SHENANDOAH COLLEGE
LIBRARY
WINCHESTER, VA.

PRACTICAL MATHEMATICS

SHENANDOAH COLLEGE
LIBRARY
WINCHESTER, VA.

PRACTICAL MATHEMATICS

RUSSELL V. PERSON
PROFESSOR OF MATHEMATICS
CAPITOL INSTITUTE OF TECHNOLOGY

VERNON J. PERSON, CPA

JOHN WILEY & SONS
NEW YORK SANTA BARBARA LONDON SYDNEY TORONTO

Copyright © 1977, by John Wiley & Sons, Inc.

All rights reserved. Published simultaneously in Canada.

No part of this book may be reproduced by any means, nor transmitted, nor translated into a machine language without the written permission of the publisher.

Library of Congress Cataloging in Publication Data:
Person, Russell V
   Practical mathematics.
   Includes index.
   1. Mathematics—1961–   I. Person, Vernon J.,
   joint author. II. Title
QA39.2.P474   513'.14   76-21732
ISBN 0-471-68216-0

Printed in the United States of America

10 9 8 7 6 5 4 3 2 1

```
QA              Person, Russell V.
39.2
.P474
1977            Practical
                   mathematics
513 P431p
```

PREFACE

This book is intended for students who need a review of general mathematics as a preparation for future study in technical fields, and for adults who have a need for the practical mathematics of everyday life. The book presents a practical working knowledge of arithmetic, geometry, algebra, and trigonometry to meet these needs.

The book is written especially as a refresher course in general mathematics for beginning students in community colleges, junior colleges, and technical institutes. Many students entering such schools as freshmen need a thorough review of some basic elementary mathematics. Some may have had little use for arithmetic or geometry for years. Some may have never studied algebra or trigonometry at all. For mature adult students, the approach to these subjects must be on an adult level. It cannot be exactly the same as for students in grade school and high school.

ORGANIZATION

The book contains four sections: *Arithmetic, Geometry, Algebra,* and *Trigonometry.* The first section, Arithmetic, contains a review of the operations with whole numbers, common fractions, decimal fractions, and percentage, as well as a chapter on the square roots of numbers.

Many adults who have used little mathematics for years need a review of the operations in arithmetic. They may have forgotten some of the rules for working with common fractions as well as with decimal fractions. The same is true regarding percentage. The entire section on arithmetic is designed as a preparation for students planning future study in any field of technology as well as for adults who have daily use for some practical mathematics.

The geometry of the second section of the book is not the demonstrative geometry of high school courses. Instead, it is better understood as conceptual and computational geometry. The purpose is to explain geometric concepts, such as various plane and solid figures. The student is shown how to compute length, area, and volume. Formulas are given, but the formulas must seem reasonable to the student, not by way of complicated axioms, postulates, and theorems, but through general understanding.

The geometry of this section might be called "Practical Measurements." Those who wish to find the perimeter of a room, the area of a garden, or the volume of a swimming pool do not need demonstrative geometry. The theorems and proofs of geometry have a definite place in mathematics, but they are not essential in a book of this kind.

Geometry is placed before algebra for a reason. In beginning with geometry before algebra, people are already familiar with many geometric forms. They already have some idea of the meaning of area and volume. Such concepts are more meaningful and more tangible than some of the ideas of algebra. In this way, we begin with the known and lead to the unknown, a basic principle of teaching and learning.

Moreover, when computational geometry precedes algebra, the formulas for areas and volumes in geometry form an excellent introduction to algebra. From such formulas, students can easily see how letters are used to represent arithmetic numbers, an idea that is the basis of algebra.

For students who have never studied algebra, the subject is like an entirely new language. A good introduction is provided through the formulas of geometry. Then it is through algebra that the student acquires a new look into the field of mathematics. He is gradually led to see the power of the equation in getting answers to problems.

The equation is introduced in Chapter 16, a very short chapter. This introduction is only for the purpose of showing the meaning and the importance of the equation. At this time, a negative number as an answer to a problem might have little meaning to many people. The formal solution of the equation is taken up later, after the explanation of positive and negative numbers.

In connection with verbal or word problems, the student is shown how such problems can be solved through the use of the equation. The various types of word problems are explained, and the steps in their solution are summarized and emphasized in the "Five Golden Rules for Solving Problems." The entire section on algebra is designed as a thorough review for those who may have forgotten much of their high school algebra. It is also an approach to algebra for those who have never studied the subject.

In trigonometry, the approach is again by way of the practical. Anyone beginning the study of trigonometry would probably like to see a use for the subject. He would probably see little use for such topics as the graphs of the trigonometric functions. Instead, he might at once appreciate the power of trigonometry in finding the height of a flagpole or the distance across a river.

For this reason, the trigonometric ratios are approached through the right triangle instead of through the general angle in standard position on the coordinate system. It is only when a student begins the study of the engineering of motion that he can see the reason for, and the meaning of, an angle greater than 180 degrees.

APPROACH

The approach and content of each subject emphasize the practical use of mathematics. Motivation is established and maintained when the student sees a need and a use for what he is learning. Moreover, the book is based on the theory that understanding must come before memorizing. A mathematical concept or process must first be understood before rules are memorized.

In developing a mathematical procedure, a specific example is used to introduce the general idea or rule. For example, the many rules in algebra are developed by beginning with examples. When students see what happens in particular examples, then, by inductive reasoning, they can usually formulate the rule themselves. This approach follows a basic principle of teaching: "Begin with the specific and lead to the general."

After a general rule or principle has been formulated through one or more examples, the principle is applied immediately to new examples. This approach can be summarized as follows:

1 Specific examples that clearly show a principle or procedure.
2 The formulation of the concept or rule in general terms.
3 The application of the rule to new situations.

The explanation of each new topic is followed by many worked-out examples that show every step in a process. The various illustrative examples point out every difficulty that the student might encounter. The student thus gets a feeling of mastery and gains confidence in his work.

EXERCISES

The exercises for assignment practice contain many examples and problems that serve to fix in mind the essential mathematical processes involved. Exercises begin with many easy routine problems and then gradually move to those that are more difficult and challenging to the better students.

Exercises are, in general, arranged so that instructors can assign either the odd-numbered or the even-numbered exercises, or every third problem, and still get a good review of all types of problems.

I wish to express my appreciation to the reviewers of the manuscript for their excellent comments and suggestions, and to the editors of John Wiley & Sons for their part in the production of a book that it is hoped will be a help to the many people who seek to enlarge their knowledge of general mathematics.

My son, Vernon J. Person, has served as assistant author. He is a CPA who has had much experience in accounting and auditing. He has contributed most of the material on percentage and other topics related to business mathematics as well as minor contributions to other parts of the book.

Silver Spring, Maryland   Russell V. Person

CONTENTS

I ARITHMETIC

1 Whole Numbers 1
1.1 Our Common Number System 1, 1.2 Kinds of Numbers 2, 1.3 Addition of Whole Numbers 3, 1.4 Subtraction of Whole Numbers 6, 1.5 Horizontal Addition and Subtraction 8, 1.6 Multiplication of Whole Numbers 10, 1.7 Division of Whole Numbers 15.

2 Common Fractions 21
2.1 Need for Fractions 21, 2.2 Definition of a Fraction 22, 2.3 Equivalent Fractions 23, 2.4 Fundamental Principle of Fractions 24, 2.5 Changing the Form of Numbers Containing Fractions 26, 2.6 Addition and Subtraction of Fractions 29, 2.7 Finding the Lowest Common Denominator 32, 2.8 Multiplication of Fractions 35, 2.9 Reciprocal of a Number 37, 2.10 Division Involving Fractions 37.

3 Decimal Fractions 43
3.1 Definition 43, 3.2 Changing a Decimal Fraction to a Common Fraction 44, 3.3 Addition of Decimal Fractions and Mixed Decimals 46, 3.4 Subtraction of Decimal Fractions and Mixed Decimals 46, 3.5 Rounding off Numbers 47, 3.6 Multiplication Involving Decimal Fractions 50, 3.7 Division Involving Decimal Fractions 53, 3.8 Changing a Common Fraction to a Decimal Fraction 55, 3.9 A Few Shortcuts 57.

4 Percentage 63
4.1 Definition 63, 4.2 Relation Between Per Cents and Fractions 63, 4.3 Rate: Base: Percentage 66, 4.4 Finding the Percentage When the Rate and the Base Are Known 66, 4.5 Finding the Rate When the Base and the Percentage Are Known 67, 4.6 Finding the Base When the Rate and the Percentage Are Known 68, 4.7 Increases and Decreases 72, 4.8 Finding the Original Amount When It Has Been Increased or Decreased by a Per Cent 73, 4.9 Buying and Selling; Gain and Loss 75,

X CONTENTS

4.10 Discount 77, 4.11 Interest 82, 4.12 Compound Interest 85, 4.13 Bank Discount 88, 4.14 Commission 91, 4.15 Taxation 92.

5 Powers and Roots 97

5.1 Definitions 97, 5.2 Roots of Numbers 98, 5.3 Finding the Root of a Number 98, 5.4 Irrational Numbers 99, 5.5 Use of a Table of Powers and Roots 101, 5.6 A Short Method for Some Square Roots 101.

II GEOMETRY

6 Plane Figures 107

6.1 The Elements of Geometry 107, 6.2 Lines: Intersecting and Parallel 110, 6.3 Angle 110, 6.4 Perpendicular Lines 112, 6.5 Kinds of Angles 112, 6.6 Measurement of Angles 113, 6.7 Polygons 115, 6.8 Triangles 116, 6.9 Quadrilaterals 117.

7 Measurement of Plane Figures 121

7.1 The Need for Measurement 121, 7.2 The Unit of Measure 121, 7.3 Perimeter 122, 7.4 Area 122, 7.5 The Rectangle 122, 7.6 The Square 125, 7.7 The Parallelogram 125, 7.8 The Triangle 127, 7.9 The Trapezoid 130.

8 The Right Triangle 135

8.1 Definition 135, 8.2 The Pythagorean Rule 135, 8.3 The Isosceles Right Triangle 139, 8.4 The Diagonal of a Square 140, 8.5 The 30°–60° Right Triangle 142, 8.6 The Equilateral Triangle 143.

9 The Circle 147

9.1 Definitions 147, 9.2 Measurements in a Circle 149, 9.3 Area of a Circle 151, 9.4 Area of a Ring 153, 9.5 Area of a Sector 154, 9.6 Area of a Segment 155.

10 Prisms 157

10.1 Geometric Solids 157, 10.2 Polyhedrons 157, 10.3 Prisms 157, 10.4 Volume of a Prism 158, 10.5 Lateral Area of a Prism 160, 10.6 The Cube 161, 10.7 Volume of Irregular Solids 164.

11 Cylinders 167

11.1 Definitions 167, 11.2 Volume of a Cylinder 167, 11.3 Lateral Area of a Cylinder 171, 11.4 Hollow Cylinder 172.

12 Pyramids and Cones 177

12.1 The Pyramid 177, 12.2 Volume of a Pyramid 177, 12.3 Lateral Area of a Regular Pyramid 178, 12.4 Frustum of a Pyramid 179, 12.5 The Cone 182, 12.6 Volume of a Cone 183, 12.7 Lateral Area of a Cone 184, 12.8 Frustum of a Cone 184.

13 The Sphere 187

13.1 Definitions 187, 13.2 Measurements in a Sphere 187, 13.3 Surface Area of a Sphere 188, 13.4 The Unit Sphere 189, 13.5 Volume of a Sphere 189.

14 The Metric System of Measurement 195
14.1 Definition 195, 14.2 Origin of the Metric System 196, 14.3 Units in the Metric System 196, 14.4 Conversion Factors Between the English and the Metric Systems 198, 14.5 Square Measure 200, 14.6 Cubic Measure: Measures of Volume 203, 14.7 Metric System of Weights 205

III ALGEBRA

15 Introduction to Algebra 209
15.1 Symbols in Arithmetic 209, 15.2 Symbols in Algebra 210, 15.3 Literal Numbers 211, 15.4 Numerical Value of an Algebraic Expression 213.

16 The Equation 217
16.1 Definition of the Equation 217, 16.2 The Solution of an Equation 220.

17 Signed Numbers 223
17.1 Definitions 223, 17.2 Absolute Value of a Number 224, 17.3 Operations with Signed Numbers: Addition 225, 17.4 Subtraction of Signed Numbers 227, 17.5 Horizontal Addition and Subtraction 229, 17.6 Multiplication of Signed Numbers 231, 17.7 Division of Signed Numbers 234.

18 Algebraic Expressions 237
18.1 Definitions 237, 18.2 Addition and Subtraction of Monomials 239, 18.3 Horizontal Addition and Subtraction 241, 18.4 Addition and Subtraction of Polynomials 243, 18.5 Horizontal Addition and Subtraction of Polynomials 243.

19 Monomials: Multiplication and Division 247
19.1 Exponents 247, 19.2 Exponents in Multiplication 247, 19.3 Multiplication of Monomials 248, 19.4 Exponents in Division 251, 19.5 Division of Monomials 252.

20 Polynomials: Multiplication and Division 253
20.1 Multiplication of a Polynomial by a Monomial 253, 20.2 Division of a Polynomial by a Monomial 255, 20.3 Multiplication of a Polynomial by a Polynomial 257, 20.4 Division of a Polynomial by a Polynomial 259.

21 Solving Equations 265
21.1 Definitions 265, 21.2 The Equation as a Balanced Scale 265, 21.3 The Axioms 266, 21.4 Transposing 268, 21.5 Parentheses in Equations 272.

22 Stated Word Problems 277
22.1 The Importance of Stated Problems 277, 22.2 The Approach to Problem Solving 277, 22.3 The "Five Golden Rules" for Solving Stated Problems 278, 22.4 General Problems 279, 22.5 Age Problems 281, 22.6 Consecutive Number Problems 283, 22.7 Coin and Other Problems About Money 286, 22.8 Problems in Uniform Motion 288.

23 Special Products and Factoring 293
23.1 Definitions 293, 23.2 Multiplication of Monomials by Inspection 293,

23.3 Factoring Monomials 294, 23.4 Prime Factors 296, 23.5 A Special Product: A Monomial Times a Polynomial 297, 23.6 Factoring by Taking Out a Common Factor 298, 23.7 A Special Product: The Sum of Two Numbers Times Their Difference 301, 23.8 Factoring the Difference Between Two Squares 303, 23.9 The Square of a Binomial: A Special Product 309, 23.10 Factoring Trinomials That Are Perfect Squares 312, 23.11 Finding the Unknown Middle Term of a Perfect Square Trinomial 314, 23.12 Completing a Square by Adding a Third Term 315, 23.13 A Special Product of the Form: $(x+a)(x+b)$ 317, 23.14 Factoring Trinomials of the Type: x^2+px+q 319, 23.15 A More Difficult Special Product: $(ax+b)(cx+d)$ 322, 23.16 Factoring a Trinomial of the Type: ax^2+bx+c 324, 23.17 Factoring the Sum and the Difference of Two Cubes 326, 23.18 Factoring by Grouping Terms 327.

24 Fractions 331

24.1 Definition 331, 24.2 Fundamental Principle of Fractions 332, 24.3 Reducing Fractions 332, 24.4 Multiplication of Fractions 337, 24.5 Division of Fractions 342, 24.6 Addition and Subtraction of Fractions 345, 24.7 Fractions with Polynomial Denominators 352.

25 Fractional Equations 361

25.1 Definition 361, 25.2 Eliminating the Denominator of a Fraction 362, 25.3 Solving a Fractional Equation 363, 25.4 Fractions and Fractional Equations: The Difference 368, 25.5 A Fraction as an Indicated Quantity 372, 25.6 Fractional Equations with Variable Denominators 374, 25.7 Literal Equations 378, 25.8 Stated Problems Involving Fractional Equations 381.

26 Systems of Equations 389

26.1 Indeterminate Equations 389, 26.2 A System of Equations 390, 26.3 Solving a System of Equations by Addition or Subtraction 391, 26.4 Solving Systems of Equations by Substitution 396, 26.5 Solving a System of Equations by Comparison 399, 26.6 Systems of Equations in Three or More Unknowns 401, 26.7 Equations: Dependent; Derived; Inconsistent; Independent 406, 26.8 Solving Word Problems by Use of Two or More Unknowns 408.

27 Graphing 421

27.1 Importance of Graphs 421, 27.2 The Rectangular Coordinate System 421, 27.3 Graph of an Equation 424, 27.4 Solving Systems of Equations by Graphing 428.

28 Exponents, Powers, and Roots 433

28.1 Multiplication 433, 28.2 Division 435, 28.3 Power of a Power 436, 28.4 Power of a Product 438, 28.5 Power of a Fraction 440, 28.6 Zero Exponent 441, 28.7 Negative Exponent 443, 28.8 Roots of Numbers 446, 28.9 Fractional Exponents 446, 28.10 Scientific Notation 449.

29 Radicals 453

29.1 Square Roots 453, 29.2 Any Root of a Number 454, 29.3 Rational and Irrational Numbers 456, 29.4 Simplifying Radicals 457, 29.5 Addition and Subtraction of Radicals 460, 29.6 Multiplication of Radicals 461, 29.7 Division Involving Radicals 466.

30 Quadratic Equations 471

30.1 Definitions 471, 30.2 Solving a Pure Quadratic Equation 473, 30.3 Solving Quadratic Equations by Factoring 475, 30.4 Solving Quadratic Equations by Completing a Square 479, 30.5 Solving Quadratic Equations by Formula 481, 30.6 Word Problems Involving Quadratic Equations 485, 30.7 Imaginary Numbers 490, 30.8 Complex Numbers 493.

31 Ratio and Proportion 499

31.1 Ratio 499, 31.2 Proportion 501, 31.3 Mean Proportional 503, 31.4 Proportional Division 505.

IV TRIGONOMETRY

32 The Trigonometric Ratios 511

32.1 Importance of Trigonometry 511, 32.2 Direct and Indirect Measurement 511, 32.3 Angles 511, 32.4 Kinds of Angles 512, 32.5 Measurement of Angles 513, 32.6 Triangles 513, 32.7 The Right Triangle 514, 32.8 The Trigonometric Ratios 516, 32.9 Definitions of the Trigonometric Ratios 517.

33 Tables of Trigonometric Ratios 521

33.1 The Trigonometric Ratios of an Angle 521, 33.2 Tables of Trigonometric Ratios 523, 33.3 Arc-Functions, or Inverse Functions 525.

34 Solving Right Triangles 529

34.1 Solving Stated or Word Problems by Trigonometry 529, 34.2 Solving a Right Triangle 534.

35 Introduction to Set Theory 541

35.1 Introduction 541, 35.2 Meaning of a Set 541, 35.3 Notation for Sets and Elements 542, 35.4 Equal Sets 544, 35.5 Subsets 545, 35.6 All Possible Subsets of a Given Set 546, 35.7 Union and Intersection of Sets; Overlapping of Sets 547, 35.8 The Venn Diagram 549.

Appendix 551

Table 1. Amount at Compound Interest 553, Table 2. Squares, Square Roots, Cubes, Cube Roots 558, Table 3. Natural Trigonometric Functions 559, Answers to Odd-Numbered Exercises 565.

Index 583

I ARITHMETIC

1 WHOLE NUMBERS

1.1. OUR COMMON NUMBER SYSTEM

Our common system of numbers makes use of only ten symbols called *digits*: 0, 1, 2, 3, 4, 5, 6, 7, 8, 9. By means of these ten symbols, we can express a number of any size, no matter how large.

When early man first began counting, he perhaps put down a mark for each object. For one object he made a mark, such as /. When he counted one more, he made another mark. Then he had two marks, //, as in the Roman notation. After many years, some early people decided to call this *two*, and for this many they made the symbol 2. After many more years, when someone counted *one more than two*, he called the number *three*, and for *three* he made the symbol 3. One more than three was called *four* and was represented by the symbol 4. The other names and symbols up to nine (9) were invented in the same way. The zero (0) was the last symbol invented and was used to represent no objects. Of course, the names and symbols from 0 to 9 were not exactly the same as we know them in English.

Some nations used other names and symbols. For example, the Romans used the symbol *V* for five, *X* for ten, *L* for fifty, *C* for one hundred, and *M* for one thousand. With these symbols it was cumbersome to write very large numbers. However, with the ten symbols in our common number system, we can write any number, however large, because of the *place value* of the symbols, which we shall now explain.

Counting up to nine, we have a symbol for each number. When counting continued to one more than nine, it was decided to lump all these objects together as one *bunch* and to write *1* for the bunch. The *1* was placed in a position a little farther to the left to indicate a bunch of ten. A zero (0) was placed at the right of the *1*. The number was written *10* to denote one bunch of ten and no single objects more. For *two* more than ten, we write *1* for the bunch of *ten*, and then write a *2* for the extra units or ones, as *12*.

2 ARITHMETIC

When we write the number 37, we mean three bunches of ten each and seven single objects more, or 30+7. In the notation 37, the 3 is in the position we call *tens'* place, and the 7 is in the position we call *units'* or *ones'* place.

This place value for the digits was one of the great inventions in mathematics. By the position of the digits, we can tell at a glance the size of a number. For example, 94 means 9 tens and 4 units more. When we count to 99, we have 9 bunches of ten, and 9 units more. In the number 44, the digit 4 in tens' place means ten times as many as the 4 in the units' place.

Now, if we have one more than 99, we have another bunch of ten; that is, we have ten bunches. However, we have no symbol for ten, so we lump the ten bunches together and call it a large bunch. It might be called a bunch of the *second order*. We write a 1 for the large bunch, and then follow it with two zeros: 100, which we call one hundred. The *1* is then said to be in *hundreds'* place. When we write the number 637, we mean 6 large bunches of one hundred each, 3 smaller bunches of ten each, and 7 ones more; that is, 600+30+7.

As we get larger bunches, or bunches of higher orders, we move the digits farther to the left. We might say 1,000,000 (one million) represents a bunch of the sixth order.

In our common number system, the first place starting at the right represents ones (or units). As we move toward the left, the second place represents tens, the third place hundreds, the fourth place thousands, the fifth place ten-thousands, the sixth place hundred-thousands, the seventh place millions. In the United States, the tenth place represents billions. That is, a billion is a thousand millions. (However, in England and some other countries, a billion means a million millions.) When we write the number 3685204, we mean 3 millions, 6 hundred-thousands, 8 ten-thousands, 5 thousands, 2 hundreds, 0 tens, and 4 ones (or units).

1.2. KINDS OF NUMBERS

Numbers may be classified as *integers* and *fractions*. An *integer* is usually called a whole number. The positive integers, 1, 2, 3, 4, and so on, are called *counting numbers* because they are used in counting. A *fraction* indicates a part of a whole number, such as 1/2, 3/5, 7/4, and so on. Fractions will be studied in Chapter 2. In the present chapter, we deal only with whole numbers.

An *even* number is a number that can be exactly divided by 2, such as 8, 26, 354, 9710. Numbers that are not exactly divisible by 2 are called *odd* numbers, such as 3, 7, 29, 481. Whether an integer is even or odd is determined by the units' digit. In any whole number, if the units' digit is even, as in 5376, then the number itself is even. If the units' digit is odd, as in 4683, then the number itself is odd.

1.3. ADDITION OF WHOLE NUMBERS

When numbers are added, the result is called the *sum* of the numbers. The sum of 7, 12, and 9 is 28. The numbers added are called *addends*. Addition is usually indicated by the plus sign (+).

In the addition of numbers, two laws are useful. We probably use these laws constantly without being aware of their names. The *commutative law* for addition states that two numbers may be added in either order. For example, to add 5 and 3, we can write

$$5+3 \quad \text{or} \quad 3+5$$

That is, we may begin with 5 objects and add 3 to this number. However, we may begin with 3 objects and add 5 to them. The result is 8 in both instances. In general., if a and b represent any two numbers, respectively, then, by the *commutative law*,

$$a+b=b+a$$

Another law useful in adding several addends is the *associative law* for addition. The associative law states that numbers to be added may be associated in any order. For example, suppose we wish to add

$$16+3+7$$

We might first say, $16+3=19$. Then we add 7 to 19 and get 26. However, we know it is convenient to add 10 to any number. Now we might recognize that $3+7=10$. Then we can say, $16+10=26$. The two ways of associating addends can be shown by parentheses. We enclose in parentheses the numbers to be added first. Then by the *associative law*, we have

$$(16+3)+7=16+(3+7)$$

In general, if a, b, and c, respectively, represent any three addends, then by the *associative law*,

$$(a+b)+c=a+(b+c)$$

The commutative and associative laws enable us to pair up any combination of addends for easy calculation. For example, note the pairs of addends that make 10 in the following example:

$$6+\underline{8+2}+\underline{7+5+3}+5=36$$

$$101010$$

4 ARITHMETIC

This pairing of addends that make 10 is very useful in column addition.

In adding numbers, we usually place one addend below the other, with units below units, tens below tens, and so on, as shown in the following example.

Example 1. Add 435 + 523.

```
 435
 523
 ───
 958
```

Solution. The work is shown at the left. First we add the units' column: 5 units + 3 units = 8 units. We place the sum, 8, below, in the units' column. Next we add 3 tens and 2 tens, and place the sum 5 in the tens' column. Finally, we add 4 hundreds and 5 hundreds, and place the sum 9 in the hundreds' column. The sum, 958, represents 9 hundreds, 5 tens, and 8 units.

Example 2. Add 549 + 734.

```
  549
  734
 ────
 1283
```

Solution. The work is shown at the left. In addition it is sometimes necessary to use a "trick" called *carrying*. When we add 9 units and 4 units, we get 13, which is 1 ten and 3 units. We write the 3 in the units' column and then add the 1 ten in the tens' column. That is, we carry the 1 over to the tens' column. Sometimes this 1 is written above the tens' column. Then we shall have 1 ten, 4 tens, and 3 tens, which makes a total of 8 tens in the answer. Finally, we add 5 hundreds and 7 hundreds, and get 12 hundreds, or 1 thousand and 2 hundreds. The *carrying* should be done metally, but it is sometimes shown.

Example 3. Add 4356 + 9678.

```
  4356
  9678
 ─────
 14034
```

Solution. The work is shown at the left except for the carrying. Adding the units' column, we get 14, which is 1 ten and 4 units. We place the 4 in the units' column and carry the 1 ten to the tens' column. Adding the tens, we get 13, which is 1 hundred and 3 tens. We place the 3 in the tens' column and carry the 1 to the hundreds' column. Adding the hundreds' column, we get 10, which is 1 thousand and no hundreds. We carry the 1 to the thousands' column and place a zero (0) in the hundreds' column. Adding the thousands, we get 14, which is 1 ten-thousand and 4 thousands. The answer is read, "fourteen thousand thirty-four," without the word *and*.

Example 4. Add 798 + 4685 + 17,893 + 77.

```
   798
  4685
 17893
    77
 ─────
 23453
```

Solution. The work is shown at the left except for the carrying. The numbers are written in column form with units' digits and the other digits in proper alignment. We begin by adding the units' digits as usual. The sum of the units' column is 23, which is 2 tens and 3 units. The 3 is placed in the units' column, and the 2 tens are carried over to the tens' column. We continue in the same manner, adding each column as we move toward

WHOLE NUMBERS 5

the left. The carrying process should be done mentally, although in long columns, the carried digits are sometimes written at the top of the proper column.

To be sure your work is correct, you should develop the habit of checking it in some way. A good way to check addition is simply to add the numbers a second time. Another way is to reverse the order of adding. If you add from top to bottom the first time, then check by adding upward starting at the bottom.

One of the best ways to avoid errors in addition is to be sure you know instantly the sum of each of the following sets of numbers. They are called the 45 addition combinations.

2	3	4	8	9	5	7	3	4	9	1	5	6	4	9
1	2	5	3	9	2	1	4	4	3	8	3	5	4	1

4	6	7	1	9	7	6	8	5	6	4	5	6	9	1
2	6	3	6	7	5	2	8	5	7	8	1	8	2	4

7	2	4	1	5	2	3	8	7	9	2	9	6	3	5
8	2	6	1	9	7	3	9	7	6	8	4	3	1	8

You should practice the above combinations until you can add them all orally in 30 seconds or write the answers in 50 seconds.

Exercise 1.1

Addition: add the following.

1. 43   45   52   51   74   65   87   74   62   85
    62   31   72   66   42   64   72   94   97   83

2. 51   65   73   54   82   93   46   84   65   92
    98   72   73   81   46   94   93   65   52   87

3. 27   54   35   56   64   45   64   49   67   35
    34   48   36   24   39   25   76   35   18   58

4. 69   98   83   23   48   78   98   87   76   56
    29   19   17   79   43   18   72   95   76   78

5. 535   334   438   625   367   549   416   827   958
    627   856   734   845   928   937   956   927   835

6 ARITHMETIC

6.	259	876	834	649	738	948	967	738	938
	829	817	457	924	747	743	913	832	658

7.	384	269	876	674	385	697	984	469	678
	598	963	579	659	925	807	676	875	598

8.	397	872	835	768	998	789	897	696	768
	748	749	598	698	909	419	919	498	958

9.	864	475	536	728	10.	349	977	471	218
	247	739	478	859		695	361	349	872
	953	351	854	381		738	183	368	867
	158	156	632	252		415	746	742	333
	632	994	259	523		872	528	637	746

11. $498 + 1534 + 23 + 43{,}576$

12. $9164 + 37 + 12{,}986 + 423$

13. $2056 + 304 + 74{,}607 + 53$

14. $43 + 3862 + 298 + 14{,}317$

15. $24 + 5986 + 307 + 63{,}364$

16. $96 + 453 + 7 + 349{,}708$

1.4. SUBTRACTION OF WHOLE NUMBERS

Subtraction is the process of taking one number from another. For example, if you have 15 dollars and then spend 7 dollars, how many dollars have you left? To find the *remainder*, we use *subtraction*. As another example requiring subtraction, suppose in a basketball game, one team makes a score of 89 and the other team makes a score of 64. What is the difference between the scores? To find the *difference*, we use subtraction.

Subtraction is the inverse of addition. When one number is subtracted from another, the result is called the *remainder* or the *difference*. The number subtracted is called the subtrahend. The number from which the subtrahend is subtracted is called the *minuend*. When we take 7 dollars from 15 dollars and get 8 dollars, the minuend is 15, the subtrahend is 7, and the remainder is 8.

The symbol for subtraction is the minus sign, a short horizontal line $(-)$. To subtract 7 from 15, we write: $15-7$. This is read: "15 minus 7." The minus sign means that the second number is to be taken from the first, not the reverse. For this reason, in subtraction, the numbers cannot be reversed, as they can be in addition. In addition, if a and b represent two numbers, respectively, then $a+b$ equals the same as $b+a$. That is, $9+5=5+9$. But in subtraction, $a-b$ does not equal the same as $b-a$. In other words, the commutative law does not hold true for subtraction.

In subtraction, we usually place the subtrahend below the minuend. We begin the subtraction with the units' digits as in addition.

WHOLE NUMBERS 7

Example 1. Subtract 36 from 78.

78 minuend
36 subtrahend
42 remainder

Solution. The work is shown at the left. We begin by subtracting 6 units from 8 units, and place the remainder 2 below in the units' place. Next, we subtract 3 tens from 7 tens, leaving 4 tens, which is placed in the tens' column. The remainder is 42.

Subtraction can be checked by addition. If the remainder is added to the subtrahend, the result should be the minuend. Checking should be done mentally.

In subtraction, it is often necessary to use a "trick" called *borrowing*. There are other methods of subtraction, but the borrowing method is probably most common. The actual borrowing should be done mentally, though at first it may be shown.

Example 2. Subtract 57 from 93.

```
         8
93     9 ¹3
57     5  7
       3  6
```

Solution. The work is shown at the left. Since we cannot subtract 7 from 3, we borrow 1 ten from the 9 tens, leaving 8 tens. The 1 ten borrowed is changed to 10 units, which we combine with the 3 units, making 13 units. This is often shown by drawing a slanted line through the 9, writing 8 above, and then writing a small 1 before the 3. Now we subtract 7 units from 13 units, leaving 6 units, which we place below in the units' place. Then we subtract 5 tens from 8 tens, leaving 3 tens. The remainder is then 36. As a check, we add mentally: $36 + 57 = 93$.

Example 3. Subtract 287 from 836.

```
8 3 6
2 8 7
5 4 9
```

Solution. The work is shown at the left except for the borrowing. We cannot subtract 7 units from 6 units. We borrow 1 ten from the 3 tens, leaving 2 tens in the minuend. The 1 ten borrowed is changed to 10 units which is combined with the 6 units, making 16 units. Now we subtract 7 units from 16 units, leaving 9 units in the remainder. As a second step, we borrow 1 hundred from the 8 hundreds leaving 7 hundreds. The 1 hundred borrowed is changed to 10 tens and combined with the 2 tens, making 12 tens. Then we take 8 tens from 12 tens, leaving 4 tens in the tens' place. Finally, we take 2 hundreds from 7 hundreds, leaving 5 hundreds in the remainder. The remainder is 549. As a check, we add: $549 + 287 = 836$, the minuend.

Example 4. Subtract 478 from 803.

```
8 0 3
4 7 8
3 2 5
```

Solution. The work is shown at the left except for the borrowing. Here we cannot borrow 1 ten from no tens. We cannot borrow 1 from 0. Then we must go to the hundreds and borrow 1 hundred from the 8 hundreds, leaving 7 hundreds. The 1 hundred borrowed is changed to 10 tens. Now,

8 ARITHMETIC

one of these tens is borrowed, leaving 9 tens in the tens' column. The 1 ten borrowed is changed to 10 ones and combined with the 3 ones, making 13 ones. Now we can subtract 8 units from 13 units, leaving 5 units, or ones, in the units' column of the remainder.

As a second step, we take 7 tens from 9 tens, leaving 2 tens for the remainder. Finally, we take 4 hundreds from 7 hundreds, leaving 3 hundreds. The remainder is then 325. As a check, we mentally add the remainder to the subtrahend. The result should be the minuend.

1.5. HORIZONTAL ADDITION AND SUBTRACTION

In horizontal addition and subtraction, the operations are performed in the order in which they occur. If any quantity is enclosed in parentheses, brackets, or any form of grouping, the operations within the parentheses must be performed first.

Examples. (a) $9-3+5-4=7$
(b) $12+(8-3)-(4+2)-(5-4)=10$

Exercise 1.2

Subtract the bottom number from the top number. Try to do all borrowing mentally. Place a blank paper below each row and then try to write the answers for the problems in each row in the time indicated at the right of each row.

									seconds
1.	84	86	95	63	95	58	47	48	
	51	43	24	53	31	13	36	25	15
2.	76	58	89	97	69	94	96	87	
	41	34	76	41	34	23	84	23	15
3.	56	86	78	97	69	59	75	79	
	46	35	61	65	42	21	35	23	15
4.	51	81	62	64	85	90	71	82	
	26	38	43	37	46	54	47	54	20
5.	63	73	84	75	85	71	90	83	
	25	48	49	37	19	59	29	24	20
6.	76	92	80	74	86	32	80	93	
	28	45	36	58	49	16	28	59	20

WHOLE NUMBERS 9

| 7. | 56 | 94 | 92 | 40 | 93 | 78 | 77 | 82 | |
| | 17 | 66 | 37 | 17 | 16 | 19 | 38 | 18 | 20 |

| 8. | 93 | 90 | 85 | 90 | 62 | 84 | 98 | 97 | |
| | 67 | 72 | 58 | 71 | 19 | 55 | 69 | 79 | 20 |

| 9. | 974 | 798 | 956 | 947 | 689 | 895 | 964 | 798 | |
| | 640 | 153 | 535 | 745 | 263 | 475 | 564 | 588 | 25 |

| 10. | 951 | 831 | 930 | 944 | 652 | 712 | 741 | 823 | |
| | 279 | 656 | 694 | 346 | 159 | 135 | 458 | 594 | 35 |

| 11. | 930 | 713 | 824 | 645 | 962 | 860 | 934 | 715 | |
| | 546 | 249 | 487 | 286 | 568 | 582 | 439 | 359 | 35 |

| 12. | 721 | 630 | 973 | 831 | 613 | 640 | 826 | 983 | |
| | 377 | 385 | 478 | 365 | 427 | 369 | 269 | 684 | 35 |

| 13. | 502 | 703 | 901 | 803 | 804 | 902 | 605 | 903 | |
| | 268 | 586 | 356 | 329 | 496 | 576 | 236 | 697 | 35 |

| 14. | 908 | 805 | 807 | 506 | 805 | 901 | 807 | 404 | |
| | 749 | 579 | 279 | 368 | 657 | 429 | 49 | 69 | 35 |

| 15. | 7128 | 9156 | 5137 | 4179 | 9285 | 8233 | 5267 | 1922 | |
| | 5841 | 5186 | 3662 | 1786 | 3295 | 4350 | 2487 | 937 | 50 |

| 16. | 5684 | 1923 | 8488 | 5635 | 3785 | 7043 | 4239 | 6104 | |
| | 4309 | 1042 | 619 | 978 | 3159 | 3897 | 3684 | 2698 | 50 |

| 17. | 9804 | 4132 | 8423 | 9006 | 5904 | 2973 | 3420 | 5600 | |
| | 2973 | 3901 | 7138 | 5842 | 4917 | 584 | 986 | 1682 | 50 |

| 18. | 3601 | 6500 | 1032 | 4000 | 7005 | 7015 | 6032 | 2341 | |
| | 2973 | 2769 | 579 | 1809 | 6738 | 3960 | 5473 | 1678 | 50 |

19. Combine: $24+(8-5)-(7+2)-(6-4)$.

20. Combine: $30-(8+3)-(9-4)+(4-1)$.

21. Combine: $27-(16-9)+(3+2)-(6+8)$.

22. Combine: $20-(4+7)-(6-5)+(8-3)$.

23. A family on an automobile tour takes a trip of 1680 miles to be completed in five days, Monday through Friday. Each day, they travel the following number of miles: Monday: 380; Tuesday: 410; Wednesday: 335; Thursday: 271. How many miles must be driven on Friday?

24. On an automobile trip from Washington, D.C. to San Francisco, you might travel the following distances between stops: D.C. to Pittsburgh, 229 mi.; Pittsburgh to Chicago, 459 mi.; Chicago to Minneapolis, 418 mi.; Minneapolis to Omaha, 364 mi.; Omaha to Denver, 537 mi.; Denver to Salt Lake City, 512 mi.; Salt Lake City to San Francisco, 754 mi. What is the total length of the trip?

25. A college has 15,382 students enrolled. The number of majors is as follows: Math and Science, 2975; Humanities, 5653; Art and Music, 2564. The remainder are undecided. How many are undecided?

26. Four voltages in series have a combined voltage of 1220 volts. If three of the voltages are, respectively, 160 volts, 348 volts, and 465 volts, find the fourth voltage. (In an electric circuit, voltages arranged in series have a combined voltage equal to the sum of the voltages.)

27. Three resistors have the following ratings, respectively: 452 ohms, 524 ohms, and 621 ohms. What additional resistance is required if the necessary resistance in series is 2100 ohms? (If resistors in an electric circuit are arranged in series, the total resistance is found by adding the resistances).

28. Three capacitors arranged in parallel in an electric circuit have a total capacitance of 92 microfarads. If two of them have ratings of 29 and 38 microfarads, respectively, find the third. (In an electric circuit, if capacitors are arranged in parallel, the total capacitance is found by adding the capacitance of the separate capacitors.)

1.6. MULTIPLICATION OF WHOLE NUMBERS

Multiplication is the process of taking a number two or more times. Multiplication may be considered as a shortened form of addition. The symbol for multiplication is the cross ($\times$). The symbol is read "times." For example, if we wish to add three 7's, we can write: 3×7. The expression is read: "3 times 7," and means the same as $7 + 7 + 7 = 21$.

Multiplication is sometimes indicated by a raised dot placed between the numbers. Then $3 \cdot 7$ means the same as 3×7. We can also enclose each number in parentheses without a sign between them: $(3)(7) = 21$. We sometimes indicate multiplication by placing one number above the other with the direction to multiply.

When we add three 7's, we have seen that we get 2 tens and 1 unit more. Then $3 \cdot 7 = 21$.

In the multiplication of two numbers, one number is called the *multiplicand* and the other is called the *multiplier*. The multiplicand is the number taken a given number of times. The multiplier tells how many times the multiplicand is to be taken. In the multiplication, 3×7, we might think of the 7 as the multiplicand and the 3 as the multiplier.

When we multiply two numbers together, the answer is called the *product* of the numbers. In mathematics, a product always implies multiplication. The numbers multiplied together are called *factors*. For example, in the multiplication, $3 \times 7 = 21$, the factors are 3 and 7; the product is 21. In the example, $2 \times 3 \times 5 = 30$, the factors of 30 are 2, 3, and 5; the product is 30.

A *composite number* is a number that is the product of two or more factors other than 1 and the number itself. For example, 15 is a composite number because it can be separated into the two factors, 3 and 5.

A *prime number* is a number that cannot be separated into any whole number factors other than itself and 1, such as the numbers 7, 13, 29, and so on.

In the multiplication, $5 \times 7 = 35$, we may think of the 5 as the multiplier and the 7 as the multiplicand. That is, five 7's added together equals 35. However, if we interchange the factors, we write 7×5. Then we can think of the problem as seven 5's. But the product is the same: 35. Therefore, we can say that 5×7 is the same as 7×5. This example is an illustration of the following law:

The Commutative Law for Multiplication states that the multiplication of any two numbers may be done in any order. In general, if a and b represent any two numbers, respectively, then by the commutative law,

$$a \times b = b \times a$$

We often make use of this law to check multiplication by reversing the order of multiplication. For example, $32 \times 47 = 47 \times 32$.

The Associative Law for Multiplication is another useful law. This law enables us to associate certain factors in such a way as to simplify multiplication. For example, suppose we have the problem in multiplication,

$$13 \times 5 \times 2$$

We might first say: $13 \times 5 = 65$. Then we multiply: $65 \times 2 = 130$. However, we know it is convenient to multiply any number by 10. Now, we might recognize that $5 \times 2 = 10$. Then we can first multiply: $5 \times 2 = 10$, and then say

$$13 \times 10 = 130$$

The two ways of associating the factors can be shown by parentheses. We enclose in parentheses the numbers to be multiplied first. Then

$$(13 \times 5) \times 2 = 13 \times (5 \times 2)$$

The answer is 130 in both cases. In general, if a, b, and c, represent any three numbers, respectively, then, by the associative law,

$$(a \times b) \times c = a \times (b \times c)$$

The associative and commutative laws for multiplication enable us to pair up any combination of factors for each computation. For example, note the multiplication by 10's in the following:

$$8 \times 2 \times 5 \times 7 \times 5 \times 2 = 5600$$

12 ARITHMETIC

To get the answer, we first take 8×7, and then multiply the result by 100.

The best way to avoid errors in multiplication is to know instantly the product of the following sets of numbers. They are called the 45 multiplication combinations. Multiply the two numbers in each set. Practice until you can state the products orally in 30 seconds, and write the answers in one minute.

2	3	4	3	8	5	9	2	7	2	8	9	3	7	6
1	2	5	4	8	3	1	6	3	4	7	9	3	4	3

9	4	8	6	5	8	7	8	1	2	8	7	5	6	3
5	4	1	4	5	2	5	9	7	2	6	9	2	1	8

9	1	7	4	6	6	3	5	9	8	1	7	8	5	7
2	1	2	9	6	9	1	6	3	5	4	6	4	1	7

In multiplying a number of more than one digit, we begin with the units' digit, as in addition and subtraction.

Example 1. Multiply 3×82.

```
 82 multiplicand
  3 multiplier
───
246 product
```

Solution. The work is shown at the left. We begin by multiplying 3×2. The product, 6, is placed below in units' place. Then we multiply, 3×8, which is really 3 times 8 tens. The product is 24, or 2 hundreds and 4 tens. The complete product is 246.

Example 2. Multiply 4×69.

```
 69 multiplicand
  4 multiplier
───
276 product
```

Solution. In this example, we use the trick of carrying as we do in addition. The work is shown at the left except for the carrying. Beginning with the units' digits, we get: $4 \times 9 = 36$, which is 3 tens and 6 units. We write the 6 below in units' place, and carry the 3 tens above the tens' place. The carrying should be done mentally, but sometimes it may be shown at the top of the tens' place. Multiplying 4 times 6 tens, we get 24 tens. Now we add the 24 tens to the 3 tens carried, which makes 27 tens, or 2 hundred and 7 tens. The product is 276.

Example 3. Multiply 7×586.

```
 586 multiplicand
   7 multiplier
────
4102 product
```

Solution. The work is shown at the left except for the carrying. In this example, it is necessary to do the carrying more than once, just as is sometimes done in addition. We begin with the multiplication: 7×6, which is 42, or 4 tens and 2 units. We place the 2 below in units' place and carry the 4 to the top of the tens' place. As a second step, we take 7×8 and get 56 tens, which, added to the 4 tens carried, makes 60 tens, or 6 hundreds and 0 tens. We place the zero (0) in tens' place and carry the 6 hundreds to the top of hundreds' place. Finally, we take 7×5, and get 35 hundreds, which, added to the 6 hundreds carried, makes 41 hundreds, or 4 thousands and 1 hundred. The final product is 4102.

Example 4.   Multiply 9×807.

```
  807 multiplicand
    9 multiplier
 ────
 7263 product
```

Solution. The work shown at the left except for the carrying. First, we take $9 \times 7 = 63$, which is 6 tens and 3 units. We place the 3 below in units' place, and carry the 6 to the top of the tens' place. Next, we take 9×0, which is 0. Then, for the tens' place we have only the 6 tens carried, which we place in tens' place below. Finally, we take 9×8, which is 72 hundreds, or 7 thousands and 2 hundreds. The complete product is 7263.

Example 5.   Multiply 48×573.

Solution. The work is shown at the right except for the carrying. First, we multiply the multiplicand, 573, by 8, the same as with a one-digit multiplier. Next, we multiply by the 4 in the multiplier. However, this 4 represents 4 tens. Therefore, when we say, $4 \times 3 = 12$, we really mean: 40×3. The product, $40 \times 3 = 120$, is really 12 tens, or 1 hundred and 2 tens. We carry the 1 hundred to the hundreds' place. After multiplying by each digit in the multiplier, we add the products.

```
   573 multiplicand
    48 multiplier
 ─────
  4584 multiplication by 8
  2292 multiplication by 40
 ─────
 27504 product
```

Example 6.   Multiply 487 by 769.

Solution. The statement really implies that the multiplier is 769. The work is shown at the right except for the carrying. The work can be checked by reversing the order or multiplication; that is, by using 487 as the multiplier.

```
    487 multiplicand
    769 multiplier
 ──────
   4383 multiplication by 9
   2922 multiplication by 60
   3409 multiplication by 700
 ──────
 374503 product
```

Example 7.   Multiply 5497 by 306.

Solution. The work shown at the right except for the carrying. If the multiplier contains a zero digit, the multiplication by this digit need not be shown, since the product of zero times any number is 0. However, the multiplication by the next digit must be properly placed.

```
    5497 multiplicand
     306 multiplier
 ───────
   32982 multiplication by 6
   16491 multiplication by 300
 ───────
 1682082 product
```

14 ARITHMETIC

Example 8. Multiply 896 by 670.

Solution. The work is shown at the right except for the carrying. Here the multiplier ends in zero. Then the multiplier is often placed on one digit farther toward the right, and a zero (0) is placed below the zero in the multiplier to indicate multiplication by the zero.

```
   896   multiplicand
   670   multiplier
 62720   multiplication by 70
  5376   multiplication by 600
600320   product
```

Example 9. Multiply 70648 by 6050.

Solution. The work is shown at the right except for the carrying. In this example, note especially the zeros in the multiplicand and in the multiplier. Such zeros often cause students much trouble.

```
    70648   multiplicand
     6050   multiplier
  3532400   multiplication by 50
  423888    multiplication by 6000
427420400   product
```

CHECKING MULTIPLICATION. The best way to check multiplication is to go back over each separate step in the multiplication to be sure that no errors have been made. In some examples, it may be practical to reverse the order of multiplicand and multiplier.

Exercise 1.3

Multiply the two numbers in each set. Try to write the answers for the sets in each row in the time indicated at the right for each row.

								Seconds
1. 96	47	43	79	56	95	83	79	30
3	6	8	4	9	8	7	7	
2. 76	93	97	48	59	85	79	69	30
8	9	6	8	7	9	9	4	
3. 786	934	863	658	678	986	896	976	45
7	9	8	7	9	7	9	8	
4. 754	879	567	758	487	978	859	789	45
8	8	9	8	7	9	7	9	
5. 73	34	65	76	59	95	23	96	80
86	97	48	59	94	29	67	86	
6. 83	76	95	89	97	58	95	68	80
54	74	73	29	98	78	87	83	

WHOLE NUMBERS **15**

7.	67	78	98	46	95	89	98	86	
	52	48	84	74	47	93	72	92	80
8.	74	87	56	97	67	62	85	68	
	94	79	58	67	59	89	75	79	80
9.	423	546	657	746	839	706	409	608	
	97	67	85	98	58	79	98	87	110
10.	357	518	419	617	823	903	607	509	
	96	67	58	96	49	78	69	78	110
11.	382	926	3807	5008	4029	3016	8314	9216	
	60	80	58	76	98	85	407	305	120
12.	459	679	6203	8206	7009	7094	7625	6342	
	90	40	67	64	48	86	708	906	120

1.7. DIVISION OF WHOLE NUMBERS

Division is the process of finding how many times one number is contained in another. The symbol for division is (÷). The expression 42 ÷ 6 is read, "42 divided by 6." It indicates that we are to find the number of times that 6 is contained in 42. The answer is 7. The number divided by another is called the *dividend*. The number divided into the dividend is called the *divisor*. The answer in division, is called the *quotient*. A quotient always implies division.

In the example, 42 ÷ 6 = 7, the dividend is 42, the divisor is 6, and the quotient is 7. In many instances, there is a remainder after the division. For example, 38 ÷ 5 = 7, with a remainder of 3.

Division is often indicated by a horizontal line with the dividend above the line and the divisor below, as $\frac{42}{6}$, which means 42 ÷ 6. In typing, division is often indicated by a slanted line between dividend and divisor, as 42/6.

Example 1. Divide 3194 ÷ 7.

$$\begin{array}{r} 456 \\ 7\overline{)3194} \\ \text{remainder} = 2 \end{array}$$

Solution. In this example, the dividend is 3194 and the divisor is 7. When the divisor is a single digit, we usually use a form called *short division*, as shown at the left. We begin by dividing 7 into 31, which is 4, with a remainder of 3. Here we are really dividing 7 into 3100, which is 400, with a remainder of 300. The quotient 4 is written directly above the 1 in 31, and the remainder 3 is carried over to the next following digit, 9, making 39; that is, the 3 remaining after the first division represents 30 tens and is combined with the 9 tens in the dividend. The result, 39, is now divided by 7. From here on, the division follows the same procedure. Whenever there is a remainder, it is carried over and placed next to the

16 ARITHMETIC

following digit. We divide 7 into 39, which is 5, with a remainder of 4. The 4 is placed next to the 4 in the dividend, making the next division, 7 into 44, which is 6, with a remainder of 2. The entire quotient is 456, with a remainder of 2.

Division can be checked by multiplication. The quotient times the divisor should equal the dividend, when there is no remainder. If there is a remainder after division, the division can be checked by multiplying the quotient by the divisor and adding the remainder. The result should equal the dividend. To check the answer in Example 1, we can multiply the quotient, 456, by the divisor, 7, and then add the remainder. The result should be the dividend.

Example 2. Divide $243765 \div 8$.

$$\begin{array}{r} 30470 \\ 8\overline{)243765} \\ \text{remainder} = 5 \end{array}$$

Solution. The work is shown at the left. In division, a digit is placed in the quotient for each division or attempted division. Zeros will sometimes appear in the quotient. In this example, notice that a zero (0) is placed in the quotient when each step in the division process involves dividing 8 into a number less than 8. However, it is not necessary to write a zero for the very first division of 8 into the first 2.

Long division is the procedure of showing the subtraction and the remainder in each step of the division. The following examples show a common form for long division. Each digit of the quotient should be placed directly above the last digit of the dividend that is used in each separate step. In Example 3, notice that the 6, the first digit in the quotient, is placed directly above the 9 in the dividend because the first division is 43 into 249.

Example 3. Divide $26975 \div 43$.

$$\begin{array}{r} 627 \\ 43\overline{)26975} \\ 258 \\ \hline 117 \\ 86 \\ \hline 315 \\ 301 \\ \hline 14 = \text{remainder} \end{array}$$

Solution. The work is shown at the left.

Step 1. Divide 43 into 269. Place the 6 in the quotient directly above the 9 of the dividend.

Step 2. Multiply the divisor, 43, by the 6 in the quotient, and write the product, 258, below the 269.

Step 3. Subtract, leaving 11, as a first remainder.

Step 4. Bring down the 7, the next digit in the quotient, and place the 7 next to the remainder 11, making 117 to be divided by 43.

Step 5. Divide 43 into 117, and write the result, 2, in the quotient directly above the 7 in the dividend.

Step 6. Multiply the divisor, 43, by the 2, and write the product, 86, below the 117.

WHOLE NUMBERS 17

Step 7. Subtract, leaving 31, as a second remainder.

Step 8. Bring down the 5, the next digit in the dividend, placing it next to the remainder 31, making 315 to be divided by 43.

Step 9. Divide 43 into 315, and write the answer, 7, in the quotient directly above the 5 in the dividend.

Step 10. Multiply the divisor, 43, by the 7, and write the product, 310, below the 315.

Step 11. Subtract, leaving a final remainder of 14.

Many students have trouble with zeros in the quotient. Sometimes, when a number is brought down from the dividend, the resulting number is too small to contain the divisor. Then a zero (0) must be placed in the quotient and the next digit brought down. Remember, every time a number is brought down from the dividend, some number must be placed in the quotient to represent the next division, even though that number is zero (0).

Although long division in arithmetic is often performed correctly by habit, it is well to consider carefully each of the steps, because the same procedure is followed in long division in algebra. In the following problem, Example 4, follow the same steps as outlined in Example 3. Notice, moreover, that in Example 4, we get a zero (0) in the quotient.

Example 4. Divide $219367 \div 54$.

```
        4062
54 ) 219367
     216
     ———
      336
      324
      ———
       127
       108
       ———
        19 = remainder
```

Solution. The work is shown at the left. In this example, notice that after the first division by the divisor 54, and the subtraction, the first remainder is only 3. When the next digit is brought down, the number 33 is not divisible by 54. Since one digit of the dividend has already been brought down at this point, a digit must be placed in the quotient to represent this division. In this case, we write a zero (0) in the quotient, and then bring down the next digit, 6, of the dividend.

Long division is used chiefly when the divisor contains two or more digits. In long division, we show the remainder after each division. However, when the divisor contains only a single digit, as in Examples 1 and 2, the remainders are carried along mentally. Of course, it is possible to use the form of long division even when the divisor contains a single digit.

In division, if the divisor contains two or more digits, it is not always easy to determine at a glance each digit of the quotient. Let us consider a two-digit divisor. If the units' digit is a 1 or a 2, as in a divisor such as 31, then it is fairly easy to guess at a particular digit for the quotient. For any step in the division, we can usually think of the tens' digit only, although even then the trial digit for the quotient may be too large. In long division, determining any digit of the quotient is a bit of guesswork.

18 ARITHMETIC

Example 5. Divide $784 \div 21$.

$$\begin{array}{r} 37 \\ 21\overline{)784} \\ 63 \\ \hline 154 \\ 147 \\ \hline 7 = \text{remainder} \end{array}$$

Solution. The work is shown at the left. First we try the tens' digit 2 of the divisor into the 7, the first digit of the dividend. That is, $7 \div 2$. The answer, 3, is placed in the quotient directly above the 8 of the dividend. We cannot be sure that the 3 is the correct digit for the quotient until we multiply back, $3 \times 21 = 63$. The 63 is placed directly below the 78 of the dividend. Subtracting, we get 15, and then bring down the 4, making 154 for the next number to be divided by 21. Next, we try dividing the 2 of the divisor into 15 and get 7 as a whole number. The 7 is placed in the quotient directly above the 4 of the dividend. Multiplying back, we get $7 \times 21 = 147$, which is placed below the 154. Subtracting, we get 7 as the remainder.

Example 6. Divide $1846 \div 31$.

$$\begin{array}{r} 59 \\ 31\overline{)1846} \\ 155 \\ \hline 296 \\ 279 \\ \hline 17 = \text{remainder} \end{array}$$

Solution. The division is shown at the left. First, we think of 3 into 18 of the dividend, which becomes 6 as a trial digit for the quotient. However, when we multiply back, 6×31, and place the product 186 below the divident, 184, we cannot subtract. Therefore, the 6 is too large for the first digit of the quotient. Then we must use a 5 for the first digit of the quotient. The final quotient is 59, with a remainder of 17.

If a two-digit divisor ends in 7, 8, or 9, such as in the number 29, it is well to try the next higher tens' digit into the first part of the dividend. Even so, one cannot be sure of the corresponding digit for the quotient.

Example 7. Divide $738 \div 29$.

$$\begin{array}{r} 25 \\ 29\overline{)738} \\ 58 \\ \hline 158 \\ 145 \\ \hline 13 = \text{remainder} \end{array}$$

Solution. The work is shown at the left. If we first think of taking 2 into 7, we might try 3 as the first digit of the quotient. However, it is best to remember that 29 is close to 30. Then we think of taking 3 into 7. If we take 2 into 7, we get 3 for the first digit of the quotient. However, when we multiply back, we get $3 \times 29 = 87$, which is too large to be subtracted from 73. Instead, if we think of taking 3 into 7, we get 2 for the first digit of the quotient. Multiplying back, we get $2 \times 29 = 58$. Now we can subtract 58 from 73, and get a remainder of 15. The next division is to take 29 into 158. Again, if we try 2 into 15, we get 7, which is too large, because $7 \times 29 = 203$. Instead, if we think of 3 into 15, we get 5, which is the correct answer for the next digit of the quotient.

Example 8. Divide $132 \div 28$.

Solution. In this example, we might first try taking 2 into 13, which is 7. However, we shall find that 7 is too large for the first digit of the quotient. Instead, we think of the divisor 28 being not far from 30. Then we think of taking 3 into 13, and get 4 as the first digit of the quotient. The work is left to the student.

WHOLE NUMBERS

Example 9. Divide $2152 \div 37$.

Solution. Here, we might first try taking 3 into 21. We get 7. However, again we shall find that 7 is too large for the first digit of the quotient. Instead, since 37 is not far from 40, we try 4 into 21, and get 5 as the first digit of the quotient. The work is left to the student.

Exercise 1.4

1. $32317 \div 4$
2. $45382 \div 5$
3. $58253 \div 6$
4. $42680 \div 7$
5. $38450 \div 8$
6. $26170 \div 9$
7. $72583 \div 9$
8. $63230 \div 7$
9. $48099 \div 8$
10. $86300 \div 8$
11. $93400 \div 9$
12. $73900 \div 7$
13. $84206 \div 7$
14. $96302 \div 8$
15. $94504 \div 9$
16. $7439 \div 21$
17. $7638 \div 31$
18. $17052 \div 41$
19. $32700 \div 61$
20. $13941 \div 51$
21. $51543 \div 81$
22. $58520 \div 71$
23. $24236 \div 91$
24. $10009 \div 23$
25. $20088 \div 32$
26. $31660 \div 43$
27. $40900 \div 52$
28. $29658 \div 42$
29. $21620 \div 53$
30. $25704 \div 63$
31. $19428 \div 24$
32. $22746 \div 74$
33. $58515 \div 83$
34. $36400 \div 64$
35. $30100 \div 73$
36. $29910 \div 84$
37. $68560 \div 92$
38. $65000 \div 19$
39. $12200 \div 28$
40. $24940 \div 37$
41. $26818 \div 46$
42. $25920 \div 59$
43. $34376 \div 68$
44. $58927 \div 77$
45. $32510 \div 86$
46. $65270 \div 95$
47. $55091 \div 89$
48. $37350 \div 75$
49. $58565 \div 85$
50. $444,444,444 \div 36$
51. $777,777,777 \div 63$
52. $555,555,555, \div 45$
53. $999,999,999 \div 81$
54. $12,345,679 \div 37$

55. An automobile trip of 576 miles required 32 gallons of gasoline. What was the average number of miles per gallon?
56. A trip of 850 miles required 53 gallons of gasoline. What was the average number of miles per gallon? Disregard the remainder.
57. An automobile trip of 1220 miles was made in 23 hours of travel time. What was the average number of miles per hour? Disregard the remainder.
58. A board 11 feet long is cut into pieces 16 inches long. How many such pieces can be cut from the board and how much remains?

Exercise 1.5

Review of Whole Numbers

1. A man weighs 148 pounds. How many ounces is that? (1 pound = 16 ounces.)
2. If you travel at an average of 378 miles per day for 6 days, find the total distance traveled in the 6 days.

20 ARITHMETIC

3. A touring trip of 2094 miles required 6 days. What was the average number of miles per day?
4. Find the number of hours, minutes, and seconds in 365 days.
5. One acre of land contains 43,560 square feet. How many square yards are there in 1 acre? (1 sq. yd. = 9 sq. ft.)
6. Find the number of square rods in 35 acres. (1 acre contains 160 sq. rd.)
7. One section of land contains 640 acres. How many square rods is that?
8. Find the cost of 38 acres of land at a price of $645 an acre.
9. Sound travels approximately 1080 feet per second. If the sound of thunder is heard 12 seconds after the flash is seen, how far away is the lightning?
10. Light travels 186,000 miles per second. How long does it take the light from the sun, about 93,000,000 miles away, to reach the earth? (omit the remainder.)
11. Find the length of a light-year, the distance light travels in 1 year.
12. In 22 hours of driving time, the odometer on a car changed from a reading of 35,479 to 36,722 miles. Find the average rate in miles per hour.
13. For the Apollo 15 flight, the J-1 rocket, first stage of V-5, had five engines, each with a thrust of 1,500,000 pounds. Find the total thrust.
14. The J-2 rocket, second stage of V-5, Apollo 15, had five engines, each with a thrust of 200,000 pounds. Find the total thrust of these engines.
15. At one point in the flight of Apollo 15, the fuel consumption was 28,000 pounds per second. If the burn continued for 150 seconds at this rate, what was the total fuel consumption, in tons, for the 150 seconds? (1 ton = 2000 lb.)
16. During one month, a family used the following appliances, with the indicated wattage, for the time indicated. Find the total number of *kwh* (kilowatt-hours) for the month. ("kilo" means "thousand"; 1 kwh = the use of 1000 watts for 1 hour.)

frying pan—1175 watts, 12 hours grill—1450 watts, 7 hours
toaster—850 watts, 9 hours lights—480 watts, 95 hours
other appliances—250 watts, 6 hours

2 COMMON FRACTIONS

2.1. NEED FOR FRACTIONS

As long as people used numbers only for counting, they had no need for fractions. A fraction indicates a part or parts of a whole number. In counting his sheep, early man could not say: $\frac{1}{2}$, $\frac{1}{4}$, and so on.

Fractions arose in connection with measuring. When we measure anything, whether length, weight, time, or other quantity, we use a *unit of measure*. The unit is simply a definite amount of the same kind of thing as the quantity that is to be measured. The measurement is made by noting how many times the quantity measured contains the unit.

When early man wished to measure a particular length (for example, the length of his hut), he may have picked up a stick of convenient length to use as a unit of measure. As he laid off the unit (the stick) along the wall of his hut, he may have laid off the unit five times and then had about a half unit left over. In this way, fractions came into arithmetic. In all measurement, we are faced with a *continuum* rather than separate and *discrete* objects. If arithmetic had been used only for counting "discrete" objects, fractions would never have been needed.

Counting is exact. All measurement is only approximate.

Exercise 2.1

Tell whether numbers are used for counting or measurement in each of the following examples.

1. The number of hairs on your head.
2. The length of a single hair.
3. The number of leaves on a certain tree.
4. Your weight.
5. The number of hours in a day.
6. The time length of a year.
7. The number of printed characters in a particular book.

8. The amount of ink required to print a particular book.
9. The area of a page of a book.
10. The number of grains of sand on a beach.

2.2. DEFINITION OF A FRACTION

There are two ways of looking at a fraction. That is, we can define the fraction $\frac{3}{4}$ in two ways. As a first way, consider a circle divided into four equal parts (Fig. 2.1). Each part is called a "fourth" and is denoted by the fraction $\frac{1}{4}$. The number below the line is called the *denominator* because it *denominates* or *names* the part (*denominate* means to *name*). It may be thought of as the *namer*. Now, if we take 3 of these "fourths," shown by the shaded portion of the circle we denote this amount by the fraction $\frac{3}{4}$. The number 3 above the line is called the *numerator* because it *enumerates* or *counts* the number of parts taken. It may be thought of as the *counter*.

There is a second way of looking at a fraction. A fraction may be considered as an indicated division. The fraction $\frac{3}{4}$ can be taken to mean 3 divided by 4, or $3 \div 4$. The horizontal line separating the numerator from the denominator can be taken as a symbol for division.

To show the difference between the two approaches, let us consider the meaning of $\frac{3}{4}$ of a dollar. First, we take the denominator 4 to mean that one dollar is first divided into four parts, each part called a "fourth" or a quarter, worth 25 cents. Then the numerator 3 means that we take 3 of the quarter. The value is 75 cents.

In the alternate approach, we look upon the fraction $\frac{3}{4}$ as $3 \div 4$; that is, 3 dollars divided into 4 parts. Then we have: $\$3.00 \div 4$. The answer is again $\$0.75$, which is 75 cents, the same as in the first approach. The answer is the same, but the approach is different.

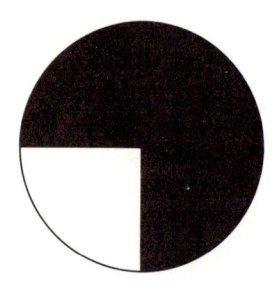

FIGURE 2.1

Note.   For convenience in typing and printing, a fraction is often indicated by a slanted line. The fraction $\frac{3}{4}$ is often written as 3/4.

As another example to illustrate the two different approaches, consider the meaning of $\frac{2}{3}$ of a foot. One-third of a foot is 4 inches; two-thirds of a foot is 8 inches. This is one approach. In the second approach, the fraction $\frac{2}{3}$ can be taken to mean 2 feet divided by 3; that is, 24 inches divided by 3. The answer is 8 inches, the same as before, but the approach is different.

There is often an advantage in considering a fraction as an indicated division. The fraction $\frac{5}{8}$ means $5 \div 8$. Moreover, division may be written as a fraction. For example, $211 \div 16$ can be written $\frac{211}{16}$.

In any fraction, the *numerator* and the *denominator* are called the *terms* of the fraction.

COMMON FRACTIONS 23

A *proper fraction* is a fraction in which the *numerator is less than the denominator*; that is, a proper fraction has a value *less than 1*, such as the following: $\frac{1}{3}, \frac{3}{4}, \frac{5}{8}, \frac{11}{16}, \frac{31}{32}$.

An *improper fraction* is a fraction in which the *numerator is equal to or greater than the denominator.* An improper fraction has a value *equal to or more than 1*, such as: $\frac{5}{3}, \frac{7}{7}, \frac{13}{8}, \frac{59}{12}, \frac{72}{18}$.

A *mixed number* is a number consisting of a *whole number and a fraction*, such as: $3\frac{1}{2}, 8\frac{2}{3}, 12\frac{5}{8}, 326\frac{13}{16}$.

It should be noted that a mixed number, such as $12\frac{5}{8}$, is really the *sum* of a whole number and a fraction. That is,

$$12\tfrac{5}{8} \quad \text{means} \quad 12+\tfrac{5}{8}$$

When we say that the length of a room is $15\frac{3}{4}$ feet, we mean that the length of the room is 15 feet and $\frac{3}{4}$ of a foot more, or $15+\frac{3}{4}$.

2.3. EQUIVALENT FRACTIONS

Two fractions having the same value may look different. The terms of one may be different from the terms of the other. The circle in Figure 2.2 is divided into 8 equal parts, each part called an *eighth*. Four of the parts are shaded. The 4 eighths can be expressed by the fraction $\frac{4}{8}$. However, notice that the shaded portion is one-half of the circle. Then we can say

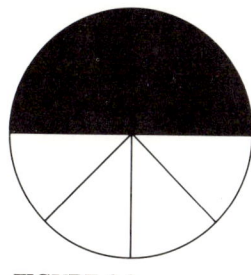

$$\frac{4}{8} \quad \text{is equivalent to} \quad \frac{1}{2}$$

FIGURE 2.2

As another example, a foot rule is divided into 12 inches, each inch equal to $\frac{1}{12}$ of a foot. If we mark off 9 inches, we have $\frac{9}{12}$ of a foot. However, note that $\frac{9}{12}$ of a foot is the same as $\frac{3}{4}$ of a foot. That is,

$$\frac{9}{12}=\frac{3}{4}$$

Although the fractions $\frac{9}{12}$ and $\frac{3}{4}$ have the same value, we often prefer to use $\frac{3}{4}$ because the terms are smaller. In the same way, if we mark off 8 inches on a foot rule, we have $\frac{8}{12}$ of a foot, but we might prefer to work with the fraction $\frac{2}{3}$, which has the same value as $\frac{8}{12}$.

A fraction can often be changed in form to an equivalent fraction having lower terms. Changing a fraction to lower terms is called *reducing* the fraction. Reducing a fraction makes the terms of the fraction smaller, but it does not change the value of the fraction.

24 ARITHMETIC

A fraction can also be changed to an equivalent fraction having *higher* terms. This is often necessary. For example, $\frac{5}{6}$ of a foot is equal to 10 inches, which is equal to $\frac{10}{12}$ of a foot. Then we can say that

$$\frac{5}{6} = \frac{10}{12}$$

2.4. FUNDAMENTAL PRINCIPLE OF FRACTIONS

In changing a fraction to higher or lower terms without changing the value of the fraction, we make use of the following principle:

FUNDAMENTAL PRINCIPLE OF FRACTIONS. *If the numerator and the denominator of any fraction are multiplied or divided by the same number (other than zero), the value of the fraction will not be changed.*

Example 1. Change the fraction $\frac{4}{7}$ to a new fraction having a denominator of 42.

Solution. To get a new denominator of 42, we must multiply the original denominator by 6. Then we also multiply the original numerator by 6. Then we get

$$\frac{4}{7} = \frac{4 \times 6}{7 \times 6} = \frac{24}{42}$$

The new fraction, $\frac{24}{24}$, has the same value as the original fraction, $\frac{4}{7}$.

You may wonder why the fraction $\frac{4}{7}$ has the same value when we multiply both numerator and denominator by a number such as 6. If we multiply the fraction $\frac{4}{7}$ by 1, the value will not be changed. In fact, any number multiplied by 1 is equal to the original number. However, the number 1 may be expressed as $\frac{3}{3}$, or $\frac{5}{5}$, or $\frac{6}{6}$, or any fraction equal to 1. Then, when we multiply the fraction $\frac{4}{7}$ by $\frac{6}{6}$, we are simply multiplying it by 1. Therefore, the value will not be changed.

Example 2. Reduce the fraction $\frac{42}{56}$ to lowest terms.

Solution. The fundamental principle also says that we may divide both numerator and denominator of any fraction by the same number. In the fraction $\frac{42}{56}$, we can divide both numerator and denominator by 2 and get the new fraction, $\frac{21}{28}$, which is equal to the original fraction. However, we can further divide both numerator and denominator by 7, and the fraction becomes $\frac{3}{4}$, which has the same value as the original fraction $\frac{42}{56}$. That is, dividing both numerator and denominator of any fraction by the same number (not zero) does not change the value of the fraction.

Incidentally, if we first divide both terms of a fraction by 2, and we could have divided both terms by 14, which is 2 times 7.

In reducing fractions to lower terms, we divide numerator and denominator by any number contained in both a whole number of times. Now the question arises: How can we tell what number can be divided into both numerator and denominator? Sometimes we can determine the number rather easily by inspection. In many cases the problem is more difficult. At this point, we need to know how to determine the *divisibility of numbers.* Here are some of the common tests. Let us test the number: 131,736.

1. A number is divisible by 2 if the last digit at the right is divisible by 2; that is, if the number is even. In this example, 6 is divisible by 2. Therefore, the entire number is divisible by 2.

2. A number is divisible by 4 if the number represented by the last two digits is divisible by 4. In the example, 36 is divisible by 4.

3. A number is divisible by 8 if the number represented by the last three digits is divisible by 8. In the example, 736 is divisible by 8.

4. A number is divisible by 3 if the sum of its digits is divisible by 3. In the example, $1+3+1+7+3+6=21$, divisible by 3.

5. A number is divisible by 9 if the sum of its digits is divisible by 9. In the example, 21 is not divisible by 9; hence the number itself is not.

6. A number is divisible by 10 if it ends in zero. This number does not.

7. A number is divisible by 5 if it ends in 5 or zero. This number does not.

8. For divisibility by 7, there is no simple test. Try it out.

9. A number is divisible by 11 if the difference between the sum of the alternate digits is zero or divisible by 11. In the example, for one set of digits we have: $1+1+3=5$; for the alternate digits: $3+7+6=16$. The difference is 11; therefore the number is divisible by 11.

Note. For divisibility by 3 or 9, the digits may be added until the result is only one digit. First, we get 21. Then we can add: $2+1=3$.

Exercise 2.2

Reduce the following fractions to lowest terms.

1. $\dfrac{6}{12}$ $\dfrac{5}{15}$ $\dfrac{7}{35}$ $\dfrac{9}{54}$ $\dfrac{10}{15}$ $\dfrac{14}{20}$ $\dfrac{16}{24}$

2. $\dfrac{4}{14}$ $\dfrac{9}{18}$ $\dfrac{4}{20}$ $\dfrac{9}{63}$ $\dfrac{12}{15}$ $\dfrac{24}{36}$ $\dfrac{15}{35}$

3. $\dfrac{13}{52}$ $\dfrac{18}{54}$ $\dfrac{11}{88}$ $\dfrac{14}{28}$ $\dfrac{15}{24}$ $\dfrac{20}{25}$ $\dfrac{14}{21}$

26 ARITHMETIC

4. $\dfrac{17}{51}$ $\dfrac{26}{39}$ $\dfrac{19}{76}$ $\dfrac{18}{45}$ $\dfrac{24}{30}$ $\dfrac{25}{35}$ $\dfrac{33}{36}$

5. $\dfrac{39}{52}$ $\dfrac{24}{60}$ $\dfrac{36}{45}$ $\dfrac{18}{63}$ $\dfrac{70}{84}$ $\dfrac{24}{40}$ $\dfrac{36}{63}$

6. $\dfrac{32}{48}$ $\dfrac{15}{50}$ $\dfrac{30}{42}$ $\dfrac{42}{56}$ $\dfrac{40}{64}$ $\dfrac{48}{72}$ $\dfrac{36}{78}$

7. $\dfrac{63}{84}$ $\dfrac{36}{96}$ $\dfrac{30}{42}$ $\dfrac{35}{77}$ $\dfrac{65}{91}$ $\dfrac{33}{110}$ $\dfrac{63}{108}$

8. $\dfrac{54}{66}$ $\dfrac{63}{81}$ $\dfrac{30}{48}$ $\dfrac{36}{96}$ $\dfrac{51}{85}$ $\dfrac{48}{132}$ $\dfrac{96}{240}$

9. $\dfrac{70}{168}$ $\dfrac{91}{117}$ $\dfrac{75}{165}$ $\dfrac{180}{252}$ $\dfrac{105}{189}$ $\dfrac{108}{120}$ $\dfrac{120}{192}$

10. $\dfrac{96}{108}$ $\dfrac{85}{119}$ $\dfrac{90}{252}$ $\dfrac{108}{144}$ $\dfrac{105}{168}$ $\dfrac{192}{288}$ $\dfrac{132}{143}$

11. $\dfrac{135}{360}$ $\dfrac{140}{224}$ $\dfrac{462}{616}$ $\dfrac{231}{528}$ $\dfrac{135}{225}$ $\dfrac{168}{378}$ $\dfrac{357}{595}$

12. $\dfrac{175}{225}$ $\dfrac{168}{840}$ $\dfrac{210}{672}$ $\dfrac{144}{192}$ $\dfrac{375}{600}$ $\dfrac{273}{351}$ $\dfrac{546}{637}$

13. $\dfrac{504}{1155}$ $\dfrac{1782}{2673}$ $\dfrac{3003}{8008}$ $\dfrac{1780}{9072}$ $\dfrac{6174}{6468}$ $\dfrac{1020}{1224}$ $\dfrac{2772}{3276}$

14. $\dfrac{735}{1764}$ $\dfrac{7182}{7938}$ $\dfrac{2310}{3234}$ $\dfrac{1260}{1512}$ $\dfrac{4165}{7595}$ $\dfrac{1092}{1456}$ $\dfrac{2574}{3762}$

Change the fractions in each of the following sets to new fractions with higher terms and with the indicated denominator.

15. $\frac{3}{4}$; $\frac{2}{3}$; $\frac{3}{8}$; $\frac{7}{12}$; denominator: 48.

16. $\frac{3}{5}$; $\frac{7}{8}$; $\frac{5}{12}$; $\frac{11}{40}$; denominator: 120.

17. $\frac{2}{3}$; $\frac{6}{7}$; $\frac{4}{5}$; $\frac{8}{15}$; denominator: 105.

18. $\frac{4}{9}$; $\frac{8}{27}$; $\frac{7}{12}$; $\frac{23}{36}$; denominator: 108.

2.5 CHANGING THE FORM OF NUMBERS CONTAINING FRACTIONS

Before we can work problems involving fractions, we must know certain rules for operating with such numbers. One thing we should know is how to change the form of a fraction.

An improper fraction can be changed to a whole number or a mixed number. For example, the improper fraction $\frac{18}{3}$ can be changed to a whole number by dividing the numerator 18 by the denominator 3. The result is

6. Then $\frac{18}{3} = 6$. As another example, the improper fraction $\frac{23}{4}$ can be changed to a mixed number by dividing the numerator 23 by the denominator 4. We know there are 4 fourths in a whole unit. To find the number of whole units in 23 fourths, we divide 23 by 4 and get 5 whole units and 3 fourths over. Therefore,

$$\frac{23}{4} = 5 + \frac{3}{4}, \quad \text{which is written } 5\frac{3}{4}$$

RULE 1. To change an improper fraction to a whole number or a mixed number, divide the numerator by the denominator. The result is a whole number and a possible remainder. Any remainder is placed over the original denominator as a proper fraction.

Sometimes it is desirable to change a whole number or a mixed number to an improper fraction. For example, we may wish to change the whole number 7 into thirds. Since there are 3 thirds in the whole number 1, then in the number 7 there are 7 times 3 or 21 thirds. That is,

$$7 = \frac{21}{3}$$

RULE 2. To change a whole number to an improper fraction with any desired denominator, multiply the whole number by the desired denominator and place the result over the desired denominator.

In some cases we may wish to change a mixed number such as $5\frac{3}{4}$ to an improper fraction. We first change the 5 to $\frac{20}{4}$, and then add the $\frac{3}{4}$. Then

$$5\frac{3}{4} = \frac{20}{4} + \frac{3}{4} = \frac{23}{4}$$

RULE 3. To change a mixed number to an improper fraction, multiply the whole number by the denominator, and to the result add the numerator of the fractional part. Then place the total over the given denominator.

Exercise 2.3

Change the following improper fractions to whole numbers or mixed numbers:

1. $\frac{15}{4}$
2. $\frac{16}{3}$
3. $\frac{40}{5}$
4. $\frac{53}{6}$
5. $\frac{31}{7}$
6. $\frac{52}{5}$

28 ARITHMETIC

7. $\dfrac{65}{6}$
8. $\dfrac{100}{7}$
9. $\dfrac{45}{8}$
10. $\dfrac{72}{9}$
11. $\dfrac{83}{10}$
12. $\dfrac{50}{11}$
13. $\dfrac{97}{12}$
14. $\dfrac{70}{13}$
15. $\dfrac{83}{8}$
16. $\dfrac{72}{12}$
17. $\dfrac{96}{16}$
18. $\dfrac{113}{10}$
19. $\dfrac{149}{12}$
20. $\dfrac{115}{15}$
21. $\dfrac{205}{16}$
22. $\dfrac{117}{13}$
23. $\dfrac{255}{17}$
24. $\dfrac{290}{19}$
25. $\dfrac{140}{15}$
26. $\dfrac{150}{18}$
27. $\dfrac{133}{20}$
28. $\dfrac{2607}{30}$
29. $\dfrac{4515}{24}$
30. $\dfrac{5467}{66}$

Change each of the following integers to thirds, fifths, and twelfths:

31. 5
32. 7
33. 9
34. 10
35. 15
36. 18

Change each of the following mixed numbers to improper fractions:

37. $4\tfrac{1}{3}$
38. $6\tfrac{3}{4}$
39. $7\tfrac{2}{5}$
40. $5\tfrac{1}{6}$
41. $8\tfrac{4}{7}$
42. $10\tfrac{5}{8}$
43. $12\tfrac{2}{9}$
44. $13\tfrac{3}{5}$
45. $9\tfrac{3}{8}$
46. $14\tfrac{5}{6}$
47. $16\tfrac{7}{8}$
48. $25\tfrac{5}{9}$
49. $13\tfrac{7}{11}$
50. $20\tfrac{4}{15}$
51. $12\tfrac{13}{16}$
52. $18\tfrac{11}{32}$
53. $30\tfrac{11}{12}$
54. $36\tfrac{9}{10}$
55. $33\tfrac{1}{3}$
56. $66\tfrac{2}{3}$
57. $124\tfrac{2}{7}$
58. $316\tfrac{12}{13}$
59. $411\tfrac{7}{18}$
60. $122\tfrac{13}{24}$
61. $118\tfrac{4}{7}$
62. $211\tfrac{1}{9}$
63. $99\tfrac{1}{11}$
64. $214\tfrac{2}{7}$
65. $142\tfrac{6}{7}$
66. $12\tfrac{17}{64}$

2.6. ADDITION AND SUBTRACTION OF FRACTIONS

The addition or subtraction of two or more fractions having the same denominator is simply a matter of adding or subtracting the numerators. In such problems, the fractions have the same name. For example, adding 4 dollars and 3 dollars, we get 7 dollars. In the same way, 4 fifths + 3 fifths = 7 fifths. Stated as fractions, we have

$$\frac{4}{5}+\frac{3}{5}=\frac{7}{5}$$

That is, if the fractions have the same denominator, we add or subtract the numerators and place the result over the common denominator.

Example 1

$$\frac{7}{8}+\frac{5}{8}-\frac{3}{8}=\frac{7+5-3}{8}=\frac{9}{8}=1\frac{1}{8}$$

Example 2

$$\frac{8}{9}+\frac{7}{9}+\frac{2}{9}-\frac{5}{9}=\frac{8+7+2-5}{9}=\frac{12}{9}=1\frac{3}{9}=1\frac{1}{3}$$

If the fractions have different denominators, we must first change the form of some or all of the fractions so that all have the same denominator.

First, we must find the *lowest common denominator* (LCD). The lowest common denominator is the smallest number that can be divided by all the given denominators. Then we change each fraction to a new fraction having the LCD as its denominator. After making this change, we proceed as usual for fractions with the same denominator.

Example 3. Add or subtract as indicated:

$$\frac{2}{3}+\frac{7}{8}-\frac{5}{12}=$$

Solution. The denominators are 3, 8, and 12. By inspection, we find that the LCD is 24. Now, each fraction is expressed as a fraction having a denominator of 24. To change the form of each fraction, we use the *fundamental principle of fractions*. Let us first write the denominator for each fraction:

$$\frac{}{24}+\frac{}{24}-\frac{}{24}$$

The fraction $\frac{2}{3}$ is changed to $\frac{16}{24}$ by multiplying the numerator and the denominator by 8. The fraction $\frac{7}{8}$ is changed to $\frac{21}{24}$ by multiplying the numerator and denominator by 3. The fraction $\frac{5}{12}$ is changed to $\frac{10}{24}$ by multiplying numerator and denominator by 2. Now the fractions can be combined:

$$\frac{16}{24} + \frac{21}{24} - \frac{10}{24} = \frac{16+21-10}{24} = \frac{27}{24} = 1\frac{3}{24} = 1\frac{1}{8}$$

In adding mixed numbers, we might use any one of several methods. Probably the most practical way is to place the numbers in column form. Then we add the whole numbers and fractions separately. To add $3\frac{1}{2}$, $9\frac{1}{4}$, $27\frac{4}{5}$, and $43\frac{7}{10}$, we place them in column form as follows:

		Short Form		or	20 (LCD)	
$3\frac{1}{2} = 3\frac{10}{20}$		$3\frac{1}{2}$	$\frac{10}{20}$		3	10
$9\frac{1}{4} = 9\frac{5}{20}$		$9\frac{1}{4}$	$\frac{5}{20}$		9	5
$27\frac{4}{5} = 27\frac{16}{20}$		$27\frac{4}{5}$	$\frac{16}{20}$		27	16
$43\frac{7}{10} = 43\frac{14}{20}$		$43\frac{7}{10}$	$\frac{14}{20}$		43	14
82 + $\frac{45}{20} = 84\frac{5}{20} = 84\frac{1}{4}$					82 +	$\frac{45}{20}$

Each fraction has been expressed as a new fraction with the lowest common denominator. Then the fractions and the whole numbers are added separately. The sum of the whole numbers is 82. The sum of the fractions is $\frac{45}{20}$ which reduces to $2\frac{1}{4}$. This quantity is combined with the 82, and the total becomes $84\frac{1}{4}$.

In some instances it is convenient, when adding fractions, to change mixed numbers to improper fractions. This method is not practical if the whole numbers are large. However, when the whole numbers are small, this method is often simple.

Example 4. Add $\frac{3}{4} + \frac{5}{6} + 2\frac{1}{12} + 3$.

Solution. We can change all the numbers to twelfths. Then we have

$$\frac{9}{12} + \frac{10}{12} + \frac{25}{12} + \frac{36}{12} = \frac{80}{12} = 6\frac{2}{3}$$

Subtraction of numbers involving fractions and mixed numbers is usually done by placing the subtrahend below the minuend as with whole numbers. However, in subtraction we often run into difficulties. The following examples show the subtraction of fractions and mixed numbers.

COMMON FRACTIONS 31

Example 5. Subtract $92\frac{7}{8} - 54\frac{1}{8}$.

This problem presents no difficulty:

Subtracting,
$$\begin{array}{r} 92\frac{7}{8} \\ 54\frac{1}{8} \\ \hline 36\frac{6}{8} = 36\frac{3}{4} \end{array}$$

Example 6. Subtract $56\frac{5}{9} - 12\frac{5}{9}$.

Subtracting
$$\begin{array}{r} 56\frac{5}{9} \\ 12\frac{5}{9} \\ \hline 44 \end{array}$$

In this example, we subtract $\frac{5}{9}$ from $\frac{5}{9}$, which is zero. However, the zero (0) must *not* be written down below the fractions. If it is so written, the entire answer would appear to be 440, which is not correct.

Example 7. Subtract $36\frac{5}{8} - 23$.

$$\begin{array}{r} 36\frac{5}{8} \\ 23 \\ \hline 13\frac{5}{8} \end{array}$$

In this example, we subtract nothing from $\frac{5}{8}$. We simply bring down the $\frac{5}{8}$ as the fractional part of the remainder.

Example 8. Subtract $36 - 23\frac{5}{8}$.

$$\begin{array}{rr} \overset{5}{3\cancel{6}\frac{8}{8}} & 35\frac{8}{8} \\ 23\frac{5}{8} & 23\frac{5}{8} \\ & \hline 12\frac{3}{8} \end{array}$$

In this example, we must borrow one whole unit from the 6 and change it to $\frac{8}{8}$ before we can subtract the $\frac{5}{8}$.

Example 9. Subtract $82\frac{4}{7} - 58\frac{6}{7}$.

$$\begin{array}{rr} 82\frac{4}{7} & 81\frac{11}{7} \\ 58\frac{6}{7} & 58\frac{6}{7} \\ & \hline 25\frac{5}{7} \end{array}$$

32 ARITHMETIC

In this example, we cannot subtract $\frac{6}{7}$ from $\frac{4}{7}$. Therefore, we borrow one whole unit from the 2 in 82, and change this unit to $\frac{7}{7}$. The quantity $\frac{7}{7}$, combined with the $\frac{4}{7}$ in the minuend, makes $\frac{11}{7}$. Now we subtract $\frac{6}{7}$ from $\frac{11}{7}$, leaving $\frac{5}{7}$ as the fractional part of the remainder.

Example 10. Combine the fractions $10\frac{5}{8} - 4\frac{2}{3} - 1\frac{7}{12}$.

We first change the fractions to 24ths; we get

$$10\frac{15}{24} - 4\frac{16}{24} - 1\frac{14}{24}$$

We must borrow 1 from 10 and change it to 24ths. Combining this with the $\frac{15}{24}$, we have

$$9\frac{39}{24} - 4\frac{16}{24} - 1\frac{14}{24} = 4\frac{9}{24} = 4\frac{3}{8}$$

2.7. FINDING THE LOWEST COMMON DENOMINATOR

We have seen that if fractions are to be added or subtracted, they must have the same denominator. If the denominators are different, then the form of the fractions must be changed. The fractions must all have the same denominator. The first step in changing fractions to higher terms is to determine some number that is divisible by all of the given denominators.

In some examples, the lowest common denominator (LCD) can be determined by inspection, which is simply a good guess. However, in many instances, we need a more systematic method for determining the LCD of two or more fractions.

It is first necessary to understand what is meant by a *multiple* of a number. A *multiple* of any given number is some number that is exactly divisible by the given number. The following numbers are all multiples of 8 because they are all exactly divisible by 8: 8, 16, 24, 32, 40, 48, 56, and so on. The following numbers are all multiples of 12 because they are all exactly divisible by 12: 12, 24, 36, 48, 60, 72, and so on.

A *common multiple* of two or more numbers is some number that is exactly divisible by each of the given numbers. For example, the following numbers are common multiples of 8 and 12 because all can be divided exactly by 8 and 12: 24, 48, 72, 96, and so on. Notice that the number 24 is the smallest number that is exactly divisible by both 8 and 12. Such a number is called the *lowest common multiple* (LCM) of the two numbers 8 and 12. The *lowest common multiple of two or more numbers is the smallest number that is exactly divisible by each of the given numbers*. When we add

COMMON FRACTIONS 33

or subtract fractions, we first find the lowest common multiple of the denominators. This number is the *lowest common denominator* of the fractions.

There are several ways of determining the lowest common multiple of two or more numbers. One method is shown in the following example.

Example. Find the lowest common multiple of the numbers: 24, 45, 60, 75.

Step 1. We write each number as a product of its prime factors. To find the prime factors of a number, we begin dividing the number by the smallest number contained in the given number. For example, to find the prime factors of 24, we divide 24 by 2, then the result by 2, and so on until the quotient is not divisible by any other factor. The result of these divisions shows that the prime factors of 24 are 2, 2, 2, and 3. We do the same for the other numbers and then write each as a product of its factors.

$$24 = 2 \cdot 2 \cdot 2 \cdot 3$$

$$45 = 3 \cdot 3 \cdot 5$$

$$60 = 2 \cdot 2 \cdot 3 \cdot 5$$

$$75 = 3 \cdot 5 \cdot 5$$

Step 2. Now, for the LCM, we write down each of the prime factors as many times as it is contained in any one of the given numbers. The factor 2 is contained three times in 24, which is the greatest number of times we find the factor 2 in any one number. Therefore, we set down: $2 \cdot 2 \cdot 2$, as part of the LCM. The factor 3 is contained twice in one of the given numbers. Therefore, we use two 3's as part of the LCM. The factor 5 is found twice in one of the numbers. The LCM is therefore the product of the following factors:

$$2 \cdot 2 \cdot 2 \cdot 3 \cdot 3 \cdot 5 \cdot 5$$

The product, 1800, is the lowest common multiple of 24, 45, 60, and 75; that is, 1800 is the smallest number that will contain all these numbers. This means that if we have the denominators, 24, 45, 60, and 75, of fractions, the lowest common denominator is 1800.

34 ARITHMETIC

Exercise 2.4

Find the lowest common multiple of each of these numbers:

1. 3, 4, 6
2. 2, 3, 5
3. 2, 5, 6
4. 6, 8, 12
5. 3, 5, 6
6. 3, 8, 12
7. 3, 4, 5
8. 2, 3, 7
9. 2, 6, 7
10. 2, 3, 4, 6
11. 2, 3, 4, 5
12. 3, 4, 6, 9
13. 4, 6, 9, 24
14. 5, 10, 15, 30
15. 2, 5, 15, 20
16. 30, 42, 60
17. 24, 54, 60, 90
18. 20, 84, 90, 180
19. 180, 210, 315
20. 120, 180, 300
21. 120, 270, 600, 800
22. 54, 324, 225, 900

23. Add the fractions in each set:

$$\frac{14}{15} \quad \frac{13}{16} \quad \frac{17}{18} \quad \frac{7}{9}$$
$$\frac{5}{12} \quad \frac{1}{12} \quad \frac{20}{27} \quad \frac{5}{8}$$

24. Add the fractions in each set:

$$\frac{11}{12} \quad \frac{13}{18} \quad \frac{17}{20} \quad \frac{29}{36}$$
$$\frac{7}{18} \quad \frac{11}{30} \quad \frac{37}{50} \quad \frac{17}{54}$$

25. In No. 23, subtract the bottom fraction from the top in each set.
26. In No. 24, subtract the bottom fraction from the top in each set.
27. Add the mixed numbers in each set:

$$17\tfrac{2}{3} \quad 78\tfrac{3}{4} \quad 49 \quad 78\tfrac{1}{4} \quad 6\tfrac{3}{4} \quad 9\tfrac{2}{3} \quad 17\tfrac{5}{8} \quad 48\tfrac{4}{9}$$
$$14\tfrac{2}{3} \quad 35 \quad 16\tfrac{5}{8} \quad 77\tfrac{3}{4} \quad 2\tfrac{5}{8} \quad 6\tfrac{3}{4} \quad 9\tfrac{7}{10} \quad 27\tfrac{11}{12}$$

28. Add the mixed numbers in each set:

$$56\tfrac{7}{9} \quad 16 \quad 14\tfrac{5}{8} \quad 46\tfrac{2}{5} \quad 8\tfrac{5}{6} \quad 15\tfrac{3}{4} \quad 6\tfrac{4}{9} \quad 51\tfrac{7}{15}$$
$$46\tfrac{7}{9} \quad 15\tfrac{3}{5} \quad 14 \quad 13\tfrac{3}{8} \quad 8\tfrac{3}{8} \quad 8\tfrac{2}{5} \quad 5\tfrac{5}{6} \quad 41\tfrac{13}{18}$$

29. In No. 27, subtract the bottom number from the top in each set.
30. In No. 28, subtract the bottom number from the top in each set.

Add and subtract the following as indicated. Reduce results if possible.

31. $7\tfrac{2}{5} + 4\tfrac{7}{8}$
32. $4\tfrac{5}{6} + 8\tfrac{4}{7}$
33. $6\tfrac{3}{4} + 5\tfrac{5}{9}$
34. $2\tfrac{7}{12} + 6\tfrac{5}{16}$
35. $6\tfrac{3}{8} + 8\tfrac{5}{12}$
36. $5\tfrac{7}{24} + 3\tfrac{13}{36}$
37. $\tfrac{7}{8} - \tfrac{5}{6}$
38. $\tfrac{5}{8} - \tfrac{3}{10}$
39. $\tfrac{11}{15} - \tfrac{7}{10}$
40. $12\tfrac{5}{16} - 8$
41. $9 - 5\tfrac{3}{8}$
42. $23 - 14\tfrac{7}{12}$
43. $38\tfrac{7}{10} - 26\tfrac{4}{15}$
44. $53\tfrac{5}{12} - 13\tfrac{8}{9}$
45. $91\tfrac{5}{24} - 48\tfrac{9}{16}$
46. $52\tfrac{7}{20} - 32\tfrac{7}{20}$

47. $11\frac{5}{12} - 10\frac{5}{12}$
48. $7\frac{4}{15} - 7$
49. $5\frac{4}{9} + 2\frac{7}{12} - \frac{5}{8}$
50. $7\frac{4}{15} - 4\frac{5}{6} + 6\frac{7}{10}$
51. $3\frac{11}{15} + 6\frac{9}{10} - 5\frac{3}{8}$
52. $9\frac{4}{9} - 3\frac{8}{15} - 4\frac{11}{18}$
53. $5\frac{8}{15} - 1\frac{5}{12} - 2\frac{7}{8}$
54. $7\frac{8}{45} - 5\frac{7}{30} - 1\frac{5}{24}$

55. Four bars of iron weigh, respectively, $5\frac{3}{4}$ lb, $6\frac{2}{3}$ lb, $8\frac{5}{6}$ lb, and $10\frac{4}{9}$ lb. Find the total weight of the four bars.

56. A strip of iron was cut into four pieces whose lengths were as follows, respectively: $13\frac{3}{8}$ in., $17\frac{5}{16}$ in., $21\frac{1}{4}$ in., and $22\frac{1}{2}$ in. What was the length of the original strip?

57. A rectangle is $7\frac{3}{8}$ inches long and $3\frac{5}{16}$ inches wide. What is the total distance around the rectangle, including both ends and both sides?

58. Three pieces measuring, respectively, $4\frac{3}{8}$ in., $5\frac{1}{4}$ in., and $5\frac{5}{16}$ in., were cut from a strip of aluminum 20 inches long. How much remained?

59. Two pieces measuring $7\frac{3}{8}$ inches and $9\frac{7}{16}$ inches, respectively, were cut from a strip of brass 30 inches long. If the amount of waste for each cut was $\frac{1}{32}$ of an inch, how much of the strip remained?

60. In a certain electric circuit, three resistors, testing $20\frac{1}{2}$ ohms, $26\frac{3}{4}$ ohms, and $38\frac{7}{8}$ ohms, respectively, are arranged in series. Find the resistance of a fourth resistor if the total resistance in series must be 120 ohms. (In an electric circuit, if resistors are arranged in series, the total resistance is found by adding the resistances.)

2.8. MULTIPLICATION OF FRACTIONS

Most people who read this will probably recall the following rule:

RULE. In multiplying fractions, multiply the numerators together for the numerator of the product, and multiply the denominators together for the denominator of the product.

Example 1

$$\frac{3}{5} \cdot \frac{2}{7} = \frac{6}{35}$$

Example 2

$$\frac{5}{\cancel{6}_2} \cdot \frac{\cancel{3}}{4} = \frac{5}{8}$$

In Example 2, if the numerators are first multiplied together, and then the denominators multiplied together, we shall find that a factor "3" can be divided into both numerator and denominator of the answer. This "3" can be divided into a numerator and a denominator before the terms are multiplied together. To show this division, we draw a slanted line through the 3 and the 6, to show that each has been divided by 3.

36 ARITHMETIC

Example 3. Multiply $\frac{8}{15} \cdot \frac{5}{12} \cdot \frac{6}{7}$.

Solution. In this example, three numbers can be divided into numerators and denominators; namely 3, 4, and 5.

$$\frac{\overset{2}{\cancel{8}}}{\underset{3}{\cancel{15}}} \cdot \frac{\cancel{5}}{\underset{\cancel{3}}{\cancel{12}}} \cdot \frac{\overset{2}{\cancel{6}}}{7} = \frac{4}{21}$$

Example 4. Multiply $1\frac{7}{8} \cdot 3\frac{5}{9}$.

Solution. The best way to multiply mixed numbers is to change the mixed numbers to improper fractions. Then this problem becomes

$$\frac{\overset{5}{\cancel{15}}}{\cancel{8}} \cdot \frac{\overset{4}{\cancel{32}}}{\underset{3}{\cancel{9}}} = \frac{20}{3} = 6\frac{2}{3}$$

Example 5. Find $\frac{2}{3}$ of $2\frac{5}{8}$.

Solution. The word "of" can be replaced by the multiplication sign. Then we have

$$\frac{\cancel{2}}{\cancel{3}} \cdot \frac{\overset{7}{\cancel{21}}}{\underset{4}{\cancel{8}}} = \frac{7}{4} = 1\frac{3}{4}$$

Example 5. Multiply $(\frac{5}{8})(376\frac{3}{4})$.

Solution. Changing a mixed number to an improper fraction sometimes leads to a large improper fraction. In some problems, the multiplication may instead be done in parts, although this method is probably susceptible to error. First we arrange the numbers in column form. We call $\frac{5}{8}$ the multiplier, and the $376\frac{3}{4}$ the multiplicand.

$$376\frac{3}{4}$$
$$\frac{5}{8}$$

First we multiply the fractions and get $\frac{15}{32}$. Now we multiply $\frac{5}{8}$ times 376 and get 235. Adding the fraction to 235, we get the product, $235\frac{15}{32}$.

2.9. RECIPROCAL OF A NUMBER

Before beginning division involving fractions, we need to know the meaning of the *reciprocal* of a number. The *reciprocal* of any number is defined as "1 divided by the number." For example, the reciprocal of 7 is $\frac{1}{7}$. The reciprocal of 25 is $1 \div 25$, which is $\frac{1}{25}$.

The reciprocal of a fraction is found in the same way. For example, the reciprocal of $\frac{3}{5}$ is $1 \div \frac{3}{5}$. This can be written as

$$\frac{1}{\frac{3}{5}}$$

In this fraction, the numerator is 1; the denominator is $\frac{3}{5}$. By use of the fundamental principle of fractions, we multiply numerator and denominator by 5, and get

$$\frac{5 \times 1}{5 \times \frac{3}{5}} = \frac{5}{3}$$

Therefore, the reciprocal of $\frac{3}{5}$ is $\frac{5}{3}$. Then we have the rule:

RULE. To find the reciprocal of a fraction, simply invert the fraction.

As an example of this rule, the reciprocal of $\frac{4}{7}$ is $\frac{7}{4}$.

By definition, then, the product of any number and its reciprocal is 1. For example,

$$\frac{3}{8} \cdot \frac{8}{3} = 1; \quad 4 \cdot \frac{1}{4} = 1; \quad \frac{7}{9} \cdot \frac{9}{7} = 1$$

2.10. DIVISION INVOLVING FRACTIONS

Most people are probably familiar with the following rule:

RULE. In the division involving fractions, multiply the dividend by the reciprocal of the divisor. In other words, invert the divisor and then multiply.

Example 2. Divide $\frac{4}{7} \div \frac{3}{5} = \frac{4}{7} \cdot \frac{5}{3} = \frac{20}{21}$.

Let us see why this rule is true. The rule depends on two facts: (a) *the product of a number and its reciprocal is* 1; (b) *any number divided by 1 is equal to the number itself.*

38 ARITHMETIC

Now let us write the foregoing problem with the dividend $\frac{4}{7}$ as the numerator, and the divisor $\frac{3}{5}$ as the denominator:

$$\frac{\frac{4}{7}}{\frac{3}{5}}$$

This arrangement is called a *complex fraction* because the numerator and the denominator themselves are fractions. The numerator is $\frac{4}{7}$, and the denominator is $\frac{3}{5}$. Now let us multiply the numerator and the denominator by $\frac{5}{3}$. We get

$$\frac{\frac{4}{7} \cdot \frac{5}{3}}{\frac{3}{5} \cdot \frac{5}{3}} = \frac{\frac{4}{7} \cdot \frac{5}{3}}{1} = \frac{4}{7} \cdot \frac{5}{3}$$

Notice that the result is exactly what the rule says. The answer is $\frac{20}{21}$.

The following examples show the application of this rule:

(a) $4 \div \frac{3}{5} = 4 \cdot \frac{5}{3} = \frac{20}{3} = 6\frac{2}{3}$ (b) $\frac{3}{5} \div 4 = \frac{3}{5} \cdot \frac{1}{4} = \frac{3}{20}$

(c) $6\frac{3}{4} \div 2\frac{5}{8} = \frac{27}{4} \div \frac{21}{8} = \frac{27}{4} \cdot \frac{8}{21} = \frac{18}{7} = 2\frac{4}{7}$

A problem involving several multiplications and divisions can be simplified by using the rules for multiplication and division of fractions. In some problems, we may find it necessary to multiply several numbers together, and then divide the product by several other numbers. For example, we may need to multiply the following: $14 \times 18 \times 20$; and then divide the product by 12, 15, and 35.

Instead of doing all the separate operations, we can arrange the problem as a single fraction. All the numbers to be multiplied together are written in the numerator. All the divisors are written as a product in the denominator. Then the problem becomes

$$\frac{14 \times 18 \times 20}{12 \times 15 \times 35}$$

Now the work can be simplified by first dividing numerators and denominators by common factors before multiplying. The following factors can be divided into numerators and denominators: 2, 2, 3, 3, 5, and 7. Then the answer simplifies to $\frac{4}{5}$.

Warning. The student is warned: This method of simplifying an expression cannot be used when additions or subtractions appear in the numerator or the denominator or both.

COMMON FRACTIONS 39

For example, in the following fraction form, the numerator and the denominator must be expanded first:

$$\frac{6 \times 10 + 1}{8 \times 15 - 2} = \frac{61}{118}, \quad \text{which cannot be reduced}$$

A problem can often be set up in a form suitable for reducing. All numbers that are to be multiplied together are placed as factors in the numerator. All numbers that are to be used as divisors are written as a *product in the denominator.*

Example 2. A man drives 360 miles in 9 hours. What is his average speed in feet per second?

Solution. A distance of 360 miles is changed to feet by multiplying by 5280; that is, 360×5280. The total number of feet is left in this form, not multiplied together. The number of feet is now divided by the number of seconds in 9 hours, which is $(9)(60)(60)$. Now we set up the problem:

$$\frac{360 \times 5280}{9 \times 60 \times 50} = 58\tfrac{2}{3} \text{ feet per second.}$$

The work is simplified by dividing numerator and denominator by any numbers found in both. Both can be divided by 10, 10, 9, 6, and 2.

When multiplications and divisions of fractions occur in the same problem, these operations are performed *in the order in which they occur*, unless otherwise indicated by parentheses or some other form of grouping.

Example 3. Simplify $\tfrac{3}{5} \cdot \tfrac{5}{6} \cdot \tfrac{8}{9}$.

Solution. We can divide numerators and denominators by 2, 3, and 5. Then the answer reduces to $\tfrac{4}{9}$.

Example 4. Simplify $3\tfrac{3}{8} \times 6\tfrac{2}{3} \div 1\tfrac{4}{5}$.

Solution. We change the numbers to improper fractions and invert only the last fraction:

$$\frac{27}{8} \times \frac{20}{3} \times \frac{5}{9} = \frac{25}{2} = 12\tfrac{1}{2}$$

Example 5. Simplify $3\tfrac{3}{4} \div 2\tfrac{2}{3} \times 3\tfrac{1}{5}$.

Solution. We change the numbers to improper fractions and invert only the second fraction:

$$\frac{15}{4} \times \frac{3}{8} \times \frac{16}{5} = \frac{9}{2} = 4\tfrac{1}{2}$$

40 ARITHMETIC

Example 6. Simplify $8\frac{3}{4} \div 3\frac{1}{3} \div 1\frac{2}{5}$.

Solution. We change to improper fractions and invert the fractions after the division signs:

$$\frac{35}{4} \times \frac{3}{10} \times \frac{5}{7} = \frac{15}{8} = 1\frac{7}{8}$$

Exercise 2.5

Multiply or divide as indicated:

1. $\dfrac{2}{5} \times \dfrac{4}{7}$
2. $\dfrac{5}{9} \times \dfrac{7}{8}$
3. $\dfrac{32}{45} \times \dfrac{35}{48}$
4. $\dfrac{12}{35} \times \dfrac{25}{36}$
5. $18 \times \dfrac{4}{9}$
6. $\dfrac{5}{8} \times 24$
7. $6\frac{3}{8} \times 9\frac{1}{3}$
8. $3\frac{5}{9} \times 5\frac{1}{4}$
9. $\dfrac{4}{15} \times 33$
10. $36 \times \dfrac{5}{16}$
11. $4\frac{5}{8} \times 36$
12. $5\frac{3}{16} \times 24$
13. $17\frac{3}{5} \times 2\frac{5}{8}$
14. $13\frac{5}{7} \times 3\frac{7}{8}$
15. $26\frac{4}{7} \times 1\frac{3}{4}$
16. $16\frac{4}{9} \times 4\frac{1}{8}$
17. $15\frac{6}{7} \times 5\frac{4}{9}$
18. $14\frac{3}{8} \times 8\frac{2}{7}$
19. $12\frac{16}{21} \times 4\frac{7}{8}$
20. $13\frac{7}{16} \times 5\frac{3}{5}$
21. $\dfrac{3}{4} \div \dfrac{5}{8}$
22. $\dfrac{7}{12} \div \dfrac{5}{6}$
23. $12 \div \dfrac{9}{10}$
24. $27 \div \dfrac{18}{25}$
25. $5\frac{1}{3} \div 24$
26. $11\frac{2}{3} \div 15$
27. $\dfrac{4}{9} \div 2\frac{2}{3}$
28. $\dfrac{16}{25} \div 1\frac{3}{5}$
29. $2\frac{5}{8} \div \dfrac{3}{10}$
30. $2\frac{4}{15} \div \dfrac{8}{9}$
31. $3\frac{13}{24} \div 7\frac{3}{16}$
32. $5\frac{11}{16} \div 9\frac{11}{12}$
33. $\dfrac{14}{25} \times \dfrac{10}{21} \div \dfrac{8}{15}$
34. $\dfrac{8}{15} \times \dfrac{5}{12} \div \dfrac{7}{6}$
35. $\dfrac{15}{16} \div \dfrac{5}{12} \times \dfrac{5}{18}$
36. $\dfrac{15}{32} \div \dfrac{5}{8} \div \dfrac{9}{16}$
37. $11\frac{2}{3} \div \dfrac{14}{15} \div 10$
38. $\dfrac{14}{25} \div \left(\dfrac{7}{5} \div \dfrac{15}{16}\right)$

39. $6\frac{2}{3} \times 5\frac{1}{4} \div 4\frac{2}{3}$
40. $3\frac{3}{4} \times 4\frac{2}{3} \div 5\frac{1}{4}$
41. $5\frac{1}{3} \div 3\frac{1}{5} \times 2\frac{1}{4}$
42. $4\frac{3}{8} \div 5\frac{1}{4} \times 3\frac{1}{5}$
43. $8\frac{3}{4} \div 4\frac{1}{6} \div 3\frac{1}{2}$
44. $5\frac{5}{6} \div 2\frac{5}{8} \div 5\frac{1}{3}$
45. $\dfrac{28 \times 75 \times 66}{125 \times 44 \times 27}$
46. $\dfrac{54 \times 55 \times 98}{35 \times 18 \times 88}$
47. $\dfrac{18 \times 30 \times 77}{22 \times 42 \times 45}$
48. $\dfrac{12 \times 60 + 1}{48 \times 30 - 2}$
49. $\dfrac{2 + 18 \times 45}{5 + 27 \times 50}$
50. $\dfrac{40 \times 50 \times 60}{40 + 50 + 60}$

Exercise 2.6

1. A motor makes 2350 revolutions per minute (rpm). How many revolutions will it make in $5\frac{1}{4}$ minutes?
2. If a man works $7\frac{3}{4}$ hours per day, how many hours does he work in $21\frac{1}{2}$ days?
3. A particular alloy contains $\frac{4}{5}$ copper, $\frac{1}{9}$ tin, and $\frac{1}{36}$ part zinc. How many pounds of each are used to make 250 pounds of the alloy?
4. A house cost $36,500. If the cost of the lot was $\frac{3}{20}$ as much as the cost of the house, find the cost of both house and lot together.
5. Five pieces, measuring, respectively, $3\frac{3}{8}$, $5\frac{1}{4}$, $4\frac{5}{16}$, $2\frac{9}{32}$, and $3\frac{15}{16}$ inches are cut from a strip of brass 2 feet 6 inches long. How much of the strip remains if the amount of waste in cutting is $\frac{1}{32}$ of an inch per cut?
6. A strip of aluminum is cut into seven pieces measuring, respectively $2\frac{3}{8}$, $3\frac{1}{16}$, $3\frac{5}{8}$, $4\frac{5}{64}$, $2\frac{17}{64}$, $3\frac{3}{4}$, and $2\frac{1}{2}$ inches. If the waste in each cut is $\frac{3}{32}$ of an inch, find the length of the original strip.
7. Twelve pieces, each $1\frac{9}{16}$ inches long, are cut from a strip of copper 24 inches long. How much of the strip remains if $\frac{1}{32}$ inch is wasted for each cut?
8. How many pieces, each $3\frac{5}{16}$ inches long, can be cut from a strip of metal that is 28 inches long if $\frac{1}{16}$ of an inch is allowed as waste in each cut?
9. How many pieces, each $2\frac{1}{4}$ feet long, can be cut from a board 18 feet long, if $\frac{1}{8}$ inch is the waste in sawing off each piece? How much is left?
10. How many cans, each holding $1\frac{1}{8}$ quarts, can be filled from a barrel of oil holding 50 gallons? (1 gallon = 4 quarts.)
11. Find the average speed of a car if it travels 273 miles in $6\frac{1}{2}$ hours. What is the average speed in miles per hour and in feet per second?
12. When potatoes sell at 3 pounds for 47 cents, what is the price per bushel? (1 bushel of potatoes weighs 60 pounds.)

3 DECIMAL FRACTIONS

3.1. DEFINITION

The decimal fraction was one of the major advances in mathematics. For a long time, people had used the place value of the digits to indicate the number 10 and numbers larger than 10. For example, when we write the number such as 4444, the 4 at the extreme right indicates units, or ones. The 4 at the left of units' place indicates *tens*; that is, the second 4 from the right has a value ten times as much as the first 4. The third 4 from the right has a value ten times as much as the second 4, and so on.

Placing a digit in the second, third, fourth, or fifth place from the right multiplies its value by 10, 100, 1000, or 10000. This place value was early recognized as one of the most important ideas in mathematics.

However, for a long time no one thought of placing digits at the right of units' place. Then, less than 400 years ago, decimal fractions were invented. It was seen that a digit could be written one place to the right of units' place with some mark, such as a period, between them. In this position a digit could represent one-tenth as much as in units' place. Thus, a 4 written one place to the right of units' place, as 0.4, means $\frac{4}{10}$.

The point separating the units' place from the fractional part is called the *decimal point*. A fraction written with digits at the right of the decimal point is called a *decimal fraction*. If no decimal point is shown in a number, the number is understood to be a whole number, and the decimal point is understood to be at the right of the number. When we write the number 63, the decimal point is understood to be just at the right of the 3.

Notice the difference in value of each of the 1's in these numbers:

1000.	one thousand	0.1	one tenth
100.	one hundred	0.01	one one-hundredth
10.	one ten	0.001	one one-thousandth
1.	one unit	0.0001	one ten-thousandth

44 ARITHMETIC

For each place that a digit is moved to the right, the digit represents another division by 10, or one-tenth as much as in the preceding place.

Annexing zeros to the right of a decimal fraction does not change the value. The following fractions have the same value:

$$0.3 = \frac{3}{10}; \quad 0.30 = \frac{30}{100}; \quad 0.300 = \frac{300}{1000}$$

In the same way, annexing zeros at the left of a whole number does not change the value. The number 000934.5 is the same as 934.5.

A number consisting of a whole number and a decimal fraction is sometimes called a *mixed decimal*. Thus,

the number 0.352 is a decimal fraction
the number 614.54 is a mixed decimal

The number of decimal places in a number means the number of digits at the right of the decimal point. The number 0.352 has three decimal places. The number 0.0086 has four decimal places; and 614.54 has two decimal places.

To summarize, whenever a digit is moved one place farther to the left, its value is multiplied by 10. Whenever a digit is moved one place farther to the right, its value is divided by 10. The number 56034.6028 means

5 ten-thousands
6 thousands
0 hundreds
3 tens
4 units (or ones)
6 tenths
0 hundredths
2 thousandths
8 ten-thousandths

In reading a number, the word "and" is used *only at the decimal point*. The foregoing number is read, "fifty-six thousand thirty-four *and* six thousand twenty-eight ten-thousandths."

3.2. CHANGING A DECIMAL FRACTION TO A COMMON FRACTION

It is sometimes desirable to change a fraction from decimal form to common fraction form. This is done simply by writing the entire decimal fraction as a common fraction and then reducing it to lowest terms. The following examples show how this is done.

DECIMAL FRACTIONS

Example 1. (a) $0.25 = \frac{25}{100} = \frac{1}{4}$ (b) $0.625 = \frac{625}{1000} = \frac{5}{8}$

Example 2. In changing a mixed decimal, change only the fraction part:

(a) $6.8 = 8\frac{8}{10} = 6\frac{4}{5}$ (b) $12.5 = 12\frac{5}{10} = 12\frac{1}{2}$

(c) $18.75 = 18\frac{75}{100} = 18\frac{3}{4}$ (d) $9.24 = 9\frac{24}{100} = 9\frac{6}{25}$

If a decimal fraction ends with a common fraction, the common fraction can sometimes be changed to a convenient decimal and annexed to the given decimal.

Example 3. (a) $0.87\frac{1}{2} = 0.875 = \frac{875}{1000} = \frac{7}{8}$ (b) $0.06\frac{1}{4} = 0.0625 = \frac{1}{16}$

Example 4. Change to common fraction form: $0.16\frac{2}{3}$.

Solution. This can be written as a common fraction:

$$\frac{16\frac{2}{3}}{100}$$

We multiply numerator and denominator by 3:

$$\frac{3 \times 16\frac{2}{3}}{3 \times 100} = \frac{50}{300} = \frac{1}{6}$$

Exercise 3.1

Change the following decimal fractions to common fraction form:

1. 0.6
2. 0.45
3. 0.24
4. 0.52
5. 3.2
6. 0.125
7. 0.375
8. $0.62\frac{1}{2}$
9. 9.875
10. 4.75
11. 6.05
12. 8.025
13. $4.07\frac{1}{2}$
14. $7.06\frac{1}{4}$
15. 3.04
16. $3.02\frac{1}{2}$
17. $0.33\frac{1}{3}$
18. $2.66\frac{2}{3}$
19. 4.0375
20. $0.043\frac{1}{8}$
21. $0.14\frac{2}{7}$
22. $0.11\frac{1}{9}$
23. $3.3\frac{1}{3}$
24. $6.6\frac{2}{3}$
25. $0.42\frac{6}{7}$
26. 3.004
27. $9.62\frac{1}{2}$
28. $4.30\frac{1}{2}$
29. $6.4\frac{3}{4}$
30. 7.005
31. $8.12\frac{1}{2}$
32. $0.008\frac{1}{3}$
33. 0.032
34. $5.01\frac{7}{8}$
35. $0.4\frac{4}{9}$
36. $4.05\frac{5}{6}$
37. $1.34\frac{1}{6}$
38. $7.61\frac{4}{5}$
39. $4.04\frac{3}{8}$
40. $2.09\frac{1}{11}$
41. $3.006\frac{3}{7}$
42. $5.002\frac{2}{9}$

3.3. ADDITION OF DECIMAL FRACTIONS AND MIXED DECIMALS

In adding decimal fractions or mixed decimals, we place the numbers in column form with the decimal points in line. Then we add the digits representing like denominations.

```
    25.
    48.7
     5.724
   864.1
     0.0372
   ─────────
   943.5612
```

Example. Add $25 + 48.7 + 5.724 + 864.1 + 0.0372$. The work is shown at the left. Note that the decimal points are in line. Sometimes zeros are annexed to the numbers to make it easier to add corresponding digits. Adding, we get the sum:

3.4. SUBTRACTION OF DECIMAL FRACTIONS AND MIXED DECIMALS

In subtracting decimal fractions, or mixed decimals, we place one number below the other with decimal points in line. In some cases, it may be necessary to annex zeros in the minuend. This does not change the value of a decimal fraction.

Example 1. Subtract $145.397 - 31.729$.

```
145.397
 31.729
────────
113.668, remainder
```

Solution. The work is shown at the left. Borrowing is done in the same way as with whole numbers.

Example 2. Subtract $56.4 - 13.6537$.

```
56.4000
13.6537
────────
42.7463, remainder
```

Solution. The work is shown at the left. Note that zeros are annexed to the minuend.

Example 3. Subtract $2 - 0.53728$.

```
2.00000
0.53728    add to
────────
1.46272 ← check
```

Solution. The work is shown at the left. Note the zeros annexed to the minuend.

Note. Check subtraction by adding the remainder to the subtrahend. The sum of these two should be the minuend.

DECIMAL FRACTIONS 47

Exercise 3.2

Add or subtract as indicated in each exercise:

1. $567.43 + 79.96 + 869.58$
2. $573.9 + 95.8 + 1957.4$
3. $7.084 + 287.6 + 0.851$
4. $4.956 + 198.3 + 0.407$
5. $789.5 + 5.89 + 9000 + 1.975$
6. $10.4 + 0.298 + 495.8 + 0.0948$
7. $4.8 + 69.64 + 39 + 909.468$
8. $89.93 + 4.9 + 0.0478 + 791$
9. $4.76 + 69.64 + 0.0649 + 802$
10. $53.75 + 85.3 + 0.0047 + 3.47895$
11. From 74.62 subtract 33.56.
12. From 871.6 subtract 316.8.
13. From 3580 subtract 18.36.
14. From 780.5 subtract 21.727.
15. From 608.2 subtract 29.853.
16. From 576.3 subtract 17.375.
17. From 477.3 take 188.324.
18. From 70.71 take 32.9072.
19. From 3.96 take 2.84926.
20. From 96.34 take 7.06071.
21. From 3 subtract 1.357.
22. From 7 subtract 2.498.
23. From 2 subtract 0.48698.
24. From 5 subtract 0.91798.
25. From 1 subtract 0.05869.
26. From 5 subtract 1.39874.
27. Find $6 - 5.48706$.
28. Find $3 - 1.29304$.
29. Find $2 - 0.61825$.
30. Find $1 - 0.71045$.

31. A family goes on a tour and drives the indicated distances during each of the first four days: Monday, 245.3 miles; Tuesday, 325.3 miles; Wednesday, 337.6 miles; Thursday, 294.1 miles. If the entire trip is 1450 miles, how many miles must they drive the fifth day to complete the trip in five days?

32. The following lengths (in inches) were cut from a 24-inch strip of copper: 3.75; 4; 4.25; 4.5; and 4.75. If the waste in each cut was 0.05 inch, what was the length of the remaining piece?

3.5. ROUNDING OFF NUMBERS

All measurement is approximate. When we say that the airline distance from New York to San Francisco is 2571 miles, we mean that the measurement is precise to the nearest mile. The word *precision* refers to the smallest unit of measure used in stating any measurement. In stating the distance as 2571 miles, we understand that this number may differ from the true distance by some small fraction of a mile.

Now we could also state that the distance is approximately 2570 miles. In this case, the measurement is precise to the nearest ten miles. The number 2571 is "rounded off" to 2570. The numbers 2, 5, and 7 are called *significant digits*. We can also say that the distance is approximately 2600 miles, precise to the nearest hundred miles. Then we have rounded off the number to two significant digits. Note that when we round off the number 2571 to 2570, we drop the "1" because it is less than 5, and we replace it with a zero (0).

To take another example, the diameter of the earth at the equator is approximately 7927 miles. This number contains four significant digits, and measurement is precise to the nearest mile. The number may be rounded to three significant digits as 7930 miles, in which the measurement is precise to the nearest ten miles. In this case, we drop the 7 but increase the 2 and make it 3, because the 7, which was dropped, is more than 5. We might further state that the distance is approximately 7900 miles, precise to the nearest hundred miles. In this case, we drop the 2 because it is less than 5, and we have rounded the number to two significant digits. We might go one step further and round the number to 8000, in which case we have one significant digit, and the measurement is precise to the nearest thousand miles.

As another example, suppose we say that the distance from the earth to the sun at a particular instant is approximately 93,162,718 miles. In most cases, we are not interested in such a high degree of precision. We would say that the distance is approximately 93,000,000 miles. That is, we round off the original number and use only the two digits, 93. We say the number 93,000,000 is stated in "round numbers." The expression "rounding off" comes from the words "round numbers," which refer to the zeros. When we round off a number, we imply that the result is only approximate, yet precise enough for our purpose.

Suppose we measure a length and state the answer to the nearest inch at 5836 inches. If we wish to state the length to the nearest ten inches, we call it 5840 inches. We drop the 6 and replace it with a zero. Since the number dropped is more than 5, we increase the preceding digit by 1 and call it 4. We can also express the length to the nearest hundred inches as 5800. This number contains two significant digits.

It is important to understand exactly what is meant by a *significant* digit. Significant digits are digits that determine the accuracy or precision of a measurement. If we say the average distance from the earth to the moon is approximately 238,900 miles, correct to the nearest hundred, then the significant digits are the 2, 3, 8, and 9, or the number 2389. The only purpose of the two zeros at the right is to place the number 2389 in the proper position. In the number 0.00042073, the significant digits form the number 42073. The only purpose of the three zeros at the left of the significant digits is to place the number 42073 in the proper position to indicate the degree of precision.

In the number, 72608000, the last three zeros are not significant unless they are definitely stated as determining the precision of the measurement. However, the zero between the 6 and 8 is significant because it is involved in the precision of measurement. A zero is always significant if it occurs between two non-zero digits.

Consider another example. A zero at the right of an integer may or may not be significant. Suppose we measure a length and find that it is 3802 feet, approximately. The measurement is precise to the nearest foot, and the number contains four significant digits. Now, if we state the length to the nearest ten feet, we call the number 3800, dropping the 2 and replacing it with a zero. However, since the number is meant to indicate precision to the nearest ten feet, then the zero after the 8 is significant, and we have three significant digits, 3, 8, and 0. If we round off the number to the nearest hundred, we still get 3800, and we have two significant digits. If we say the measurement 25,000 miles, is precise to the nearest mile, then the three zeros are significant, because they indicate the degree of precision.

In a decimal fraction, any zeros between the decimal point and the non-zero digit are not significant. In the number 0.0052, the zeros are not significant. However, suppose we measure a line segment and find that it is 2.4 inches long. This number implies precision to the nearest tenth of an inch. Now, if we write the measurement as 2.40 inches, we imply precision to the nearest hundredth of an inch. A zero should not be so annexed at the right of a decimal fraction unless we wish to indicate the implied degree of precision.

To summarize, for rounding off numbers, we have the following rules.

1. If the digit dropped is less than 5, then the last digit kept is left as it is. Then the number is rounded *downward*.

2. If the digit dropped is more than 5, then the last digit kept is increased by 1. Then the number is rounded *upward*.

3. If the digit dropped is exactly 5, followed only by zeros in the original number, then we often use a special rule. When the 5 is dropped, the last digit kept is not changed if it is *even*. If the last digit kept is odd, it is changed to the next higher digit, which makes it even. That is, if the digit dropped is 5, the last digit kept is made an even digit. Remember, this special rule applies only when the digit dropped is 5 followed by only zeros in the original number.

As a result of this last rule, if we round off many numbers ending in 5, the result will be that we shall round off such numbers *upward* about as many times as we round *downward*. If numbers ending in 5 were always rounded *upward*, the result would be an accumulation of errors *upward*.

An actual experience will show the reason for the special rule for numbers ending in 5. A lady bought some canned vegetables at the supermarket. One kind was marked "2 for 59 cents." The other kind was marked "2 for 39 cents." She bought one can of each kind. For the first can, the clerk rang up 30 cents. For the second, he rang up 20 cents. The lady said, "You charged a half-cent extra for the first can. Then you should drop the half-cent for the second can." Of course, the clerk could not do so, but the incident led to a ten-minute argument, and the lady lost.

Exercise 3.3

Round off each of the following numbers to five, four, three, and two significant digits:

1. 471528
2. 28.5916
3. 4071382
4. 194.835
5. 852.7129
6. 358391.4
7. 39.25004
8. 0.08297486
9. 0.1829651
10. 9182.753
11. 688535
12. 6285316
13. 7290304
14. 0.2910852
15. 3.259175
16. 0.000419965
17. 0.00574046
18. 0.0960108
19. 87.0048
20. 0.06084545

3.6. MULTIPLICATION INVOLVING DECIMAL FRACTIONS

In multiplication involving decimal fractions, we can look upon the decimals as common fractions. Then the decimal part of the number will indicate the denominator of the common fraction. Consider the following example.

Example 1. Multiply 0.4×0.7.

Solution. We first take 4 times 7 and get 28. Now our problem is to place the decimal point in the proper position in the product. Let us see what we get when we state the numbers as common fractions:

$$\frac{4}{10} \times \frac{7}{10} = \frac{28}{100} = 0.28$$

This example is an illustration of the rule for the decimal point in a product:

RULE: The number of decimal places in a product is equal to the number of places in the multiplier and the multiplicand combined.

In Example 1, there is *one* decimal place in each of the factors. Then there are *two* decimal places in the product: that is, $1+1=2$.

DECIMAL FRACTIONS

Example 2. Multiply 3.2×4.78.

```
  4.78
   3.2
  ----
   956
  1434
 -----
 15296
```

Solution. The multiplication is shown at the left. In this example involving decimals, we write one number below the other with the right-hand digits in line, just as we do with whole numbers. Then we perform the multiplication without regard to the decimal points, just as though the numbers were whole numbers. Our problem now is to place the decimal point in the proper position in the product.

To place the decimal point in the product, we count as many decimal places as there are decimal places in the two numbers combined. In this example, we count *three* places from the right. The product is 15.296.

To see the reason for the rule for the decimal point in a product, suppose we write the numbers in Example 2 as improper fractions. Note that the number 3.2 indicates a denominator of 10, and the number 4.78 indicates a denominator of 100. Then the multiplication

$$3.2 \times 4.78 \quad \text{means the same as} \quad \frac{32}{10} \times \frac{478}{100} = \frac{15295}{1000} = 15.296$$

The product of the numerators is 15296. The product of the denominators is 1000, which means that we divide the numerator by 1000. Therefore, we count off three decimal places in the numerator and get the answer.

In some problems in the multiplication involving decimal fractions, it is necessary to annex zeros at the left of the answer we get.

Example 3. Multiply 0.04×0.023.

```
 0.023
  0.04
 -----
    92
```

Solution. Multiplying 4 times 23, we get 92. Now we must have *five* decimal places in the product. Therefore, it is necessary to annex three zeros at the left of the 92. Then the product is 0.00092. The answer now has five decimal places as required.

Example 4. Multiply 0.765×38.4.

```
   38.4
  0.765
  -----
   1920
   2304
   2688
  ------
 293760
```

Solution. The work is shown at the left. First we multiply without regard to the decimal points. Now we count off *four* places from the right and we get the correct answer: 29.3760. In this example, the answer can be rounded off to 29.376, since the zero does not affect the answer unless we wish to indicate six-place accuracy. If we wish to round off the answer to four significant digits, it can be written as 29.38. Moreover, the answer correct to the nearest tenth is 29.4.

Example 5. Multiply 4.32×6.827 and express the answer to the nearest thousandth, to the nearest hundredth, and to the nearest tenth.

```
   6.827
   4.32
  ------
  13654
  20481
  27308
  -------
  2949264
```

Solution. The work is shown at the left. Here we must count off *five* decimal places in the answer, which then becomes 29.49264. The answer:

correct to 5 significant digits, is 29.493, to the nearest thousandth;
correct to 4 significant digits, is 29.49, to the nearest hundredth;
correct to 3 significant digits, is 29.5, to the nearest tenth.

Example 6. Multiply 0.697×592.4.

```
   592.4
   0.697
  ------
   41468
   53316
   35544
  -------
  4129028
```

Solution. The work is shown at the left. In this example, we must count off *four* decimal places from the right in the product. The answer then becomes 412.9028. Rounding off, we get the answer

correct to the nearest thousandth: 412.903
correct to the nearest hundredth: 412.90
correct to the nearest tenth: 412.9
correct to the nearest unit: 413.

In connection with multiplication involving decimal fractions, we can formulate two special rules that are useful. From the definition of a decimal fraction, and the place value of a digit, we have seen that if a digit is moved one place farther toward the left, it is multiplied by 10. That is, *multiplying a digit by* 10 *moves it one place toward the left*, as shown in these examples:

$$10 \times 6 = 60 \qquad\qquad 10 \times 0.006 = 0.06$$
$$10 \times 60 = 600 \qquad\qquad 10 \times 0.3 = 3.$$

Multiplying a digit by 100 *moves it two places toward the left:*

$$100 \times 7 = 700 \qquad\qquad 100 \times 0.0004 = 0.04$$

RULE 1. To multiply any number by 10, 100, 1000, and so on, move the decimal point toward the right as many places as there are zeros in the multiplier.

Since division is the reverse of multiplication, we have a second rule.

RULE 2. To divide any number by 10, 100, 1000, and so on, move the decimal point toward the left as many places as there are zeros in the divisor.

Examples

$$10 \times 37.62 = 376.2$$
$$100 \times 37.62 = 3762.$$
$$1000 \times 37.62 = 37620.$$
$$100000 \times 0.004963 = 496.3$$

$$37.62 \div 10 = 3.762$$
$$37.62 \div 100 = 0.3762$$
$$37.62 \div 1000 = 0.03762$$
$$593000 \div 1000000 = 0.593$$

3.7. DIVISION INVOLVING DECIMAL FRACTIONS

In division involving decimal fractions, if the divisor is a whole number, the decimal point in the answer is placed directly above the decimal point in the dividend, as shown in the following examples.

Example 1. Divide $39.41 \div 7$.

$$7\overline{)39.41}$$

$$7\overline{)39.41} \quad \text{5.63}$$

Solution. Here the divisor is the whole number 7. First, we place the decimal point in the answer directly above the decimal point in the dividend, as shown. Then we divide as with whole numbers. The work is shown at the left. The first division is 7 into 39, which is 5, with a remainder of 4. Then we continue as with whole numbers.

$$53\overline{)171.72}$$

$$\begin{array}{r} 3.24 \\ 53\overline{)171.72} \\ \underline{159} \\ 127 \\ \underline{106} \\ 212 \\ \underline{212} \\ 0 \end{array}$$

Example 2. Divide $171.72 \div 53$.

Solution. If the divisor is a whole number and consists of two or more digits, we follow the same procedure. The first step is to place the decimal point for the answer directly above the decimal point in the dividend. Then we divide as usual with whole numbers. The first division is 53 into 171, which is 3, with a remainder. Now the steps are the same as with whole numbers. The answer, 3.24, can be checked by multiplication: $53 \times 3.24 = 171.72$

If the divisor contains a decimal fraction, we use the following procedure. We mentally move the decimal point in the divisor and the dividend to the right as many places as there are decimal places in the divisor. This makes the divisor a whole number. Then we proceed as in Example 2, above.

$$0.21\overline{)90.78\,3}$$

$$\begin{array}{r} 4\,32.3 \\ 0.21\overline{)90.78\,3} \\ \underline{84} \\ 67 \\ \underline{63} \\ 48 \\ \underline{42} \\ 63 \\ \underline{63} \\ 0 \end{array}$$

Example 3. Divide $90.783 \div 0.21$.

Solution. After setting up the form for division, we first mentally move the decimal point in the *divisor* and the *dividend,* as shown. This makes the divisor a whole number. Then we really have the division: $9078.3 \div 21$. The decimal point for the answer is placed directly above the new position of the decimal point in the dividend. This should be done before the actual division is begun. Now, in order to be sure that the answer will have the decimal point in the proper place, we place each digit of the quotient *directly above the last digit used in each step of the division.* Then the division is performed just as with whole numbers.

Example 4. Divide $297.56 \div 0.215$.

Solution. The work is shown at the left. In this example, we move the decimal point of the *divisor* and the *dividend three* places to the right. Notice that it is necessary to annex a zero (0) to the dividend in order to move the decimal point three places. Now the divisor is a whole number, and we proceed as with whole numbers.

```
                1 384.
        0.215/297.560
              215
              ---
              825
              645
              ---
             1806
             1720
             ----
              860
              860
              ---
                0
```

Example 5. Divide $138.4 \div 0.32$.

Solution. The work is shown at the left. We first move the decimal point two places to the right in the divisor and the dividend. To do so, we annex a zero (0) to the dividend. If we annex another zero to the dividend, we divide to the first decimal place of the answer, and get the answer 432.5.

```
              4 32.5
        0.32/138.40 0
             128
             ---
             104
              96
              --
              80
              64
              --
             160
             160
             ---
               0
```

Example 6. Divide $78.69 \div 0.243$; continue to divide to five decimal places in the answer by adding zeros to the dividend. Then express the answer to the nearest thousandth, the nearest hundredth; the nearest tenth; and the nearest whole numbers.

Solution. The work is shown at the left. We move the decimal point three places to the right in the divisor and the dividend. Then we add zeros to the dividend to get five decimal places in the answer. The answer is expressed as follows:

```
               323.82716
     0.243/78.690 00000
           72 9
           ----
            5 79
            4 86
            ----
              930
              729
              ---
             2010
             1944
             ----
              660
              486
              ---
             1740
             1701
             ----
              390
              243
              ---
             1470
             1458
             ----
               12
```

   323.827   to the nearest thousandth
   323.83   to the nearest hundredth
   323.8   to the nearest tenth
   324.   to the nearest whole number.

3.8. CHANGING A COMMON FRACTION TO A DECIMAL FRACTION

```
  0.75
4/3.00
  2 8
    20
    20
     0
```

The principle of division involving decimals can be used to change a common fraction to an equivalent decimal fraction. The fraction, $\frac{3}{4}$, can be understood to mean $3 \div 4$. To find the decimal fraction equivalent to $\frac{3}{4}$, we place a decimal point after the 3 and then annex zeros at the right of the decimal point as we divide. We place the decimal point for the answer directly above the decimal point in the dividend as usual. After two divisions the remainder is zero, as shown at the left. Then the result is called a *terminating decimal*. The answer, 0.75, can be checked by writing 0.75 as a common fraction, $\frac{75}{100}$, and reducing it to $\frac{3}{4}$.

When a common fraction is changed to a decimal fraction, if the remainder is never zero, the result is a *repeating decimal*. For example,

$$\frac{1}{3} = 0.3333 \cdots \qquad \frac{3}{7} = 0.4285714285714 \cdots$$

Whenever a common fraction is changed to a decimal fraction, the result is either a *terminating* decimal or a *repeating* decimal, as shown in these examples:

$$\frac{2}{5} = 0.4; \qquad \frac{1}{4} = 0.25; \qquad \frac{3}{8} = 0.375; \qquad \frac{4}{25} = 0.16$$

$$\frac{7}{32} = 0.21875; \qquad \frac{2}{3} = 0.666666 \cdots \qquad \frac{5}{11} = 0.45454545 \cdots$$

By a *repeating decimal*, we mean a decimal in which the number after a certain point has the same set of consecutive digits. In the fraction, $\frac{1}{3}$, the decimal will consist of a repetition of 3's. In the fraction, $\frac{5}{11}$, the decimal will consist of a repetition of the two digits, 45.

Since a decimal fraction always indicates a denominator of 10, 100, 1000, or some power of 10, it follows that many common fractions cannot be stated exactly as decimals. If the common fraction has a denominator that can be changed to 10, 100, 1000, and so on, then the fraction can be stated exactly as a decimal. For example, $\frac{3}{5}$ can be changed to $\frac{6}{10}$, or 0.6. The fraction $\frac{7}{25}$ can be changed to $\frac{28}{100}$, or 0.28. However, the fraction $\frac{4}{7}$ cannot be changed to tenths, or hundredths, and therefore cannot be written exactly as a decimal.

Exercise 3.4

Perform the following multiplications. Round off answers to four significant digits:

1. 91.7×4.8 **2.** 63.9×0.58
3. 0.948×8.6 **4.** 8.56×3.8

56 ARITHMETIC

5. 7.85×0.75
7. 7.36×0.089
9. 0.748×3.9
11. 0.0293×0.054
13. 98.5×0.0087
15. 8.963×3.21
17. 697.82×0.408
19. 0.05796×1.007
21. 0.8507×0.0513
23. 4965.2×0.204

6. 9.28×59
8. 84.9×0.58
10. 0.0863×0.074
12. 0.00628×0.39
14. 0.0279×0.067
16. 0.09378×0.516
18. 0.75806×0.083
20. 1087.4×0.816
22. 0.00126×1.034
24. 0.00769×3500

In the following divisions, divide to five significant digits and then round off the answers to four significant digits:

25. $2.1044 \div 26$
27. $0.11862 \div 46$
29. $68.83 \div 0.019$
31. $0.03301 \div 3.9$
33. $48613 \div 0.065$
35. $0.07814 \div 0.057$
37. $55.413 \div 1.87$
39. $458.21 \div 0.673$
41. $0.00402 \div 58.1$

26. $3621.5 \div 54$
28. $141.38 \div 3.9$
30. $234.91 \div 0.28$
32. $6.1392 \div 0.76$
34. $0.47602 \div 0.492$
36. $118.6 \div 0.0019$
38. $36435 \div 38.2$
40. $4394.2 \div 0.048$
42. $0.02002 \div 0.037$

Exercise 3.5

Change the following common fractions and mixed numbers to decimal form:

1. $\frac{5}{8}$
2. $\frac{9}{16}$
3. $\frac{4}{11}$
4. $1\frac{1}{4}$
5. $3\frac{3}{16}$
6. $5\frac{6}{7}$
7. $6\frac{7}{8}$
8. $4\frac{5}{16}$
9. $8\frac{2}{5}$
10. $7\frac{5}{6}$
11. $3\frac{7}{16}$
12. $1\frac{3}{32}$
13. $4\frac{7}{20}$
14. $9\frac{8}{9}$
15. $\frac{31}{15}$
16. $5\frac{31}{40}$
17. $9\frac{5}{32}$
18. $\frac{127}{8}$
19. $\frac{185}{128}$
20. $\frac{135}{64}$
21. $\frac{77}{18}$
22. $\frac{432}{125}$
23. $\frac{105}{32}$
24. $\frac{181}{48}$

DECIMAL FRACTIONS 57

3.9. A FEW SHORT-CUTS

There are many so-called *short-cuts* for fast computation in arithmetic. The following five rules cover a few of the most useful in multiplication and division. The first two have been previously mentioned.

RULE 1. To multiply any number by 10, 100, 1000, and so on, move the decimal point toward the right as many places as there are zeros in the multiplier.

Examples. $1000 \times 38.92 = 38920$; $100000 \times 0.0045 = 450$.

RULE 2. To divide any number by 10, 100, 1000, and so on, move the decimal point toward the left as many places as there are zeros in the divisor.

Examples. $38.92 \div 1000 = 0.03892$; $0.045 \div 100000 = 0.00000045$

RULE 3. To multiply any number by 5, move the decimal point one place toward the right and divide by 2.

Examples. $5 \times 28 = 140$; $5 \times 379 = \frac{1}{2}$ of $3790 = 1895$.

Rule 3 means that we actually multiply by 10 and then divide by 2, which is equivalent to multiplying by 5. The rule can be extended to multiplying by 50, 500, 5000, and so on.

RULE 4. To multiply any number by 25, move the decimal point two places toward the right and divide by 4.

Examples. $25 \times 36.8 = \frac{1}{4}$ of $3680 = 920$; $25 \times 87.3 = \frac{1}{4}$ of $8730 = 2182.5$.

Rule 4 means that we actually multiply by 100 and then divide by 4, which is equivalent to multiplying by 25. The rule can be extended to multiplying by 250, 2500, 25000, and so on.

RULE 5. To multiply a number ending with the fraction $\frac{1}{2}$ by itself, multiply the whole number by one more than itself and then annex the fraction $\frac{1}{4}$.

This is a useful rule for rapid calculation in arithmetic. It is especially useful in engineering and statistics. We show several examples.

Example 1. $6\frac{1}{2} \times 6\frac{1}{2}$. Take 6×7 and annex $\frac{1}{4}$. Answer: $42\frac{1}{4}$.

Example 2. $19\frac{1}{2} \times 19\frac{1}{2}$. Take 19×20, and annex $\frac{1}{4}$. Answer $380\frac{1}{4}$.

Example 3. Rule 5 can also be used when $\frac{1}{2}$ is written as a decimal, 0.5. 8.5×8.5. Take 8×9, and annex the fraction, 0.25. Answer: 72.25.

Example 4. 24.5×24.5. Take 24×25, and annex 0.25. Answer: 600.25.

Example 5. Rule 5 can also be used for any number ending in 5, even though the number does not contain a decimal fraction. 85×85. Take 8×9, and annex the digits 25. Answer: 7225.

Example 6. 125×125. Take 12×13 and annex 25. Answer: 15625

If a problem requires several multiplications and divisions involving decimal fractions, it can often be simplified in the same manner as with whole numbers. We have seen how the following problem can be reduced:

$$\frac{24 \times 25 \times 49}{35 \times 28 \times 45} = \frac{2}{3}$$

In this example, the entire numerator and denominator can be divided successively by 3, 4, 5, 5, 7, and 7. The fraction reduces to $\frac{2}{3}$.

If the numbers in the numerator and/or in the denominator contain decimals, the computation can be simplified in the same way. However, in that case it is best to multiply the numerator and the denominator by 10, 100, 1000, or by some number of 10's so that the decimals will disappear. This can be done simply *by moving the decimal point the same number of places in the numerator and the denominator.*

Example 7. Suppose, in arranging the work for some problem, we get the expression

$$\frac{0.6 \times 1.75 \times 32}{0.075 \times 4 \times 2.8}$$

The decimal fractions can be eliminated by multiplying the numerator and the denominator by 10 a sufficient number of times. In this example, we multiply the numerator and the denominator by 10 four times, that is, by 10000. Then all decimals can be made to disappear. To multiply both terms by 10000, we simply move the decimal point a total of four places in the numerator, and a total of all four places in the denominator. Then the problem becomes

$$\frac{6 \times 175 \times 320}{75 \times 4 \times 28}$$

All numbers are now integers, and the fraction reduces to 40.

DECIMAL FRACTIONS

Exercise 3.6

Multiply each of the following numbers by 10, by 100, and by 1000:

1. 47.3
2. 8.53
3. 0.0076
4. 0.0527
5. 62
6. 2900
7. 0.638
8. 0.00006
9. 185
10. 3.4

11–20. Divide each of the numbers in Exercises 1–10 by 10, by 100, and by 1000:

Multiply each of these numbers by 5 and by 25, using the short method:

21. 28
22. 72
23. 34
24. 93
25. 312
26. 23.6
27. 56.8
28. 25.3
29. 15.7
30. 7.31

Multiply by a short method:

31. $7\frac{1}{2} \times 7\frac{1}{2}$
32. $11\frac{1}{2} \times 11\frac{1}{2}$
33. $5\frac{1}{2} \times 5\frac{1}{2}$
34. 23.5×23.5
35. 31.5×31.5
36. 62.5×62.5
37. $18\frac{1}{2} \times 18\frac{1}{2}$
38. 95×95

Simplify each of the following expressions:

39. $\dfrac{240 \times 7.2 \times 0.36 \times 8.4}{4.8 \times 0.15 \times 168.96}$
40. $\dfrac{42 \times 62.5 \times 2.43 \times 0.7854}{2.64 \times 0.175 \times 270 \times 432}$
41. $\dfrac{3.1416 \times 6.45 \times 62.5 \times 8.8}{2.2 \times 1728 \times 31.5 \times 2.54}$
42. $\dfrac{6.28 \times 377 \times 60 \times 0.0128}{0.7854 \times 39.37 \times 62.4 \times 60}$

Exercise 3.7

Assume that measurements are approximate.

1. What is the total thickness of a pile of 12 metal sheets of iron, if each sheet is 0.045-inch thick?
2. A pile of 15 sheets of metal has a total thickness of 2.14 inches. What is the approximate thickness of each sheet?
3. Five bars of iron weigh, respectively, $7\frac{3}{4}$, $5\frac{1}{8}$, $4\frac{9}{16}$, $6\frac{1}{2}$, and $5\frac{13}{32}$ pounds. Change these weights to decimal fractions, and then find the total weight of the five bars and the average weight of each bar. Round off the final answer to three significant digits.
4. Six strips of brass have, respectively, the following measurements in inches: 3.275; 4.35; 2.625; 0.875; 5.125; and 4.6. If these six strips were cut from a strip 32 inches long, how long a piece is left if 0.045 of an inch was wasted in each cut?

60 ARITHMETIC

5. A strip of copper was cut into five strips of the following lengths (in inches), respectively: 4.25; 3.95; 7.3; 6.485; and 5.74. If the waste for each cut was 0.035 of an inch, how long was the original strip?

6. Five pieces, each measuring 1.375 inches long, were cut from a strip of silver alloy 12 inches long. If the waste in each cut was 0.018 of an inch, what was the length of the remaining piece?

7. How many pieces, each 3.125 inches long, can be cut from a strip of brass if the waste per cut is $\frac{1}{16}$ of an inch and the strip is 25 inches long?

8. How many pieces, each 2.225 inches long, can be cut from a strip of aluminum 1 yard long if the waste is 0.04 inch per cut?

9. Six pieces measuring, respectively, $2\frac{7}{16}$, $5\frac{1}{4}$, $6\frac{1}{2}$, $7\frac{5}{8}$, $3\frac{1}{8}$, and $3\frac{5}{16}$ inches are cut from a piece of aluminum 30 inches long. If the waste is $\frac{1}{16}$ of an inch per cut, find the length of the remaining strip.

10. A bar of copper is cut into 8 pieces, each $2\frac{7}{16}$ inches long. If the waste per cut is $\frac{1}{32}$ of an inch, find the length of the original bar.

11. An automobile trip of 400 miles required 24 gallons of gasoline. Find the average number of miles per gallon.

12. What is the average number of miles per gallon if 94.8 gallons of fuel are required for a trip of 1620 miles?

13. Find the average number of miles per gallon if 68.5 gallons of fuel are required for a trip of 1245 miles.

14. Find the average number of miles per gallon of fuel if 21.4 gallons are required for a trip of 363.2 miles.

15. If a trip of 490 miles is made in 9 hours and 15 minutes, find the average speed in miles per hour.

16. Find the average speed if a trip of 350.8 miles is made in 6 hours and 45 minutes travel time.

17. If a motorcycle averages 52.9 miles per gallon of fuel, how far can it travel on 8.6 gallons?

18. If a motorcycle travels 312.5 miles and uses 5.9 gallons of fuel, what is the average number of miles per gallon, to the nearest tenth of mile?

19. A motorcycle made a trip of 246.8 miles. If it averaged 51.5 miles per gallon of fuel, how many gallons were required for the trip? Round off the answer to the nearest tenth of a gallon.

20. A certain motorcycle tank holds 3.2 gallons of fuel. On a trip of 625 miles, the tank was filled 4 times. At the end of the trip, the amount of fuel left in the tank was 1.2 gallons. What was the average number of miles per gallon of fuel?

21. In October 1936, the CB & Q railroad made a record run from Chicago to Denver, a distance of 1017.23 miles, in 12 hours, 12 minutes, 27 seconds. What was the average speed in miles per hour? In feet per second?

22. In 1959, the Burlington Zephyr train had a scheduled run of 54.6 miles to be traversed in 43 minutes. What was the rate in miles per hour?

23. Japan has a train that made a run of 320.1 miles in 3 hours, 10 minutes. What is the rate of speed in miles per hour?

24. Charles Lindbergh made his historic flight from New York to Paris, a distance of 3610 miles, in 33 hr. 29 min. 30 sec. in May 1927. What was his speed in miles per hour?

25. In November 1957, Captain Sweet of the USAF flew from Los Angeles to New York and back to Los Angeles, a distance of 4891.8 miles, in 6 hr, 40 min. 36.23 sec. What was his average speed in miles per hour (mph)?

26. If 3.6 pounds of beef is priced at $1.24 a pound, find the total cost.

27. A turkey, weighing 15.3 pounds, cost $18.50. What was the price per pound to the nearest cent?

28. The weights of four chickens, in pounds, respectively, were 3.4; 4.1; 2.7; and 4.3. If the price was $0.63 per pound, find the cost (nearest cent).

29. Five brands of breakfast cereal, respectively, cost as follows: brand A: 18 oz. for 69¢; brand B: 12 oz. for 79¢; brand C: 16 oz. for 67¢; brand D: 9 oz. for 73¢; brand E: 15 oz. for 69¢. Find the cost of each per ounce. Which was the best buy?

4 PERCENTAGE

4.1. DEFINITION

The expression *per cent* is so common in everyday speech that most people probably know what is meant by "one hundred per cent," "50 per cent," or other similar expressions. The statement, "You are one hundred per cent right," means "You are completely right." When we say, "He spent 50 per cent of his money," we mean he spent half of it.

The words "per cent" mean "hundredths" or "by the hundred." The expression *25 per cent* means "25 hundredths" or "25 out of every hundred." The symbol for *per cent* is %. When we say "25%," we mean "25 out of every hundred." It means $\frac{25}{100}$ as a common fraction, or 0.25 as a decimal fraction.

The expression *per cent*, such as 25%, means a part of some quantity. The expression 25% is equivalent to the common fraction $\frac{1}{4}$. For example, 25% of $48 means 0.25 of $48, or $\frac{1}{4}$ of $48, which is $12.

Note especially in the example, 25%, that the number 25 by itself is a whole number, or integer; 25 alone is not a fraction. The expression becomes a fraction when we attach the per cent sign (%), as in 25%. It also becomes a fraction when we place the denominator 100 below the 25, as $\frac{25}{100}$; or when we place the decimal point before the 25, as in 0.25.

4.2. RELATION BETWEEN PER CENTS AND FRACTIONS

In multiplication and division involving per cents, it is first necessary to change the per cent to a decimal or common fraction. On the other hand, when we have found an answer in the form of a decimal fraction, we often wish to state it as a per cent. The following rules and examples show how these changes are made.

64 ARITHMETIC

RULE 1. To change a per cent to a common fraction, omit the per cent sign (%) and write 100 below the number of per cent. Then reduce the fraction if possible.

Examples.

$$25\% = \frac{25}{100} = \frac{1}{4}; \qquad 20\% = \frac{20}{100} = \frac{1}{5}; \qquad 15\% = \frac{15}{100} = \frac{3}{20}$$

$$32\% = \frac{32}{100} = \frac{8}{25}; \qquad 47\% = \frac{47}{100}; \qquad 5\% = \frac{5}{100} = \frac{1}{20}$$

$$125\% = \frac{125}{100} = \frac{5}{4}; \qquad 200\% = \frac{200}{100} = 2; \qquad 1\% = \frac{1}{100}$$

$$12\tfrac{1}{2}\% = \frac{12\tfrac{1}{2}}{100} = \frac{25}{200} = \frac{1}{8}; \qquad 66\tfrac{2}{3}\% = \frac{66\tfrac{2}{3}}{100} = \frac{200}{300} = \frac{2}{3}$$

$$0.5\% = \frac{0.5}{100} = \frac{5}{1000} = \frac{1}{200}; \qquad 0.04\% = \frac{0.04}{100} = \frac{4}{10000} = \frac{1}{2500}$$

RULE 2. To change a per cent to a decimal fraction, move the decimal point two places toward the left and then omit the per cent sign.

We can see that this rule is reasonable if we first change the per cent to a common fraction.

Examples.

$$37\% = \frac{37}{100} = 0.37; \qquad 20\% = 0.20; \qquad 8\% = 0.08$$

$$125\% = 1.25; \qquad 4.2\% = 0.042; \qquad 6.25\% = 0.0625$$

$$0.03\% = 0.0003; \qquad 8\tfrac{1}{2}\% = 0.085; \qquad 300\% = 3.00$$

RULE 3. To change a decimal fraction to per cent, move the decimal point two places toward the right and annex the per cent sign (%).

Note. A common fraction can be changed first to a decimal fraction and the result changed to per cent.

Examples.

$$0.24 = 24\%; \qquad 0.0375 = 3.75\%; \qquad 4 = 400\%$$

$$2.5 = 250\%; \qquad 0.615 = 61.5\%; \qquad 10 = 1000\%$$

$$\frac{3}{4} = 0.75 = 75\%; \qquad 0.14\tfrac{2}{7} = 14\tfrac{2}{7}\%; \qquad 0.5 = 50\%$$

$$\frac{5}{16} = 0.3125 = 31.25\%; \qquad \frac{1}{32} = 0.03125 = 3.125\%$$

Exercise 4.1

Change the following per cents to decimal fractions and to common fractions or mixed numbers:

1. 45%
2. 24%
3. 6%
4. 12%
5. 75%
6. 60%
7. 10%
8. 35%
9. 125%
10. 275%
11. $37\frac{1}{2}$%
12. $8\frac{3}{4}$%
13. $28\frac{4}{7}$%
14. $22\frac{2}{9}$%
15. 520%
16. 450%
17. 0.8%
18. 0.12%
19. 22.5%
20. 3.75%
21. $4\frac{1}{5}$%
22. $6\frac{1}{4}$%
23. $1\frac{7}{8}$%
24. $\frac{5}{16}$%
25. $\frac{1}{2}$%
26. $\frac{3}{4}$%
27. $1\frac{1}{3}$%
28. $\frac{7}{8}$%

Change the following common fractions, mixed numbers, and decimal fractions to per cents:

29. $\frac{2}{5}$
30. $\frac{7}{16}$
31. $\frac{9}{64}$
32. $\frac{25}{32}$
33. $4\frac{5}{8}$
34. $2\frac{3}{25}$
35. $4\frac{1}{4}$
36. $5\frac{1}{2}$
37. $\frac{2}{3}$
38. $\frac{5}{11}$
39. $4\frac{4}{9}$
40. $3\frac{3}{8}$
41. 0.075
42. 0.0045
43. 3.36
44. 1.65
45. 1.08
46. 4.3
47. 0.08
48. 0.36
49. 0.005
50. 0.0125
51. 2
52. 3.5

4.3. RATE: BASE: PERCENTAGE

In any problem involving per cent, there are three quantities that must be considered. These quantities are called the *rate*, the *base*, and the *percentage*. In order to see the relation between these quantities, consider the following example:

$$25\% \text{ of } \$48 = \$12$$

In this statement, the *rate* is 25%; the *base* is $48; and the *percentage* is $12.

In any percentage problem, the *rate* is the indicated fractional part of a particular quantity. For example, 25% means $\frac{25}{100}$ or $\frac{1}{4}$ of a quantity. The *base* is the quantity of which a fractional part is taken. The base usually has a name or denomination of some kind, such as feet, pounds, or dollars. The *percentage* is a portion of the base and has the *same denomination as the base*. Note that if the base is in dollars, the percentage is also in dollars.

Let us write the foregoing example by using a *decimal fraction* in place of *per cent* for the rate:

$$0.25 \times \$48 = \$12$$

From this example we see that the general percentage statement can be shown as

$$\text{rate} \times \text{base} = \text{percentage}$$

Note that this is a problem in multiplication. When we multiply two numbers together, the answer is called the *product*. The numbers multiplied together are called *factors*. For example, in the problem, $5 \times 7 = 35$, the factors are 5 and 7, and the product is 35. Similarly, in the percentage statement,

$$0.25 \times \$48 = \$12$$

the factors are 0.25 and 48, and the product is 12.

4.4. FINDING THE PERCENTAGE WHEN THE RATE AND THE BASE ARE KNOWN

In the percentage statement,

$$\text{rate} \times \text{base} = \text{percentage}$$

note that the *rate* and the *base* are *factors*. The *percentage* is a *product*. If we know the rate and the base, we have a problem in which the two factors are known. Then we multiply the *rate times the base* to find the *percentage*. Therefore, we have the following rule.

RULE 4. To find the percentage when the rate and the base are known, multiply the rate times the base.

The rule may be stated as

$$\text{rate} \times \text{base} = \text{percentage}$$

The percentage has the same denomination as the base.

We have said that 100% means $\frac{100}{100}$, or all of a particular quantity. If the rate is 100% (which is equal to 1), then the percentage is equal to the base. If the rate is less than 100% (i.e., less than 1), then the percentage is less than the base. If the rate is more than 100%, (i.e., more than 1), then the percentage is more than the base. These three conditions are illustrated in the following examples.

Example 1.   100% of $60 = $60.

Example 2.   20% of 60 ft = 12 ft.

Example 3.   150% of 60 pounds = 90 pounds.

Example 4.   A family buys a house priced at $36,000. The down payment is 15% of the cost. What is the amount of the down payment?

Solution.   In this problem, the base is the price of the house, $36,000. The rate of down payment is 15%. To find the percentage, we use the formula

$$\text{rate} \times \text{base} = \text{percentage}$$

or

$$0.15 \times 36{,}000 = 5400$$

Then the down payment is $5400.

4.5. FINDING THE RATE WHEN THE BASE AND THE PERCENTAGE ARE KNOWN

If we know the base and the percentage, we have a problem in which the product of the two factors is known. Consider the following problem:

Example 5.   Suppose a man receives a salary of $950 a month. One month he spends $304 for food. What rate of his income did he spend that month for food?

68 ARITHMETIC

Solution. In this problem, the rate is unknown. We know the base, $950, and the percentage, $304. The percentage, $304, is the product of the two factors. The two factors are the *rate* and the *base*. One factor, the *base*, is known. Our problem is to find the other factor, the *rate*.

In a problem in multiplication, whenever we know the product and one factor, we can find the other factor by division. For example, if we know the product of two factors is 91, and one factor is 7, then we find the other factor by dividing the product 91 by the known factor 7. That is,

$$91 \div 7 = 13$$

To find the rate in the foregoing problem, we divide the percentage, $304, by the base, $950. That is,

$$304 \div 950 = 0.32, \text{ or } 32\%$$

We can check the answer by multiplying the rate times the base:

$$0.32 \times 950 = 304$$

Then, we have the following rule.

RULE 5. To find the rate when the base and the percentage are known, divide the percentage by the base.

The rule may be stated as

$$\text{percentage} \div \text{base} = \text{rate}$$

Example 6. A family has a yearly income of $12,800 and spends $3360 a year for rent. What per cent of their income do they spend for rent?

Solution. The base is $12,800. The percentage is $3360. Dividing, we get

$$3360 \div 12,800 = 0.2625$$

Thus, they spend 26.25% of their income for rent.

4.6. FINDING THE BASE WHEN THE RATE AND THE PERCENTAGE ARE KNOWN

A third type of problem involves finding the base when the rate and the percentage are known. In this case, we again know the product, which is

PERCENTAGE 69

the percentage; and we know one factor, the rate. To find the unknown factor, the base, we divide the percentage by the rate. Consider the following problem:

Example 7. A man spends $2668 for a car, which is 18.4% of his yearly income. How much is his yearly income?

Solution. In this problem, his income is the base, which is unknown. We know the rate and the percentage. The percentage, $2668, is the product of two factors. One factor is the rate, 18.4%. To find the other factor, the base, we divide the product, $2668, by the known factor, the rate:

$$2668 \div 0.184 = 14{,}500$$

His yearly income is $14,500.
For this type of problem, we have the following rule.

RULE 6. To find the base when the rate and the percentage are known, divide the percentage by the rate.

The rule may be stated as

$$\text{percentage} \div \text{rate} = \text{base}$$

Example 8. A family spends $295 a month for food. If this is 23.6% of their monthly income, how much is their monthly income?

Solution. In this problem, we know the product, which is the percentage, $295. The rate, 23.6%, is one factor. To find the base, which is the other factor, we divide

$$295 \div 0.236 = 1250$$

Therefore, their monthly income is $1250.

Sometimes the answer must be rounded off, as in the next example.

Example 9. One year a family spent $3780 for rent. If this was 27.3% of their income for the year, find their income.

Solution. In this problem the percentage and the rate are given. The percentage, $3780, is the product of two factors, one of which is the rate, 0.273. To find the base, the yearly income, we divide:

$$\text{percentage} \div \text{rate} = \text{base}$$

or

$$3780 \div 0.273 = 13{,}846.1538\ldots$$

The answer is approximately $13,850, which is their yearly income.

Note. *In a problem of this kind, the answer is rounded off. The answer, $13,850, therefore, is approximate. However, it should be remembered that it is not his income that is approximate. His income is an exact number of dollars and cents. The same is true with regard to the amount spent for rent. The part that is approximate is the rate, 27.3%, or 0.273.*

In working problems in percentage, probably the greatest difficulty is in identifying the *rate*, the *base*, and the *percentage*. The following three statements may be helpful in working problems in percentage.

(1) *The percentage is a product of two factors.* Therefore, whenever the percentage is given, the process will be *division*; that is, *the percentage must be divided by the given factor*, whether base or rate.

(2) *The base and the percentage have the same name or denomination*, such as dollars, feet, miles, pounds, or objects of any kind.

(3) *The rate has no name or denomination but is simply an indicated fractional part of some quantity. The rate may be more than 100%.*

Exercise 4.2

Find the percentage in each of the following problems:

1. 8% of $2300
2. 7.5% of $480
3. 12.5% of $5600
4. 5.2% of $650
5. 9.6% of $350
6. 6.2% of 85 feet
7. 22.5% of 264 feet
8. 32.5% of 420 miles
9. 1.6% of 300 pounds
10. $1\frac{1}{2}$% of 260
11. $9\frac{3}{4}$% of $1600
12. $6\frac{2}{5}$% of $4500
13. 0.5% of 300
14. 0.21% of 500
15. 0.65% of $4800
16. 0.09% of $45,000
17. 0.01% of $32,000
18. 0.3% of $65,800
19. $\frac{1}{4}$% of $3600
20. $\frac{5}{8}$% of $360,000
21. 12.5% of 560 miles

22. During one year, a family received an income of $12,400, and spent the following portions for each of the items shown. Find the amount spent for each item:

    Food 19%   Clothing 10.5%
    Insurance and savings 12%   Housing 18%
    Medical 8.5%   Utilities 15%
    Taxes 9.7%   Miscellaneous 7.3%

23. A man bought a house for $45,000 and paid 15% of the price as a first down payment. How much did he pay and how much was left to be paid?
24. A cattle rancher had 25,600 cattle. He sold 15% to one buyer, 30% to another, and 45% to a third buyer. If he kept the rest, how many did he sell to each buyer, and how many did he have left?
25. On an automobile trip of 1250 miles, a family drove 26% of the distance the first day, 28% the second day, 24% the third day, and finished the trip the fourth day. How many miles did they travel each of the four days?
26. A college has a total student body of 8400 students. Of the entire student body, 19% are seniors, 23% are juniors, 26% are sophomores, and the remainder are freshmen. Find the number of students in each year.
27. A farmer raised 3500 bushels of corn on a certain field last year. This year, due to excellent growing conditions, he expects a crop 112% as much as last year. How many bushels does he expect this year?
28. A contractor agreed to build a road 42 miles long. When he had finished 35% of the road, he was unable to complete the rest. How many miles did he finish and how many miles were left to be built?
29. A man has a total of $56,000 invested as follows: 23% in stocks; 35% in bonds; and the rest in savings and loan companies. How much has he invested in each type of investment?
30. Last year, a house was priced at $36,500. This year, the price of the house is 120% as much as last year. What is the present price?

Find the rate (in per cent) in each of the following problems:

31. $480 is what % of $3200?
32. 936 is what % of 7800?
33. $405 is what % of $360?
34. $636 is what % of $480?
35. $18 is what % of $720?
36. $54.40 is what % of $850?
37. 6.3 lb is what % of 120 lb?
38. $84 is what % of $84?
39. A contractor agreed to build a road 520 miles long. After finishing 167 miles, he was unable to do more. What per cent had he finished?
40. A family receiving a yearly income of $12,500 spent the following amounts for the items listed. What per cent was spent for each?

   food $2800   clothing $950
   travel and amusement $930 housing $2340
   medical $1025   utilities $1080
   insurance $520   miscellaneous (remainder)

Find the base in each of the following problems:

41. 25% of what number is 16?
42. $36 is 15% of what amount?
43. $42\frac{1}{2}$% of what number is 40.63?
44. $156 is 32.5% of what amount?

Fill in the correct number for each blank in each of the following:

45. 12.5% of $68 = ____
46. 67.5% of ____ = $216
47. 87.5% of 360 miles = ____
48. 135% of $24 = ____
49. ____% of $720 = $108
50. 351 cm is ____% of 5560 cm

51. 135% of ____ = $837   **52.** 32.5% of ____ = $53.30
53. $66 is ____% of $48   **54.** 0.045 ft. = ____% of 0.72 ft.

4.7. INCREASES AND DECREASES

In many problems a particular quantity is to be increased or decreased by a certain per cent of the quantity. In such problems we first find the amount of increase or decrease and then add or subtract to find the new amount.

Example 1. One year the population of a certain town was 42,800. The following year the population had increased by 6.5%. What was the population after the increase?

Solution. To find the amount of increase, we take

$$6.5\% \text{ of } 42,800$$

or

$$0.065 \times 42,800 = 2782, \text{ (the increase)}$$

Since the population increased by 2782, the population the second year was

$$42,800 + 2782 = 45,582$$

Example 2. One week a merchant's sales amounted to $36,800. The next week it had decreased by 4.5%. What was the amount of his sales after the decrease?

Solution. First we find the amount of the decrease:

$$0.045 \times \$36,800 = \$1656$$

To find the amount of sales after the decrease, we subtract:

$$\$36,000 - \$1656 = \$35,144$$

In the case of increases or decreases, the base should always be taken as the original number; that is, the *first number in time*.

Example 3. If the population of a town increases from 40,000 to 50,000, what is the rate of increase?

Solution. For the increase, we have

$$50,000 - 40,000 = 10,000, \text{ (increase)}$$

To find the *rate* of increase, we take the first number, 40,000, as the base.

Then, for the rate, we have

$$10{,}000 \div 40{,}000 = 0.25, \text{ or } 25\%, \text{ (rate of increase)}$$

Example 4. If the population of a town decreases from 50,000 to 40,000, find the rate of decrease.

Solution. For the decrease, we have

$$50{,}000 - 40{,}000 = 10{,}000, \text{ (decrease)}$$

To find the rate of decrease, we take the *first number*, 50,000, as the base. Then, for the rate, we have

$$10{,}000 \div 50{,}000 = 0.20, \quad \text{or } 20\%, \text{ (rate of decrease)}$$

Exercise 4.3.

1. One year the population of a town was 23,400. The following year it had increased by 7.5%. Find the population after the increase.
2. One year the population of a town was 32,000. Five years later it had increased by 125%. Find the population after the increase.
3. One year a man received a salary of $9600. The following year he received an increase of $12\frac{1}{2}\%$. Find his salary after the increase.
4. A TV set was originally priced at $360. During a sale, the price was reduced by 15%. Find the selling price.
5. One year a merchant's sales amounted to $53,200. The following year the amount of sales was $51,870. Find the rate of decrease in sales.
6. During a five-year period, the population of a town increased from 24,800 to 65,100. What was the rate of increase for the five-year period?
7. During a strenuous exercise period, the weight of an athlete decreased from 234 pounds to 219 pounds. What was the rate of loss in weight?
8. During one month, a merchant's sales amounted to $21,000. The next month his sales amounted to $63,000. Find the rate of increase in sales. Would you say that the amount of his sales the second month was three times *more than the first*?
9. During a sale, the price of a TV set was reduced from $456 to $399. Find the *amount* and the *rate* of reduction.

4.8. FINDING THE ORIGINAL AMOUNT WHEN IT HAS BEEN INCREASED OR DECREASED BY A PER CENT

This type of problem is probably one of the most difficult in percentage. In such a problem, the original number has been increased or decreased by a

74 ARITHMETIC

given per cent. We are given the rate of increase or decrease and the amount after the change. However, the base, which is the original number, is unknown. Consider the following example.

Example 1. After the price of an article had been increased by 25%, the new price is $60. What was the original price?

Solution. We cannot take 25% of $60. Instead, it is understood that the 25% was taken on the original price, which is unknown. The original price must be taken as the base. The problem is analyzed in the following way. We take the original price to be 100% of itself. Then, when 25% is added to the original price, the result is 125% of the original price. That is,

$$\$60 \text{ is } 125\% \text{ of the original price}$$

or

$$\$60 = 1.25 \times \text{base}$$

Then

$$\text{base} = \$60 \div 1.25 = \$48, \text{ the original price}$$

To check the correctness of the answer, we can now take

$$25\% \text{ of } \$48 = \$12, \text{ the increase}$$

and

$$\$48 + \$12 = \$60, \text{ the price after the increase}$$

In the case of decrease, we proceed in a similar manner.

Example 2. After a decrease of 15%, the price of an article was $35.70. What was the original price?

Solution. We cannot take 15% of $35.70. Instead, the base must be taken as the original price. Again, we take the original price to be 100% of itself. Then, when 15% is subtracted from this, the result is 85% of the original price. That is,

$$\$35.70 \text{ is } 85\% \text{ of the original price}$$

or

$$\$35.70 = 0.85 \times \text{base}$$

Then

$$\text{base} = \$35.70 \div 0.85 = \$42, \text{ original price}$$

The problem can be checked by taking 15% of $42, and subtracting the result.

Exercise 4.4

1. After an increase of 7.5% in his salary, a man receives $10,320 a year. What was his salary before the increase? How much was the increase?
2. After a lady's weight decreased by 10%, it was 127.8 pounds. What was her weight before the decrease? What was the loss in weight?
3. After a man's weight increased by 6%, he weighed 174.9 pounds. How much did he weigh before the increase? What was his increase in weight?
4. An article was sold for $144 at a gain of $12\frac{1}{2}$% on the cost. What was the cost? What was the gain?
5. An article was sold for $162.80 at a loss of $7\frac{1}{2}$% of the cost. What was the cost? How much was the loss?
6. A boat was sold for $410, which was 18% less than the cost. What did the boat cost?
7. A sports car supply store one month had sales amounting to $20,580. If this was an increase of 22.5% over the previous month, find the amount of sales for the previous month. Find the increase in sales.

4.9. BUYING AND SELLING; GAIN AND LOSS

A merchant buys goods from a *wholesaler* and then sells the goods to his customers. The merchant is then called the *retailer*. The merchant is said to buy goods at *wholesale* and sell them at *retail*.

A merchant who buys and sells goods is concerned with the profit or gain he makes on the articles he buys and sells. The profit or gain is the difference between the cost and the selling price. For example, he may buy a radio for $36 and sell it for $50. His profit on the radio is $14. Of course, in some cases, he may have to sell an article for less than it cost him. In that case, he takes a loss.

If an article is bought for $40 and sold for $50, the gain is $10. If an article is bought for $50 and sold for $40, the loss is $10.

In determining what is to be the selling price of an article, the merchant will usually decide on a price that is some per cent above the cost. For example, he may have paid $84 for a men's suit. He decides that he must make a profit or gain of 25% on the cost. To find the gain, we take

$$25\% \text{ of } \$84 = \$21, \text{ gain or profit}$$

If he wishes to make a profit of $21, he marks the suit to sell for $105. This is called the *marked price*.

Example 1. Find the selling price of a desk bought for $60 and sold at a gain of 25%.

Solution. For the gain, we have

$$0.25 \times \$60 = \$15, \text{ gain}$$

Then we get the selling price:

$$\$60 + \$15 = \$75, \text{ selling price}$$

The desk must be sold for $75 if the gain is to be 25% of the cost.

Example 2. If an article is bought for $140 and sold at a loss of 15%, find the loss and the selling price.

Solution.
$$0.15 \times \$140 = \$21.00, \text{ loss}$$

For the selling price, we have

$$\$140 - \$21.00 = \$119.00, \text{ selling price}$$

To find the *rate* of gain or loss on an article, we first find the gain or loss, which is the percentage. To find the rate, we take the cost as the base.

Example 3. Find the rate of gain on an article bought for $72 and sold for $90.

Solution. For the amount of gain, we have

$$\$90 - \$72 = \$18, \text{ gain}$$

To find the rate of gain, we take the cost, $72, as the base. Then we have

$$\$18 \div \$72 = 0.25, \quad \text{or } 25\% \text{ gain}$$

Note. Some large stores compute the rate of gain or loss with reference to the selling price instead of the cost. That is, the selling price is taken as the base. In Example 3, if the gain is computed with reference to the selling price, we should have

$$\$18 \div \$90 = 0.20, \quad \text{or } 20\% \text{ gain}$$

However, the rate of gain or loss should always be computed using the cost as the base because the cost always occurs *before* the selling price.

Example 4. Find the rate of loss on a boat that is bought for $300 and sold for $252.

Solution. For the loss, we have

$$\$300 - \$252 = \$48, \text{ loss}$$

To find the rate of loss, we take the cost, $300, as the base. Then

$$\$48 \div \$300 = 0.16, \quad \text{or } 16\% \text{ rate of loss}$$

Note. If the rate of loss is computed on the selling price as the base, we should have

$$\$48 \div \$252 = 0.19 \text{ (approximately)}$$

Then the rate of loss is approximately 19%.

Note. In Example 3, if the gain is computed on the selling price as the base, the rate of gain *appears to be* less *than when computed on the cost. In Example 4, if the loss is computed on the selling price as the base, the* rate of loss *appears to be* greater *than when computed on the cost.*

Exercise 4.5

Find the gain or loss and the selling price for each of the following items:

1. TV set, cost $350; rate of gain, 15%.
2. Radio, cost $54; rate of gain, 15%.
3. Car, cost $2500; rate of gain, 18%.
4. House, cost $35,000; rate of gain, 12%.
5. Car, cost $3200; rate of loss, $12\frac{1}{2}\%$.
6. Davenport, cost $280; rate of loss, $8\frac{1}{2}\%$.
7. Rod and reel, cost $56; rate of loss, 6.25%.
8. Air conditioner, cost $180; rate of loss, 7.5%.

Find the rate of gain or loss on the following items, using the cost as the base: (first find the amount of gain or loss).

	Cost	Selling price		Cost	Selling price
9.	$120	$150	10.	$150	$120
11.	$144	$165.60	12.	$176	$165
13.	$75	$60	14.	$60	$75
15.	$100	$150	16.	$150	$100

4.10. DISCOUNT

The price at which a merchant marks an article to be sold is called the *marked price*. During a sale, merchants usually offer goods at a reduction from the marked price. A reduction from the marked price is called a *discount*. The discount is usually stated as a *per cent* of the marked price.

Example 1. A man's suit is marked to sell for $120. During a sale, the suit was priced to sell at a discount of 20% off the marked price. Find the discount and the selling price.

Solution. For the discount, we have

$$20\% \text{ of } \$120 = \$24, \text{ discount}$$

Then the selling price becomes

$$\$120 - \$24 = \$96, \text{ selling price}$$

At a sale various items may be sold at different discount rates. For the amount of deduction, we take the *rate* of discount times the marked price. Then we subtract the discount from the marked price. The result is the selling price.

A merchant may offer a discount for various reasons, such as sales, special events, partly damaged goods, articles out of date, or because of immediate payment in cash.

Example 2. A radio is marked to sell for $45. If a discount of 2% is allowed for payment in cash, find the net cost.

Solution

$$0.02 \times \$45 = \$0.90, \text{ discount for cash}$$

For the net cost, we have

$$\$45 - \$0.90 = \$44.10, \text{ net cost}$$

If we have given the marked price and the discount itself, our problem is to find the *rate* of discount. If the marked price and the selling price are given, we must first find the amount of discount or reduction. Here, again, we have the problem of finding the *rate* in the percentage statement. The discount itself is computed as a per cent of the marked price, which is taken as the base.

Example 3. A tire regularly priced at $52 is sold for $44 during a sale. Find the rate of discount.

Solution. For the amount of discount, we have

$$\$52 - \$44 = \$8, \text{ discount}$$

Now the $8 is computed as a per cent of the marked price, $52.

$$\$8 \div \$52 = 0.154 \text{ (approximately)}$$

Then the rate of discount is approximately 15.4%.

Example 4. A merchant buys a record player at wholesale for $140. He marks it to sell at a gain of 25% of the cost. When it is sold, the purchaser is given a discount of 2% for cash payment. Find the marked price, the selling price, and the gain. What is the rate of gain on the cost?

Solution. To find the markup above the cost, we take

$$25\% \text{ of } \$140 = \$35.00$$

Then the marked price is

$$\$140 + \$35.00 = \$175.00$$

Now the 2% discount is computed on the marked price as the base. Then we have

$$0.02 \times \$175 = \$3.50, \text{ discount off marked price}$$

The selling price becomes

$$\$175.00 - \$3.50 = \$171.50$$

The actual gain becomes

$$\$171.50 - \$140 = \$31.50, \text{ gain}$$

For the *rate* of gain, we have

$$\$31.50 \div \$140 = 0.225, \quad \text{or } 22.5\%$$

The rate of gain on the cost is 22.5%.

Sometimes, two or more discounts are offered on the same article. Then the first discount is first computed and deducted. The second discount is computed on the remainder after the first deduction. When two or more discounts are given, they are called *successive discounts*. When successive discounts are offered, the first is computed and deducted, and the next discount is computed on the remainder.

80 ARITHMETIC

Example 5. At a sale, a TV set, marked at a regular price of $275, is to be sold at a 20% discount off the marked price. An additional discount of 3% is allowed for cash payment. Find the net cost of the TV set.

Solution. First we take

$$20\% \text{ of } \$275 = \$55, \text{ sale discount}$$

For the sale price, we have

$$\$275 - \$55 = \$220, \text{ sale price}$$

For the discount for cash,

$$0.03 \times \$220 = \$6.60, \text{ discount for cash}$$

Then we subtract

$$\$220 - \$6.60 = \$213.40, \text{ net cost}$$

When two successive discounts are allowed, it is necessary to compute each discount in succession. Each discount is computed and deducted in order. The two discounts cannot be first added and treated as a single discount. For example, two successive discounts of 25% and 5% do not have the same result as a single discount of 30%, as the following example shows.

Example 6. A typewriter is listed at $160 and is sold at discounts of 25% and 5%. Find the net cost of the typewriter.

Solution. For the first discount,

$$0.25 \times \$160 = \$40$$

Now we deduct the first discount:

$$\$160 - \$40 = \$120$$

For the second discount, we have

$$0.05 \times \$120 = \$6.00$$

Deducting the second discount, we get

$$\$120 - \$6.00 = \$114.00$$

PERCENTAGE 81

In this problem, if we had used a single discount of 30%, the net cost would have been $112.00.

It is sometimes desirable to find a single discount that would be equivalent to two or more successive discounts. For example, in the case of the typewriter in Example 6, the net cost is $114. To find the equivalent single discount, we can proceed as follows. We first find the total discount, which is $46. Then we divide:

$$\$46 \div \$160 = 0.2875$$

Therefore, the two successive discounts of 25% and 5% are equivalent to a single discount of 0.2875, or 28.75%.

Exercise 4.6

Find the net cost of each of the following items, discounted as shown.

Item	Marked price	Rate of discount	Item	Marked price	Rate of discount
1. Piano	$1225	20%	2. Car	$3284	15%
3. House	$43,500	10%	4. Tire	$48.50	20%
5. TV set	$320	30%	6. Bicycle	$82	25%
7. Battery	$42.50	35%	8. Stereo	$285	45%

At a furniture sale, the following items, with list price shown, were sold at 30% off the list price and an additional discount of 5% for cash. Find the cash price of each item.

Item	List price	Item	List price
9. Davenport and chair	$480	10. Table and chairs	$245
11. Writing desk	$95	12. Bed and stand	$368
13. Dining table and chairs	$520	14. Dresser and vanity	$425

At a clothing store sale, the following items, with list price shown, were sold at 35% off the list price, and an additional discount of 2% for cash. Find the cash price of each item.

Item	List price	Item	List price
15. Men's suits	$125	16. Men's slacks	$48
17. Men's ties	$6.50	18. Dresses	$65

Find the rate of discount on each of the following items, showing the list price and the sale price.

82 ARITHMETIC

Item	List price	Sale price		Item	List price	Sale price
19. Bicycle	$79.99	$66.00		20. Stereo	$99.99	$85.00
21. 8-track player	$49.99	$39.00		22. Automobile	$2850	$2550
23. Suitcase	$28.95	$11.99		24. Tire	$34.95	$29.95

Find the rate of discount on each of the following items. The regular price and the sale price are shown for each item.

Item	Regular price	Sale price		Item	Regular price	Sale price
25. Men's suits	$115	$95		26. Slacks	$14	$9.99
27. Sport coats	$39.99	$19.89		28. Shoes	$35	$26.50
29. Tennis racket	$45.50	$32		30. Tire	$29	$23
31. Tape recorder	$65.50	$42.50		32. Battery	$27.95	$21.95
33. Portable TV	$85.60	$55		34. Car	$4160	$3760

The following items were purchased by dealers at the cost shown. For the selling price, the items were marked up at the rates shown. Later the items were sold at the indicated discount off the marked price. Find the net cost of each item and the rate of gain on the cost as a base.

35. Men's suits; cost $60 each; marked up 35%; sold at a discount of 10%.
36. Tape recorder; cost $45; marked up 40%; sold at a discount of 20%.
37. Bicycle; cost $50; marked up 30%; sold at a discount of 15%.
38. TV set; cost $220; marked up 50%; sold at a discount of 25%.
39. Dresses; cost $36; marked up 45%; sold at a discount of 40%.
40. Stereo; cost $180; marked up 60%; sold at a discount of 50%.

A catalog showed the following list prices for various articles. They were sold at the indicated successive discounts. Find the single discount equivalent to the given successive discounts for each article.

41. List price: $400; discounts: 30% and 20%.
42. List price: $650; discounts: 40% and 20%.
43. List price: $80; discounts: 30% and 10%.
44. List price: $360; discounts: 40% and 10%.
45. List price: $60; discounts: 20% and 20%.
46. List price: $180; discounts: 25%, 10%, and 2%.
47. List price: $460; discounts: 30%, 20%, and 5%.
48. List price: $240; discounts: 30%, 15%, and 5%.

4.11. INTEREST

Interest is a charge for the use of money. Interest may be called the *rent* for the use of money, just as we pay rent for the use of a boat or a house.

Interest involves a *borrower* and a *lender*. The money borrowed is called

PERCENTAGE 83

the *principal*. In borrowing money, the borrower agrees to pay the lender *interest* for the use of the money borrowed. The *rate* of interest is usually expressed as a per cent of the principal for each year.

The borrower usually signs a note stating the principal, the rate of interest to be paid, and the length of time the money is to be used. Sometimes the due date is stated on the note. The interest paid will depend on the rate and also on the length of time of the note.

Example 1. A man borrows $500 at an interest rate of 8% per year. That is, if he has the use of the money for one year, he pays 8% of $500 as interest. If he has the money two years, how much interest does he pay, and what is the total amount he must pay back at the end of two years?

Solution. First we find the interest for 1 year:

$$0.08 \times \$500 = \$40.00$$

If he has the use of the money for 2 years, the interest is

$$2 \times \$40 = \$80$$

The total amount paid back is

$$\$500 + \$80 = \$580, \text{ called the } amount$$

To find the interest on a particular principal for a certain time, we have the rule:

$$\text{interest} = \text{principal} \times \text{rate} \times \text{time}$$

The time is stated in years.

To show the rule in a more concise form, we may use letters to represent the quantities mentioned in the rule. We use p to represent the principal; r to represent the *rate*; t to represent the *time* (in years); and i to represent the *interest*. Then we can state the rule by using letters. The rule becomes

$$i = p \times r \times t; \quad \text{often written simply:} \quad i = prt$$

Such a simple rule, as $i = prt$, is called a *formula*. A formula is a rule stated in letters that represent quantities. If A represents the amount to be paid back, including principal and interest, then we have

$$A = p + i$$

or

$$A = p + prt$$

Example 2. A man borrows $800 at 6% interest per year for 3 years. Find the interest and the amount.

Solution

$$0.06 \times \$800 = \$48, \text{ interest for one year}$$

$$3 \times \$48 = \$144, \text{ interest for 3 years}$$

$$A = \$800 + \$144 = \$944, \textit{ amount}$$

Using the formula, we write

$$i = 800 \times 0.06 \times 3 = 144$$

$$A = 800 + 144 = 944$$

Note. When money is borrowed and used for several years, the interest due is often paid at the end of each year.

Example 3. Find the interest and amount on a loan of $120 borrowed for 3 months at an interest rate of 7.5% per year.

Solution. We express 3 months as $\frac{1}{4}$ year. Using the formula, we have

$$i = \$120 \times 0.075 \times \tfrac{1}{4} = \$2.25, \text{ interest for 3 months}$$

The amount is

$$\$120 + \$2.25 = \$122.5, \quad \text{amount}$$

Example 4. A man needs to borrow $3600 for only 15 days. The interest rate is 10% per year. Find the interest and the amount due in 15 days.

Solution. We express 15 days as $\frac{15}{360}$ year. Then

$$i = \$3600 \times 0.10 \times \tfrac{15}{360} = \$15.00, \text{ interest}$$

To find the amount, we take

$$A = \$3600 + \$15 = \$3615, \text{ amount due}$$

Sometimes when money is borrowed, the note will state the date on which it is due, that is, the *maturity date*. If the maturity date is given but not the duration of the note, then it is necessary to compute the time from the date of borrowing to the maturity date.

In determining the time of a note, it is customary to take one month as 30 days, and 360 days as one year. The time from a particular date of one month to the same date of the following month is taken as one month of 30 days, regardless of the exact number of days. The time from July 15 to August 15 of the same year is taken as one month. The time from March 12, 1974 to June 12, 1975 is called 1 year and 3 months.

However, sometimes the exact number of days is computed between dates. This is especially true if the maturity date falls on a day of the month different from the day of the month of borrowing. For example, the time from April 20 to June 7 of the same year is 48 days. Then the time is taken as $\frac{48}{360}$ of a year. In some instances, as in calculating interest on government bonds or where required by law, one year is taken as 365 days. However, in most cases, a year is taken as 360 days.

Exercise 4.7

Find the interest and the amount for each of the following loans.

	Principal	Rate of interest	Time		Principal	Rate of interest	Time
1.	$200	8%	3 years	2.	$50	10%	6 months
3.	$80	10%	3 months	4.	$150	8%	4 months
5.	$3200	6.5%	8 months	6.	$15,200	5.5%	18 months
7.	$60	12%	2 months	8.	$480	7.5%	15 months
9.	$360	$7\frac{1}{4}$%	10 months	10.	$250	$6\frac{1}{4}$%	$2\frac{1}{2}$ years
11.	$2400	9%	15 days	12.	$840	11%	20 days
13.	$4500	10%	45 days	14.	$6000	12%	10 days

15. $300, borrowed March 20, 1975, due July 24, 1975, at 8%.
16. $480, borrowed July 7, 1975, due October 20, 1975, at 10%.
17. $560, borrowed April 15, 1975, due June 4, 1975, at 9.5%.
18. $840, borrowed May 20, 1975, due August 20, 1975, at 7.5%.
19. $680, borrowed February 15, 1975, due November 15, 1975, at 8.5%.
20. $2500, borrowed August 5, 1975, due August 20, 1975, at 10%.

4.12. COMPOUND INTEREST

Let us first consider a problem in simple interest. If you deposit $100 in a savings account at 6% interest per year, the interest amounts to $6 in one year. If you withdraw the $6 interest at the end of the one year, you will have $100 on deposit at 6% for the second year. If you withdraw the interest at the end of each year, you will always have a principal of $100 on which you will receive interest of $6 each year. Then at the end of 10 years, you will have received $60 as interest, and the principal will remain at $100. The $60 interest received for the 10 years is called *simple interest*.

Now, suppose, at the end of 1 year, you leave the $6 interest with the $100 principal. Then the total amount on deposit for the second year will be $106. During the second year, your principal of $106 earns interest at 6%. Then the interest for the second year will be

$$0.06 \times \$106 = \$6.36$$

If this amount of interest is then left with the principal of $106, the principal for the third year will be $112.36. The original principal of $100 has now earned $12.36 as interest. The $12.36 is called *compound interest.* Compound interest represents not only the interest on the original principal but also interest on the accumulated interest.

If $100 is placed at 6% interest, and the interest is withdrawn at the end of each year, then at the end of 10 years, the simple interest will be $60. However, if the interest is *compounded annually* (that is, the interest is added to the principal at the end of each year), then the total amount at the end of ten years will be $179.08, of which $79.08 is the compound interest. At the end of 12 years, the simple interest would be $72, whereas the compound interest would be 101.22. That is, at the end of 12 years, the $100 principal would have more than doubled itself.

In the example of $100 deposited as principal at 6% interest, if the interest due is added on every half-year, we say that the interest is *compounded semiannually.* At 6% interest per year, the interest for 6 months will be one-half as much as the interest for 1 year. For $100 principal at 6% interest per year, the interest for one-half year is $3.00. Now, if the $3 interest is added to the principal at the end of 6 months, then the principal for the second half-year is $103. The interest on the $103 for the second 6 month period will be $3.09. At the end of the year, the amount will be $106.09.

It should be clear that if the interest is compounded semiannually (that is, every 6 months), then the accumulation or amount on deposit at the end of 10 years will be greater than when the interest is compounded annually. We have seen that, when the interest is compounded annually, then, at the end of 10 years, the interest will be $79.08. Now, if the interest is compounded semiannually, then at the end of 10 years, the interest will be $80.61.

A compound interest table (see Appendix, Table 1) shows the amount or accumulation of $1 at various rates of interest when the interest is compounded for certain numbers of periods. For example, the table shows that the amount of $1 at 6%, compounded annually for 10 years is $1.79085. Since this amount includes the $1 principal, the compound interest alone is $0.79085, or $0.79.

To find the amount of $1 at 6% interest compounded semiannually, we take *one-half the yearly rate* and *double the number of interest periods.* For example, at an annual rate of 6% compounded semiannually, for 10 years, we take a rate of 3% for 20 interest periods. Then we find that the amount is $1.80611, of which the interest alone is $0.80611.

If the earned interest on a deposit is added to the principal every three months, then we say the interest is *compounded quarterly*. Then we take one-fourth of the annual rate, and multiply the number of years by 4. For example, for $1 at an annual rate of 6% for 5 years, we take $\frac{1}{4}$ of the yearly rate, or 1.5% as the rate for 3 months. Then for 5 years, we have 20 interest periods. That is, the interest is added to the principal 20 times.

The values shown in the table are based on $1 as the original principal. To find the amount or accumulation for any principal, we look for the amount of $1 at the proper interest rate and for the proper number of interest periods. Then we multiply this amount by the original principal.

Example 1. Find the compound amount and the compound interest on a principal of $300 at an annual rate of 7%, compounded quarterly for a total of 8 years.

Solution. Using the table, we take a tax rate of 1.75%, which is $\frac{1}{4}$ of 7%, since the interest is compounded each 3 months, or $\frac{1}{4}$ year. Then, for 8 years, the number of interest periods is 32. For 1.75% and 32 interest periods, the table shows $1.74221 as the amount for $1. For the principal of $300, we multiply

$$300 \times 1.74221 = 522.663$$

Then the amount is $522.66 for the $300 principal for 8 years. Of this amount, the interest is found by subtracting:

$$\$522.66 - \$300 = \$222.66, \quad \text{compound interest}$$

Exercise 4.8

Find the amount and the compound interest on each of the following principals at the given rate of interest per year, and for the given number of years. (Use the table.)

1. A principal of $3000 at 6% per year for 12 years, compounded annually.
2. Same as No. 1, compounded semiannually.
3. Same as No. 1, compounded quarterly.
4. A principal of $2400 at 8% per year for 6 years, compounded annually.
5. Same as No. 4, compounded semiannually.
6. Same as No. 4, compounded quarterly.
7. A principal of $600 at 5% for 10 years, compounded annually.
8. Same as No. 7, compounded semiannually.
9. Same as No. 7, compounded quarterly.

10. A principal of $80 at 7% per year for 10 years, compounded annually.
11. Same as No. 10, compounded semiannually.
12. Same as No. 10, compounded quarterly.
13. From the table, we find that $1 at 6% interest per year, compounded annually, will accumulate an amount of $2.0122 in 12 years. That is, the amount at the end of 12 years will be 2.0122 times the principal at the beginning of the 12-year period. Then what would be the amount at the end of 24 years (that is, after another 12 years)? After 36 years? After 48 years? After 96 years?
14. Approximately what would be the amount of $1 at 6% interest per year, compounded annually, for 240 years if the principal of $1 could be permitted to remain at compound interest for that length of time? (Consider that the principal approximately doubles itself every 12 years.)
15. Work out without the use of the table the compound interest on $100 at 6% interest per year, compounded annually, for a period of 5 years.

4.13. BANK DISCOUNT

Suppose you borrow $200 from a bank at 8% interest for one year. The interest is $16. If the interest is paid at the end of the year, you receive $200 at the time of borrowing, and the total amount to be paid back at the end of the year is $216.

However, in some cases, the bank deducts the interest at the time of borrowing. Interest collected by the bank at the time of borrowing is called *bank discount*. If you borrow $200, and the bank at once collects the interest of $16, then the actual amount you receive is $184. The amount received by the borrower after the bank has deducted the interest is called the *proceeds*.

It is often said that interest should be paid at once when money is borrowed. Since interest is money paid for the use of money, then it is like rent paid for the use of a boat or a house. Such rent is usually paid in advance. Therefore, it is argued, interest for the use of money should also be paid in advance.

When bank discount is deducted from a loan in advance, the amount the borrower receives is less than the amount of the loan. For example, when you borrow $200 at 8% for 1 year, if the bank deducts the discount, then you receive only $184. At the end of the year you pay back $200. If the interest of $16 is based on the $184 used for 1 year, then the *rate* of interest is greater than 8%. That is,

$$\$16 \div \$184 = 0.087 \text{ (approximately)}, \quad \text{or} \quad 8.7\%$$

It often happens that a bank discounts a note that already bears a rate of interest. Consider the following example.

Example 1. Mr. Smith lends Mr. Jones $600 at 7% interest for 8 months. Mr. Jones gives Mr. Smith a note for $600. At the end of 8 months, Mr.

Smith will receive $28 interest and the $600 principal from Mr. Jones. The $628 is called the *maturity value* of the note. We say the note *matures* in 8 months. Mr. Jones need not pay the amount until the maturity date.

Now, two months after lending the money, Mr. Smith finds that he needs money. He takes the note to a bank and asks the bank to give him money for the note. If the bank accepts, then it will receive the principal and the interest due at the end of 6 months. The bank will receive the maturity value, $628. Now, the bank does not give Mr. Smith the entire $628, since the bank will not receive that amount for another 6 months. Instead, the bank will discount the note at some rate, such as 8.5%. The discount is based on the maturity value, $628. To find the discount, we take 8.5% of the maturity value for the 6 months remaining:

$$\text{discount} = \$628 \times 0.085 \times \frac{1}{2} = \$26.69, \text{ discount}$$

Subtracting, we get the proceeds:

$$\$628 - \$26.69 = \$601.31$$

Then Mr. Smith receives the proceeds, $601.31, from the bank. Actually, he has then received only $1.31 as interest on $600 for 2 months.

There is another type of problem, slightly more difficult, involving bank discount.

Example 2. Mr. Brown needs $1500 to pay a hospital bill. He goes to a bank to borrow the money at 8% for 6 months. The bank makes out a note for $1500, and then discounts the note at 8%. That is, the interest is deducted at once, and Mr. Brown is to get the proceeds. If we take

$$\$1500 \times 0.08 \times \frac{1}{2} = \$60.00, \text{ discount}$$

then Mr. Brown is to get the proceeds, $1440. "But," says Mr. Brown, "I need $1500." Now, our question is this: What amount should the note state as principal to be paid in 6 months so that Mr. Brown can now receive the $1500 he needs, after the bank has deducted the discount?

Solution. The discount at 8% per year is to be computed on the maturity value of the note. The 8% per year becomes 4% for 6 months. Then the $1500 is 4% less than the maturity value. Therefore, we take

$$\text{maturity value} = 100\% \text{ of maturity value}$$

Deducting 4%,

$$\text{present value} = 96\% \text{ of maturity value}$$

That is,

$$\$1500 = 0.96 \text{ of maturity value, amount to be paid}$$

Then

$$\text{maturity value} = \$1500 \div 0.96 = \$1562.50$$

Therefore, the note should state $1562.50 as the amount borrowed.

We check the answer as follows.
For the bank discount, we have

$$0.08 \times \$1562.50 \times \frac{1}{2} = \$62.50$$

Deducting the discount, we get

$$\$1562.50 - \$62.50 = \$1500, \text{ proceeds}$$

Exercise 4.9

Find the bank discount and the proceeds of the following loans:

	Face of note	Rate of discount	Time		Face of note	Rate of discount	Time
1.	$300	8%	6 months	2.	$480	7%	4 months
3.	$240	$8\frac{1}{2}$%	3 months	4.	$520	7.5%	8 months
5.	$800	7.5%	3 months	6.	$1200	8%	2 months
7.	$2500	6%	2 years	8.	$3600	6.5%	$1\frac{1}{2}$ years

Find the bank discount, the proceeds, and the actual rate of interest paid on the money used during the time period for the following loans:

9. $400 at 8% for 1 year.
10. $600 at 8% for 6 months.
11. $800 at 7% for 1.5 years.
12. $1200 at 6% for 2 years.

Find the bank discount and the proceeds for each of the following interest bearing notes:

13. A note for $80 with interest at 8% for 3 months is discounted at 10% one month after the date of the note.
14. A note for $450 with interest at 7%, dated March 4, 1975 for 6 months, is discounted on April 4, 1975 at 8%.
15. A note for $2000 with interest at 5.5% for 1 year, dated November 20, 1974, is discounted on February 20, 1975 at 6%.

16. A note for $120 with interest at 8%, dated April 16, 1974, for 9 months, is discounted on May 16, 1974 at 8.5%.

17. A note for $60 with interest at 7.5% for 120 days, is discounted 15 days after the date of the note at 8%.

Find the face of the note (principal) for each of the following in order that the stated amount shall remain after discount has been deducted:

18. $300 at 8% for 6 months.
19. $600 at 8% for 3 months.
20. $1500 at 6% for 4 months.
21. $1200 at 7% for 1 year.
22. $2400 at 6% for 2 months.
23. $1800 at 8% for 9 months.
24. $3000 at 5% for 3 years.
25. $48 at 7% for 8 months.

4.14. COMMISSION

An agent who buys or sells goods for another is paid a *commission* for his services. The commission is the amount of money he receives as his pay. The commission is usually based as a rate per cent on the money value of the goods bought or sold. The agent may sometimes also be paid a regular salary or wage.

Example 1. A man sells a house for the owner and receives $31,500 from the buyer. The agent's commission is 6% of the sale. How much is the commision?

Solution

$$0.06 \times \$31,500 = \$1890, \text{ amount of commission}$$

Example 2. A saleslady sells 4 sets of encyclopedias at $165 each and receives a commission of 30% on the amount of the sales. How much is her commission?

Solution

$$4 \times \$165 = \$660, \text{ amount of sales}$$

$$0.30 \times \$660 = \$198, \text{ amount of commission}$$

Example 3. A man receives a commission of $527 for selling a second-hand automobile for $4250. What is his rate of commission?

Solution. Here the rate is unknown. Then we use division.

$$\$527 \div \$4250 = 0.124, \quad \text{or } 12.4\%, \text{ rate of commission}$$

Example 4. A salesman receives a commission of 24% on the amount of his sales. One month he received $1644 as his commission. What was the amount of his sales?

Solution. Here the base is unknown. Then we use division.

$$\$1644 \div 0.24 = \$6850, \text{ amount of his sales}$$

Exercise 4.10

Find the amount of commission on the following sales (No. 1–5):

1. A car sold for $3150; rate of commission: 10%.
2. A house sold for $52,000; rate of commission: 6%.
3. Forty acres of land, worth $650 an acre; rate of commission: 5.2%.
4. One hundred fifty shares of stock at $85 a share; rate of commission: $\frac{1}{2}$%.
5. Insurance policy for $12,500; rate of commission: $\frac{1}{4}$%.

Find the rate of commission on the following sales (No. 6–10):

6. A car sold for $2800; commission: $420.
7. Men's suits at $135 each; commission: $29.70.
8. Set of encyclopedias sold for $180; commission: $63.
9. Magazines sold for $85; commission: $23.80.
10. A motorboat sold for $2100; commission: $315.
11. A store clerk receives a salary of $180 a week plus 5% of the amount of sales. What is the total amount he receives for a week if his sales that week amount to $1150?
12. A saleslady receives a commission of 20% on all sales of cosmetics less than $100, and 40% on amounts over $100 per week. One week her sales amounted to $165. How much did she receive as her commission?
13. A broker buys 200 shares of stock at $120 a share for a customer. The broker receives a commission of $\frac{5}{8}$%. How much is his commission?
14. An agent buys 2500 bushels of potatoes for a dealer at $6.50 a bushel. The agent receives 6.5% commission. How much is his commission?

4.15. TAXATION

Taxes are levied on various items such as sales, income, homes, land, and other kinds of property. Money collected through taxation by governments is used to provide services that cannot be so well provided by individuals, such as eduction, police protection, highway construction, and other government services.

One of the basic kinds of tax is the tax on property. This tax is based on the value of property. The value placed on property as a basis for the tax is called the *assessed valuation*. The assessed value is usually less than the actual market value of the property. In most cases, the assessed value is a certain stated per cent of the market value. When the assessed value has been computed, then the tax is taken as a stated rate of the assessed value.

The tax rate may be stated in several ways:

1. As a per cent of the assessed value.
2. As a rate per $100.
3. As a rate per $1000.
4. As a number of mills per dollar. (1 mill = 0.1 of a cent).

As an example, a rate of 2% on the assessed valuation is the same as $2.00 per $100; $20.00 per $1000; or 20 mills on $1. However, in most cases we shall state the rate in per cent of the assessed value.

Example 1. A house has a market value of $45,000. Find the tax if the assessed value is 50% of the market value, and the rate of tax is 3.6% of the assessed value. (3.6% is the same as 36 mills per dollar.)

Solution. First we find the assessed value. That is,

$$0.50 \times \$45,000 = \$22,500, \text{ assessed value}$$

For the tax, we have

$$0.036 \times \$22,500 = \$810, \text{ amount of tax}$$

The difference in rates of assessed valuation and in rates of tax may lead to misunderstanding. A tax rate alone does not determine the amount of tax to be paid. Compare the tax in Examples 2 and 3 below.

Example 2. In one community, a house whose market value is $40,000 is assessed at 60% of its market value. Find the amount of tax if the tax rate is 2.5% of the assessed value. (2.5% = 25 mills on a dollar.)

Solution

$$0.60 \times \$40,000 = \$24,000, \text{ assessed value}$$

For the tax, we have

$$0.025 \times \$24,000 = \$600, \text{ amount of tax}$$

Example 3. In another community, a house having a market value of $40,000 is assessed at 40% of its market value. Find the amount of tax if the tax rate is 3.5% of the assessed value.

Solution

$$0.40 \times \$40{,}000 = \$16{,}000, \text{ assessed value}$$

For the tax, we have

$$0.035 \times \$16{,}000 = \$560, \text{ amount of tax}$$

Note. In Example 3, the tax rate is higher than in Example 2, but the amount of tax is less, even though the market value of each house is the same.

Example 4. In a certain community where the assessed value is 55% of the market value, a house is assessed at $23,100. What is the market value?

Solution. Here we have the rate and the percentage given. We must find the base, which is the market value. Therefore, the problem involves division:

$$\$23{,}1000 \div 0.55 = \$42{,}000, \text{ market value}$$

Example 5. In a certain community, the market value of property subject to taxation is $26,800,000. If the assessed value is 45% of the market value, and the community requires a total of $422,100 in taxes, what must the tax rate be on the assessed value?

Solution. First we find the assessed value of the taxable property:

$$0.45 \times \$26{,}800{,}000 = \$12{,}060{,}000, \text{ assessed value}$$

For the rate, we take

$$\$422{,}100 \div \$12{,}060{,}000 = 0.035$$

The tax rate is then 3.5% of the assessed value, or 35 mills on a dollar.

Exercise 4.11

Find the amount of tax on the following properties:

	Property	Market value	Rate of assessed value	Rate of tax
1.	House	$44,000	45%	3.4%
2.	House	$36,000	50%	36 mills per $1
3.	House	$58,000	55%	3.5%
4.	Store	$127,500	60%	3.8%

5. Find the tax on 80 acres of farmland worth $450 per acre, assessed at 30% of market value, with a tax rate of 3.5%.

Find the tax rate on each of the following properties:

	Property	Market value	Rate of assessed value	Amount of tax
6.	House	$48,000	50%	$864
7.	Store	$124,000	65%	$3465.80

8. In a certain community, the market value of taxable property was $28,000,000. The property was assessed at 45% of the market value. If the community needed a total of $441,000 in taxes, what must be the tax rate? Find the tax on a house worth $52,000.

9. In one community, the assessed valuation was $7,290,000, which was 45% of the market value of the property. If the community needed to raise a total of $255,150 through property taxation, what must be the tax rate? Find the market value of the property, and find the tax on a house that has a market value of $64,500.

10. In one community, the market value of taxable property was $25,200,000. Property was assessed at 50% of its market value. If the community needed to raise $567,000 by the property tax, what must be the tax rate. Find the tax on a store whose market value was $136,000.

5 POWERS AND ROOTS

5.1. DEFINITIONS

If we multiply a number by itself, we call the product the *square* of the number. For example, consider the multiplication,

$$6 \times 6 = 36$$

In this example, 36 is called the *square* of 6. It is also called the *second power* of 6.

To indicate the square or second power of a number, we use a small number called an *exponent*. To show the second power of 6, we write

$$6^2 = 36$$

In this expression, the *exponent* is 2. We call 6 the *base*. The exponent 2 means that two 6's are to be multiplied together.

If we multiply three 6's together, the product is called the *cube* of 6. For example,

$$6 \times 6 \times 6 = 216$$

In this example, 216 is called the *cube* of 6. It is also called the *third power* of 6. To indicate the cube or *third power* of a number, we use the *exponent* 3. For the third power of 6, we write

$$6^3 = 216$$

The expression is read, "6 cubed equals 216." In this case, the exponent is 3. The small 3 means that three 6's are to be multiplied together.

> *Definition.* An exponent is a small number placed at the right and a little above another number, called the base, to show how many times the base is to be used as a factor.

Example. 5^3 means $5 \times 5 \times 5 = 125$. In this example, the base is 5. The exponent 3 means that three 5's are to be multiplied together. Then 125 is called the *third power* of 5, or the *cube* of 5.

The second power of a number, as we have said, is called the *square* of the number. The third power is called the *cube* of the number. If the exponent is greater than 3, we name the power by the exponent. Thus:

5^4 is called the 4th power of 5; the 4th power of 5 is 625
2^5 is called the 5th power of 2; the 5th power of 2 is 32
10^6 is called the 6th power of 10; the 6th power of 10 is 1,000,000

5.2. ROOTS OF NUMBERS

The *root* of a number is the opposite of the power. As an example,

$$6^2 = 36$$

Then the square of 6 is 36. Then we say that the *square root* of 36 is 6. In the same way, the *square root* of 49 is 7, because the square of 7 is 49.

Other examples

The square root of 100 is 10, because the square of 10 is 100;
The square root of 169 is 13, because the square of 13 is 169;
The cube root of 125 is 5, because the cube of 5 is 125;
The fourth root of 81 is 3, because the fourth power of 3 is 81.

To indicate a root of a number, we use a symbol called the *radical sign*, $\sqrt{}$. This symbol, placed over a number, means that a certain root is to be found. For example, $\sqrt[2]{49}$ means that we are to find the square root of 49. The number, 49, under the radical sign, is called the *radicand*. The small 2, called the *index*, placed in the notch of the radical sign, indicates the square root is to be found. Then $\sqrt[2]{49} = 7$.

For the cube root of a number, we use the index number 3. For example $\sqrt[3]{125}$ means to find the cube root of 125. Then

$$\sqrt[3]{125} = 5, \quad \text{because} \quad 5^3 = 125$$

Also, $\sqrt[4]{81} = 3$; that is, the fourth root of 81 is 3, because $3^4 = 81$. The index 2 is usually omitted because the square root is used so often.

5.3. FINDING THE ROOT OF A NUMBER

Finding a root of a number is the opposite of finding a power. The problem of finding a root is more difficult than finding a power. To find a power of a

POWERS AND ROOTS

number, we simply multiply the number by itself. For example, to find the second power of 87, we simply multiply: $87 \times 87 = 7569$. To find the third power of 28, we multiply: $28 \times 28 \times 28 = 21,952$. To find the fifth power of 12, we multiply: $12 \times 12 \times 12 \times 12 \times 12 = 248,832$.

Now, suppose we wish to find the square root of, say, 784. We indicate the problem by the symbol: $\sqrt{784}$. This problem is more difficult than finding a power. In this case it happens that: $\sqrt{784} = 28$. However, there is no simple way to determine the square roots of large numbers.

Many square roots of numbers can be found by inspection. For example, if we write the square of each of the whole numbers from 1 to 12, we get the squares: 1; 4; 9; 16; 25; 36; 49; 64; 81; 100; 121; 144. The square roots of these numbers can then be determined from memory. Here are some examples:

$$\sqrt{1} = 1; \quad \sqrt{9} = 3; \quad \sqrt{36} = 6; \quad \sqrt{64} = 8; \quad \sqrt{121} = 11; \quad \sqrt{144} = 12$$

Some other square roots might easily be memorized:

$$\sqrt{169} = 13; \quad \sqrt{196} = 14; \quad \sqrt{225} = 15; \quad \sqrt{256} = 16; \quad \sqrt{289} = 17; \quad \sqrt{361} = 19$$

Some cube roots and other roots can be given from memory:

$$\sqrt[3]{8} = 2; \quad \sqrt[3]{64} = 4; \quad \sqrt[3]{125} = 5; \quad \sqrt[3]{1000} = 10; \quad \sqrt[4]{81} = 3; \quad \sqrt[5]{32} = 2; \quad \sqrt[6]{64} = 2$$

At this time we shall be concerned with only square roots and cube roots. When we come to larger numbers, we cannot tell the square roots or cube roots from memory. Moreover, in most cases, the roots do not come out to be whole numbers or fractions. Then we need to know what is meant by *irrational numbers*.

5.4. IRRATIONAL NUMBERS

In connection with finding roots of numbers, we must first understand what is meant by a *perfect square* or a *perfect power*. A number such as 25 is a perfect square because the square root of 25 is exactly 5. The square root of 169 is exactly 13 because the square of 13 is 169. Then 169 is a perfect square. As another example, the number 6.25 is a perfect square because it is the square of 2.5; that is, the square root of 6.25 is 2.5. A *perfect square, by definition, is a number whose square root can be stated exactly as a whole number or as a common fraction or decimal fraction.*

On the other hand, the number 17 is not a perfect square. There is no whole number or fraction whose square is exactly 17. It is close to 4.1. If we multiply 4.1 by itself, we get a number slightly less than 17. If we multiply 4.2 by itself, we get a number slightly more than 17. If we use several decimals, we still do not get exactly 17. That is,

$(4.1)^2 = 16.81$ $(4.2)^2 = 17.64$
$(4.12)^2 = 16.9744$ $(4.13)^2 = 17.0569$
$(4.123)^2 = 16.999129$ $(4.124)^2 = 17.007376$

The exact square root of 17 cannot be stated as a whole number or decimal fraction, no matter how many decimal places we use. Then we say that the square root of 17 is *irrational*

> **Definition.** *An irrational number is a number that cannot be stated exactly as a whole number or common fraction or decimal fraction.*

In contrast, a *rational number* is a number that can be stated as a whole number, or a common fraction, or an exact decimal fraction.

The square roots and other roots of most numbers are irrational. They do not come out exactly, no matter how many decimal places we take. However, it must be understood that the number 17 is *rational*. It is the *square root of* 17 that is *irrational.*

A *perfect cube* is a number such as 8, because the cube root of 8 is exactly 2. The number 125 is a perfect cube because the cube root of 125 is exactly 5. The number 729 is a perfect cube because $\sqrt[3]{729} = 9$.

On the other hand, the cube roots of most numbers are irrational. For example, the cube root of 50 is a little more than 3.6, but it cannot be stated exactly as a decimal, no matter how many decimal places we use. That is,

$(3.6)^3 = 46.656$ $(3.7)^3 = 50.653$
$(3.68)^3 = 49.836032$ $(3.69)^3 = 50.243409$
$(3.684)^3 = 49.998717 \cdots$ $(3.685)^3 = 50.039444 \cdots$

If the square root or the cube root of a number is irrational, we use as many decimal places as we think best for the desired degree of accuracy.

There is a long method of finding square roots and cube roots of numbers. However, in most cases in actual practice, we find square roots and cube roots by looking in a table, or, in some cases, we use a slide rule or a calculator. A table of squares, square roots, cubes, and cube roots should be available. (See Appendix, Table 2).

5.5. USE OF A TABLE OF POWERS AND ROOTS

To find the square, the square root, the cube, or the cube root of any given number, we first find the given number in the first column of the table under the heading "N". Then, opposite the given number, we find the desired power or root in the proper column. For example, to find the square of 56, we find the number 56 in the first column, N. Under the heading "N^2", we find the number 3136. That is, $56^2 = 3136$.

Using the same column, we can also find the squares of any number, including decimal fractions, that have the two digits, 56, in succession. For example,

$$(5.6)^2 = 31.36; \quad (0.056)^2 = 0.003136; \quad (5600)^2 = 31360000$$

Of course, we must observe the correct placing of the decimal point.

For the square root of a number, such as 56, we look under the heading, $\sqrt{N}$. For $\sqrt{56}$, we find the number, 7.483 (rounded off to three decimal places). That is, the square root of 56 is approximately 7.483.

For the cube or third power of a number, we look under the heading, "N^3". Opposite the number 56, we find the number 175,616. That is,

$$(56)^3 = 175{,}616; \quad \text{also,} \ (5.6)^3 = 175.616; \quad (0.056)^3 = 0.000175616$$

For the cube root of a number, we look under the heading, "$\sqrt[3]{N}$". Opposite the number 56, we find the number 3.825862. That is,

$$\sqrt[3]{56} = 3.826 \quad \text{(rounded off to three decimal places)}$$

5.6. A SHORT METHOD FOR SOME SQUARE ROOTS

We can find the square roots of some numbers by use of a short method. To do so, we make use of the following principle:

Principle. If a radicand can be separated into factors, the radical can be written as the product of two radicals.

The use of this principle depends on our knowing the approximate square

102 ARITHMETIC

roots of a few small numbers, such as these approximate square roots:

$\sqrt{2} = 1.4142$ (approx.); $\qquad \sqrt{3} = 1.732$ (approx.);

$\sqrt{5} = 2.236$ (approx.); $\qquad \sqrt{6} = 2.449$ (approx.);

$\sqrt{7} = 2.646$ (approx.); $\qquad \sqrt{10} = 3.162$ (approx.)

We show the method by several examples.

Example 1. Find, without using a table, $\sqrt{18}$.

Solution. Here we have *one radical*, the entire expression: $\sqrt{18}$. We separate the radicand, 18, into two factors such that one factor is a perfect square:

$$\sqrt{18} = \sqrt{(9) \cdot (2)}$$

Now we write the radical as the *product* of *two radicals*:

$$\sqrt{(9) \cdot (2)} = \sqrt{9} \cdot \sqrt{2}$$

Note that we can now take the square root of one radical, and get

$$\sqrt{9} \cdot \sqrt{2} = 3 \cdot \sqrt{2}$$

Now we make use of the fact that we *remember* that $\sqrt{2} = 1.4142$ (approx). Then we write

$$3 \cdot (1.4142) = 4.2426 \text{ (approx.)}$$

To check our work we can consult a table showing the square root of 18.

Example 2. Find $\sqrt{48}$ by the short method.
Solution. $\sqrt{48} = \sqrt{(16) \cdot (3)} = \sqrt{16} \cdot \sqrt{3} = 4 \cdot (1.732) = 6.928$ (approx.).

Example 3. Find $\sqrt{180}$ by the short method.
Solution. $\sqrt{180} = \sqrt{36} \cdot \sqrt{5} = 6 \cdot (2.236) = 13.416$ (approx.), or 13.42.

Example 4. Find $\sqrt{150}$ by the short method.
Solution. $\sqrt{150} = \sqrt{25} \cdot \sqrt{6} = 5 \cdot (2.449) = 12.245$, or 12.25 (approx.).

Exercise 5.1

By use of the table, find the following powers:

1. $(78)^2$ **2.** $(6.7)^3$
3. $(0.073)^2$ **4.** $(0.086)^3$
5. $(482)^2$ **6.** $(960)^3$
7. $(0.831)^2$ **8.** $(0.934)^3$
9. $(76.2)^2$ **10.** $(2140)^3$
11. $(0.0653)^3$

Find the square roots of these numbers:

12. 65   **13.** 560
14. 394   **15.** 632
16. 912   **17.** 784

Find the cube roots of these numbers:

18. 57   **19.** 684
20. 343   **21.** 792
22. 876   **23.** 729

Use the short method for finding the approximate square roots of the following numbers. Do not use the table. Refer to Section 5.6 for the approximate square roots of 2, 3, 5, 6, 7, and 10.

24. 12   **25.** 24
26. 28   **27.** 32
28. 45   **29.** 54
30. 72   **31.** 75
32. 80   **33.** 90
34. 96   **35.** 98
36. 125   **37.** 147
38. 160   **39.** 162
40. 175   **41.** 196
42. 200   **43.** 242
44. 243   **45.** 250
46. 300   **47.** 1800
48. 3200   **49.** 18000

II GEOMETRY

6 PLANE FIGURES

6.1. THE ELEMENTS OF GEOMETRY

Geometry is the study of figures of various shapes and sizes that we see around us every day, such as circles, squares, rectangles, triangles, lines, angles, and points. These are called *geometric figures*. Geometry also includes the study of the shapes and sizes of objects that *take up space*, such as boxes, balls, cylinders, cones, pyramids, and other geometric solids. The word *geometry* comes from two words that mean "*earth measurement*": "geo" meaning *earth* and "meter" meaning *measurement*.

Geometry deals with the form and size of geometric figures. Whenever we use such expressions as the following, we are conscious of form: a *square* table; a *round* plate; a *rectangular* rug; a *straight* road; a *curved* path; a *crooked* trail. Such expressions all refer to the form of geometric figures.

We are also conscious of *size*. Whenever we use expressions such as these, we have a feeling of *size* or *measurement*: a *big* man; a *little* child; a *small* box; a *wide* river; a *thin* board; a *thick* rug.

To understand geometry and the study of geometric figures, we must understand just what is meant by *points, lines,* and *surfaces* that make up geometric figures. Exactly what is meant by a point, a line, and a surface? First, what is meant by a *point*? The word is not easy to define exactly. In geometry, we say a point is an *undefined element*. A true geometric point has no size, length, width, or thickness. It indicates *position* only. We usually make a dot to indicate the location of a point, but such a dot is not a true point because every dot we might make has some length, width, and thickness. You might get some idea of the meaning of a geometric point if you think of the tip of a needle just at the place where you leave the needle.

A line is sometimes called another *undefined element*. However, we can say a line is the *path of a moving point*. Then the line has no width or thickness. It has *length* only; that is, *one dimension*. Again, we often indicate a line by drawing a mark with a pen or pencil. However, such a mark is not a true geometric line because it has some width and thickness. A line is

108 GEOMETRY

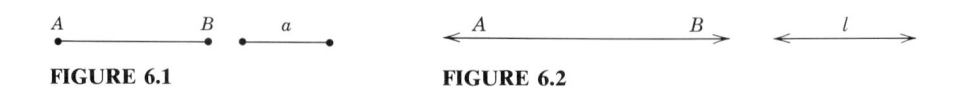

A B a

FIGURE 6.1

A B   l

FIGURE 6.2

sometimes said to consist of an infinite number of points. A line might be defined as the intersection of two surfaces.

In geometry, a line is understood to have *unlimited extent;* that is, it has no end points. A limited portion of a line is called a *line segment.* In many cases, when we use the word *line,* we mean a *line segment.* To indicate a line segment, we draw a mark with a pen or pencil (Fig. 6.1). Then we place a letter at each end of the line segment, such as *A* and *B*, to mark the end points. A line segment can also be named by a single letter, as line segment *a*.

To indicate a *line,* we often draw a figure showing a line segment. (Fig. 6.2). Then we place an arrowhead at each end of the segment to indicate that the line has no end. The line can then be named in various ways. We can name the line by naming a couple of points on the line, as *A* and *B*. Another way often used is to place the letter *l* near the line. If we have more than one line to consider in the same figure or problem, we often use subscripts to the letter *l*, each subscript referring to a different line, as l_1, l_2, l_3, and so on. These are read, "*l* sub 1, *l* sub 2, *l* sub 3," and so on.

In some problems we are concerned with what is called a *half-line,* or a *ray.* A ray or *half-line* is a portion of a line that is understood to have one endpoint and unlimited extent in the opposite direction. (Figure 6.3). If we take a point on a line and then consider only the portion of the line on one side of the point, we have a *half-line* or *ray.* When we speak of a ray of light, we think of the ray as originating at a point (the source of light) and then extending outward from the point infinitely far. The ray can be named by naming its endpoint and any other point on the ray, as the ray *OA*.

A *straight line* (Fig. 6.4*a*) is the path of a moving point that always moves in the same direction. A *curved line* (Fig. 6.4*b*) is a line having no straight portion. As a point moves to form a curve, its direction is always changing. A *broken line* (Fig. 6.4*c*) is a line consisting of straight line segments. A *closed curve* encloses a definite amount of area. (Fig. 6.4*d*). In everyday conversation, the word *line* refers to a line that is *straight.* However, a line can also be curved, as the equator, or a hanging telephone wire.

A *surface* has two dimensions: *length* and *width.* It has no thickness. For the meaning of a surface, we might think of the paint on a wall. Yet even the

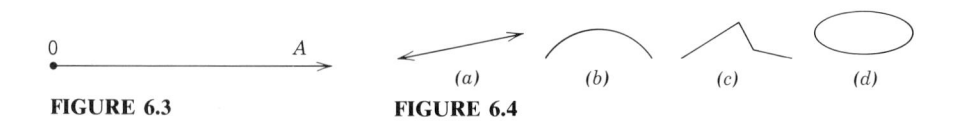

0   A

FIGURE 6.3

(a) (b) (c) (d)

FIGURE 6.4

paint has thickness. When we speak of any kind of surface, as the surface of a tabletop, we are not concerned with the thickness of the tabletop. A surface is sometimes defined as the *path of a moving line* that does not move in its own direction. The surface of the water in a pool is not concerned with the depth of the water.

A surface may be *plane* or *curved*. A *plane* surface is one usually called "flat." To be technical, we say that a plane surface is a surface such that if a straight line has any two of its points in the surface, then the entire line lies in the surface. If we connect any two points in a plane surface with a straight line, then the entire line will lie in the surface. Carpenters test a surface to see if it is flat by placing a straight edge in several positions on the surface. Figure 6.5 shows a *curved surface*. A *curved surface* is a surface having no flat part. Here are some examples of curved surfaces.

(a) the outside of a round silo.
(b) the outside of a round pencil;
(c) the outside or inside of the round portion of a round stew pan;
(d) the outside of a wire.

A geometric solid has three dimensions: length, width, and thickness (or height). The word *solid* in geometry has a meaning different from its meaning in everyday conversation. We often say a ball is *solid wood*, or *solid iron*, or *solid ivory*. We are then thinking of the material that makes up the object. A geometric solid is not concerned with the material of which an object is composed. In fact, it need not be any material at all. A geometric solid refers only to a particular *amount of space*. When we say a geometric solid is 6 inches long, 4 inches wide, and 2 inches high (or thick), we mean just that amount of space. Of course, in most cases, a solid with which we are concerned is actually composed of some kind of material substance.

A plane surface is usually called simply a *plane*. A *plane figure* is a figure that can be drawn on a plane or flat surface. Such figures have no thickness, only length and width. For example, on a flat piece of paper, we can show plane figures such as a line, an angle, a triangle, a circle, a square, and other plane figures. (Fig. 6.6). When we draw such figures, we are not concerned with any thickness. For example, when we paint the top of a floor, we do not think of the thickness of the floor.

Plane geometry deals with figures on a plane; that is, with plane figures. *Solid geometry* is the study of figures that take up space, such as boxes, cylinders, cones, and so on. In solid geometry, we deal with geometric solids. Solid geometry is sometimes called *space geometry*.

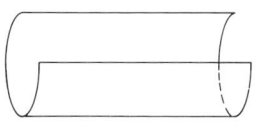

FIGURE 6.5

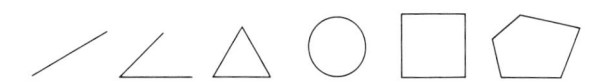

FIGURE 6.6. Plane figures.

110 GEOMETRY

6.2. LINES: INTERSECTING AND PARALLEL

Two straight lines in the same plane are either *parallel* or *intersecting* lines. If the lines cross each other, they are called *intersecting* lines and they intersect at one point only (Fig. 6.7a). If two straight lines in the same plane do not intersect no matter how far they are extended, they are called *parallel* lines (Fig. 6.7b). According to this definition, two parallel lines never intersect.

In space, however, two straight lines can be nonparallel, yet not intersect. Such lines are called *skew* lines. Two telephone wires extending in different directions, one above the other, form skew lines.

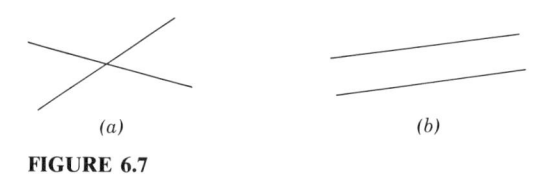

FIGURE 6.7

6.3. ANGLE

Most people know what is meant when we use the word *angle* as referred to a geometric figure. However, to define an angle precisely is not so simple as it might seem. For a definition, we might use various approaches. [The word angle is used so frequently in geometry that it is often represented by a small angle, ∠, as a symbol (plural: ∠s).]

An angle is sometimes defined as the figure formed by two *rays* drawn from the same point. For example, the rays *OA* and *OB* drawn from the same point *O* form an angle. (Fig. 6.8). According to this definition, we think of the angle as the *configuration* or *picture* formed. The two rays, *OA* and *OB*, are called the *sides* of the angle. The point *O* were they meet is called the *vertex* of the angle (plural: *vertices*). A small curved line is often drawn between the sides of the angle near the vertex to indicate the angle, as ⊿.

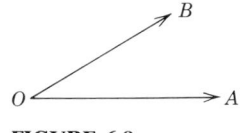

FIGURE 6.8

An angle is sometimes defined as the *amount of opening* between two rays drawn from the same point, as *OA* and *OB*. (Fig. 6.9a). Then the amount of opening refers to the *size* of the angle.

Another definition especially useful in engineering is the following: beginning with a ray, *OA*, with its end point at *O*, we define an angle as the amount of turning or rotation of the ray *OB* about its end point (Fig. 6.9b). Then we think of only one ray that moves from one position to another position. Then we have two *different positions* of the *same ray*. For example, the ray *OA* can rotate about point *O* until it reaches the new position *OB*. The curved arrow shows that the ray *OA* has rotated to the position *OB*.

FIGURE 6.9

This view is similar to considering the angle formed by the two positions of the short hand of a clock as it moves from 1 o'clock to 2 o'clock. This definition of an angle looks upon the angle as the result of *motion*. It may be called a *dynamic* definition. This definition is especially useful in any engineering where motion is involved. If we think of an angle only as the figure formed, then there is no thought of motion.

Notice that an angle can be indicated by two straight lines on a plane or flat sheet of paper. It might be well to call this kind of angle a *plane angle*.

Wrong definitions of an angle sometimes lead to misunderstanding and confusion about terms. We might say carelessly, "An angle is two lines drawn from the same point." But the angle is *not* the two lines. The two lines are the *sides* of the angle. Again, we might say, "The angle is the point where the two lines meet." But a point cannot be an angle. Besides, the point where the lines meet is the *vertex* of the angle.

If we say "The angle is the space between the two rays," we are no better off. Since the sides of the angle are understood to extend infinitely far, then the space between them is unlimited. Even if we consider the sides limited in extent, the space included between them is no measure of the size of the angle. In Figure 6.10, the angle at B is larger than the angle at A. The size of an angle is not determined by the lengths of the sides.

An angle is named in various ways. It is often named by a single capital letter at the vertex, as $\angle B$ (Fig. 6.11a). It is sometimes named with three capital letters as $\angle ABC$. If three letters are used to name an angle, the *vertex letter is mentioned second*. The angle should be named in the same order as you would draw it without lifting the pencil from the paper. An angle may also be named by a small letter or number within the angle, as ∡ or △. If there is any danger of confusion as to which angle is meant, the angle should always be named by the use of three letters. In Figure 6.11d, the meaning is not clear if we say simply $\angle O$. In fact, there are six angles with point O as a vertex.

FIGURE 6.10

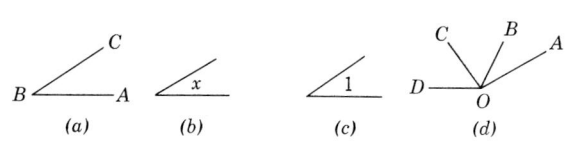

FIGURE 6.11

112 GEOMETRY

Adjacent angles are two angles that have the *same vertex and a common side between them*, as angles *x* and *y*, (Fig. 6.12). The two adjacent angles are $\angle AOB$ and $\angle BOC$.

If two straight lines intersect, four angles are formed, as shown in Figure 6.13. The following are pairs of adjacent angles: $\angle 1$ and $\angle 2$; $\angle 2$ and $\angle 3$; $\angle 3$ and $\angle 4$; $\angle 4$ and $\angle 1$. Opposite angles, such as $\angle 1$ and $\angle 3$, are called *vertical angles*. (It can be proved that *vertical angles are equal to each other*.)

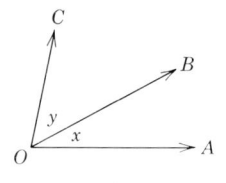

FIGURE 6.12

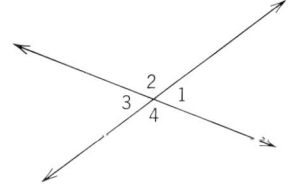

FIGURE 6.13

6.4. PERPENDICULAR LINES

If two straight lines intersect to that two adjacent angles are equal, then the lines are said to be *perpendicular to each other.* (Figure 6.14). The *symbol for perpendicular* is $\perp$. Then $AB \perp CD$. Also, $CD \perp AB$.

If two straight lines are perpendicular to each other, then the two adjacent angles are called *right angles*. A right angle is what we often call a square corner. Figure 6.14 shows four right angles. A right angle is often indicated by the symbol ∟.

6.5. KINDS OF ANGLES

As we have said, a *right angle* is an angle formed by two intersecting lines that are perpendicular to each other (Fig. 6.15a). An *acute angle* is an angle smaller than a right angle (Fig. 6.15b). An *obtuse angle* is an angle greater than a right angle but less than two right angles, (Figure 6.15c). If two right angles are placed adjacent to each other, two of the sides form a straight line. An angle formed as the sum of two adjacent right angles is called a *straight angle*, (Fig. 6.15d). Actually, a straight angle forms a straight line.

If four right angles are placed with all the vertices at the same point and each right angle is adjacent to two others, the four right angles can be made to fill exactly the entire space around the point (Fig. 6.16). The four adjacent right angles represent one complete rotation of a ray around a point. That is, if we take a ray and rotate it through four right angles, we come back to the starting position.

A square corner, or right angle, can easily be formed with any irregular sheet of paper. Fold the paper once at any place. The fold represents a straight line. Now fold the paper again so that the straight line is folded over

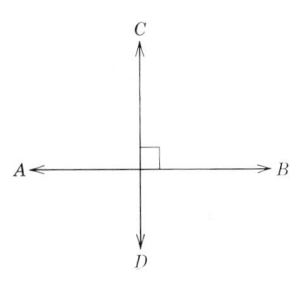

FIGURE 6.14

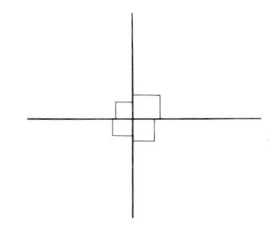

FIGURE 6.16

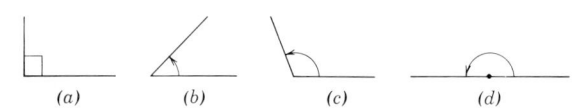

(a) (b) (c) (d)

FIGURE 6.15. (*a*) Right angle. (*b*) Acute angle. (*c*) Obtuse angle. (*d*) Straight angle.

PLANE FIGURES 113

exactly along itself. The result is a square corner because the two adjacent angles are equal and together they form a straight line or straight angle.

6.6. MEASUREMENT OF ANGLES

In order to measure the size of angles, the early Babylonians decided to divide a complete circle or one rotation into 360 equal parts. (They might just as well have divided it into 100 parts or 60 parts or some other number.) Each part they called *one degree*. This system of measuring angles is still in common use. The unit of measurement, one *degree* (written 1°) is therefore $\frac{1}{360}$ of a complete rotation. (Fig. 6.17).

Since one complete rotation is called 360 degrees (360°), then one right angle is called 90 degrees (90°), and one straight angle is called 180 degrees (180°). An angle greater than 180° but less than 360° is called a *reflex* angle.

The size of an angle can be measured by an instrument called a protractor (Fig. 6.18). The protractor is a semicircle, with the curved portion representing the circumference of the circle, and the straight portion representing the diameter. The center of the circle is the midpoint of the diameter, shown as point O. The curved portion, representing the circumference is marked off in 180 degrees for the semicircle. To measure an angle, we place the point O of the protractor at the vertex of the angle, and one side of the angle along the straight portion of the protractor. One side of the angle will then lie along the line from O to the zero mark of the protractor. The other side of the angle will fall along a mark on the curved portion of the protractor. This mark will show the size of the angle in degrees.

Some protractors have two rows of numbers along the semicircumference. One row of numbers is read from right to left, the other from left to right.

The protractor can also be used to draw an angle of any required size. We first draw a line to represent one side of the angle, with an end point for the vertex of the angle. Then we place the protractor with the point O at the vertex of the required angle, and the line along the straight edge of the

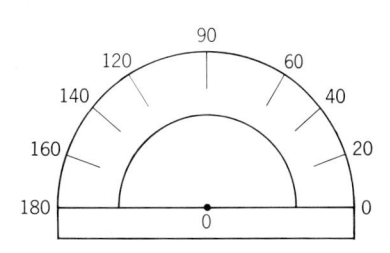

FIGURE 6.17. An angle of approximately one degree. **FIGURE 6.18**

protractor. We make a mark at the required number of degrees, and then connect this point with the vertex. For example, if we wish an angle of 30°, we place a mark at 30° on the protractor, and then connect this mark with the end point of the line we have drawn. The resulting angle will be 30°.

For very fine measurements, (that is, for more precision), the degree is divided into 60 smaller parts or angles, called *minutes*. A minute of rotation is therefore $\frac{1}{60}$ of one degree. For extreme precision, such as that required in astronomy, the minute is further divided into 60 parts called *seconds*. One second of rotation is a very small angle. In fact, an angle of one second is so small that the distance between its sides is only about $\frac{3}{10}$ of an inch at a distance of a mile from the vertex.

Two angles whose sum is equal to a right angle, or 90°, are called *complementary*. Two complementary angles need not be adjacent. An angle of 32° and an angle of 58° are complementary since their sum is 90°. Two angles whose sum is equal to a straight angle, or 180°, are called *supplementary*. Two supplementary angles need not be adjacent. An angle of 46° and an angle of 134° are supplementary, since their sum is 180°.

Exercise 6.1

1. Draw two intersecting lines that are not perpendicular to each other. Measure the four angles formed. What can you conclude about a pair of vertical angles? What can you conclude concerning adjacent angles?
2. Draw two parallel lines approximately 2 inches apart. Now draw another line, called a *transversal*, intersecting the two parallel lines, but not perpendicular to them. Measure the eight angles thus formed. What can you conclude about the eight angles formed? What angles are equal to each other? Mark each of four equal angles with the letter x, and mark the other four equal angles with the letter y.
3. Show by a drawing that if a transversal is perpendicular to one of two parallel lines, then it is also perpendicular to the other.
4. Name several different examples of the use of parallel lines in everyday life. Which ones are necessary?
5. Name several examples of intersecting lines in everyday life. Why are such lines necessary?
6. Name several examples of right angles in your immediate surroundings as you look from reading this page.
7. Name some examples of skew lines in everyday life. Why are skew lines necessary?
8. Draw two complementary angles that are not adjacent.
9. Draw two supplementary angles that are not adjacent.
10. From a given point, draw seven lines outward in different directions. Measure each of the seven angles formed and add them together. What do you conclude about the total number of degrees in all angles around a point?

6.7. POLYGONS

A *polygon* may be defined as a *plane closed figure bounded by straight line segments*. If the figure is not closed, it is not a polygon. A polygon therefore encloses a definite amount of area on a flat surface. The boundary of a polygon consists of any number of straight line segments. These line segments are called the *sides* of the polygon. The *perimeter* of the polygon is the sum of the sides, or the distance around the polygon. The points where the sides meet are called the *vertices* of the polygon.

Polygons are named according to the number of sides. If a plane closed figure is to be bounded by straight line segments, it must have at least three sides. A polygon of three sides is called a *triangle*. The prefix "tri" means *three*. Notice that the name *triangle* really means "three angles." Yet we define a triangle in terms of sides instead of angles. Figure 6.19 shows triangles. However, remember that the sides must be straight. Perhaps no figure that we draw is a perfect polygon, yet we can imagine that the sides are perfectly straight.

A polygon of four sides is called a *quadrilateral* (Fig. 6.20). The prefix "quad" means *four*, and the word "lateral" refers to *side*. Then the word really means *four sides*. If a triangle were named after its three sides, it might be called a "trilateral." Moreover, a quadrilateral can also be named after the angles and called a "quadrangle."

Figure 6.21 shows several kinds of polygons. A polygon of five sides is called a *pentagon*. A *hexagon* has six sides; a *heptagon* has seven sides; an *octagon* has eight sides; a *decagon* has ten sides.

A *regular* polygon is a polygon having all sides equal and all angles equal. Figure 6.22 shows some regular polygons.

FIGURE 6.19

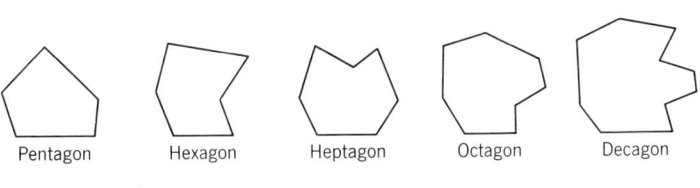

FIGURE 6.20

Pentagon   Hexagon   Heptagon   Octagon   Decagon

FIGURE 6.21

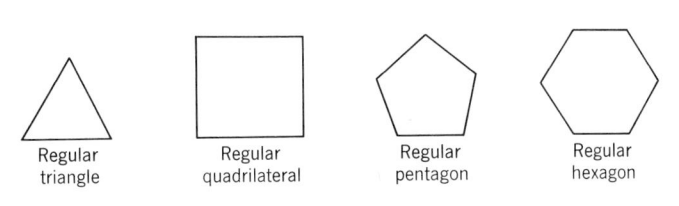

Regular triangle   Regular quadrilateral   Regular pentagon   Regular hexagon

FIGURE 6.22

6.8. TRIANGLES (Symbols: △, △)

Triangles are named according to their general shape. A triangle may have all three sides equal, two sides equal, or no two sides equal. A triangle having all three sides equal is called an *equilateral* triangle. ("equi" is a prefix meaning *equal*) (Fig. 6.23a). It can be proved that if the three sides of a triangle are equal, then all three angles are equal; that is, an equilateral triangle is also equiangular. This condition is not necessarily true for other polygons. A quadrilateral may have all its four sides equal, yet not have all its angles equal.

A triangle having two sides equal is called an *isosceles* triangle (Fig. 6.23b). In an isosceles triangle, it can be proved that the angles opposite the equal sides are also equal. Moreover, if a triangle has two equal angles, then the sides opposite these angles are also equal to each other. A triangle having no two sides equal is called a *scalene* triangle (Fig. 6.23c).

A triangle may have angles of various sizes. However, the following statement is true with regard to the angles of all triangles.

In any triangle, the sum of the three angles is equal to 180° (two right angles, or one straight angle.)

As an example, if one angle of a triangle is 43° and a second angle is 79°, then the third angle can be found. It must be 58° because the sum of the three angles must be 180°.

A triangle having one right angle is called a *right triangle* (Fig. 6.24a,b). Since a right triangle has one right angle, the other two angles must be *acute angles*. Moreover, the two acute angles are *complementary*; that is, their sum is 90°. The two sides forming the right angle are often called the *legs* of the right triangle. The side opposite the right angle is called the *hypotenuse* of the right triangle. If the two legs of a right triangle are equal in length, then the triangle is also *isosceles* (Fig. 6.24b).

A triangle having one obtuse angle is called an *obtuse triangle* (Fig. 6.24c). A triangle having all three angles acute is called an *acute triangle* (Fig. 6.24d). Acute triangles and obtuse triangles are often called *oblique* triangles to distinguish them from right triangles. Since the sum of all three angles of any triangle is equal to 180°, it follows that no triangle can have more than one right angle or obtuse angle.

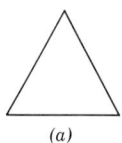

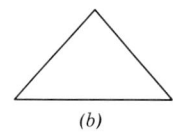

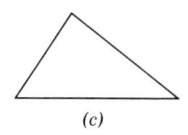

FIGURE 6.23. Triangles. (*a*) Equilateral. (*b*) Isosceles. (*c*) Scalene.

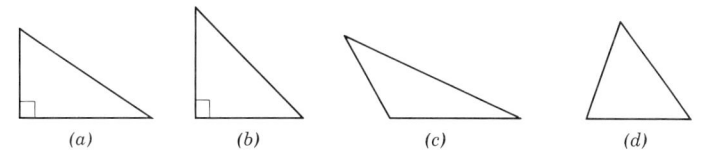

FIGURE 6.24. Triangles. (*a*) and (*b*) Right triangles. (*c*) Obtuse triangle. (*d*) Acute triangle.

The *base* of a triangle is the side on which is it understood to rest. Any side of a triangle may be considered as the base. The *vertex* of a triangle is the vertex opposite the base. The angle at the vertex is called the *vertex angle*. The *altitude* of a triangle is a straight line segment drawn from the vertex *perpendicular to the base* (Fig. 6.25.) Since any side of a triangle may be considered as the base, a triangle has three altitudes. All three altitudes meet at the same point. An altitude is often indicated by the letter *h*.

If one leg of a right triangle is taken as the base, then the other leg is the *altitude* (Fig. 6.25*b*). In an obtuse triangle two of the altitudes fall outside the triangle and must be taken to the base extended (Fig. 6.25*c*).

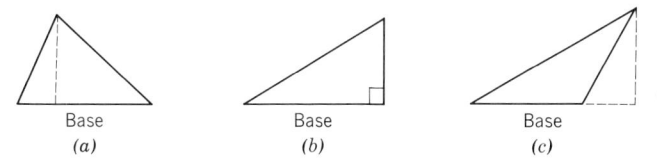

FIGURE 6.25. Triangles showing altitudes. (*a*) Acute triangle. (*b*) Right triangle. (*c*) Obtuse triangle.

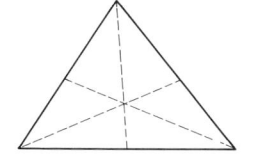

FIGURE 6.26

A *median* of a triangle is a straight line segment drawn from a vertex to the midpoint of the opposite side (Fig. 6.26). A triangle has three medians, all of which meet at a point called the *centroid*. This point is important, since it is the *center of gravity*, or *balancing point of the triangle*.

6.9. QUADRILATERALS

A *quadrilateral* may have many different shapes. Some of these are shown in Figure 6.27. A quadrilateral having no pair of opposite sides parallel is called a *trapezium* (Fig. 6.27*a*). A quadrilateral having one pair of opposite sides parallel is called a *trapezoid* (Fig. 6.27*b*). If the two nonparallel sides of a trapezoid are equal, the figure is called an *isosceles trapezoid*. The two parallel sides are called the *bases* of the trapezoid. The *altitude* of a trapezoid is the perpendicular distance between the bases.

A quadrilateral having both pairs of opposite sides parallel is called a

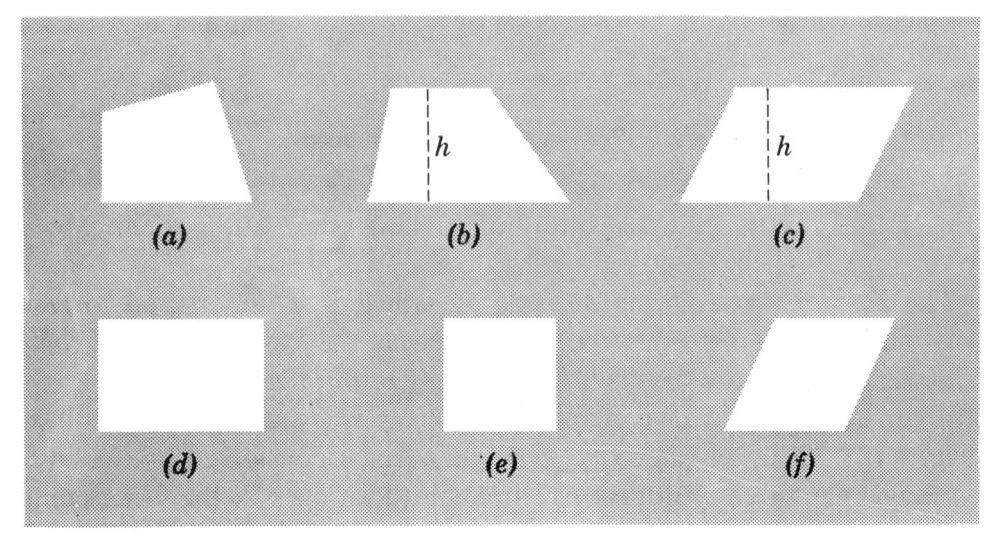

FIGURE 6.27

parallelogram (symbols: ▱, ▱). By this definition, Figures 6.27c, d, e, and f are all parallelograms, since in each of these figures both pairs of opposite sides are parallel.

Of the four parallelograms shown, we see that some of them have right angles. A parallelogram whose angles are right angles is called a *rectangle* (symbols: ▫, ▫). By this definition, Figures 6.27d and e are rectangles, since each one is a parallelogram with right angles.

Of the two rectangles shown, Figures 6.27d and e, we note that one of them has its sides all equal. This figure (e) is a square (symbols: ▫ ▫). A *square* is defined as a rectangle having equal sides.

Notice that all rectangles are parallelograms and that they are also quadrilaterals, having four sides. A square is a rectangle, a parallelogram, and a quadrilateral.

A *rhombus* is defined as a parallelogram having equal sides. By this definition, a square is also a rhombus. However, the word *rhombus* is usually restricted to mean an equilateral parallelogram that is not a square (Fig. 6.27f).

The following facts concerning quadrilaterals can be shown to be true. The student should make a drawing or sketch of a figure and try to understand clearly the meaning of each statement.

1. The sum of all the interior angles of a quadrilateral is 360°.

2. Any two opposite sides of a parallelogram are equal.

3. If the opposite sides of a quadrilateral are equal, the figure is a parallelogram.

PLANE FIGURES 119

4. The opposite angles of any parallelogram are equal.

5. If the opposite angles of a quadrilateral are equal, the figure is a parallelogram.

6. The diagonals of any rectangle (including a square) are equal.

7. The diagonals of any rhombus (including a square) are perpendicular to each other.

Exercise 6.2

1. If two straight lines intersect and form four angles, and if one of the angles is 58°, what is the size of each of the other three angles?
2. What is the complement of an angle of 61°?
3. What is the supplement of an angle of 74°?
4. How many minutes are there in an angle of 24°? How many seconds?
5. Two angles of a triangle are 48° and 75°, respectively. How large is the third angle?
6. The two equal angles of a certain isosceles triangle are each 67°. How large is the vertex angle?
7. The vertex angle of an isosceles triangle is 38°. Find the size of each of the base angles. (The base angles are the two equal angles.)
8. What is the size of each angle of an equilateral triangle?
9. If a right triangle has one angle equal to 34°, find the other acute angle.
10. If one angle of a parallelogram is 64°, find the other three angles.
11. Carefully draw a quadrilateral that has no two sides parallel. Measure the interior angle at each corner and find the total number of degrees in the four angles. How near is the result to 360°?
12. If you start at any point on one side of a quadrilateral and walk around the quadrilateral until you arrive at your starting point, through how many degrees will you turn?
13. Draw a hexagon, *not* necessarily a *regular* hexagon. Measure the interior angle at each corner and find the total number of degrees in the six angles of the hexagon. How near is your total to 720°? It can be proved that the six interior angles of any hexagon have a total of 720°.
14. If you start at any point on the side of a hexagon and walk around the figure until you arrive at your starting point, through how many degrees will you turn? Through how many degrees would you turn if the figure were a decagon?
15. How many diagonals can be drawn in (a) a square, (b) a pentagon, (c) a hexagon, (d) a triangle?
16. Draw a right triangle, an acute triangle, and an obtuse triangle, and then measure the three angles in each triangle. Do your measurements confirm the statement that the sum of the three angles of any triangle is 180°?
17. Draw an obtuse triangle that is also scalene. Carefully draw the three altitudes. If the altitudes are extended, do they intersect at the same point?
18. Draw an obtuse triangle that is also isosceles. Measure the angles opposite the two equal sides. Are these angles equal?

19. Why can a triangle not have more than one obtuse angle?
20. Draw a right triangle and show the three altitudes.
21. Why must the two acute angles of a right triangle be complementary?
22. Draw a right triangle with one acute angle equal to 35°. Draw another, larger right triangle with one angle equal to 35°. Are all the angles of one triangle equal, respectively, to the angles of the other? Such figures are called *similar* figures because they have the *same shape*.
23. Draw two acute triangles, one larger than the other, but with the angles of each equal, respectively, to 42°, 67°, and 71°. Are these two triangles similar?
24. Draw a quadrilateral having sides equal to 2 inches, 1 inch, $1\frac{1}{2}$ inches, and $2\frac{1}{4}$ inches, respectively. Can you draw another quadrilateral having sides of the same lengths as those mentioned, but of a different shape?
25. Draw a triangle having its sides equal to 2 inches, $2\frac{1}{4}$ inches, and $1\frac{1}{2}$ inches, respectively. Can you draw another triangle having sides of the same lengths as those mentioned in this problem but of an entirely different shape?
26. Draw a quadrilateral having two opposite sides each equal to 1 inch, and the other two sides each equal to 2 inches. After this is done, try to determine whether the opposite sides are also parallel.
27. Draw a quadrilateral, making the opposite sides parallel without trying to make them equal. Then measure the sides and determine whether they are also equal.
28. Draw a rectangle and then draw the diagonals. Are the diagonals equal to each other? Are they perpendicular to each other?
29. Draw a rectangle much longer than it is wide. Then draw one diagonal and try to determine whether the diagonal divides the right angle into two equal angles.
30. Draw a parallelogram that is not a rectangle. Then draw the diagonals. Are the diagonals equal to each other?

7 MEASUREMENT OF PLANE FIGURES

7.1. THE NEED FOR MEASUREMENT

Whenever we use such words as *long, short, large, small,* or *wide,* we are conscious of *size.* Every day we are faced with some problem that has to do with *measurement.* We want to know *how long, how heavy, how fast, how much, how far,* and so on. When we ask the time of day, we are really asking about measurement of *time.* It is hard to realize what our lives would be like if it were not for measurement.

It has been said that man first became civilized when he started to measure things. All scientific ideas must be translated in measurement before they can become useful. Measurement is necessary, whether in making a cake or building a house. Thousands of precise measurements must be made in the building of a single airplane or a satellite.

In the measurement of plane figures, we are interested in two measurable quantities: the lengths of line segments, and the areas of limited portions of surfaces. For example, we may need to find the length of a house, the width of a room, or the distance around a circular race track. We may also need to find the area of a wall, the area of a rug, or the area of a circular table top.

7.2. THE UNIT OF MEASURE

To measure anything we must use a *unit of measure.* The unit is simply a definite amount of the same kind of thing as the quantity to be measured.

To measure a length, we use a *unit of length* such as an inch, a foot, or a mile. Units of length are called *linear units.* To measure area, we use a *unit of area* such as a square inch, a square foot, or an acre. To measure volume, we use a *unit of volume* such as a cubic inch, a cubic foot, a pint, or a gallon. To measure weight, we use a *unit of weight* such as an ounce, a gram, a pound, or a ton. To measure time, we use a *unit of time* such as a second, a minute, an hour, a day, or a year. To measure angles, we use a small angle called a *degree,* or a larger unit called a *radian.*

122 GEOMETRY

In the study of electricity, we use several units. To measure current, we use the *unit of current* called an *ampere*. Voltage is measured by the unit *volt*. Electrical resistance is measured by the *unit of resistance*, called the *ohm*.

In measuring any quantity, we see how many times the unit is contained in the quantity measured. Therefore, the unit of measure must be of the same kind of thing as the quantity to be measured. For example, we cannot measure length in pounds; we cannot measure an angle in inches; we cannot measure electrical current in feet, inches, pounds, or minutes.

7.3. PERIMETER

The *perimeter* of any geometric figure is the distance around it. The perimeter of a polygon is found by adding the lengths of the sides. For example, if the sides of a triangle are 5 inches, 6 inches, and 8 inches, respectively, then the perimeter of the triangle is 19 inches. If a room is 18 feet long and 12 feet wide, the perimeter of the room is 60 feet. If a polygon is equilateral (having equal sides), then the perimeter can be found by multiplication. If one side of a square is 7 inches, the perimeter of the square is 28 inches. If one side of a regular hexagon is 8 inches, the perimeter is 48 inches.

Notice that the perimeter of a polygon is expressed in *linear units*, since the perimeter is a *length*.

7.4. AREA

To measure area, we must use a *unit of area*. To measure a given area, we lay off the unit area as many times as it takes to cover the given area. For example, to measure the area of this page, we might take a coin such as a nickel and see how many times the area of the page contains the unit, the nickel. However, a circular unit of area is not convenient because the units do not fit together. Instead, we use a rectangular unit, or, best of all, a *square unit*, such as a square inch or a square foot.

We shall see that there are special rules for finding the areas of rectangles, squares, triangles, and other plane figures. However, it is well to learn where these special rules come from. We shall also find perimeters of figures.

7.5. THE RECTANGLE

Suppose we have a rectangle that is 5 inches long and 3 inches wide (Fig. 7.1). To find the area of the rectangle, we usually say: *multiply the length*

MEASUREMENT OF PLANE FIGURES

times the width. That is,

$$(5 \text{ inches}) \times (3 \text{ inches}) = 15 \text{ square inches}$$

The answer is correct. However, is it not strange that we can multiply two line segments together and get an area? If you will take a moment to analyze this problem, you may better understand later problems in areas and volumes. Let us see why we get the correct answer.

An area cannot be measured in linear units such as an inch. We must use a unit of area. We shall use a unit 1 inch long and 1 inch wide. This unit is called one *square inch.*

Now, starting at one corner of the rectangle (Fig. 7.2), we lay off the unit, 1 square inch, as many times as possible along one side of the rectangle. There will be 5 units of area in 1 row, because the unit is 1 inch long and 1 inch wide, and there are 5 inches in the length of the rectangle.

After laying off one row of 5 square inches, we continue with the next row and so on until the rectangle is covered. There will be 3 rows because the rectangle is 3 inches wide. At this point, we can find the total number of units (square inches) by multiplication:

$$3 \times (5 \text{ sq. in.}) = 15 \text{ sq. ins.}$$

The advantage of using a square unit of area is that we can measure the length in linear units (inches), the width in the same linear units (inches), and then easily find the area by the following rule.

RULE. Multiply the number of linear units in the length by the number of the same kind of linear units in the width. The result is the number of corresponding square units in the area.

Stated in short form, the rule is

$$\text{area} = (\text{length})(\text{width})$$

The rule may be stated as a *formula.* A formula is a rule stated in letters and signs to represent the words. If we use the letter l to represent the length (in linear units), use w to represent the width (in linear units), and A

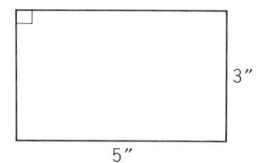

FIGURE 7.1

FIGURE 7.2

to represent the area (in corresponding square units), then we can write the rule as

$$A = l \times w, \quad \text{or simply} \quad A = lw$$

(When two letters are written without a sign between them, they indicate *multiplication*.) The formula can be used for all kinds of rectangles.

Example 1. Find the area of a rectangular room, 18.5 ft. long and 12 ft. wide.

Solution. Area = (18.5)(12) = 222, number of square feet.

The rule can also be used in reverse.

Example 2. If a rectangle is 6 inches wide and has an area of 48 square inches, find the length.

Solution. We find the length by division.

length = area ÷ width; or length = 48 ÷ 6 = 8, length (inches)

Example 3. A field is 20 rods long and contains 280 square rods. Find the width of the field.

Solution. Since the area is given, we use division:

width = area ÷ length; or width = 280 ÷ 20 = 14, width (in rods)

In the three foregoing examples we have used three formulas:

$$A = lw; \quad l = A \div w; \quad w = A \div l$$

The *perimeter* of a rectangle is the distance around it along the 2 ends and 2 sides. Then it is equal to *twice the length plus twice the width*, or, as a formula,

$$P = 2l + 2w$$

The perimeter, of course, is expressed in linear units. Whenever we *add linear units*, we get an answer representing *linear units*. When we *multiply linear units* by *linear units*, we get a product that represents the number of *square units*.

7.6. THE SQUARE

A square is a special rectangle in which all sides are equal (Fig. 7.3); that is, $l = w$. If we represent the number of linear units in one side by the letter s, we have the formula for the area:

$$A = s \times s; \quad \text{or} \quad A = s^2$$

Since all four sides are equal in length, we can find the perimeter of a square by taking 4 times the length of a side; that is,

$$P = 4s.$$

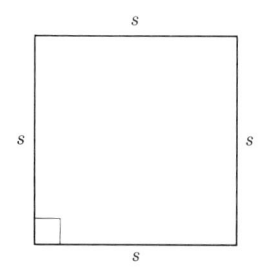

FIGURE 7.3

Example 4. Find the area and perimeter of a square rug measuring 12 feet on a side.

Solution. For the area, we have

$$A = s^2 = (12)(12) = 144, \text{ area (in sq. ft.)}$$

For the perimeter, we have

$$P = 4(s) = 4(12) = 48, \text{ perimeter (in ft.)}$$

Example 5. Find the perimeter of a square whose area is 162 sq. in.

Solution. To find the perimeter, we first find one side. For the side, we take the square root of the area. Then we have

$$s = \sqrt{162} = 9\sqrt{2} = 12.726 \text{ (approx.), (inches)}$$

For the perimeter, we take 4 times the length of a side and get

$$P = (4)(9\sqrt{2}) = 36\sqrt{2} = 50.9 \text{ (approx.)}$$

7.7. THE PARALLELOGRAM

If we know the four sides of a parallelogram, we can find the perimeter but we cannot find the area from this information alone. For example, in Figure 7.4, the parallelogram has its 4 sides equal to 12 inches, 8 inches, 12 inches, and 8 inches, respectively. The perimeter is 40 inches.

However, to find the area we must know the *altitude* of the parallelogram. The *altitude*, h, is the perpendicular distance between two sides. One side is called the *base* instead of the length.

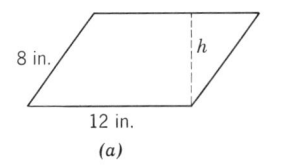

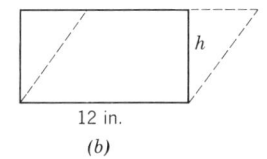

FIGURE 7.4

To find a formula for the area, let us assume that a triangle is cut off at one end of the parallelogram, as shown in Figure 7.4b, and attached to the other end of the parallelogram. The result is a rectangle having the same base and altitude as the parallelogram. Therefore, it has the same area as the rectangle. Then, for the area of a parallelogram, we have the formula

$$A = \text{base} \times \text{altitude}, \quad \text{or} \quad A = bh$$

The formulas can also be used in reverse, as with a rectangle. If the area and the base of a parallelogram are known, the altitude can be found by division, as with a rectangle. If the area and the altitude are known, the base can be found by division. That is,

$$h = A \div b, \quad \text{and} \quad b = A \div h$$

Example 6. Find the area of a parallelogram with base = 15 inches and altitude = 7 inches.

Solution. $A = bh$; then $A = (15)(7) = 105$, area (in sq. in.)

Example 7. A parallelogram has an area of 152 sq. in. and a base of 16 in. Find the altitude.

Solution. Here we use the formula:

$$h = A \div b$$

Then we have
$$h = 152 \div 16 = 9.5, \text{ altitude (inches)}$$

Exercise 7.1

Find the area and perimeter of each of the following rectangles:

1. Length, 16.5 inch width, 10 in.
2. Length, 48 ft., width 5.4 ft.
3. Length, 3 ft., width, 15 inches.
4. Length, 42 rods; width, 23.5 rods.
5. Find the perimeter and the area of a room, 16 ft. long, and $13\frac{1}{2}$ ft. wide.
6. Find the area of a sidewalk, 6.5 ft. wide, and 100 yards long.
7. Find the area of a concrete road 32 ft. wide and $\frac{1}{2}$ mile long.

8. Find the area of a small rug, 2.5 ft. wide and 4.5 ft. long.
9. Find the area of a wall, 18 ft. long and 9.5 ft. high.
10. Find the area of a TV screen, 24 inches by 18.5 inches.
11. Find the area and perimeter of a square room, 18.5 ft. on a side.
12. Find the area and perimeter of a square tabletop, 5.5 ft. on a side.
13. Find the area and perimeter of a parallelogram having sides of 7 in., 4 in., 7 in., and 4 in., and altitude of 3 inches to the 7-inch side.
14. Find the area and perimeter of a parallelogram having sides, 8.5 in., 5 in., 8.5 in., and 5 in., respectively, and altitude of 4.5 in., to the 8.5-inch side.
15. A rectangle has an area of 90 sq. in. and a width of 7.5 in. Find the length and the perimeter of the rectangle.
16. A rectangle has an area of 121.8 sq. in. and a length of 14.5 in.; find the width and the perimeter of the rectangle.
17. A rectangle has an area of 273 sq. ft. and a length of 6.5 yards. Find the width and the perimeter of the rectangle.
18. A square has an area of 225 sq. ft. Find the length of a side.
19. A square has an area of 576 sq. ft. Find the side and the perimeter.
20. A square has an area of 552.25 sq. ft. Find the side and the perimeter.
21. Find the side and area of a square whose perimeter is 52 inches.
22. Find the side and area of a square field whose perimeter is 98 rods.
23. A baseball diamond is a square with 90 ft. between the bases. Find the area. What part of an acre is that? (One acre = 43,560 sq. ft.)
24. A rectangle has a perimeter of 65 inches. The width is 14 inches. Find the length and then find the area of the rectangle.
25. The perimeter of a rectangle is 54 ft. The length is 15 ft. Find the width and then find the area of the rectangle.
26. A piece of plywood is 4 ft. wide and 12 ft. long. What is the cost of the plywood at 23.5 cents per sq. ft.?
27. Find the cost of paving 10 miles of road 36 ft. wide at a cost of 40¢ a sq. yd.
28. A blackboard is 3 ft. 4 in. high. How long must it be to contain 100 sq. ft.?

7.8. THE TRIANGLE

To derive the rule for the area of a triangle, let us suppose that we have a triangle, *ABC*, with the base *b* equal to 11 inches, and the altitude *h* equal to 8 inches (Fig. 7.5). Now we construct another congruent triangle adjacent

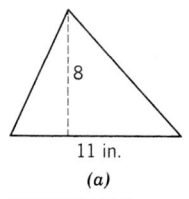

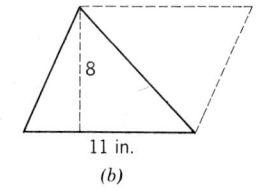

FIGURE 7.5

128 GEOMETRY

to the given triangle but in an inverted position. (A *congruent* triangle is another triangle of the same shape and size as the given triangle.) We have now formed a parallelogram that has the same base and the same altitude as the given triangle; that is, $b = 11$, and $h = 8$. For the parallelogram, we have the formula

$$A = bh$$

Then the area of the parallelogram is $(8)(11) = 88$ (square inches).

We also see that the area of the original triangle is exactly *one-half* the area of the parallelogram. Therefore the area of the triangle is 44 square inches. This is *one-half* of the *base times altitude* of the triangle. Now we can state the rule for the area of a triangle. As a formula, we have

$$A = \tfrac{1}{2}bh, \quad \text{or} \quad A = \frac{bh}{2}$$

This is the formula for the area of any triangle.

Example 8. A triangle has a base of 15 in. and an altitude of 11 in. Find the area.

Solution. Here we use the formula:

$$A = (\tfrac{1}{2})bh; \quad \text{or} \quad A = \frac{bh}{2}$$

Then we get

$$A = (\tfrac{1}{2})(15)(11) = 82.5, \text{ area (sq. in.)}$$

In a right triangle, either leg can be taken as the base and the other leg as the altitude. The *area of a right triangle* is therefore *one-half the product of the two legs*.

Example 9. A right triangle has one leg equal to 12.5 inches, and the other leg equal to 16.4 inches. Find the area of the triangle.

Solution. Here, either leg may be called the base and the other leg may be called the altitude of the triangle. For the area we have

$$A = (\tfrac{1}{2})(16.4)(12.5) = 102.5, \text{ area (sq. in)}$$

The formula for the area of a triangle can also be used in reverse. For example, if we know the area and either the base or the altitude, we can find the unknown dimension. We use the following formulas:

$$h = \frac{2A}{b} \quad b = \frac{2A}{h}$$

MEASUREMENT OF PLANE FIGURES 129

Example 10. A triangle has a base of 16 in. and an area of 60 sq. in. Find the altitude.

Solution. Here we use the formula:

$$h = \frac{2A}{b}$$

Then

$$h = \frac{(2)(60)}{16} = 7.5, \text{ altitude (inches)}$$

To find the perimeter of a triangle, we must know the lengths of the three sides. Then, for the perimeter, we simply add the lengths of the sides. For example, if a triangle has its sides equal, respectively, to 7 inches, 9 inches, and 12 inches, the perimeter becomes (Fig. 7.6)

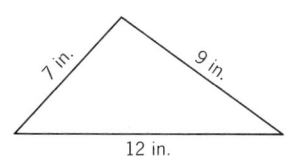

$$7 + 9 + 12 = 28 \text{ (inches)}$$

FIGURE 7.6

Note that the perimeter is stated in *linear units* because it is a length. If we let *p* represent the number of units in the perimeter, and let *a*, *b*, and *c*, respectively, represent the number of units in the three sides, we have the formula

$$p = a + b + c$$

In the triangle in Figure 7.6, since the sides are 7, 9, and 12 inches, respectively, the shape and size of the triangle are definitely determined. That is, if the three sides of any triangle are given, then there is *only one shape and size* of the triangle. This is the principle that forms the basis for the use of triangles in bracing. *Triangular bracing* produces *rigidity* in buildings, bridges, furniture, etc. This is not true with respect to quadrilaterals or other polygons.

Now, since the triangle in Figure 7.6 has a definite shape and size, it should be possible to compute the area by using only the lengths of the sides. The method was explained almost two thousand years ago by a mathematician named Hero (Heron). Hero showed that the area of a triangle can be computed by the following formula. We let *a*, *b*, and *c*, respectively, represent the three sides. Then we let *s* represent one-half the sum of the three sides. Then the formula for the area is

$$A = \sqrt{s(s-a)(s-b)(s-c)}$$

130 GEOMETRY

In the foregoing triangle, $a = 7$, $b = 9$, $c = 12$, $s = 14$. Then we have

$$A = \sqrt{14(14-7)(14-9)(14-12)}$$

$$= \sqrt{(14)(7)(5)(2)} = \sqrt{980} = 31.3 \text{ (approximately)}$$

7.9. THE TRAPEZOID

A trapezoid is a quadrilateral having two sides parallel and the other two sides not parallel. The two parallel sides are called the *bases*. The *altitude* of the trapezoid is the perpendicular distance between the bases. Now we shall derive a formula for the area of a trapezoid.

Suppose we have given the trapezoid in Fig. 7.7*a*. Let us call the bottom base B and the top base b. Then, in the given trapezoid, $B = 8$ in., and $b = 5$ in. The altitude $h = 4$ in.

To derive the rule for the area, we first construct a congruent trapezoid adjacent to the given trapezoid but in an inverted position (Fig. 7.7*b*). Note that a parallelogram is formed by the two trapezoids. The parallelogram has an altitude of 4 inches, the same as the given trapezoid. The base of the parallelogram is equal to the *sum* of the bases of the trapezoid, that is, $8 + 5$, or $B + b$. The base of the parallelogram is 13 in., and the altitude is 4 in. Therefore, the area of the parallelogram is 52 sq. in.

Now we see that the area of the original trapezoid is *one-half* the area of the parallelogram, or 26 sq. in.

In general terms, we see that the base of the parallelogram is $(B + b)$, and the altitude is h. Then the area of the parallelogram is given by

$$A = h(B + b) \quad \text{(area of parallelogram)}$$

Then the formula for the area of a trapezoid becomes:

$$A = \tfrac{1}{2}h(B + b) \quad \text{(area of a trapezoid)}$$

In words, *to find the area of a trapezoid, add the two parallel sides; multiply the sum by the altitude; finally, divide by 2.* If convenient, we can first divide the altitude by 2.

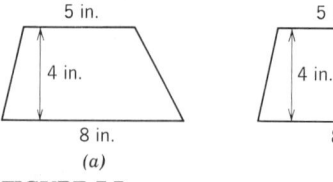

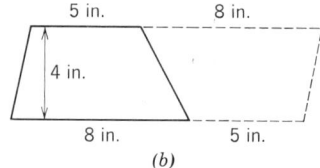

FIGURE 7.7

MEASUREMENT OF PLANE FIGURES 131

Example 11. A lawn has the shape of a trapezoid. One base is 48 ft. and the other base is 35ft. If the altitude is 26 ft., find the area.

Solution. We use the formula:

$$A = (\tfrac{1}{2})(h)(B+b)$$

We get

$$A = (\tfrac{1}{2})(26)(48+35) = (13)(83) = 1079, \text{ area, (sq. ft.)}$$

The altitude or either base, B or b, can be found by the following formulas if the other measurements are given:

$$h = \frac{2A}{B+b}; \quad B = \frac{2A}{h} - b; \quad b = \frac{2A}{h} - B$$

Note that the second and third formulas are essentially the same.

Example 12. In a certain trapezoid, one base $B = 18.2$ in., the other base $b = 14.3$ in., and the area is 136.5 sq. in. Find the altitude h.

Solution. Here we use the foregoing formula for the altitude h. We get

$$h = \frac{(2)(136.5)}{18.2+14.3} = \frac{273}{32.5} = 8.4, \text{ altitude, (inches)}$$

Example 13. A window in the shape of a trapezoid has a bottom base B of 42 in. and an altitude of 24 in. The area of the window is 936 sq. in. Find the top base b.

Solution. To find an unknown base we use the formula:

$$b = \frac{2A}{h} - B$$

We get

$$b = \frac{(2)(936)}{24} - 42 = 78 - 42 = 36, \text{ one base, (inches)}$$

Example 14. In a certain trapezoid, base $b = 18.5$ in., the altitude $h = 12.5$ in., and the area is 280 sq. in Find the other base B.

Solution. Using the formula for base B, we get

$$B = \frac{(2)(280)}{12.5} - 18.5 = \frac{560}{12.5} - 18.5$$

$$= 44.8 - 18.5 = 26.3, \text{ (inches)}$$

Exercise 7.2

Find the areas of the following triangles (No. 1–4).

1. Base 15 in., altitude 18 in.
2. Base 22 in., height 13.6 in.
3. Altitude 26.5 in., base 36 in.
4. Altitude 25 in., base 25 in.

Find the areas of the following trapezoids: (No. 5–8)

5. Bases 12 in. and 7 in., altitude 8 in.
6. $B = 13$ in., $b = 7$ in., $h = 5$ in.
7. Bases 15 in. and 9 in., height 7 in.
8. $B = 8.5$ in., $b = 6.3$ in., $h = 4.5$ in.
9. A triangle has an area of 52 sq. in. and base of 13 in. Find the altitude.
10. A triangle has an area of 72 sq. in. and altitude of 9 in. Find the base.
11. A triangle has an area of 90 sq. in. and altitude of 16 in. Find the base.
12. A triangle has an area of 63.4 sq. in. and base of 17.6 in. Find the altitude.
13. A trapezoid has an area of 63 sq. in. and bases of 8 in. and 13 in. Find the altitude.
14. A trapezoid has an area of 86 sq. in. The altitude is 8 inches, and one base is 9 inches. Find the other base.
15. A triangular lawn at the corner of a rectangular lot has one side 40 feet and the other side 35 feet. Find the area of the lawn.
16. A triangular window has a base of 52 inches and a height of 21 inches. Find the area of the window.
17. The gable of a house is in the shape of a triangle, measuring 27 feet across and a height of 11 feet. Find the area of the gable.
18. The end of a box is in the shape of a trapezoid. The distance across the top is 25 inches; across the bottom it is 16 inches, and the altitude is 13 inches. Find the area of the end of the box.
19. A farmer has a field in the shape of a trapezoid whose bases are 20 rods and 14.5 rods, respectively, and whose altitude is 52 rods. Find the area.
20. A ditch is dug in the shape of a trapezoid. The ditch is 12.5 feet across the top, 7.5 feet across the bottom, and is 6.5 feet deep. Find the cross section area of the ditch.
21. The legs of a right traingle are 7 in. and 10 in., respectively. Find the area.
22. The legs of a right triangle are 6.5 in. and 9.5 in., respectively. Find the area.
23. A triangle has sides equal to 5 in., 7 in., and 8 in., respectively. Find the area of the triangle by use of Hero's formula.
24. A triangle has sides of 6 in., 10 in., and 15 in., respectively. Find the area of the triangle by use of Heron's formula.
25. A triangle has sides of 5 in., 12 in., and 13 in., respectively. Find the area by use of Hero's formula.
26. A wire 120 inches long is to bent to form a rectangle. Find the area of each rectangle formed if the following values are taken for the width:
    (a) 30 in
    (b) 24 in.
    (c) 20 in.
    (d) 15 in.
    (e) 12 in.
    (f) 10 in.
    (g) 8 in.
    (h) 6 in.

(i) 4 in. (j) 2 in.
(k) 1 in. (l) 0.5 in.

27. A rectangle contains 576 sq. in. Find the perimeter of each of the rectangles formed if the following values are taken for the width.

(a) 24 in. (b) 18 in.
(c) 16 in. (d) 12 in.
(e) 9 in.   (f) 8 in.
(g) 6 in.   (h) 4 in.
(i) 3 in.   (j) 2 in.
(k) 1 in.   (l) 0.5 in.

28. A rectangular field bounded on one side by a river is to be fenced on the three remaining sides with a total length of 180 rods. If the length of the field is taken as the distance along the river and the width as the distance outward from the river, find the area enclosed for each of the following widths:

(a) 10 rods (b) 20 rods
(c) 30 rods (d) 40 rods
(e) 45 rods (f) 50 rods
(g) 60 rods (h) 70 rods
(i) 80 rods

8 THE RIGHT TRIANGLE

8.1. DEFINITION

A triangle having one right angle is called a *right triangle*. The right triangle is one of the most important figures in engineering and in general mathematics. Since one angle is a right angle (90°), the other two angles are acute angles.

We know the sum of the three angles of any triangle is equal to 180°. Since the right angle is equal to 90°, the sum of the two acute angles must be 90°. Then the two acute angles are complementary.

In a right triangle the side opposite the right angle is called the *hypotenuse*. The hypotenuse happens to be the longest side. The other two sides of a right triangle are sometimes called the *legs* of the right triangle. The two legs form the right angle and are therefore perpendicular to each other. If we call one leg the base of the triangle, then the other leg is the altitude.

8.2. THE PYTHAGOREAN RULE

A farmer was building a chicken house. When the sills had been laid down temporarily, he told his small son, "Now we must be sure the corners are square." So the son, wanting to be helpful, got his father's carpenter's 2-foot square and tried to test one corner of the sills. His father said, "No, we can't do it that way because the square is too small. Instead, we shall measure 8 feet along one side and make a mark. Then we measure 6 feet along the end and make a mark. Then the distance between the marks must be 10 feet if the corner is square." The little boy asked, "Why?" The father said, "Well, that is just a rule that carpenters use." The boy continued to wonder why. When he studied geometry in high school, he learned the answer. The answer is the Pythagorean rule.

A right triangle has the following important characteristic:

In every right triangle, the square of the hypotenuse is equal to the sum of the squares of the two legs.

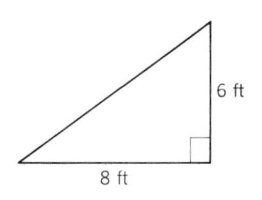

136 GEOMETRY

This principle is known as the Pythagorean rule. It is one of the most useful rules in mathematics, especially engineering. To understand the meaning clearly, consider the right triangle in Figure 8.1. The two legs of the right triangle measure 6 feet and 8 feet, respectively. To use the Pythagorean rule, we first find the square of the length of each leg:

$$6^2 = 36 \qquad 8^2 = 64$$

Now we add the squares together: $36 + 64 = 100$.

The sum of the squares is 100. Then 100 is the square of the hypotenuse. To find the hypotenuse, we take the square root of 100, which is 10. Therefore, if the two legs of the right triangle are 6 feet and 8 feet, respectively, the hypotenuse must be 10 feet, if the corner is square.

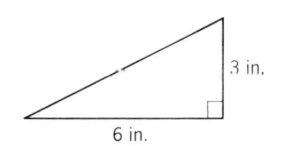

Sometimes the answer does not come out to a whole number. As another example, consider the right triangle in Figure 8.2. One leg is 3 inches long and the other leg is 6 inches long. If we square the two legs, we get

$$3^2 = 9 \qquad 6^2 = 36$$

Adding the two squares, we get $9 + 36 = 45$.

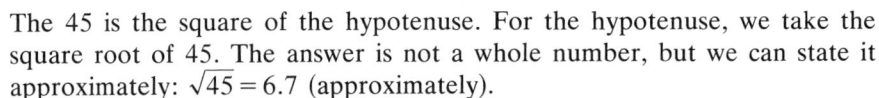

The 45 is the square of the hypotenuse. For the hypotenuse, we take the square root of 45. The answer is not a whole number, but we can state it approximately: $\sqrt{45} = 6.7$ (approximately).

The Pythagorean rule is named after Pythagoras, a Greek mathematician and philosopher, who proved the truth of this principle about 500 B.C. The idea was known and used for specific examples long before Pythagoras, but he was the first to prove that the rule is true for all right triangles. Many people use it in their work without knowing why it is true. It has been proved in many different ways.

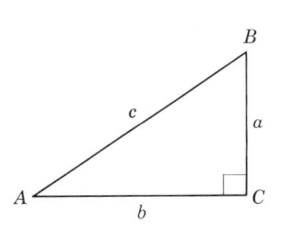

The Pythagorean rule can be stated as a formula. In the general right triangle ABC (Fig. 8.3), angle C is the right angle, and angles A and B are the acute angles. Each side is named with a small letter that corresponds to the opposite angle. For example, side a is the side opposite angle A. The hypotenuse is side c, since it is opposite angle C, the right angle. Side b is the side opposite angle B.

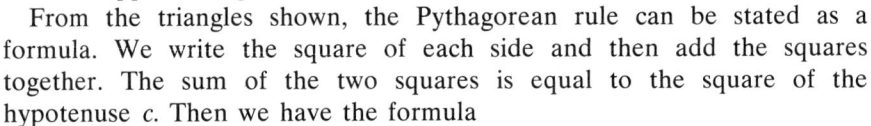

From the triangles shown, the Pythagorean rule can be stated as a formula. We write the square of each side and then add the squares together. The sum of the two squares is equal to the square of the hypotenuse c. Then we have the formula

$$a^2 + b^2 = c^2, \qquad \text{or} \qquad c^2 = a^2 + b^2$$

This formula enables us to compute the length of the hypotenuse of any

THE RIGHT TRIANGLE

right triangle if we know the lengths of the legs. For example, if the legs of a right triangle are 8 inches and 15 inches, respectively, then the length of the hypotenuse can be computed:

$$8^2 = 64 \qquad 15^2 = 225$$

Adding the squares, we get

$$64 + 225 = 289$$

Then

$$c^2 = 289, \quad \text{and } c = 17, \quad \text{the length of the hypotenuse}$$

Following the formula, we can write

$$c^2 = 8^2 + 15^2$$

$$c^2 = 64 + 225$$

$$c^2 = 289$$

$$c = 17$$

The Pythagorean formula can also be used to find the length of a leg if the lengths of the other leg and the hypotenuse are known. As an example, if we know the hypotenuse of a right triangle is 10 inches, and one leg is 8 inches, we can find the length of the other leg. We first find the square of the hypotenuse and then *subtract* the square of the given leg. In this example, we have

$$\begin{array}{rl} \text{square of hypotenuse,} & 10^2 = 100 \\ \text{square of given leg,} & 8^2 = 64 \\ \text{subtracting,} & 36 \end{array}$$

Then the square of the other leg is 36, and the length is 6 inches.

In the foregoing problem, after finding the squares of the hypotenuse and the given leg, we must *subtract* instead of adding because the hypotenuse, the longest side, is given. In any problem such as this when the hypotenuse is given, then we must use subtraction of the squares to find the unknown value.

We shall work several problems showing the use of the Pythagorean rule.

Example 1. Find the hypotenuse of a right triangle having legs of 21 inches and 28 inches, respectively (Fig. 8.4). Also find the perimeter and the area.

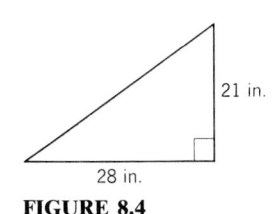

FIGURE 8.4

138 GEOMETRY

Solution. Squaring the two legs, we have

$$21^2 = 441 \qquad 28^2 = 784$$

Adding the squares, we get

$$441 + 784 = 1225$$

Then the square of the hypotenuse is 1225, and the hypotenuse is $\sqrt{1225}$, which is 35 (inches). To find the perimeter of the triangle, we add the three sides and get 84 inches. For the area, we take $\frac{1}{2}$ the product of the legs, and get 294 sq. in.

Example 2. A telephone pole, standing perpendicular to level ground, is to be braced with a wire cable. One end of the cable will be fastened to the pole 24 feet above the ground. The other end is to be fastened to a stake in the ground 15 feet from the foot of the pole. How long a cable will be required if 4 feet extra are needed for fastening to the pole and to the stake?

Solution. A right triangle is formed by the pole, the ground, and the cable (Fig. 8.5). The pole and the ground form the legs of the right triangle, and the cable forms the hypotenuse. From the information given, we know the lengths of the legs. Then we find the squares of the legs and add these squares:

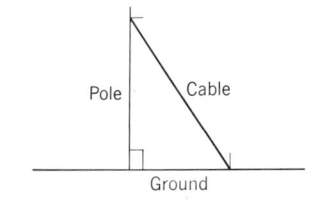

$$24^2 = 576 \qquad 15^2 = 225$$

FIGURE 8.5

Adding the squares of the legs, we get

$$576 + 225 = 801$$

Then the number, 801, represents the square of the hypotenuse, the cable. To find the hypotenuse, we take the square root of 801:

$$\sqrt{801} = 28.3 \text{ (approximately)}$$

Then the hypotenuse is approximately 28.3 feet. Since 4 feet extra are needed for fastening, the length of the cable must be about 32.3 feet.

Example 3. One leg of a right triangle measures 13.2 inches, and the hypotenuse measures 21.3 inches (Fig. 8.6). Find the length of the other leg.

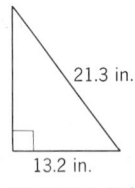

FIGURE 8.6

Solution. Since the hypotenuse, the longest side, is given, we must use subtraction after we find the squares of the two given lengths:

$$\text{square of hypotenuse,} \quad 21.3^2 = 453.69$$
$$\text{square of leg,} \quad 13.2^2 = \underline{174.24}$$
$$\text{subtracting,} \quad 279.45$$

The number, 278.85, represents the square of the unknown leg. Taking the square root of 278.85, we get

$$\sqrt{279.45} = 16.7 \text{ (approx.)}$$

Therefore, the other leg is approximately 16.7 inches long.

8.3. THE ISOSCELES RIGHT TRIANGLE

An isosceles triangle is a triangle having two equal sides. If a right triangle has its two legs equal, then the right triangle is also isosceles (Fig. 8.7). The right triangle shown has each leg equal to 6 inches. The legs are denoted by the small letters a and b, respectively. To find the length of the hypotenuse, we square each leg, add the squares, and then find the square root of the sum, as usual for finding the hypotenuse of any right triangle:

$$\text{square of one leg,} \quad 6^2 = 36$$
$$\text{square of other leg,} \quad 6^2 = \underline{36}$$
$$\text{adding the squares,} \quad 72$$

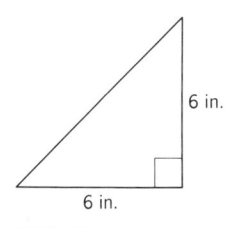

FIGURE 8.7

The number 72 represents the square of the hypotenuse. Then the hypotenuse is equal to the square root of 72. The square root of 72 can be found by the short method:

$$\sqrt{72} = \sqrt{(36)(2)} = \sqrt{36} \cdot \sqrt{2} = 6\sqrt{2}$$

To find the approximate decimal value, we take

$$6\sqrt{2} = 6(1.414) = 8.484 \cdots \quad \text{or } 8.5 \text{ (approx.)}$$

In any isosceles triangle, it can be proved that the angles opposite the equal sides are also equal. That is, if $a = b$, then angle A = angle B. In an isosceles right triangle, since the two acute angles are equal, each of the acute angles is equal to 45°. A right triangle of this particular shape is often called a "45-degree right triangle."

8.4. THE DIAGONAL OF A SQUARE

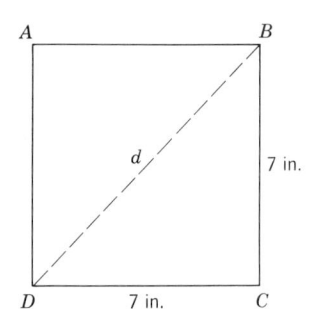

The diagonal of a square is a straight line segment joining two nonadjacent vertices (Fig. 8.8). In the square, ABCD, the diagonal is the line segment BD. The diagonal divides the square into two *congruent* right triangles. (Geometric figures that have exactly the same size and shape are called *congruent* figures.) In the triangle formed, BCD, each leg is a side of the square. Then the triangle BCD is an *isosceles* right triangle, with each leg equal to 7 inches, the side of the square. The hypotenuse of this triangle is the diagonal of the square.

To find the hypotenuse of triangle BCD, we square each leg (or side of the square) and get

$$7^2 = 49$$

Adding the squares,
$$\frac{7^2 = 49}{98}$$

The number 98 represents the square of the hypotenuse, BD, or the diagonal of the square. The hypotenuse becomes (by the short method):

$$\sqrt{98} = 7\sqrt{2}$$

Therefore, the diagonal of the square is $7\sqrt{2}$, or approximately 9.898.

If we were to begin with a square whose side is 8 inches, we should find that the diagonal is equal to $8\sqrt{2}$. If we begin with a square whose side is 10 inches, we shall find the following rule is true:

The diagonal of a square is equal to the length of a side multiplied by the square root of 2.

The rule may be stated as a formula. If we represent the side of a square by s and the diagonal by d (Fig. 8.9), then we have the formula

$$d = s\sqrt{2}$$

This formula can also be used in reverse. If we know the length of the diagonal of a square, we can find the length of one side by the formula

$$s = \frac{d}{\sqrt{2}}; \quad \text{or} \quad s = 0.707d$$

Example 4. Find the length of the diagonal of a square having a side equal to 15.4 inches.

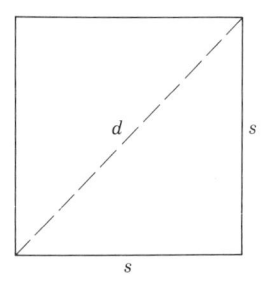

THE RIGHT TRIANGLE 141

Solution. We use the formula

$$d = s\sqrt{2}.$$

Then we have

$$d = 15.4\sqrt{2} = 15.4(1.414) = 21.7756$$

Thus the diagonal is equal to 21.8 inches, rounded off to three significant digits.

Example 5. Find the diagonal of a square whose perimeter is 50 inches.

Solution. The perimeter of a square is 4 times the length of one side. Therefore, the side of a square is $\frac{1}{4}$ of the perimeter. Since the perimeter of this square is 50 inches, the length of one side is 12.5 in. For the diagonal, we have

$$d = 12.5\sqrt{2} = 12.5(1.414) = 17.7 \text{ (approx.)}$$

Example 6. Find a side and the area of a square whose diagonal is 24 in.

Solution. To find the length of a side, we use the formula:

$$s = \frac{d}{\sqrt{2}}, \quad \text{or} \quad s = (0.707)(d)$$

We get

$$s = \frac{24}{\sqrt{2}}; \quad \text{or} \quad (0.707)(24) = 16.968, \quad \text{or} \quad 16.97 \text{ (approx.)}$$

For the area we find the square of a side. For the side, we can use

$$s = \frac{24}{\sqrt{2}}; \quad \text{then } s^2 = \left(\frac{24}{\sqrt{2}}\right)^2 = \frac{576}{2} = 288$$

The area of a square can be found directly from a given diagonal. Suppose we make a drawing of a square and show a diagonal. Now we construct a square on the diagonal. The square on the diagonal has twice the area of the original square. Therefore, to find the area of the square, we can first square the diagonal and then divide by 2. That is,

$$\text{area} = (\tfrac{1}{2})(d^2)$$

Example 7. Find the area and then the side of a square whose diagonal is equal to 32 inches.

142 GEOMETRY

Solution. First we find the area by the formula:

$$A = (\tfrac{1}{2})(d^2).$$

We get

$$A = (\tfrac{1}{2})(32^2) = (\tfrac{1}{2})(1024) = 512, \text{ area, (sq. in.)}$$

Now we can find the length of a side by taking the square root of 512. We get

$$s = \sqrt{512} = 16\sqrt{2} = 22.6, \text{ (approx.) (inches)}$$

8.5. THE 30°–60° RIGHT TRIANGLE

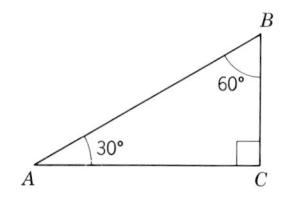

A special triangle that occurs often in mathematics is a right triangle whose acute angles are 30° and 60°, respectively (Fig. 8.10). A triangle of this particular shape is called a "30°–60° right triangle." Notice that the triangle is scalene; that is, no two sides are equal. This is the particular type of triangle that Plato, a Greek philosopher and mathematician, called "the most beautiful scalene right-angled triangle."

A very important fact can be proved concerning such a triangle:

FIGURE 8.10

In any 30°–60° right triangle, the hypotenuse is always equal to twice the shortest side.

By use of this principle, we can find all the sides of such a triangle if we know just one side.

Example 8. Find the hypotenuse and the unknown sides of the 30°–60° right triangle having the shortest side equal to 4 inches (Fig. 8.11). Also find the perimeter and the area of the triangle.

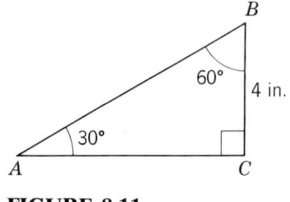

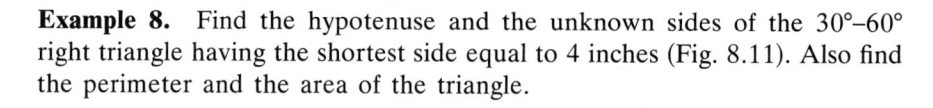

Solution. From the principle just stated, the hypotenuse AB is twice the shortest side, BC. Since $BC = 4$ inches, then $AB = 8$ inches. Now we can use the Pythagorean rule to find the other leg. We find the squares of the hypotenuse and the given leg:

FIGURE 8.11

$$\text{square of hypotenuse,} \quad 8^2 = 64$$
$$\text{square of short leg,} \quad \underline{4^2 = 16}$$
$$\text{subtracting,} \quad 48$$

Then 48 is the square of the side AC. Taking the square root, we get the

THE RIGHT TRIANGLE 143

length of the other leg:

$$AC = \sqrt{48} = 4\sqrt{3}, \quad \text{or approximately 6.9 inches}$$

Notice the longer leg is $\sqrt{3}$ times the length of the short leg. For the perimeter, we add the three sides

$$8 + 4 + 6.9 = 18.9 \text{ (in.)}$$

For the area, we take $\frac{1}{2}$ of the product of the two legs, and get

$$(\tfrac{1}{2})(4)(6.9) = 13.8 \text{ (sq in.)}$$

In this example, of course the answers are approximate values.

8.6. THE EQUILATERAL TRIANGLE

An *equilateral* triangle is a triangle having all sides equal. This is another type of special triangle that is very useful. We have said that if two sides of a triangle are equal, then the angles opposite the equal sides are also equal. That is, an *equilateral* triangle is also *equiangular*. Since the sum of the three angles of any triangle is 180°, then in any equilateral triangle, each angle is equal to 60°.

In connection with an equilateral triangle, there are two problems with which we are concerned. They are finding the altitude and the area.

In order to see how to find the altitude and the area of an equilateral triangle, suppose the triangle *ABC* in Figure 8.12 has each side equal to 10 inches. Now we wish to find the altitude and the area of the triangle. If we draw the straight line segment *CD*, bisecting the angle *C* into two equal angles, each angle at *C* is 30°. The line meets the side *AB* at the point *D*, dividing the figure into two triangles.

Now we want to show that DBC is a right triangle. We do so as follows:

angle B = 60°;   angle DCB = 30°

Therefore,   angle BDC must be 90°

Now we know that the triangle BDC is a 30°–60° right triangle.

We now use our knowledge of a 30°–60° right triangle. In the triangle *DBC*, the hypotenuse *BC* is 10 inches. Therefore, *DB* = 5 inches, since it is the shortest side of the 30°–60° right triangle, *DBC*. (In trying to prove that a statement is true in geometry, we must be careful not to assume that it is true just because it looks like it.)

Note that the altitude *DC* of the equilateral triangle is one leg of the small right triangle *DBC*. Now we have the following information: in the triangle *DBC*, the hypotenuse is 10 inches and one leg is 5 inches. Then we can find

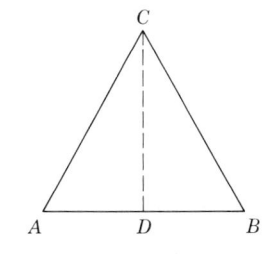

FIGURE 8.12

144 GEOMETRY

the length of DC:

$$\begin{aligned}
\text{squaring the hypotenuse,} \quad & 10^2 = 100 \\
\text{squaring the short leg,} \quad & 5^2 = \underline{25} \\
\text{subtracting,} \quad & 75
\end{aligned}$$

The number 75 represents the square of the leg DC, which is the altitude of the equilateral triangle. Then the altitude is $\sqrt{75}$, or $5\sqrt{3}$, which is approximately 8.66 inches.

Let us take another look at the altitude of the triangle in Figure 8.12. The altitude is $5\sqrt{3}$. Notice that the altitude is $\sqrt{3}$ multiplied by one-half of the side of the equilateral triangle. This is always true for the altitude of an equilateral triangle. If we use s to represent one side of an equilateral triangle, then the formula for the altitude h is

$$h = \tfrac{1}{2}(s)\sqrt{3}$$

That is, the altitude of an equilateral triangle is $\tfrac{1}{2}$ the side times $\sqrt{3}$.

To find the area of the triangle in Figure 8.12, we begin with the formula for the area of any triangle:

$$A = \tfrac{1}{2}bh$$

In the triangle shown, the base is 10 inches. Also we have found that the altitude h is equal to $5\sqrt{3}$. Then, for the area, we have

$$A = \left(\frac{1}{2}\right)(10)(5)(\sqrt{3}) = 25\sqrt{3} = 43.3 \text{ (approx.)}$$

In general, if we use s to represent one side of an equilateral triangle, we have the following: base $= s$; altitude $h = \tfrac{1}{2}s\sqrt{3}$. Using this information, and the general formula for the area of any triangle, $A = \tfrac{1}{2}bh$, we get the formula for the area of any equilateral triangle:

$$A = \left(\frac{1}{2}\right)(s)\left(\frac{1}{2}\right)(s)\sqrt{3} = \frac{1}{4}s^2\sqrt{3}$$

This formula shows that, to find the area of any equilateral triangle, we take $\tfrac{1}{4}$ of the square of a side, and then multiply the result by $\sqrt{3}$.

Example 9. Find the altitude and area of an equilateral triangle having a side equal to 12 inches.

Solution. For the altitude we have the formula: $h = \frac{1}{2}s\sqrt{3}$. Then

$$h = \left(\frac{1}{2}\right)(12)\sqrt{3} = 6\sqrt{3}, = 10.4 \text{ (in.) (approx.)}$$

For the area we have the formula: $A = (\frac{1}{4})(s^2)\sqrt{3}$. Then

$$A = \left(\frac{1}{4}\right)(12^2)\sqrt{3} = \left(\frac{1}{4}\right)(144)\sqrt{3} = 36\sqrt{3} = 62.4 \text{ (sq in.) (approx.)}$$

Exercise 8.1

Find the hypotenuse, perimeter, and area of each of these right triangles:

1. Legs: 24 inches and 32 inches.
2. Legs: 20 inches and 48 inches.
3. Legs: 12.5 cm and 30 cm.
4. Legs: 37.5 cm and 50 cm.
5. Legs: 17 yds and 23 yds.
6. Legs: 24 rd and 33 rd.
7. Legs 24.3 in. and 34.1 in.
8. Legs: 31.4 in. and 42.3 in.

Find the perimeter and area of each of the following right triangles:

9. Hypotenuse: 30 in.; a leg: 16 in.
10. Hypotenuse: 65 in., a leg: 25 in.
11. Hypotenuse: 42.5 cm, a leg: 20 cm.
12. Hypotenuse: 37.5 in., a leg: 36 in.
13. Hypotenuse: 25.1 cm., a leg: 18.2 cm.
14. Hypotenuse: 32.5 cm., a leg: 14.3 cm.

Find the length of the diagonal of each of the following squares:

15. Side = 13.2 in.
16. Side = 28.7 cm.
17. Side = 3.52 ft.
18. Side = 1000 ft.
19. Side = 4570 yd.
20. Side = 5280 ft.
21. Side = 360 rd.
22. Side = 16.5 ft.
23. Side = 5.5 yd.
24. Perimeter = 540 in.
25. Perimeter = 38 in.
26. Perimeter = 1 foot.

Find the area and perimeter of each of the following 30°–60° rt. triangles:

27. Shortest side = 7.4 inches.
28. Thy hypotenuse = 25.2 cm.
29. Hypotenuse = 480 ft.
30. Shortest side = 3.57 cm.
31. Hypotenuse = 6.73 in.
32. Hypotenuse = 21.47 cm.

Find the area of each of the following equilateral triangles:

33. A side = 18 inches.
34. A side = 13.4 cm.
35. Perimeter = 42 in.
36. A side = 72.4 ft.
37. A side = 52.4 ft.
38. Perimeter = 23.4 cm.

39. How long a wire will be needed to brace an electric light pole if the wire is fastened to the pole 35 feet above the ground and to a stake in the ground at a distance of 23 feet from the foot of the pole?
40. A TV broadcasting tower is braced with a wire cable 360 feet long. The cable is fastened in the ground at a point 270 feet from the tower. How far up on the tower is the cable fastened?
41. How long a diagonal brace is needed for a gate 5 feet high and 16 feet long.
42. A field is 60 rods long and 30 feet wide. A road extends diagonally from one corner to the opposite corner. How long is the road?
43. A ladder 32 feet long leans up against a vertical wall. How high on the wall does the ladder reach if the foot is placed 7.2 feet from the wall?
44. A ladder is 28 feet long. How far from the wall must the foot of the ladder be placed in order that the ladder will just reach a window that is 26.4 feet from the ground?
45. Two cars start at the same point, one traveling north at 45 mph and the other east at 34 mph. How far apart are they after 30 minutes?
46. A man rows his boat directly across a river at right angles to the current. The river is 800 feet wide. If he rows at 200 feet per minute, and the river flows at a rate of 150 feet per minute, how fast does he travel and in what direction? How long will it take him to cross the river?
47. The length of a rectangle is 42.3 cm. The diagonal is 53.4 cm. Find the area and the perimeter of the rectangle.
48. The width of a rectangle is 21.5 ft. The diagonal is 62.5 ft. Find the area and perimeter of the rectangle.

9 THE CIRCLE

9.1. DEFINITIONS

The circle is one of the most common as well as most useful of all geometric figures. It is used for beauty in design. It has a practical use in the wheels of machinery, industry, and most vehicles.

Although the circle is a very common figure, it is not easy to define. We may say that *the circle is a plane closed curve such that every point on the curve is the same distance from a point within called the center.* By this definition, the circle is the curve itself (Fig. 9.1). For the word circle we use the symbols: ⊙; Ⓢ.

The *circumference* of the circle is the length of the curve. Sometimes we say the curve itself is the circumference. Then the circle is taken to mean the space within the circumference.

There are several lines and line segments that refer to a circle. A line segment such as *AB* (Fig. 9.1) through the center with the end-points on the circumference is called a *diameter* of the circle. The diameter divides the circle into two equal halves, each one called a *semicircle*. A straight line segment such as *OB* connecting the center with a point on the circumference is called a *radius* of the circle (plural: *radii*). Notice that the diameter is equal in length to twice the length of a radius. If we use D to represent the number of units in the diameter, and r to represent the number of units in the radius, then we have the two formulas

$$D = 2r \quad \text{and} \quad r = \tfrac{1}{2}D, \quad \text{or} \quad r = \frac{D}{2}$$

From the definition of a circle, we see that *all radii of a circle are equal.* That is why a circle can be drawn by using a *compass*.

A *tangent* to a circle is a straight line, such as line *t* that touches a circle in only one point and does not cut through the circumference. A *secant* is a straight line, such as line *s*, that cuts through the circumference in two

147

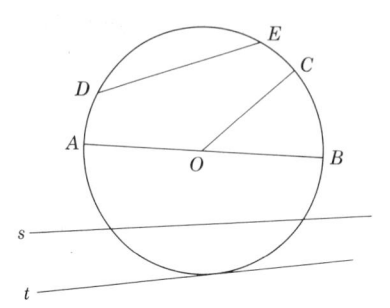

FIGURE 9.1

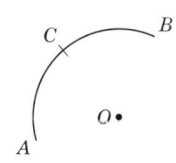

FIGURE 9.2

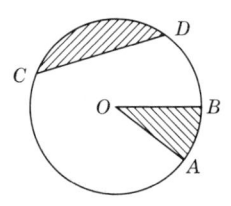

FIGURE 9.3

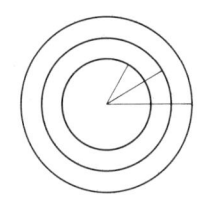

FIGURE 9.4

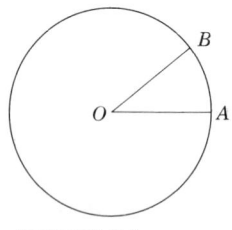

FIGURE 9.5

points. A *chord* is a straight line segment, such as *DE*, that has its end points on the circumference. A chord, therefore, has a limited length. The longest chord that can be drawn in a circle is a diameter.

A *semicircle* is a half-circle. A *semicircumference* is half of the circumference. An *arc* is a portion of a circumference, such as *arc BC*. Any two points on a circumference divide the circumference into two arcs. If one arc is longer than the other, then the longer one is called the *major* arc; the other is called the *minor* arc. The symbol for *arc BC* is $\widehat{BC}$. If a major arc is meant, it must be indicated by using three letters, such as arc *BAC* (Fig. 9.1).

The *midpoint* of an arc is a point that divides the arc into two equal arcs. In Figure 9.2, the midpoint of *arc AB* is the point *C*. However, an arc is also said to have a *center*. By the center of an arc, we mean the center of the circle of which the arc is a part. The center of *arc AB* is the point *O*.

A *sector* of a circle is a portion of the area bounded by an arc and two radii (Fig. 9.3). A *segment* of a circle is a portion of the area bounded by an arc and a chord (Fig. 9.3). *Concentric circles* are two or more circles that have the same center but different radii (Fig. 9.4).

An angle formed by two radii at the center of a circle intercepts an arc on the circumference. In Figure 9.5, the angle *AOB* at the center of the circle intercepts the arc *AB* on the circumference. Since there are 360° in one complete revolution, then an angle of 1° at the center of a circle intercepts a small arc that is $\frac{1}{360}$ of the circumference. This small arc we might call one *arc degree*. Then the angle of 1° we may call one *angle degree*. To distinguish between the two, try to think of one *angle degree* as a very small *angle*, and one *arc degree* as a very small portion of the circumference cut off by a 1° angle at the center.

An angle formed by two radii at the center of a circle is called a *central angle*, such as angle *AOB* (Fig. 9.6). If a central angle contains 10 degrees, then it cuts off an arc of 10 arc degrees on the circumference. Whatever the size of the central angle in a circle, it will intercept on the circumference an arc that contains just as many arc degrees as the angle contains angle degrees. Then we say that a central angle is equal, in the number of degrees, to the intercepted arc. This fact is usually stated as follows:

A central angle is measured by the intercepted arc.

We must be careful how we make this statement. When we say a central angle of 20 degrees intercepts an arc of 20 arc degrees, we might carelessly say that the central angle equals the arc. But an angle cannot equal an arc. The two things are not comparable. It is no more correct to say that an angle equals an arc than to say, for example, that one day equals 2 feet. One day equals 24 hours. There are 24 inches in 2 feet. The only thing equal about the two quantities is the number 24. In the same way, an angle cannot equal an arc. We can say, however, that the number of angle degrees in a central angle is the same as the number of arc degrees in the intercepted arc.

An *inscribed angle* is an angle formed by two chords drawn from the same point on the circumference of a circle, as angle *CDE* (Fig. 9.6). The inscribed angle *CDE* intercepts the arc *CE*. In this case, the inscribed angle contains only one-half as many angle degrees as the intercepted arc contains arc degrees. Then we say that *an inscribed angle is measured by one-half the intercepted arc.* For example, if an inscribed angle intercepts an arc of 50 arc degrees, then the angle contains only 25 angle degrees. An inscribed angle of 40 degrees intercepts an arc of 80 arc degrees. An angle inscribed in a semicircle (that is, with the end points of the angle at the end points of a diameter) intercepts an arc of 180 arc degrees. Then the angle contains 90°, or a right angle. Therefore, an *angle inscribed in a semicircle is a right angle.* Carpenters make use of this fact in drawing a circle by using a square.

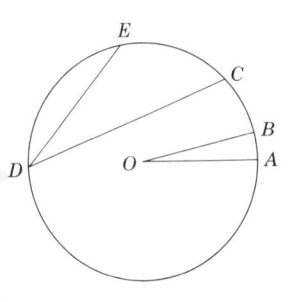

FIGURE 9.6

9.2. MEASUREMENTS IN A CIRCLE

In any circle, we are usually concerned with measurements that involve the diameter, the radius, the circumference, and the area. In the formulas concerning the circle, we use the following notations:

C represents the number of *linear* units in the circumference.

D represents the number of *linear* units in the diameter.

r represents the number of *linear* units in the radius.

A represents the number of *square* units in the area.

We have already mentioned the fact that since the diameter is equal to twice the radius, we have the two formulas:

$$D = 2r \quad \text{and} \quad r = \tfrac{1}{2}D, \quad \text{or} \quad r = \frac{D}{2}$$

It is a little difficult to measure around a circle with a straight ruler. Of course, we could use a tape measure. However, instead, we can measure the diameter. Then the circumference is a little more than three times the diameter. The circumference of any circle always has a fixed ratio to the diameter. The ratio is close to $3\frac{1}{7}$, but the true value cannot be stated exactly as a decimal or common fraction.

The exact ratio of circumference to diameter of any circle is called π (pronounced *pie*). The value of π is approximately equal to $3\frac{1}{7}$, but more nearly 3.1416. For many years, people believed that π could be written as an exact common fraction or decimal. Now it is known that the exact value of π cannot be so written. It is an unending decimal. The value to 30 decimal places is

$$3.141592653589793238462643383280\ldots \text{(rounded off)}$$

In most problems, π is rounded off to 3.14159 or 3.1416. On a slide rule, we use 3.14 for π. In an actual problem, you should use the value that you think best for that particular problem, depending on the degree of precision you want.

If we know the diameter of a circle, we find the length of the circumference by multiplying the diameter by π (3.1416 or 3.14). Representing the number by π, we have the formula:

$$C = \pi D$$

Since the diameter is twice the radius, we can replace D with $2r$, and write

$$C = \pi(2r)$$

which is usually written

$$C = 2\pi r$$

The formulas can be used in reverse; that is, if the circumference is known, then we can find the diameter and the radius. Then we have the formulas

$$D = \frac{C}{\pi} \quad \text{and} \quad r = \frac{C}{2\pi}$$

That is,

$$D = C \div \pi \quad \text{and} \quad r = C \div 2\pi$$

Example 1. The radius of a certain circle is 7.5 inches. Find the diameter and the circumference.

THE CIRCLE 151

Solution. For the diameter, we have $D = 2r$; then $D = 2(7.5) = 15$ (in.). For the circumference, we have $C = \pi D$; then $C = 15(3.1416) = 47.124$ (in.).

Example 2. Find the diameter and the radius of a circular race track if the track is $\frac{1}{4}$ mile long.

Solution. We might first change the length to feet: $\frac{1}{4}$ mile = 1320 ft. To find the diameter, we divide the circumference, 1320 ft., by 3.1416:

$$1320 \div 3.1416 = 420.168$$

The answer can be rounded off to 420.2 feet. For the radius, we take $\frac{1}{2}$ the diameter: $\frac{1}{2}$ of 420.2 ft = 210.1 ft, (radius).

9.3. AREA OF A CIRCLE

Most people probably remember from their grade school days the formula for the area of a circle: $A = \pi r^2$. The exact formula can be derived only by more advanced mathematics. However, it is not difficult to see why the rule is reasonable.

Consider a circle with center at O (Fig. 9.7) and radius $OA = 7$ inches. We circumscribe a square around the circle. The small, shaded square on the radius contains 49 square inches, or the square of the radius, r^2. Note that the large square about the circle contains 4 times as much area as the small square. That is, the large square has an area equal to $4r^2$, or 196 square inches.

Now we see that the circle contains less area than the large square. In fact, the circle has an area a little more than 3 times the square of the radius. The area of the circle is approximately 3.1416 times the square of the radius; that is, the area of a circle is given by the formula

$$A = \pi r^2$$

FIGURE 9.7

The formula for the area of a circle can be reasoned in a somewhat more logical way. We imagine a circle divided into any number of sectors, resembling triangles (Fig. 9.8). Now we can say that the altitude of these triangles is equal to the radius of the circle, and the sum of the bases is equal to the circumference. The total area of the triangles can then be found by the formula for the area of a triangle. The area is equal to $\frac{1}{2}$ the altitude times the sum of the bases; or

$$A = \left(\frac{1}{2}\right)(\text{radius})(\text{circumference})$$

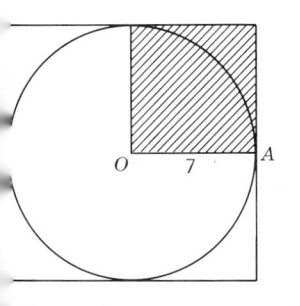

152 GEOMETRY

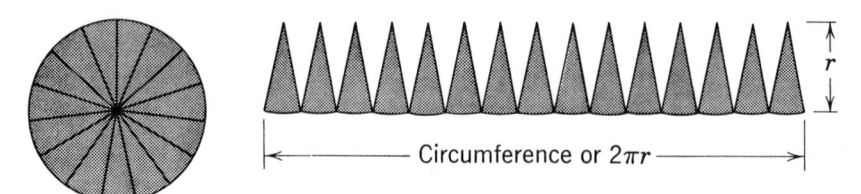

FIGURE 9.8

or

$$A = \left(\frac{1}{2}\right)(r)(2\pi r) = \pi r^2$$

Of course, each sector is not a true triangle. However, if we imagine that the number of sectors increases, the base of each sector becomes a smaller and smaller arc that approaches the condition of a straight line. The sum of the bases of the sectors is still equal to the circumference of the circle, and the altitude is still equal to the radius.

If the area of a circle is known, the formula can be reversed. Then we have the formula for the radius of the circle:

$$r^2 = \frac{A}{\pi} \quad \text{or} \quad r = \sqrt{\frac{A}{\pi}}$$

Another formula for the area of a circle is obtained by substituting the value, $D/2$, for the radius. We get the following formula

$$A = \pi \left(\frac{D}{2}\right)^2 \quad \text{or} \quad A = \frac{\pi}{4} D^2$$

This formula can be used when the diameter is known. Then the value of $\pi/4$ is usually taken as 0.7854. The area of a circle is therefore equal to 0.7854 times the area of the circumscribed square, or $A = 0.7854 D^2$.

Example 3. Find the area of a circular patio having a diameter of 26 ft.

Solution. We can first find the radius, which is $(\frac{1}{2})(26) = 13$, ft. Then we use the formula:

$$A = \pi r^2$$

We get

$$A = \pi(13^2) = 169\pi, = 530.9, \text{(sq. ft., approx.)}$$

If we wish, we can use the formula:

$$A = \frac{\pi}{4}D^2 = 0.7854(676) = 530.9$$

Example 4. Find the radius and diameter of a circle whose area is 1018 sq. ft.

Solution. To find the square of the radius, we use the formula

$$r^2 = \frac{A}{\pi}$$

We get

$$r^2 = \frac{1018}{3.1416} = 324 \text{ (approx.)}$$

for the radius, we have

$$r = \sqrt{324} = 18, \text{ (ft.)}$$

9.4. AREA OF A RING

By the area of a ring, we mean the area between two concentric circles. It is often necessary to find such an area. For example, we may wish to find the area of a circular sidewalk around a circular fountain. In Figure 9.9, a sidewalk 4 feet wide is laid around a circular fountain having a diameter of 20 feet. To find the area of the sidewalk alone, we can first find the area of the large circle, then the area of the smaller circle, and then subtract the smaller area from the larger. The large circle, which includes the fountain, has a radius of 14 feet. The small circle, the fountain, has a radius of 10 feet. For the large circle the area is $\pi(14^2) = 196\pi$. For the small circle we get the area, $\pi(10^2) = 100\pi$. Then the area of the sidewalk itself is equal to

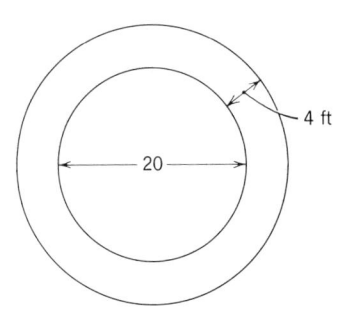

FIGURE 9.9

the difference between the areas of the two circles. That is,

$$A(\text{of walk}) = 196\pi - 100\pi = 96\pi, \quad \text{or} \quad 301.6 \text{ sq. ft (rounded off)}$$

The method for finding the area of any ring may be stated as a formula. To derive the formula, we use R to represent the radius of the large circle, and r to represent the radius of the smaller circle. We use A_L to represent the area of the larger circle, and A_S to represent the area of the smaller circle. Then we have the following formulas for the areas of the two circles:

$$A_L = \pi R^2 \quad \text{and} \quad A_S = \pi r^2$$

For the area of the ring, we take the difference between the two areas:

$$A(\text{of ring}) = A_L - A_S = \pi R^2 - \pi r^2 \quad \text{or} \quad A = \pi(R^2 - r^2)$$

The formula shows that we can square the radius of each circle, then subtract the smaller from the larger, and finally multiply by π.

Warning. We cannot first take the difference between the two radii. That is, we cannot first take $R - r$. Instead, we must square each radius, and then subtract. Finally, we multiply the difference by π.

Example 5. A circular fountain is surrounded by a walk 6 ft. wide. The inside diameter of the walk is 24 ft. Find the area of the walk.

Solution. Since the walk is 6 ft. wide, the diameter to the outside is 36 ft. Then we have the radius of the inner and the outer circle. Calling the greater radius R, and the smaller radius r, we use the formula for the area of a ring:

$$A = \pi(R^2 - r^2)$$

Then we get

$$A = \pi(324 - 144) = \pi(180) = 565.5 \text{ (sq. ft. approx.)}$$

9.5. AREA OF A SECTOR

A *sector* of a circle is a portion of the circle bounded by *two radii and an arc* of the circumference. The length of the arc and the area of the sector depend on the radius of the circle. For example, in Figure 9.10, the radius of the circle is 12 inches and the angle of the sector is 20°. The entire circumference of any circle represents 360°. Then, in the sector shown, the central angle will intercept an arc that is $\frac{20}{360}$ of the circumference. In a similar manner, the area of the sector is $\frac{20}{360}$ of the area of the entire circle. The circumference of the circle shown is $2\pi r$, or 75.3984 inches. Then the arc of the sector is $\frac{20}{360}$ of 75.3984 inches or 4.1888 inches. The area of the

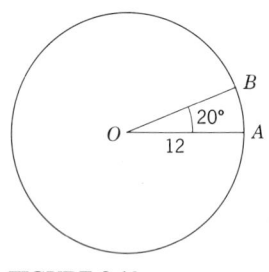

FIGURE 9.10

circle is 452.3904 square inches. Then the area of the sector is $\frac{20}{360}$ of 452.3904 square inches or 25.1328 square inches.

9.6. AREA OF A SEGMENT

A *segment* is a portion of a circle bounded by a *chord and an arc* (Fig. 9.11). There is no simple exact formula for the area of a segment. For the approximate area, we sometimes use a formula that depends on the *width* and the *height* of the segment. The *width w* is the length of the chord, and the *height h* is the perpendicular distance from the midpoint of the chord to the arc. Then we find the approximate area of a segment by the formula

$$A = \tfrac{2}{3}hw + \frac{h^2}{2w}$$

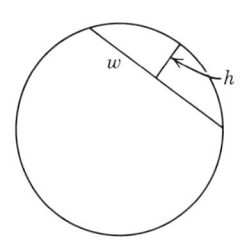

FIGURE 9.11

Example 6. A segment of a circle is bounded by a chord 20 in. long. The height *h* of the segment is 4.5 in. Find the area of the segment.

Solution. We use the formula for the approximate area. We get

$$A = (\tfrac{2}{3})(4.5)(20) + \frac{(4.5)^2}{40} = 60 + 0.506 = 60.5 \text{ (sq. in. approx.)}$$

If the height *h* of a segment is small compared with the width *w*, then the second term of the formula may be omitted.

The area of a segment can sometimes be found conveniently in another way. In Figure 9.12, the area of the segment formed by the arc *AB*, and the chord *AB* is equal to the area of the sector *AOB* minus the area of the triangle *AOB*.

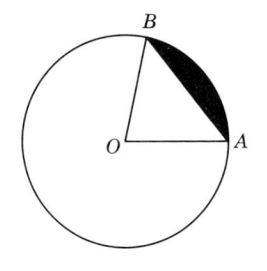

FIGURE 9.12

Exercise 9.1

Find the circumference and the area of each of the following circles:

1. Circular table, diam. = 50 in.
2. Circular window; diam. = 5.5 ft.
3. Circular mirror; diam. = 26 in.
4. Clock dial; diam. = 11.5 in.
5. Drum head; diam. = 28 in.
6. Coin; diam. = $1\tfrac{3}{16}$ in.
7. Circular clock dial; $r = 5.75$ in.
8. Circular race track; $r = 210$ ft.

9. A circle is made with a compass set at 6.3 inches. Find the circumference.
10. A circular race track has a diameter of 320 feet. How many laps must be made for a distance of 1 mile?

11. A bicycle wheel is 26 inches in diameter. How many turns does it make in going 1 mile if there is no slipping?
12. An automobile wheel 24 inches in diameter makes 3 revolutions per second. What is the speed in miles per hour if there is no slipping?
13. On a merry-go-round you sit 15 feet from the center. How far do you ride in making 20 rounds?
14. A locomotive wheel is 72 inches in diameter. How many revolutions does it make per second when the train travels 80 miles per hour?
15. A circular race track is 1320 feet ($\frac{1}{4}$ mile). Find the diameter of the circle and the area enclosed by the track.
16. The distance around a large tree is 10 ft 6 in. Find the diameter and the radius of the tree.
17. A circular skating rink having a radius of 46 feet is surrounded by a sidewalk that is 10 feet wide. Find the area of the sidewalk.
18. A circular fountain, 16 feet in diameter, is surrounded by a walk that is 7 feet wide. Find the area of the walk.
19. A circle is inscribed in a square that is 18 inches on a side. Find the area of each corner of the square that is outside the circle.
20. Four concentric circles are drawn with radii of 4 in., 6 in., 8 in., and 10 in., respectively. Find the area of the innermost circle and the area between each of the other circles; that is, the area of each ring.
21. A belt pulley has a diameter of 11.5 inches. The pulley makes 20 revolutions per second. If there is no slipping, how fast is the belt traveling in miles per hour?
22. Two circular ventilating pipes, each 8 inches in diameter, are joined to form one single, large circular pipe having the same capacity as the combined capacity of the smaller pipes. What is the diameter of the large pipe?
23. A square ventilating tube 8 inches on a side is to be replaced with a circular tube having the same ventilating capacity as the square tube. What must be the diameter of the circular tube?
24. A circular water main 18 inches in diameter branches off into two equal smaller circular mains having the same combined capacity as the large main. What is the diameter of each of the smaller mains?
25. A 24-inch circular water main branches off into 4 smaller equal circular mains having a combined capacity equal to the capacity of the 24-inch main. Find the diameter of each of the smaller mains.
26. The minute hand of a clock is 4.5 inches long. How far does a point on the tip of the hand travel in ten minutes? How much area does the hand sweep over in ten minutes? (A sector of a circle.)
27. A circle 30 inches in diameter has a chord 24 inches long. How far is the chord from the center of the circle? Find the approximate area of the segment of the circle cut off by the chord.
28. A golfer has decided to put in a circular putting green in his back yard. What will be the area of the green if the diameter is 27.5 feet?
29. A ship's rudder has jammed, and the ship is traveling in a circle with a radius of $\frac{1}{2}$ mile at a rate of 6 miles per hour. How many minutes will it take for the ship to travel once around its circular path?

10 PRISMS

10.1. GEOMETRIC SOLIDS

All the figures we have studied up to this point do not take up space, such as triangles, polygons, and circles. All of them can be drawn on a plane. They are understood to have no thickness.

Now we come to geometric figures that take up space, such as a box, a cone, and a sphere. Figures that take up space are called *geometric solids*. As we have said before, a geometric solid is concerned with only a certain amount space, not with the material of which a solid might be composed, such as wood, iron, or ivory. The study of geometric solids is called *solid geometry*, sometimes called *space geometry*.

10.2. POLYHEDRONS

A *polyhedron* is a geometric solid bounded by planes; that is, flat surfaces (Fig. 10.1). (The prefix "poly" means *many*, and the word "hedron" refers to *faces*.) The plane surfaces of a polyhedron are called *faces*. For example, a cube has six faces, all squares. Any two faces of a polyhedron intersect in a line called an *edge*. The intersection of three or more edges is a point, called a *vertex* (plural: vertices).

10.3. PRISMS

A *prism* is a special kind of polyhedron. The most common kind of prism is what we call a *rectangular solid* such as an ordinary box with square corners (Fig. 10.2a). Such a prism has six faces. All the faces are rectangles and all corners are square corners. It has 12 edges and 8 vertices. If all the six faces are squares, we call the solid a *cube* (Fig. 10.2b).

A rectangular solid and a cube are special types of prisms. In general, a prism has two faces that are congruent and parallel polygons (Fig. 10.3). The

158 GEOMETRY

FIGURE 10.1. Polyhedrons. **FIGURE 10.2**

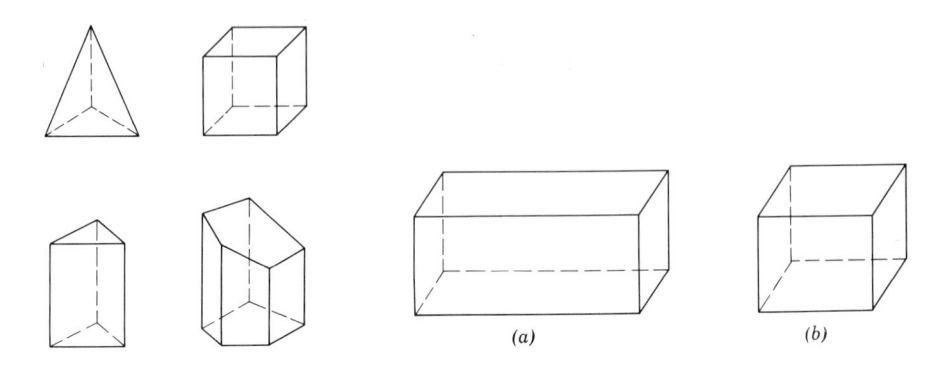

other faces are parallelograms. The two parallel equal faces are called the *bases* of the prism. The bases may be triangles, quadrilaterals, pentagons, or any other kind of polygon. The other faces of a prism are called its *lateral faces*. They meet at the lateral edges.

A *right prism* is a prism having its lateral edges perpendicular to the bases (Fig. 10.3*a,b,c*). That is, the prism is said to be upright. If the lateral edges are not perpendicular to the bases, the prism is called an *oblique* prism (Fig. 10.3*d*).

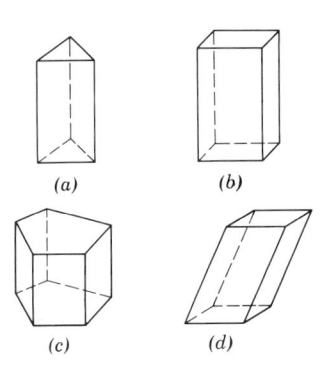

FIGURE 10.3

10.4. VOLUME OF A PRISM

Volume refers to the amount of *space* occupied by a *geometric solid*. To show the method of finding volume, we begin with a rectangular solid, (Fig. 10.4). Suppose the solid is 5 inches long, 3 inches wide, and 4 inches high. Most people who read this will remember the rule: Multiply as follows: (length) × (width) × (height). We get $5 \times 3 \times 4 = 60$. The answer, 60, shows the

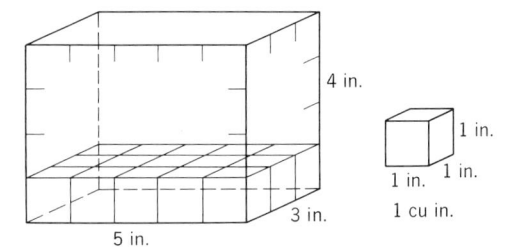

FIGURE 10.4

number of cubic inches in the volume. The answer is correct, but does it not seem strange that we can multiply three *lengths* together and get volume? If you take a moment to analyze the problem, you will get a better understanding of the method for finding the volume of all kinds of prisms, as well as cylinders and other solids.

To measure volume, we must use a *unit of volume.* A common unit for measuring the volume of a solid of this size is a *cubic inch*; that is, a cube 1 inch long, 1 inch wide, and 1 inch high. Now we try to determine how many times this unit is contained in the given rectangular solid. We begin by placing units along one side at the bottom as shown. We find that we can place 5 cubic-inch units in one row. In the bottom layer there will be 3 rows. Then the bottom layer will contain

$$3 \times 5 \text{ cu. in} = 15 \text{ cu. in.}$$

Since the solid is 4 inches high, there will be 4 layers. Then the volume of the entire solid will be

$$4 \times 15 \text{ cu. in.} = 60 \text{ cu. in.}$$

The advantage of using a cubic unit to measure volume is that we can measure the *length*, the *width*, and the *height* in *linear units*. When we multiply these *linear units* together, we get the number of corresponding *cubic units* of volume. The rule may be stated as

$$V = lwh$$

The formula for the volume of a rectangular solid, $V = lwh$, may also be used in reverse. If we know the volume and two dimensions of such a solid, we can find the unknown dimension by using the formula in *reverse*. We can use the following formulas involving *division* to find l, w, or h.

$$l = \frac{V}{wh} \qquad w = \frac{V}{lh} \qquad h = \frac{V}{lw}$$

For example, if a rectangular solid is 4 inches wide and 3 inches high, and has a volume of 96 cubic inches, we can find the length by using the formula

$$l = \frac{V}{wh}$$

then

$$l = \frac{96}{12} = 8 \text{ (length, in inches)}$$

As another formula for the volume of a prism, consider the following type of problem. Suppose we know that the base of a prism contains 48 square inches. Even if we do not know the length and width, we know that since the area of the base is 48 square inches, then 48 cubic inches can be placed in one layer. Now, if we know that the height is 5 inches, we can find the volume. We multiply the area of the base by the height, and get

$$5 \times 48 \text{ cu. in.} = 240 \text{ cu in.}$$

This example illustrates another useful formula for the volume of a prism. If we use B for the area of the base (in square units), we have the formula

$$V = Bh$$

The formula can be used to find the volume of any prism if we know the area of the base and the altitude (height).

Example 1. A triangular prism, 10 inches high, has bases that are triangles. Each triangular base has a side equal to 8 inches, and the altitude of the triangle is 5 inches (Fig. 10.5). Find the volume of the prism.

Solution. For the area of the base, we use the formula for the area of a triangle: $A = (\frac{1}{2})ab$. Then the area of the base becomes $A = (\frac{1}{2})(5)(8) = 20$. This means the area of the base is 20 sq. in. For the volume, we have

$$V = Bh = (20)(10) = 200 \text{ (cu. in. in volume)}$$

The formula, $V = Bh$, for the volume of a prism, can also be used for the volume of an oblique prism, but then the altitude must be measured perpendicular to the bases.

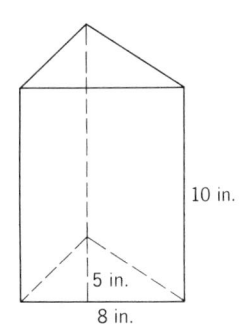

FIGURE 10.5

10.5. LATERAL AREA OF A PRISM

The lateral area of a prism is found by computing the area of each side, or lateral face, and then adding these areas. If the solid is a right prism, the

PRISMS

lateral area can be found by multiplying the perimeter by the height of the prism. For the total area of a prism, we add the areas of the two bases to the lateral area.

In the case of a rectangular solid, such as a rectangular box, which has six rectangular faces, probably the best way to get the total area is to find separately the area of the two sides, the two ends, and the top and bottom, and then add these areas together. The total area can also be found by adding the area of one side, one end, and the bottom, and then multiplying by 2.

10.6. THE CUBE

A *cube* is a rectangular solid whose length, width, and height are equal, and whose faces are squares (Fig. 10.6). If we use e to represent the length of one edge, then the formula for the volume of a rectangular solid, $V = lwh$, becomes

$$V = (e)(e)(e); \quad \text{or} \quad V = e^3$$

Each face of the cube is a square whose side is equal to the length of an edge of the cube, e. Then the area of each face is e^2. For the entire surface of the six faces of a cube, we have the formula

$$A = 6e^2$$

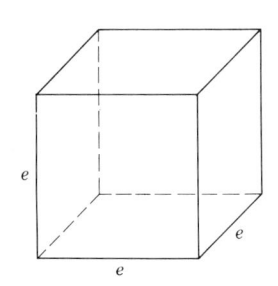

FIGURE 10.6

Example 2. A rectangular box is 36 inches long, 24 inches wide, and 15 inches deep. Find the volume in cubic feet. Find the area of the 4 sides and the top and bottom (square feet). (1 cu. ft = 1728 cu. in: 1 sq. ft = 144 sq. in.)

Solution. For the volume we have the formula: $V = lwh$. Then

$$V = (36)(24)(15) = 12960 \text{ (cu. in.)}$$

To change cubic inches to cubic feet, we divide by 1728. Then we get

$$12960 \div 1728 = 7.5 \text{ (cu. ft. in volume)}$$

For the lateral area, we have 2 sides, 2 ends, and top and bottom.
For 2 sides, we have

$$(2)(36)(15) = 1080 \text{ (sq. in.)}$$

For 2 ends, we have

$$(2)(24)(15) = 720 \text{ (sq. in.)}$$

For top and bottom,

$$(2)(36)(24) = 1728 \text{ (sq. in.)}$$

For total area, we have

$$1080 + 720 + 1728 = 3528 \text{ (sq. in.)}$$

To change square inches to square feet, we divide by 144, and get

$$3528 \div 144 = 24.5 \text{ (sq. ft in total surface area)}$$

Example 3. A rectangular solid is 12 inches long and 8 inches wide. If the volume is 624 cu. in. find the height and the total surface area.

Solution. The area of the base is

$$(l)(w) = (12)(8) = 96 \text{ (sq. in.)}$$

To find the height, we divide the volume by the area of the base. Then,

$$624 \div 96 = 6.5 \text{ (in. in height)}$$

Now we find the total surface area:

2 sides:

$$(2)(6.5)(12) = 156$$

2 ends:

$$(2)(6.5)(8) = 104$$

Top and bottom:

$$(2)(12)(8) = 192$$

Total surface area:

$$156 + 104 + 192 = 452 \text{ (sq. in. in surface area)}$$

Example 4. A rectangular solid has a height of 7.5 inches and a square base. If the volume is 1080 cubic inches, find the side of the square base.

PRISMS 163

Solution. To find the area of the base, we divide the volume by the height and get

$$1080 \div 7.5 = 144 \text{ (sq. in. in area of base)}$$

To get the side of the square base, we take the square root of 144:

$$\sqrt{144} = 12 \text{ (in. in the side of the square base)}$$

Example 5. Find the surface of a cube whose volume is 552 cu. in.

Solution. For the volume of a cube, we have the formula: $V = e^3$. To find the edge of the cube we must find the *cube root* of 552. Then, using a table, we find

$$\sqrt[3]{552} = 8.2 \text{ (approx.)}$$

For the lateral area we have the formula: $A = 6e^2$. The area of one face of the cube is the square of 8.2:

$$(8.2)^2 = 67.24 \text{ (sq. in.)}$$

For the 6 faces, we have

$$(6)(67.24) = 403.44 \text{ (sq. in.)}$$

Example 6. The total surface of a cube is 288 sq. in. Find the volume.

Solution. The 288 sq. in. represents the area of the six faces. Then the area of one face of the cube is

$$288 \div 6 = 48 \text{ (sq. in.)}$$

To find the edge of the cube, we take the square root of 48. This can be found by the short method:

$$\sqrt{48} = (\sqrt{16})(\sqrt{3}) = 4\sqrt{3} \text{ (in. in edge)}$$

Now we recall that the square root of 3 is approximately 1.732. Then

$$\sqrt{48} = 4(1.732) = 6.928 \text{ (in. in the edge of the cube)}$$

For the volume of the cube, we use the formula: $V = e^3$. Then

$$V = (4\sqrt{3})^3 = 332.5 \text{ (cu. in., approx.)}$$

10.7. VOLUME OF IRREGULAR SOLIDS

The volume of an *irregular* solid can be found by immersing the solid in a liquid and noting the rise of the liquid in the container. Of course, this cannot be done if the solid is soluble in the liquid or if the object is porous.

Exercise 10.1

1. A cardboard packing box is 20 in. long, $14\frac{1}{2}$ in. wide, and 10 in. deep. Find the number of cubic inches in the volume. Find the area of the 4 sides and the bottom.
2. A box is 14 in. long, $5\frac{1}{2}$ in. wide, and $4\frac{1}{2}$ in. deep. Find the volume in cubic inches, and the total surface area.
3. A brick is 8 in. long, 4 in. wide, and 2.5 in. thick. How many cubic inches does it contain? What is its total surface area?
4. A trash bin is a cube, 6.5 feet on a side. Find its volume and the total surface area.
5. A basement is 42 feet long, 28 feet wide, and 7 feet high. How many cubic yards does it contain? (1 cu. yd = 27 cu. ft.) Find the area of the four walls and the floor of the basement.
6. A room is 18 feet long, 12 feet wide, and 9.5 feet high. What is the volume of the room in cubic yards? Find the number of square yards in the four walls and the ceiling. (1 sq. yd = 9 sq. ft.)
7. A room is 18 feet long, 15 feet wide, and 9 feet high. Find the number of cubic yards in the room. What is the cost of painting the four walls and ceiling at 50 cents a square yard?
8. An aquarium is 30 in. long, 16.5 in. wide, and 14 in. high. How many gallons of water are required to fill the aquarium up to 2 inches from the top? Find the number of square inches of glass in the four walls of the aquarium. (1 gallon is equivalent to 231 cu. in.; then 1 cu. ft holds approximately 7.5 gal.)
9. Find the number of cubic feet of concrete in a driveway 62 feet long, 15 feet wide, and 6 inches thick. How many cubic yards of concrete are required to build the driveway?
10. Find the number of cubic yards of concrete required to build $\frac{1}{2}$ mile of road, 30 feet wide and 8 inches thick. How many truck loads will be needed if a truck load is 6 cubic yards?
11. The base of a rectangular prism contains 37 sq. in. The altitude of the prism is 7.5 in. Find the volume of the prism.
12. The base of a triangular prism contains 13.5 sq. ft. The altitude of the prism is 9 feet. Find the volume of the prism.
13. A school room is 36 feet long, 27 feet wide, and 14.5 feet high to the ceiling. If the usual number of people in the room is 40, what is the number of cubic feet of air per person?
14. A rectangular packing box is 18.5 in. long, 12 in. wide, and 8.5 in. deep. Find the volume and the total surface area. How long a string is needed to tie the box with one strand around the box in each of three dimensions?
15. A bar of iron is 18 feet long. The cross section of the bar is a square that

measures 4 inches on a side. Find the number of cubic inches and the number of cubic feet of iron in the bar. (1 cu. ft. = 1728 cu. in.) If the specific gravity of iron is 6.8, find the weight of the bar. (Water weighs 62.4 pounds per cu. ft. Specific gravity of 6.8 means that the iron weighs 6.8 times as much as water.)

16. Could you lift a block of gold, 10 in. long, 6 in. wide, and 3.5 in. thick? (The specific gravity of gold is 19.3.). What is the block of gold worth at a price of $150 an ounce (take 1 pound = 16 ounces)?

17. A triangular prism has an altitude of 16 inches. The base is a triangle having one side equal to 9 inches, and the altitude of the triangular base is 5 inches. Find the volume of the prism.

18. A prism has a base in the shape of a trapezoid whose area is 22.5 sq. in. If the altitude of the prism is 3 feet, find the volume.

19. A liquid container has the shape of a cube, 14 inches in each dimension. How many gallons does it hold? Find the total surface area. (1 gal. = 231 cu. in.)

20. A rectangular metal can is 6 in. long, 4.5 inches wide, and 9 in. high. Does it hold 1 gallon?

21. A ditch is 300 feet long. It is 8 feet wide at the top, 4 feet wide at the bottom, and 6 feet deep. How many cubic yards of dirt were taken out in digging the ditch? (The cross section has the shape of a trapezoid.)

22. A swimming pool is 24 feet wide and 60 feet long. It is 3 feet deep at one end and 9 feet deep at the other end. How many gallons of water does it hold? How long will it take to fill the pool if the water runs into the pool at a rate of 4 gallons per second? (A trapezoid is involved.)

23. A rectangular box is 12 in. long, 8 in. wide, and has a volume of 528 cubic inches. Find the altitude.

24. A rectangular box has a volume of 441 cu. in. It is 7 inches wide and 6 inches high. Find the length.

25. A rectangular solid is 18 in. long, 10 in. high, and contains 2430 cu. in. Find the width.

26. A rectangular solid contains 336 cu. in. It is 5.25 in. high. If the base is a square, find the size of the edge of the square.

27. A cube contains 343 cu. in. Find the length of the edge and surface area.

28. A cube contains 422 cu. in. Find the length of an edge and surface area.

29. A cube has a surface area of 150 sq. in. Find the volume of the cube.

30. A cube has a surface area of 432 sq. in. Find the volume of the cube.

31. The total surface area of a rectangular solid is 108 sq. in. If the length is 6 in. and the width 4 in., find the height and the volume.

32. The total surface area of a rectangular solid is 304 sq. in. The altitude is 5.5 in., and the base is square. Find the volume.

33. The total area of a rectangular solid is 384 sq. in. What is the volume if the base is a square 8 in. long and 8 in. wide? What is the volume if the base is 12 in. long and 4 in. wide. (Total area is still 384 sq. in.)

34. The volume of a rectangular solid is 216 cu. in. What is its surface area if it is a cube? What is the surface area if the base is 9 in. long and 6 in. wide?

35. What is the surface area (Exercise 34) if the base is 12 in. long and 6 in. wide? What is the surface area (Exercise 34) if the base is 12 in. long and 9 in. wide?

11 CYLINDERS

11.1. DEFINITION

A *cylinder* is a geometric solid bounded by a closed *cylindrical surface* and by portions of two parallel planes (Fig. 11.1). Technically, a cylindrical surface is a curved surface formed by a moving line that moves in such a way that it is always parallel to another fixed straight line. The ends of the cylinder are called the *bases*. They are the plane surface boundaries. If the bases are circles, the cylinder is called a *circular cylinder*. A cylinder need not be circular. The bases may be any shape, such as an ellipse or some other curve. However, according to the definition, both bases of the cylinder are exactly the same size and shape.

The most common kind of cylinder is a *right circular cylinder*, such as a tin can. The word *right* refers to *perpendicularity* (Fig. 11.1a). In a right circular cylinder, the straight line connecting the centers of the bases is *perpendicular to the bases*. Otherwise, the cylinder is called an *oblique* cylinder (Fig. 11.1b). The *radius r* of a circular cylinder is the radius of one of the circular ends. The *altitude*, or height h, is the perpendicular distance between the bases of the cylinder.

Right circular cylinders are common in our everyday life. They have many sizes and uses. Much of our canned food is put up in cylindrical cans. Storage tanks often have the shape of a right circular cylinder. We see cylinders of all sizes from the very small, such as the shaft of a small wheel in a wrist watch, to the very large, such as a large storage gas tank, perhaps 100 feet in diameter. Some cylinders, such as a coin, have a very small altitude compared with the diameter. Others, such as a round wire, have a very small diameter compared with the altitude (Fig. 11.2).

11.2. VOLUME OF A CYLINDER

To find the volume of a cylinder, we use a formula similar to that for a prism. We have used the formula, $V = Bh$, for the volume of a prism. Now,

168 GEOMETRY

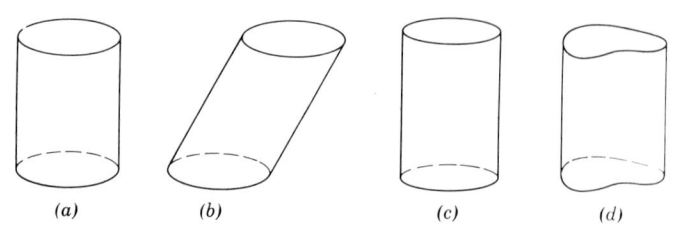

(a) (b) (c) (d)

FIGURE 11.1

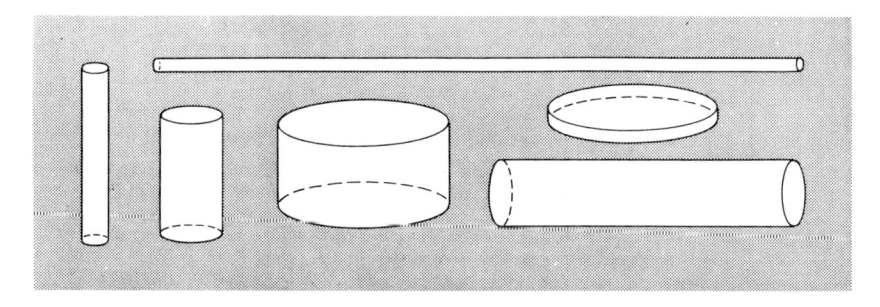

FIGURE 11.2. Circular cylinders.

if the base for a cylinder contains 20 square inches, then we can reason that 20 cubic inches can be placed in one layer on the bottom of the cylinder. If the altitude of the cylinder is 8 inches, then, for the volume, we multiply the area of the base by the altitude (height h). That is

$$V = Bh$$

$$V = 20 \times 8 = 160 \text{ (cu. in. in volume)}$$

That is, if we can find the area of the base of any cylinder, we can multiply the base by the altitude to get the volume.

In a circular cylinder, the base is a circle (Fig. 11.3). Then, for the area of the base, we use the formula for the area of a circle, $A = \pi r^2$. We have said that the volume can be found by the formula, $V = Bh$. Now, if we use the value, πr^2, for the area of the base, we have the formula for the volume of a circular cylinder:

$$V = \pi r^2 h$$

This formula can be used for the volume of all circular cylinders. It also applies to circular cylinders that are oblique, if we remember to take the altitude h as the perpendicular distance between the bases.

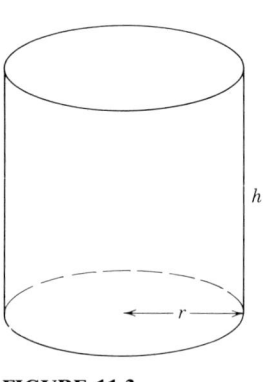

FIGURE 11.3

CYLINDERS

Example 1. A circular cylinder has a diameter of 6 in. and altitude of 8 in. Find the volume. Does it hold more or less than 1 gallon? (1 gal. holds 231 cu. in.)

Solution. We use the formula for the volume of a cylinder: $V = \pi r^2 h$ ($r = 3$). Then

$$V = (3.1416)(3^2)(8) = 226.2 \text{ (cu. in., approx.)}$$

Since 1 gallon holds 231 cu. in., the cylinder holds a little less than 1 gallon.

Example 2. A circular cylinder has a diameter of 2 feet and an altitude of 30 in. How many gallons does it hold? (The radius is 12 inches.)

Solution. We shall work the problem in two ways. First, we find the number of cubic inches.

$$V = (3.1416)(12^2)(30) = 13{,}571.7 \text{ (cu. in., approx.)}$$

To find the number of gallons, we divide by 231, and get

$$13{,}571.7 \div 231 = 58.75 \text{ (gallons, approx.)}$$

As a second method, we can express the entire problem as a fraction, with the numerator containing the numbers to be multiplied together, and the denominator containing the divisor or divisors. Then the fraction can often be reduced very easily:

$$\frac{3.1416 \times 144 \times 30}{231} = 58.752$$

Now the numerator and denominator can be divided by 3, 7, and 11. In fact, the value, 3.1416, can be divided by 231 itself. We get the answer: 58.75.

Example 3. Find the number of cubic feet of copper in a wire $\frac{3}{8}$ inch in diameter and 1 mile long. Find the weight if specific gravity of copper is 8.8.

Solution. We change all measurements to inches. 1 mile $= (5280)(12)$ in. Now we express the problem in the form of a fraction, in the formula: $V = \pi r^2 h$. For the number of cubic feet, we have ($r = \frac{3}{16}$)

$$\frac{3.1416 \times 3 \times 3 \times 5280 \times 12}{16 \times 16 \times 1728} = 4.05 \text{ (cu. ft., approx.)}$$

170 GEOMETRY

To find the weight, we multiply by the weight of water, 62.4 pounds per cu. ft. and then by 8.8, the specific gravity of copper:

$$4.05 \times 62.4 \times 8.8 = 2224 \text{ (lb, approx.)}$$

The entire problem could be set up as follows:

$$\frac{3.1416 \times 3 \times 3 \times 5280 \times 12 \times 62.4 \times 8.8}{16 \times 16 \times 1728} = 2224 \text{ (approx.)}$$

Example 4. The volume of a circular cylinder is approximately 5655 cubic inches. If the radius is 7.5 inches, find the altitude.

Solution. For the volume, we have the formula: $V = \pi r^2 h$. Now we use the formula in reverse. We first find the value of the expression, πr^2. We get

$$\pi r^2 = 3.1416 \times (7.5)^2 = 3.1416 \times 56.25 = 176.715$$

Since the entire volume is equal to 5655 cu. in., we divide the volume by the value, πr^2, or 176.715, and get

$$5655 \div 176.715 = 32 \text{ (inches in height, approx.)}$$

Example 5. The volume of a circular cylinder is 1215π cubic inches. If the altitude is 15 inches, find the radius of the cylinder.

Solution. In this problem we know the value of $\pi r^2 h$ is 1215π. Now, if we divide the volume by πh, we shall have the result, r^2. That is,

$$1215\pi \div 15\pi = 81 \qquad \text{(81 is the square of the radius)}$$

For the radius, we take the square root of 81: $\sqrt{81} = 9$ (inches in radius).

Exercise 11.1

1. A hot-water tank has a diameter of 14 inches and is 5 feet high. The shape is a circular cylinder. How many gallons will it hold?
2. An oil drum (a circular cylinder) has a diameter of 2 feet and an altitude of 40 inches. How many gallons will it hold?
3. An oil tank has a diameter of 15.6 feet and an altitude of 24 feet. How many gallons will it hold? (1 cu. ft. will hold 7.5 gallons.)
4. An oil tank-car (railroad) is 42 feet long and has a diameter of 7.4 feet. How many gallons will it hold?

5. A cylindrical oil can has a diameter of 10 in. and an altitude of 15 in. Does it hold 5 gallons?
6. A copper wire is 2 miles long and $\frac{1}{4}$ inch in diameter. Find the weight of the wire. (Water weighs 62.4 pounds per cu. ft.; specific gravity of copper is 8.8.)
7. A 5-gallon can has a radius of 4.8 inches. Find the height for 5 gallons.
8. A rectangular gasoline tank on a car is 42 inches long, 13 inches wide, and 7 inches deep. On another car the tank is a circular cylinder, 38 in. long and has a diameter of 9 inches. Which tank holds more and how much?
9. A circular cylinder has a diameter of 21.6 in. and an altitude of 33 in. How many gallons will it hold?
10. A circular cylinder is 42 inches in altitude and has a volume of 2672 cu. in. Find the approximate length of the radius.
11. Find the total piston displacement (in cubic inches) in an 8-cylinder engine if each cylinder has a diameter of 3 inches and a stroke of 3.75 in.
12. A cylindrical can has a diameter of 1.75 inches. What must be the altitude of the cylinder if it is to hold one quart? (1 gal. holds 231 cu. in.)
13. A cylindrical can has a diameter of 18.4 cm, and a height of 32 cm. How many liters does it hold? (1 liter equals 1000 cu. cm.)
14. A cylindrical container is 28.3 inches high. What must be the radius if it is to hold 60 gallons?

11.3. LATERAL AREA OF A CYLINDER

It is often necessary to compute the *lateral area* of a cylinder. By *lateral area*, we mean the area of the curved surface. We may wish to paint the outside surface of a cylindrical gasoline storage tank, to find the amount of heating surface of a steam pipe, or to find the number of square inches of metal required in the manufacture of cylindrical tin cans.

To arrive at a formula for the lateral area, consider the cylinder in Figure 11.4. Let us assume that the lateral area is cut along one side and spread out flat like a sheet of paper. Then the lateral area has the shape of a rectangle. The length of the rectangle is the circumference C of the cylinder. The width of the rectangle is the altitude h of the cylinder. We know that the

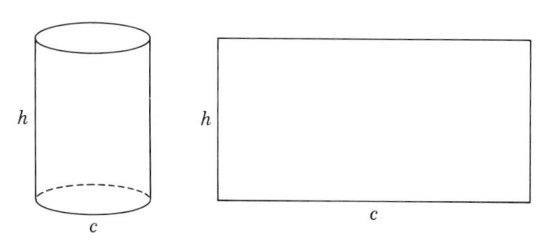

FIGURE 11.4

area of a rectangle is given by the formula

$$A = (\text{length}) \times (\text{width})$$

The lateral area (LA) of the cylinder now becomes the area of the rectangle. If we substitute the measurements of the cylinder, we get the formula

$$LA = (\text{circumference}) \times (\text{altitude}); \quad \text{or} \quad LA = 2\pi rh$$

The formula can be used to find the lateral area of any right circular cylinder, large or small. Notice that in this formula we have the *product* of *two linear measurements*. Then the answer is in *square units*.

For the *total* outside surface of a cylinder, it is also necessary to consider the area of the two ends. Each end is a circle, and its area can be found by the formula: $A = \pi r^2$. However, if the diameter is very small compared with the length, as in a wire, then the area of the ends is so small as to be negligible.

11.4. HOLLOW CYLINDER

By the volume of a hollow cylinder, we mean the volume of the solid shell between two concentric cylinders; that is, cylinders with the same axis but different radii (Fig. 11.5). As an example, we may wish to find the amount of iron in a hollow iron pipe. The problem is similar to that of finding the area of a ring.

To find the volume of a hollow cylinder, or shell, we could compute the volume of the large cylinder, then the volume of the small inner cylinder (or the hollow), and then find the difference between the two volumes. The difference will be the volume of the material in the shell.

Let us call the radius of the larger, or outer, cylinder R. Then we use small r to represent the radius of the smaller, or inner, cylinder. The altitude h is the same for both cylinders. Now, we use V_L to represent the volume of the large cylinder, and V_S to represent the volume of the small, or inner, cylinder (that is, the hollow). Then we have the formulas for the volumes of the outer (large) cylinder and the small (inner) cylinder:

$$V_L = \pi R^2 h \quad \text{and} \quad V_S = \pi r^2 h$$

The volume of the shell is the difference between the two:

$$V(\text{of shell}) = V_L - V_S = \pi R^2 h - \pi r^2 h$$

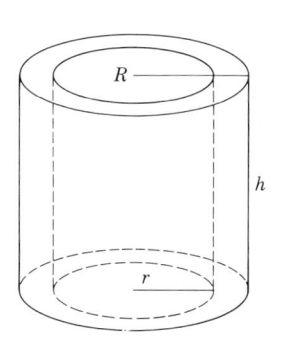

FIGURE 11.5

The formula may be written

$$V(\text{of shell}) = \pi(R^2 - r^2)h, \quad \text{or} \quad V = \pi h(R^2 - r^2)$$

Note that we first find the squares of the radii, and then subtract, just as we do in finding the area of a ring.

Example 6. A metal can in the shape of a circular cylinder is 16 in. high, and has a diameter of 12 in. Find the volume and the total surface area.

Solution. For the volume we have the formula: $V = \pi r^2 h$; $r = 6$ in. Then,

$$V = (3.1416)(36)(16) = 1809.56 \text{ (cu. in., approx.)}$$

For the lateral area, we have the formula: $LA = 2\pi rh$. Then,

$$LA = (2)(3.1416)(6)(16) = 603.19 \text{ (sq. in., approx.)}$$

For the area of the top and bottom, each a circle, we have the formula: $A = \pi r^2$. For the two circles,

$$A = (2)(3.1416)(36) = 226.20 \text{ (sq. in., approx.)}$$

Total surface area:

$$603.19 + 226.20 = 829.39 \text{ (sq. in., approx.)}$$

The answers can be rounded off to: $V = 603.2$ cu. in.; $A = 829.4$ sq. in.

Example 7. Find the number of cubic yards of concrete in a concrete culvert, a hollow cylinder, 30 feet long, with inside diameter 24 inches, if the concrete is 6 inches thick. (27 cu. ft. = 1 cu. yd.)

Solution. In this problem, it is more convenient to express all measurements in feet. Inside diameter = 2 feet; outside diameter = 3 feet. For the radii, we have

$$R = 1.5; \quad r = 1.$$

Now, we could find the volume of the culvert as if it were solid, and then find the volume of the inside hollow, and subtract. However, we can apply the formula for the volume of a hollow cylinder: $V(\text{of shell}) = \pi(R^2 - r^2)h$. Then,

$$V = (3.1416)(2.25 - 1)(30) = 117.81 \text{ (cu. ft., approx.)}$$

174 GEOMETRY

To change cubic feet to cubic yards, we divide by 27, and get

$$117.81 \div 27 = 4.36 \text{ (cu. yd., approx.)}$$

Example 8. Find the volume of an iron shell 5 feet long if the inside diameter is 8 inches and the shell itself is 2 inches thick. How much does it weigh if the specific gravity of the iron is 6.8? (Water weighs 62.4 pounds per cubic foot; specific gravity of 6.8 means 6.8 times as heavy as water.)

Solution. The radius of the inner, or small, cylinder is 4 inches. Since the metal is 2 inches thick, the radius of the large outer cylinder is 6 inches. The height or altitude is 5 feet (60 inches), for both cylinders. Now we use the formula for the volume of the shell (all measurements in inches):

$$V = \pi h (R^2 - r^2) = (3.1416)(60)(6^2 - 4^2)$$

$$= (3.1416)(60)(36 - 16) = (3.1416)(60)(20) = 3769.9$$

The answer, 3769.9, represents the number of cubic inches. To find the weight, we change the volume to cubic feet by dividing by 1728, and then multiply by the weight of water and the specific gravity of iron:

$$\frac{3769.9}{1728}(62.4)(6.8) = 925.7 \text{ (approx. number of pounds)}$$

Exercise 11.2

1. Find the total surface area (lateral area and area of both ends) of a tin can having an altitude of 10 inches and a diameter of 8 inches.
2. How many square feet of asbestos will be needed to wrap a hot-water furnace that is 5 feet high and 42 inches in diameter? (Lateral area and top.)
3. A blacktop roller is 4 feet in diameter and 12 feet long. How many rotations will the roller make in rolling a blacktop strip of road that is 32 feet wide and 1 mile long?
4. A steam boiler of a steam engine has 15 flues, each 2 inches in diameter and 12 feet long. Find the total number of square inches of heating surface of the 15 flues.
5. A large cylindrical gasoline tank is 12 feet long and holds 2545 gallons. One cubic foot holds 7.5 gallons (approximately). Find the number of cubic feet in the volume of the tank. Then find the radius of the tank and the total surface area, including lateral surface and both ends.
6. A concrete culvert is 24 feet long. The inside diameter is 18 inches, and the concrete itself is 6 inches thick. Find the number of cubic yards of concrete in the culvert.

CYLINDERS

7. A hollow iron pipe is 20 feet long and has an outside diameter of 4 inches. The iron itself is $\frac{1}{4}$ inch thick. Find the number of cubic inches and the number of cubic feet of iron in the pipe.

8. An iron pipe has an inside diameter of 1 inch, and the metal is $\frac{1}{8}$ inch thick. Find the number of cubic inches and the number of cubic feet in 100 feet of the pipe.

9. A concrete culvert contains 39.77 cubic feet of concrete. The outside diameter is 32 inches and the inside diameter is 20 inches. Find the length of the culvert. If the inside and the outside is to be painted with some water-proofing paint, how many square feet would that be? (Also include both ends.)

10. A metal cylindrical container, labeled "1 gallon," has a diameter of 5 in. and an altitude of 12 in. Another metal cylindrical container is also labeled "1 gallon." The second container has a diameter of 6.8 in. and an altitude of 6.5 in. Do they hold the same amount? What is the lateral area and the area of both ends of each container? Which has the greater total surface area?

11. A manufacturer makes some quart cans with a diameter of 4 in. and another size of can having a diameter of 3 in. Which size of can requires more metal for the total area? How many more square feet of metal are required for 10,000 cans of one size than for the other?

12. If 10 gallons of hot water are placed in a rectangular container that measures 11 in. by 14 in. by 15 in., how much is the total radiating surface? How much surface is exposed for radiation of heat if the water is placed in a cylindrical container 3 inches in diameter?

13. A rectangular block of copper is 4 ft. long and has a square cross section that measures 6 in. on a side. What is the volume in cubic feet? What is the surface area of the block? Surface area is important when exposure to air is considered. What would be the surface area exposed to the air if the block of copper is formed into a solid cube 1 foot on the edge? What would be the surface area if the block of copper were drawn into a wire $\frac{1}{4}$ inch in diameter?

12 PYRAMIDS AND CONES

12.1. THE PYRAMID

A *pyramid* is a special kind of polyhedron. The *base* of a pyramid is a polygon of some kind. It may be a triangle, a quadrilateral, or some other kind of polygon. The sides of the pyramid are called the *lateral faces*. The faces come together at a point called the *apex* of the pyramid (Fig. 12.1). The altitude h of a pyramid is the perpendicular distance from the apex to the base.

If the base of a pyramid is a regular polygon with all sides equal and all angles equal, and if the apex is directly opposite the center of the base, then the pyramid is called a *regular pyramid* (Fig. 12.2). In a regular pyramid, the line segment from the apex to the center of the base represents the *altitude* of the pyramid. All the lateral faces of a regular pyramid are *isosceles triangles*, such as the triangle ABC. The altitude of one of these triangles, FA, is called the *slant height* of the pyramid. The intersection, such as DA, of two lateral faces is called a *lateral edge* of the pyramid. The intersection, BC, of a lateral face with the base of the pyramid is called the *base edge* of the pyramid.

12.2. VOLUME OF A PYRAMID

To get an understanding of the volume of a pyramid, let us begin with a right prism, such as a rectangular solid (Fig. 12.3a). For the volume of the prism, we have the formula: $V = Bh$. Now, if we cut away part of the prism on each side at a slant from a point in the top down to each of the lower edges, as shown in Figure 12.3b, we have a pyramid left. Then our question is: how much of the prism has been cut away and how much remains? The remaining pyramid is shown in Figure 12.3c.

The pyramid (c) has the same base and the same altitude as the corresponding prism (a). It has been found and it can be proved that the volume

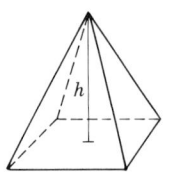

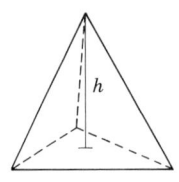

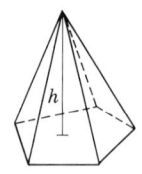

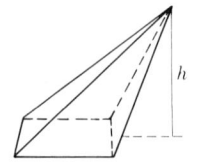

FIGURE 12.1. Pyramids.

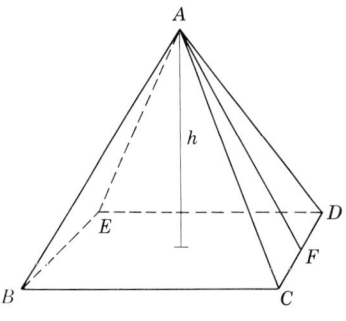

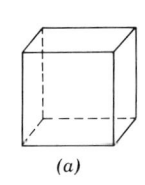

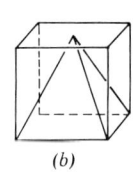

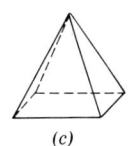

FIGURE 12.2

FIGURE 12.3

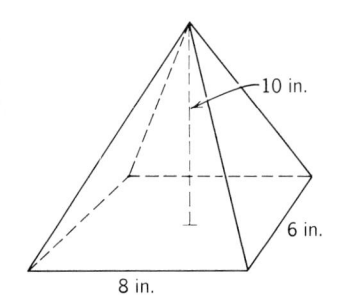

FIGURE 12.4

of the pyramid is exactly one-third as much as the volume of the corresponding prism. That is, the volume is equal to one-third of the product of the altitude h and the area of the base B. As a formula we have, for the volume of a pyramid,

$$V = \tfrac{1}{3}BH \quad \text{or} \quad V = \frac{Bh}{3}$$

Example 1. A pyramid has an altitude of 10 inches and a rectangular base that is 8 inches long and 6 inches wide (Fig. 12.4). Find the volume of the pyramid.

Solution. For the volume we have the formula

$$V = \tfrac{1}{3}Bh$$

The area of the base is

$$(6) \times (8) = 48 \text{ (sq. in.)}$$

Then

$$V = (\tfrac{1}{3})(48)(10) = 160 \text{ (cu in.)}$$

12.3. LATERAL AREA OF A REGULAR PYRAMID

It is sometimes necessary to find the area of the lateral faces of a pyramid. In a regular pyramid all the lateral faces are congruent isosceles triangles.

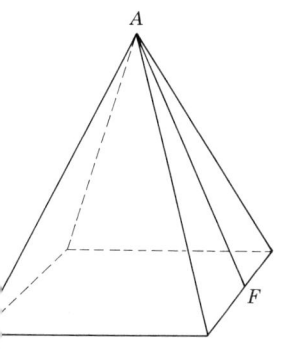

PYRAMIDS AND CONES 179

Therefore, we first find the area of one of the faces and then multiply by the number of faces. Each face, an isosceles triangle, has a base equal to the base edge of the pyramid (Fig. 12.5). The altitude of each of the triangles is the slant height of the pyramid, such as the line FA. We can find the area of a lateral face by taking one-half the product of the slant height and the base edge. Here we simply use the formula for the area of a triangle. Then we find the total lateral area by multiplying by the number of sides.

However, since all the lateral-face triangles have the same altitude (the slant height), we can first find the sum of the bases, which is equal to the perimeter of the pyramid base. Then we can find the total lateral area by the following multiplication:

$$\text{total lateral area} = \tfrac{1}{2}(\text{perimeter of base})(\text{slant height})$$

or, as a formula,

$$LA = \tfrac{1}{2}ps$$

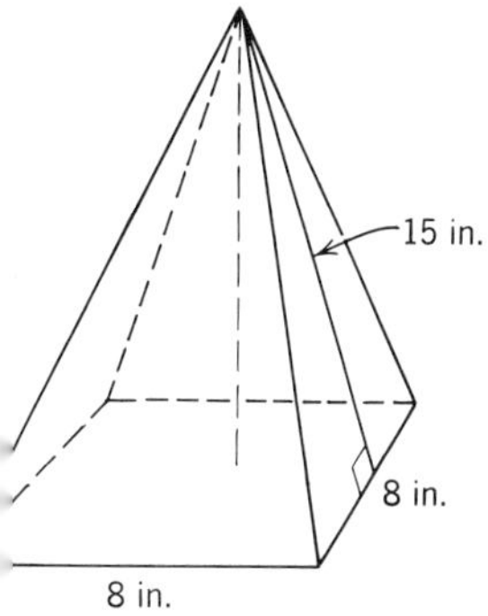

Example 2. A regular pyramid has a square base with each base edge equal to 8 inches. The slant height of the pyramid is 15 inches. Find the total surface area of the pyramid including lateral faces and the area of the base itself (Fig. 12.6).

Solution. Each side is a triangle with base equal to 8 inches and altitude equal to 15 in. Then the area of each lateral face is $(\tfrac{1}{2})(8)(15) = 60$ (sq. in.). The total area of the 4 lateral faces is 240 sq. in. The area of the base of the pyramid is $(8)(8) = 64$ (sq. in.) Then the total area of the four faces and the base of the pyramid is $240 + 64 = 304$ (sq. in.)

12.4. FRUSTUM OF A PYRAMID

The *frustum of a pyramid* is the part of a pyramid that remains when the top is cut off by a plane parallel to the base (Fig. 12.7). Such a geometric solid may appear, for example, as a box containing a loudspeaker. The top of the frustum is a polygon of exactly the same shape as the base but smaller in size. The top is sometimes called the *top base*. The two bases are often denoted by B and b, respectively.

The volume of a frustum of a pyramid can be found by the formula

$$V = \tfrac{1}{3}h(B + b + \sqrt{Bb})$$

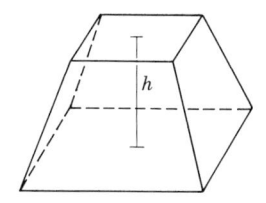

In the formula, h represents the altitude, which is the perpendicular distance between the bases; B represents the area of one base in square units; b

180 GEOMETRY

represents the area of the other base in square units; and V represents the volume in cubic units. Note that in the formula, we add the bases, $B+b$, and then add the square root of their product, $\sqrt{Bb}$. Finally we multiply this sum by $\frac{1}{3}$ of the altitude h.

Example 3. Find the volume of the frustum of a pyramid having a square bottom base, 15 inches on a side, and a square top base, 9 inches on a side. The altitude of the frustum is 8 inches (Fig. 12.8).

Solution. Since the bases are square, then $B = 15^2 = 225$; $b = 9^2 = 81$; $h = 8$; and $\sqrt{Bb} = \sqrt{(225)(81)} = 135$. Using the formula, we have

$$V = (\tfrac{1}{3})(8)(225 + 81 + 135) = 1176 \text{ (cu. in. in volume)}$$

To find the area of the lateral faces of a frustum of a pyramid, we use the formula for the area of a trapezoid, since all the faces are trapezoids. Instead of finding the area of each trapezoid separately, we can first consider the sum of the bases of the trapezoids. The bases of the trapezoids form the perimeters of the top base and the bottom base of the frustum. Recall that the area of a trapezoid is equal to one-half the altitude times the sum of the basis. For the frustum of a pyramid, we use P to represent the perimeter of one base, and p to represent the perimeter of the other base. We use these values in the formula for the area of a trapezoid. The altitude of each trapezoidal face is the slant height h of the frustum. Then, for the lateral area, we have the formula

$$LA = (\tfrac{1}{2})(s)(P + p)$$

Example 4. Find the total surface area of the frustum of a pyramid having a square bottom base 9 inches on a side, a top base 5 inches on a side, and a slant height of 7 inches (Fig. 12.9).

Solution. The square bottom base has a perimeter of 36 inches. The square top base has a perimeter of 20 inches. Using the formula, we have,

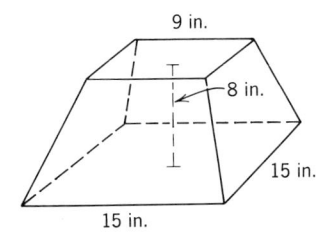

FIGURE 12.8

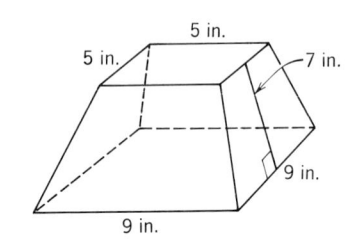

FIGURE 12.9

for the lateral area,

$$A = (\tfrac{1}{2})(7)(36+20) = 196 \text{ (sq. in.)}$$

Next, we find the area of each of the bases:

area of bottom base $= 9^2 = 81$ (sq. in.)   top base $= 5^2 = 25$ (sq. in.)

For the total surface area, we have

$$196 + 81 + 25 = 302 \text{ (sq. in.)}$$

Exercise 12.1

1. A rectangular prism and a pyramid each have square bases 12 in. on a side and each is 20 in. high. Find the volume of each.
2. A monument of granite has the shape of a regular pyramid having a base of 20 square feet and altitude of 30 feet. Find the weight of the pyramid if the granite weighs 170 pounds per cubic foot.
3. A pyramid has a rectangular base 32 in. by 24 in. The altitude is 15 ft. If the pyramid is made of solid wood that weighs 48 pounds per cu. ft., find the total weight of the pyramid.
4. A marble monument in the shape of a pyramid has a square base 18 in. on a side. The altitude is 10 feet. Find the weight of the monument if the specific gravity of the marble is 2.7 (that is, 2.7 times as heavy as water).
5. The base of a triangular pyramid is an equilateral triangle 12 in. on a side. The pyramid is 24 inches high. Find the volume of the pyramid.
6. The Great Pyramid of Egypt is approximately 480 feet high. The base is a square approximately 750 feet on a side. Find the volume of the pyramid in cubic yards.
7. The roof of a house has the shape of a regular square pyramid. A side of the house is 28 ft. and the slant height of the roof is 18 feet. Find the number of square yards in the entire roof (4 sides).
8. A tent has the shape of a regular square pyramid. A side of the tent measures 12 feet. The apex (peak) of the tent is 8 feet above the floor. Find the number of cubic feet of air in the tent; find the slant height and the number of square feet of canvas needed for the tent (walls and floor).
9. A regular square pyramid has a square base 6 in. on a side. The volume of the pyramid is 128 cu. in. Find the altitude, the slant height, and the total surface area (lateral area and base).
10. The volume of a regular square pyramid is 1526 cu. in. If the altitude is 18 in., find each side of the square base.
11. The frustum of a regular square pyramid has a square top base of 12 in., and a square bottom base of 24 in. If the altitude of the frustum is 18 in., find the volume of the frustum. Find the slant height and the total area including the four lateral surfaces and the two bases.

12. A metal container has the shape of the frustum of a regular square pyramid. The square bottom is 8 in. on a side, and the open top measures 12 in. on the side. The altitude of the container is 10 in. How many gallons of liquid will the frustum hold? How many square inches of metal are required to make the bottom and the sides of the container?

13. A box for a radio speaker has the shape of the frustum of a regular square pyramid. The front opening (one base) is 18 in. on the side. The back (the other base) is a 6-inch square. If the slant height is 11 in., find the lateral area of the box.

14. A waste basket has the shape of the frustum of a square pyramid. The bottom is 9 in. square, and the top is 12 in. square. If the slant height is 14 in., find the volume.

15. The main part of a monument is the frustum of a pyramid whose bottom base is a square 36 in. on a side. The top of the frustum is a square 12 in. on a side. The altitude of this portion of the monument is 16 feet. This frustum is surmounted by a pyramid whose altitude is 12 in. Find the weight of the monument if the granite of which it is made has a specific gravity of 2.75.

16. Imagine that you stand 8 feet from a wall on which a picture is hung flat against the wall. The picture is 24 in. high and 30 in. long (horizontally). As your eyes follow the perimeter once around the picture, what is the volume of the geometric solid that your line of sight encloses?

12.5. THE CONE

A *cone* is a geometric solid that has many applications in our everyday life as well as in science. We have conical tents, conical speakers in radios, ice-cream cones, and other conical containers.

A *cone* is a solid bounded by a plane forming the *base* and by a lateral *curved surface* that comes to a point called the *apex* of the cone. A cone might be thought of as a pyramid having an infinite number of lateral faces. Figure 12.10 shows several cones.

If the base of a cone is a circle, the cone is called a *circular cone* (Fig. 12.10a, b). The altitude of a cone is the length of a straight line segment, h, from the apex perpendicular to the base. In a *right circular cone* the base is a circle, and the altitude is measured to the center of the base. Figure 12.10a shows a right circular cone. If a right triangle is rotated around one of its legs, a right circular cone is generated.

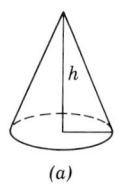

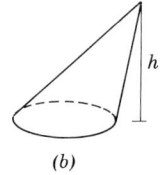

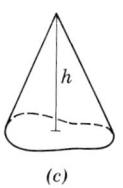

FIGURE 12.10

PYRAMIDS AND CONES 183

12.6. VOLUME OF A CONE

A cone has the same relation to a cylinder as a pyramid has to a prism. To get an understanding of the volume of a cone, let us begin with a right circular cylinder (Fig. 12.11a). For the volume of the cylinder, we have the formula: $V = \pi r^2 h$. Now, if we begin at the center of the top of the cylinder and cut away part of the cylinder sloping downward to the circumference of the base (Fig. 12.11b), the part that remains is a cone (Fig. 12.11c). Then our question is: how much of the cylinder remains?

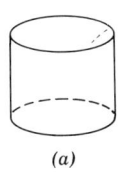

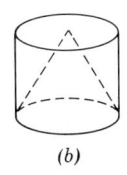

FIGURE 12.11

The cone has the same base and the same altitude as the corresponding cylinder (a). It has been found, and it can be proved, that the volume of any cone is exactly *one-third* as much as the volume of a corresponding cylinder. Therefore, the volume of the cone can be found by the formula

$$V \text{ (of cone)} = \tfrac{1}{3}\pi r^2 h, \quad \text{or} \quad V = \frac{\pi^2 h}{3}$$

Example 4. A right circular cone has an altitude of 20 inches. The diameter of the circular base is 1 foot. Find the volume of the cone. How many gallons will it hold? (1 gal. = 231 cu. in.) (Fig. 12.12).

Solution. For the volume we have the formula

$$V = \tfrac{1}{3}\pi r^2 h$$

Substituting, $r = 6$ in., $h = 20$ in. Then

$$V = \tfrac{1}{3}(3.1416)(36)(20) = 753.98$$

Then $v = 754$ cu. in. (approximately). To find the number of gallons, we divide by 231:

$$754 \div 231 = 3.26 \text{ gallons (approx.)}$$

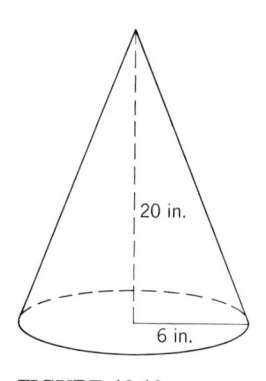

FIGURE 12.12

12.7. LATERAL AREA OF A CONE (*LA*)

The *lateral area* (the curved surface) of a right circular cone is found by the same formula as that used for a regular pyramid: $LA = \frac{1}{2}sp$. However, for a circular cone, the perimeter p is the circumference of the base of the cone, which is equal to $2\pi r$ (Fig. 12.13). Then the formula for the lateral area becomes

$$LA = \tfrac{1}{2}(\text{slant height})(\text{circumference})$$

The formula reduces to

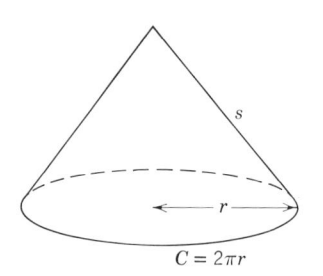

$$LA = \tfrac{1}{2}(s)(2\pi r) = \pi rs$$

FIGURE 12.13

Example 5. In Figure 12.13, suppose the cone has a slant height of 12 inches, and the diameter of the base is 16 inches. Find the total surface area of the cone including the area of the base.

Solution. $LA = \pi rs = (3.1416)(8)(12) = 301.6$ (sq. in. in lateral area).
For the area of the base, we have

$$A = \pi r^2 = (3.1416)(64) = 201.1 \text{ (area of base)}.$$

Total area = 301.6 + 201.1 = 502.7 (sq. in. in total area).

12.8. FRUSTUM OF A CONE

The frustum of a cone is the part that remains when the top of a cone is cut off by a plane parallel to the base (Fig. 12.14). Such a geometric form appears in many familiar objects, such as paper drinking cups, buckets, pails, dishes, lamp shades, megaphones, and other containers of many kinds.

The two plane boundaries of a frustum are called the *bases*. In the frustum of a right circular cone, one circular base is smaller than the other. The radius of one base may be denoted by R, and the radius of the other base by r. The altitude of the frustum is the perpendicular distance between the bases. The slant height s is the distance from one circumference to the other along the lateral surface.

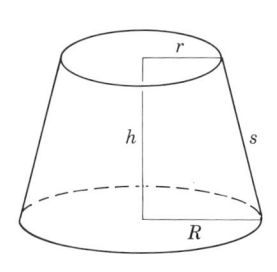

FIGURE 12.14

If we denote the area of one base by B and the area of the other base by b, then the formula for the volume is the same as that used for the frustum of a pyramid:

$$V = \tfrac{1}{3}(h)(B + b + \sqrt{Bb})$$

However, in the case of the frustum of a right circular cone, the bases are circles. Then the formula can be reduced to the form

$$V = \tfrac{1}{3}(\pi)(h)(R^2 + r^2 + Rr)$$

Example 6. In Figure 12.14, suppose the diameter of the top base is 10 in., and the diameter of the bottom base is 16 in.; suppose also that the altitude of the frustum is 7 inches. Find the volume of the frustum.

Solution. We first find the radius of the top base and the bottom base.

$$R = \tfrac{1}{2}(16) = 8 \text{ (in.)} \qquad r = \tfrac{1}{2}(10) = 5 \text{ (in.)}$$

Substituting in the formula, we have

$$V = \tfrac{1}{3}(3.1416)(7)(64 + 25 + 40) = 945.6 \text{ (cu. in. in volume)}$$

The lateral area of the frustum of a cone may be found in a way that is similar to the method used for finding the lateral area of the frustum of a pyramid. Recall that for the frustum of a pyramid, we find the lateral area by taking one-half the slant height times the sum of the perimeters; that is, for the lateral area of the frustum of a pyramid, we have the formula

$$LA = \tfrac{1}{2}(s)(P + p) \text{ (where } s = \text{slant height)}$$

However, in the case of the frustum of a right circular cone, the perimeters are now the circumferences of the circular bases. Then, for the perimeters we use the formula for the circumference of a circle: $C = 2\pi r$. Then, for the lateral area of the frustum of a circular cone, we have the formula

$$LA = \tfrac{1}{2}(s)(2\pi R + 2\pi r)$$

which reduces to

$$LA = \pi s(R + r)$$

Exercise 12.2

1. A pile of sand in the shape of a right circular cone has a base whose diameter is 8.8 feet. The altitude of the cone is 4.2 ft. How many cubic yards of sand are there in the pile?
2. An ice-cream cone has a diameter of 2 in. and an altitude of 3.6 in. Find the volume of the cone.
3. A paper cup in the shape of a right circular cone has a diameter of 3.2 in. The altitude of the cone is 4 in. Find the volume.

4. A conical tank has a diameter of 8.5 feet and an altitude of 12 feet. If the cone is inverted with the apex at the bottom, how many gallons will it hold?

5. Some grain piled in one corner of a bin has the shape of a quarter of a right circular cone with a radius of 4.2 feet and a height of 3.6 feet. How many bushels are there in the bin? (1 bushel = 1.25 cu. ft.)

6. Imagine you stand 8 feet from a vertical wall on which hangs a circular mirror 34 inches in diameter. As your eyes follow the circumference of the mirror once around the mirror, what is the volume of the geometric solid that your line of sight encloses?

7. The conical roof of a silo has a diameter of 18 feet. The slant height of the roof is 10.5 feet. Find the total area of the roof.

8. As a warning to motorists, some conical signs are set up along a road when construction work is being done. If the diameter of the cone is 16 in. and the slant height is 36 in., what is the altitude of a cone?

9. A waste basket has the shape of the frustum of a cone. The top diameter is 15 in. and the bottom diameter is 9 in. If the altitude of the basket is 18 in., find the volume and the surface area of the lateral surface and the bottom.

10. A drinking cup in the shape of the frustum of a cone has a top diameter of 2.6 in. and a bottom diameter of 2 in. If the altitude of the cup 2.8 in., how many cups can be filled from 1 gallon of lemonade?

11. A flower pot in the shape of a frustum of a right circular cone has a top diameter of 10 in. and a bottom diameter of 6 in. If the slant height is 7 inches, find the altitude, the volume, and the total surface area including the lateral surface and the bottom.

12. A megaphone in the shape of the frustum of a right circular cone has a diameter of 3 in. at one end, and a diameter of 12 in. at the other end. If the altitude of the megaphone is 24 in., find the slant height and then find the lateral area.

13. Some stands for flower displays have the shape of the frustum of a right circular cone. The bottom diameter is 22 in., and the top diameter is 12 in. If the slant height of a stand is 24 in., what is the cost of covering the lateral surface with material at a cost of 20 cents per sq. ft.?

14. A log of wood is 12 feet long. One end is 18 in. in diameter, and the small end is 9 in. in diameter. What is the weight of the log if the specific gravity of the wood is 0.40.?

13 THE SPHERE

13.1. DEFINITIONS

A *sphere* is a geometric solid bounded by a curved surface such that every point on the surface is the same distance from a point within called the *center* (Fig. 13.1). If a circle is rotated about one of its diameters, the figure formed is a sphere.

A *radius* of a sphere is a straight line segment joining the center with any point on the surface, such as *OA*. The radius is often defined as the *distance* from the center to any point on the surface. A *diameter* can be defined as a straight line segment through the center with its end points on the surface, such as *BOC. A diameter is therefore equal in length to twice the radius.*

If a plane cuts through a sphere, the cut surface is circular; that is, the section is a circle, such as *DEFG*. If a thin slice is cut off a sphere, the section is a small circle. As the intersecting plane moves closer to the center of the sphere, the sections become larger circles. The largest circle that can be cut by a plane intersecting a sphere is a section through the center of the sphere, *BHCI*. Such a circle is called a *great circle* of the sphere. The length of a minor arc of a great circle is the shortest distance between two points on the surface of a sphere. That is why airplanes often fly along the *great circle* route as the shortest distance.

The portion of a sphere cut off by a plane is called a *segment* of the sphere (Fig. 13.2). If two parallel planes cut through a sphere, the portion of the sphere between the planes is called a *segment of two bases* (Fig. 13.3).

13.2. MEASUREMENTS IN A SPHERE

In connection with a sphere, there are two measurable quantities in which we are interested. They are the *surface area* and the *volume*. In stating the size of a sphere, we need to give only one measurement. The measurement may be the radius, the diameter, or the circumference. If the radius of a

188 GEOMETRY

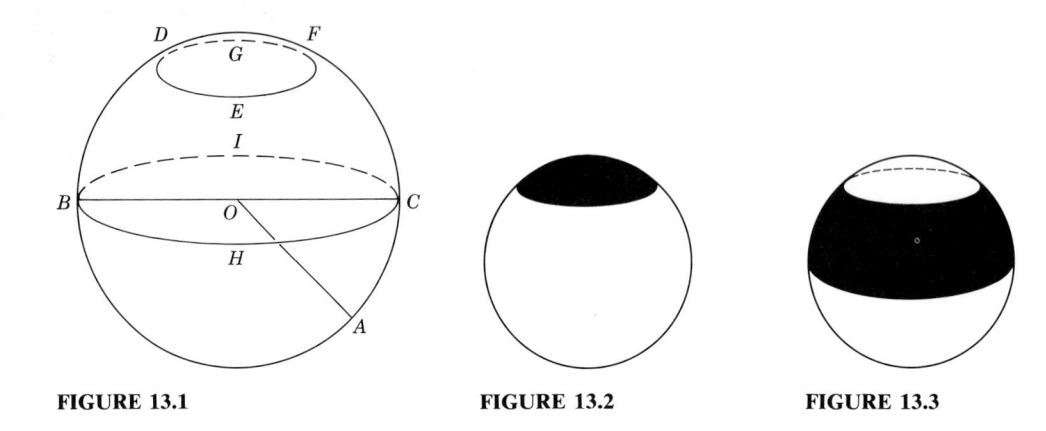

FIGURE 13.1

FIGURE 13.2

FIGURE 13.3

sphere is 5 inches, then the sphere is completely determined in size, and all measurable facts can be computed from the one measurement alone.

To measure the size of a sphere, we might place a plane (flat) surface on each side of the sphere so that the planes are parallel and touching the sphere. Then we can measure the perpendicular distance between the planes. This distance is the diameter of the sphere. The diameter may also be measured by a set of plane calipers.

Practically, it is impossible to measure the radius of a solid sphere directly because we cannot get to the center. However, the formulas for the surface area and the volume are usually stated in terms of the radius rather than the diameter.

13.3. SURFACE AREA OF A SPHERE

The formula for the surface area of a sphere can be derived only by more advanced mathematics. However, the formula itself is not difficult to understand. Suppose we have a sphere with center at point O, and a given radius r (Fig. 13.4). If we cut the sphere in two by a plane passing through the

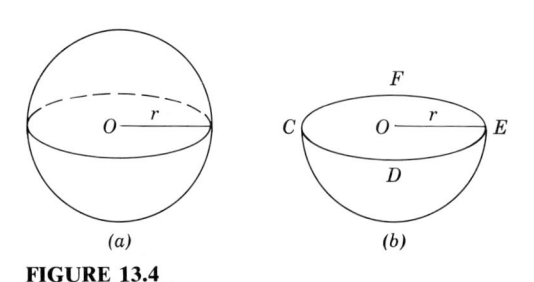

FIGURE 13.4

center, we have a hemisphere (*a*). The hemisphere resembles a kettledrum. The plane surface of the hemisphere, *CDFF* (the head of the kettledrum), is a great circle of the sphere. Its radius *r* is the radius of the sphere.

Now, we can find the area of this great circle, the head of the kettledrum. We use the formula for the area of any circle: $A = \pi r^2$. Let us look at the curved surface of the hemisphere, that is, the bottom of the kettledrum. This curved surface, clearly, has a greater area than the flat top, the head of the drum. In fact, it can be proved that the curved surface of the hemisphere (*b*) is exactly *twice* the area of the great circle of the sphere. Since the area of the great circle is πr^2, then the *area of the curved surface of the hemisphere is* $2\pi r^2$. Therefore, the area of the entire surface of the sphere is given by

$$A \text{ (of surface of sphere)} = 4\pi r^2$$

Example 1. A sphere has a diameter of 14 inches. Find the surface area (Fig. 13.5).

Solution. The radius is 7 inches. Using the formula, $A = 4\pi r^2$, we have

$$A = 4(3.1416)(49) = 615.8 \text{ (sq. in., approx.)}$$

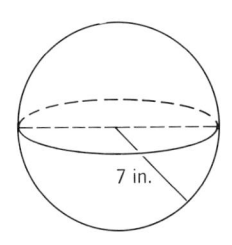

FIGURE 13.5

13.4. THE UNIT SPHERE

A *unit sphere* is a sphere whose radius is 1 unit in length. The unit may be any measurement, such as 1 centimeter, 1 inch, 1 foot, and so on. The unit sphere is an important concept in many applications of the sphere in scientific study. It has important application in the study of magnetism.

Suppose we have a sphere whose radius is 1 centimeter. This is a sphere about the size of a fairly small marble. The total area of the sphere is given by the formula: $A = 4\pi r^2$. Since $r = 1$, we have

$$A = 4\pi(1^2) = 4\pi \text{ (sq. centimeters)}$$

Whatever unit is taken for the radius of the unit sphere, the area is equal to 4π square units.

13.5. VOLUME OF A SPHERE

The volume of a sphere can be found by the formula: $V = \frac{4}{3}\pi r^3$. The formula is not difficult to memorize or to use. It can be proved only by advanced mathematics. However, let us see why it is reasonable.

You will recall that in finding the area of a circle, we imagined the circle divided into many sectors. In a similar way, we can imagine a sphere composed of many pyramids. (Figure 13.6 shows one of these pyramids.) Of course, the base of each pyramid is slightly curved. However, we can assume that the number of pyramids increases so that the base of each is practically flat.

For the volume of any pyramid, we have the formula: $V = \frac{1}{3}Bh$. Instead of finding the volume of each of the many pyramids making up the sphere, we can first add the bases of all the pyramids.

Now, we recall that the sum of the bases of the pyramids is the surface area of the sphere, that is, $4\pi r^2$. The altitude is the same for all the pyramids, that is, r, the radius of the sphere. Now, we can take the sum of the bases for B in the formula, and the radius for h. Then we have

$$V = \tfrac{1}{3} \text{ (sum of bases)(radius)}$$

If we substitute $4\pi r^2$ for the sum of the bases, and r for the radius, we get

$$V = \tfrac{1}{3}(4\pi r^2)(r)$$

which reduces to

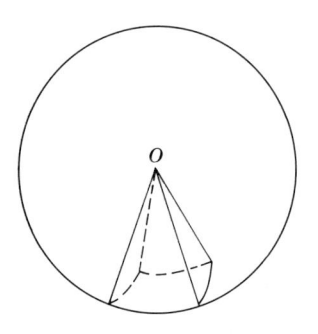

$$V = \tfrac{4}{3}\pi r^3$$

FIGURE 13.6

Example 2. A sphere 16 inches in diameter is filled with water. How many gallons will it hold? What is the surface area?

Solution. The radius of the sphere is 8 inches. Using the formula, we have

$$V = \tfrac{4}{3}(3.1415)(8^3) = 2144.7 \text{ (cu. in. in volume)}$$

For the number of gallons, we divide by 231:

$$2144.7 \div 231 = 9.28 \text{ (gallons)}$$

For the surface area, we have

$$A = 4\pi r^2 = 4(3.1416)(64) = 804.2 \text{ (sq. in.)}$$

Example 3. The approximate average radius of the earth is 3959 miles. Find the length of the circumference of a great circle of the earth. Also find the approximate area of the surface of the earth.

Solution. For the circumference, we have

$$C = 2\pi r = 2(3.1416)(3959) = 24{,}875 \text{ (miles, approx.)}$$

For the surface area, we use the formula: $A = 4\pi r^2$. Then,

$$A = 4(3.1416)(3959^2) = 196{,}962{,}000 \text{ (sq. mi., approx.)}$$

Example 4. The circumference of a sphere is 132 inches. Find the diameter, the radius, the surface area, and the volume of the sphere.

Solution. Since the circumference is equal to the diameter times π, we find the diameter by dividing the circumference by π. Then

$$\text{Diameter} = 132 \div 3.1416 = 42.0 \text{ (approx.)}$$

For the radius, we take one-half the diameter, and get: $r = 21$ (approx.) Now we use the length of the radius, 21, to find the surface and volume.

$$A = 4\pi r^2$$

Then

$$A = 4(3.1416)(21^2) = 5541.8 \text{ (sq. in., approx.)}$$

$$V = \tfrac{4}{3}\pi r^3$$

Then

$$V = \tfrac{4}{3}(3.1416)(21^3) = 38{,}792 \text{ (cu. in., approx.)}$$

Example 5. The surface area of a sphere is approximately 1810 sq. in. Find the radius, the diameter, and the volume of the sphere.

Solution. Since the surface area is given, we use the formula in reverse. The formula for the surface is $A = 4\pi r^2$. Then the quantity, πr^2, must be $\tfrac{1}{4}$ of 1810, which is equal to 452.5. Then,

$$\pi r^2 = 452.5; \quad \text{or} \quad r^2 = 452.5 \div 3.1416 = 144.03 \text{ (approx.)}$$

To get the radius, we take the square root of 144.0, which is 12 (approx.) Since the radius is 12, the diameter is 24 inches. For the volume, we have

$$V = \tfrac{4}{3}\pi r^3 = \tfrac{4}{3}(3.1416)(12^3) = 7238 \text{ (cu. in. in volume, approx.)}$$

Example 6. The surface area of a sphere is equal to 104π sq. in. Find the radius, the diameter, and the volume of the sphere.

Solution. Since we know the surface area, we can write: $104\pi = 4\pi r^2$. If we divide both sides by 4π, we know that $r^2 = 26$. From a table, we find that the square root of 26 is approximately 5.1. Then diameter = 10.2. For

192 GEOMETRY

the volume, we take

$$V = \tfrac{4}{3}(3.1416)(5.1^3) = 555.6 \text{ (cu. in., approx.)}$$

Note. The work in finding unknown measurements in connection with a sphere can be simplified if we first compute the following values:

$$4\pi = 12.5664 \quad \text{and} \quad \tfrac{4}{3}\pi = 4.1888$$

Example 7. Find the surface area of a sphere whose radius is 20 in.

Solution. We use the formula: $A = 4\pi r^2$; $4\pi = 12.5664$. Then,

$$A = (12.5664)(400) = 5026.56 \text{ (sq. in.)}$$

For the volume, we use the formula $V = \tfrac{4}{3}\pi r^3$; $\tfrac{4}{3}\pi = 4.1888$. Then,

$$V = (4.1888)(8000) = 33{,}510 \text{ (cu. in. in volume)}$$

Example 8. The volume of a sphere is 2145 cu. in. (approx.) Find the radius and the surface area of the sphere.

Solution. We first write the formula: $V = \tfrac{4}{3}\pi r^3$. If we use the value, $\tfrac{4}{3}\pi = 4.1888$, we have the formula: $V = (4.1888)(r^3)$. To find the r^3, we can divide the volume by the quantity, 4.1888. Since we know the volume is 2145, we divide as follows:

$$r^3 = 2145 \div 4.1888 = 512 \text{ (approx.)}$$

Now, using Table 2 of the Appendix, we find the cube root of 512 is 8. Then, $r = 8$. For the surface area, we use the formula: $A = 4\pi r^2$. Then

$$A = 4\pi(8^2) = (12.5664)(64) = 804.2 \text{ (sq. in., approx.)}$$

Exercise 13.1

Find the circumference, the surface area, and the volume of these spheres:

1. Iron ball, diameter = 3 inches.
2. Polo ball, diam. = $3\tfrac{1}{4}$ in.
3. Beach ball, diam. = 18.4 in.
4. Marble, diam. = $\tfrac{3}{4}$ in.

Find the diameter, the radius, surface area, and volume of these spheres:

5. Tennis ball, circumference = 8.25 in.
6. Basketball, circumference = 29.8 in.
7. Croquet ball, circumference = $8\tfrac{7}{8}$ in.
8. Baseball, circumference = 8.64 in.

THE SPHERE 193

9. A globe, 18 inches in diameter, is filled with eater. How many gallons does it hold? (1 gallon = 231 cu. in.)
10. A globe 3 feet in diameter is filled with water. Find the weight of the water in the globe if water weighs 62.4 pounds per cu. ft.

In Exercises 11–14, the surface area of each sphere is given. Find the radius and the volume of each sphere.

11. Area = 452.4 sq. in.
12. Area = 254.5 sq. in.
13. Area = 2464 sq. in.
14. Area = 98.6 sq. ft.

In Exercises 15–18, the volume of the sphere is given. Find the radius and the surface area of each sphere.

15. Volume = 523.6 cu. in.
16. Volume = 4189 cu. in.
17. Volume = 288π cu. in.
18. Volume = 113.1 cu. cm.

19. Find the weight of a hollow iron sphere if the outside diameter of the sphere is 18 inches and the metal is 2 inches thick. (The specific gravity of the iron in the hollow ball is 6.8, which means that the iron is 6.8 times as heavy as water).
20. Could you lift a sphere of gold 8 inches in diameter? (The specific gravity of gold is 19.3.)
21. You are shown a sphere of gold 4 inches in diameter, and you are told that you have a choice of taking either an outside shell $\frac{1}{2}$ inch thick or the remainder of the sphere. Which would you choose?
22. The diameter of the moon is approximately 2160 miles. Find the surface area of the moon.
23. The diameter of Jupiter is approximately 86,900 miles. Find surface area.
24. What is the diameter of a 16-pound shot if the specific gravity is 6.8?
25. The total surface area of a geometric solid is 384 sq. in. What is its volume if the figure is a cube? What is the volume if it is a rectangular solid 12 inches long and 9 inches wide? What is the volume if the solid is a sphere? (Remember, the area is still 384 sq. in.)
26. The volume of a rectangular solid is 216 cu. in. What is the surface area if the solid is 12 in. long and 9 in. wide? What is the surface area if the figure is a cube? What is the surface area of a sphere containing the same volume? Can you tell from your answer why all soap bubbles are round?

14 THE METRIC SYSTEM OF MEASUREMENT

14.1. DEFINITION

The metric system of measurement is a decimal system. It is based on the number 10, just as our number system itself is based on 10. To change the size of the units in the metric system, we simply move the decimal point to the right or to the left.

To understand the convenience of a decimal system, imagine what we might do if our common English system of measurements were a decimal system. Suppose there were 10 inches in 1 foot; 10 feet in 1 yard; 10 yards in one rod; 10 rods in 1 furlong; 10 furlongs in 1 mile. Then

23457 inches could be changed to:
2345.7 feet
234.57 yards
23.457 rods
2.3457 furlongs
0.23457 miles

Instead, in our present system, called the English system, we must perform the following divisions: for 23457 inches, we have

$23457 \div 12 = 1954.75$, number of feet
$1954.75 \div 3 = 651.58$, number of yards
$651.58 \div 5.5 = 118.47$, number of rods
$118.47 \div 40 = 2.962$, number of furlongs
$2.962 \div 8 = 0.3702$, number of miles

Note. The smallest practical unit of length in the English system is the mil, *which is equal to* $\frac{1}{1000}$ *of an inch. This paper has a thickness of approximately 3 mils. The mil is often used in stating the measurement of the diameters of wires. A diameter of 0.0264 inch is equal to 26.4 mils.*

14.2. ORIGIN OF THE METRIC SYSTEM

Our common English system of measurements grew up piece by piece without any definite plan. As a result, there was no attempt to make it easy to convert one kind of unit into another. The metric system, instead, was planned as a decimal system. It was devised by the French in 1789 at the time of the French Revolution. The French decided to make a new start in setting up a logical decimal system of measurements.

The foot measurement originally came from a length in some way related to that of a man's foot. The French decided to start with some measurement that was fixed in nature and not so variable as men's feet. They began with a measurement that they considered as one ten-millionth of the distance from the equator to the North Pole. This distance was to be the primary unit and was called one *meter* (m) meaning *measure*. The distance taken for this length is equal to 39.37 inches in the English system, although later calculations showed that this length is not exactly one ten-millionth of the distance from the equator to the North Pole. However, the relation between the English system and the metric system has been established as

$$1 \text{ meter} = 39.37 \text{ inches}$$

The meter is therefore a little longer than the English yard of 36 inches.

14.3. UNITS IN THE METRIC SYSTEM

In the metric system all measurements of length, area, volume, and even weight are based on the primary unit, the *meter*. The meter is divided into smaller units, and all smaller units are in *tenths*. The primary unit, the meter, is divided into ten parts, each part being $\frac{1}{10}$ of a meter. The prefix for *tenth* is "deci." Therefore, a length of $\frac{1}{10}$ meter is called a *decimeter* (dm). A decimeter (Fig. 14.1) is equal to 3.937 inches.

The decimeter is further divided into ten parts. Each part is called a *centimeter* (cm). The prefix "centi" means one *one-hundredth*. The word for our coin, the *cent*, means $\frac{1}{100}$ of a dollar. One centimeter is $\frac{1}{100}$ of a meter, or 0.3937 inch, which is approximately $\frac{2}{5}$ of an inch. The following relation is a very close approximation:

$$1 \text{ inch} = 2.54 \text{ centimeters (approx.)}$$

The centimeter is therefore a rather small unit of measure. However, it is much used in scientific work.

The centimeter is further divided into ten divisions, each division called one *millimeter* (mm). The prefix "milli" means one *one-thousandth*. The

THE METRIC SYSTEM OF MEASUREMENT

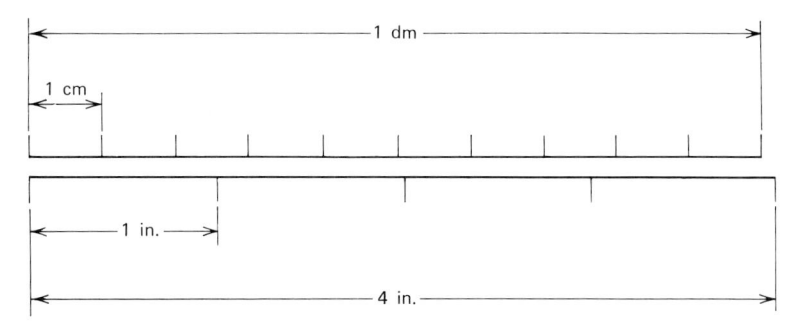

FIGURE 14.1

millimeter is therefore $\frac{1}{1000}$ of a meter. It is a very small unit of measure, approximately $\frac{1}{25}$ of an inch. The millimeter has quite common usage in our everyday speech. A 35-mm camera film is 35 millimeters wide. The sizes, 8 mm and 16 mm, also refer to the width of films. The bores of cannons and small firearms are often stated in terms of the metric system. An 8-mm rifle has approximately the same calibre as a "30–30" rifle. A 150-mm cannon is about the size of a 6-inch gun.

There is one extremely small unit of length in the metric system. It is called the *micron* and is equal to one-millionth of a meter. The micron is used in extremely fine measurement in science.

So far, in our explanation, we began with the primary unit, the meter, and moved toward smaller and smaller units, the decimeter, the centimeter, the millimeter, and the micron. Now we shall move toward larger units.

A distance of 10 meters is called 1 *decameter* (dkm), sometimes spelled *dekameter*. The prefix "deca" means *ten*. One decameter is a distance of about 32.8 feet. A distance of 10 decameters is called 1 *hectometer* (hkm). The prefix "hecto" means 100. A hectometer is equal to 100 meters. The decameter and the hectometer are seldom used in everyday speech. For example, a sprint of 1 hectometer is called the "hundred-meter dash."

A distance of 10 hectometers is called 1 *kilometer* (km). The prefix "kilo" means 1000. A kilometer is equal to 1000 meters, or approximately $\frac{5}{8}$ of a mile. One million meters is called a *megameter*.

The following table is a summary of the relations between the units of length in the metric system:

$$10 \text{ millimeters (mm)} = 1 \text{ centimeter (cm)}$$
$$10 \text{ cm} = 1 \text{ decimeter (dm)}$$
$$10 \text{ dm} = 1 \text{ meter (m)}$$
$$10 \text{ m} = 1 \text{ decameter (dkm)}$$
$$10 \text{ dkm} = 1 \text{ hectometer (hkm)}$$
$$10 \text{ hkm} = 1 \text{ kilometer (km)}$$

$$100 \text{ cm} = 1 \text{ meter (m)}$$
$$1000 \text{ mm} = 1 \text{ meter (m)}$$
$$1 \text{ cm} = \tfrac{1}{100} \text{ meter}$$
$$1 \text{ mm} = \tfrac{1}{1000} \text{ meter}$$

Summary of prefixes:

deci means $\tfrac{1}{10}$ *deca* means 10
centi means $\tfrac{1}{100}$ *hecto* means 100
milli means $\tfrac{1}{1000}$ *kilo* means 1000
micro means $\tfrac{1}{1000000}$ *mega* means 1000000

To see the convenience of the metric system, consider the following example. Suppose we begin with 1,295,860 millimeters. Then

$$\begin{aligned}1295860 \text{ mm} &= 129586 \text{ cm}\\ &= 12958.6 \text{ dm}\\ &= 1295.86 \text{ m}\\ &= 129.586 \text{ dkm}\\ &= 12.9586 \text{ hkm}\\ &= 1.29586 \text{ km}\end{aligned}$$

14.4. CONVERSION FACTORS BETWEEN THE ENGLISH AND THE METRIC SYSTEMS

To change measurements from English to metric, or metric to English units, we use chiefly these conversion facts:

1 meter = 39.37 inches
1 inch = 2.54 centimeters (approximately)
1 kilometer = 0.621 mile (approximately)

Example 1. Change 23 meters to feet.

Solution. We can first change 23 meters to inches, then to feet:

$$23 \text{ m} = 23(39.37 \text{ in.}) = 905.51 \text{ in.}$$

Dividing by 12,

$$905.51 \div 12 = 75.46 \text{ (ft)}$$

Example 2. Change 5 feet 2 inches to centimeters; then to meters.

Solution. 5 feet and 2 inches = 62 inches. Now we multiply by 2.54, the number of centimeters in 1 inch:

$$(62)(2.54) = 157.48 \text{ (cm)}$$
$$= 1.5748 \text{ m}$$

Example 3. When a car travels at a speed of 30 miles per hour, how many kilometers is that?

Solution. We use the relation: 1 kilometer = 0.621 mile. Then we divide the number of miles by 0.621:

$$30 \div 0.621 = 48.3 \text{ (km per hour)}$$

Example 4. What is the difference in length between 70 meters and 230 feet?

Solution. The answer can be expressed in various units. We shall change each measurement to inches:

$$70 \text{ m} = (70)(39.37 \text{ in.}) = 2755.9 \text{ in.}, \quad 230 \text{ ft} = 2760 \text{ in.}$$

Then the difference is found by

$$2760 - 2755.9 = 4.1 \text{ (in.)}$$

Exercise 14.1

Change each of the following measurements to the form indicated. Carry decimals out to a reasonable degree.

1. 35 in. to cm.
2. 87 cm. to in.
3. 250 m. to ft.
4. 21.3 ft. to m.
5. 31.2 ft. to cm.
6. 1 yd. to mm.
7. 1.32 miles to m.
8. 32.5 yd. to m.
9. 7 ft. 6 in. to cm.
10. 12.5 in. to mm.
11. 1600 ft. to m.
12. 1800 m. to yd.
13. 250 mils to in.
14. 16.5 ft. to cm.
15. 87 miles to km.
16. A man is 5 ft. 10 in. tall. State his height in centimeters.
17. A rectangular mirror is 26 inches wide and 4 feet long. State its width and length in centimeters.
18. A certain camera has a 60-mm lens. How many inches is this?
19. If a ball is dropped and falls without any interference, it will fall approximately 490 cm. in the first second. Change the distance to feet.
20. A certain camera takes a picture of a size stated as 2.25 in. by 3.25 in. Change these measurements to millimeters.
21. A family on a trip drove 300 miles one day. Change to kilometers.
22. A golf drive was 160 yards. Change this length to meters.
23. A baseball player hit an "inside-the-park" home run. The ball hit the surrounding wall 420 feet from home plate. How many meters was this?

24. One race in the Olympic Games is the 5000-meter run. Change this measurement to feet; then to miles.
25. A paperclip is made of wire 0.036 inches in diameter. Change to mils. How many millimeters is this?
26. An airplane flies at a height of 6.1 miles. What is the height in feet?
27. Change the following speeds to feet per second and meters per second: 15 mph; 30 mph; 45 mph; 60 mph; 80 mph.

14.5. SQUARE MEASURE

Square measures in the metric system are found in the same way as in our common English system. For example, by a square foot, we mean the amount of area contained in a square 1 foot on a side. A square inch is the area equal to that of a square 1 inch on a side. In the same way, by a square meter, we mean the area contained in a square 1 meter on a side.

To see the relation between a square inch and a square foot, we begin with the statement:

$$1 \text{ foot} = 12 \text{ inches}$$

then

$$1 \text{ square foot} = 12^2 \text{ square inches (or } 12^2 \text{ in}^2.\text{)}$$

Note. To indicate "square inches" we often write "$in.^2$". To indicate the square of any linear measure, we use the exponent 2 on the linear measure as well as on the numerical value itself.

In the same way, we have

$$1 \text{ yard} = 3 \text{ feet}$$

then

$$1 \text{ sq. yd.} = 3^2 \text{ sq. ft. or } 1 \text{ yd}^2 = 9 \text{ ft}^2$$

From these examples we can see the relation between linear units and corresponding square units. That is, we take the relation between the linear units and then square that number to get the relation between square units. As another example,

$$1 \text{ rod} = 16.5 \text{ feet}$$

then

$$1 \text{ square rod} = (16.5)^2 \text{ square feet (or } 16.5^2 \text{ ft}^2)$$

$$= 272.25 \text{ ft}^2$$

In the same way, to find the relation between the squares in the metric

THE METRIC SYSTEM OF MEASUREMENT 201

system, we square the linear relation. For example,

1 cm = 10 mm;   then   1 sq. cm. = 100 sq. mm.;   or $1 \text{ cm}^2 = 10^2 \text{ mm}^2$

1 m = 100 cm;   1 sq. m. = 10000 sq. cm;   written: $1 \text{ m}^2 = 100^2 \text{ cm}^2$

The relation between square measures in the English and the metric systems can be stated in a similar way. That is, we state the linear relation and square the number that represents this relation. For example,

1 inch = 2.54 cm;   then   1 sq. in. = $(2.54)^2$ sq. cm.

Stated in another way, $1 \text{ in.}^2 = (2.54)^2 \text{ cm}^2 = 6.45 \text{ cm}^2$ (approximately).
As another example, 1 m = 39.37 in., then $1 \text{ m}^2 = (39.37)^2 \text{ in.}^2 = 1550 \text{ in.}^2$

Example 1. A picture is 15 inches wide and 20 inches long. Change these measurements to centimeters and find the perimeter and the area in metric units.

Solution. For the linear measurements, we use the relation: 1 in. = 2.54 cm. For the width, we have

$$(15)(2.54 \text{ cm}) = 38.1 \text{ cm (approx.)}$$

For the length, we have

$$(20)(2.54 \text{ cm}) = 50.8 \text{ cm (approx.)}$$

For the perimeter, we take twice the width and twice the length:

$$2(38.1) + 2(50.8) = 177.8 \text{ cm (approx.)}$$

To check the answer, we can work the problem first in English units and then convert the answer to metric units. In English units,

$$\text{perimeter} = 2(15) + 2(20) = 70 \text{ (in.)}$$

Now we multiply this answer by 2.54, and get 177.8 (cm).
For the area, we multiply width times length, using the metric units: Then

$$(38.1)(50.8) = 1935.48 \text{ (sq. cm., approx.)}$$

To check this answer, we can first find the area in English units:

$$(15)(20) = 300 \text{ (sq. in.)}$$

Now we use the relation

$$1 \text{ sq. in.} = 6.45 \text{ sq. cm. (approx.)}$$

Then for the area, we get

$$(300)(6.45) = 1935 \text{ (sq. cm.)}$$

This answer for the area differs from the first computation by a small fraction because numbers have been rounded off.

Example 2. In the metric system, a room measures 6.8 meters wide and 7.8 meters long. Find the number of feet in the perimeter and the number of square feet in the area.

Solution. We shall first change meters to inches by use of the relation: 1 meter = 39.37 inches. Then we change the number of inches to feet by the relation: 1 foot = 12 inches.

For the width,

$$(6.8)(39.37 \text{ in.}) = 267.72 \text{ in.} = 22.3 \text{ ft}$$

For the length,

$$(7.8)(39.37 \text{ in.}) = 307.1 \text{ in.} = 25.6 \text{ ft}$$

For the perimeter, we take twice the width plus twice the length:

$$\text{perimeter} = 2(22.3) + 2(25.6) = 95.8 \text{ (ft in perimeter)}$$

For the area, we multiply width times length:

$$\text{area} = (22.3)(25.6) = 570.9 \text{ (sq. ft., approx.)}$$

Example 3. Find the number of square feet in 1 square meter.

Solution. This is a very useful relation to remember. We first find the number of square inches in 1 square meter; then we divide by 144.

$$1 \text{ m} = 39.37 \text{ in.}$$

THE METRIC SYSTEM OF MEASUREMENT

Then
$$1 \text{ sq. m} = (39.37)^2 \text{ sq. in.}$$
$$= 1550 \text{ sq. in. (approx.)}$$

Now we divide by 144, the number of square inches in 1 square foot:

$$1550 \text{ sq. in.} \div 144 \text{ sq. in.} = 10.764 \text{ (sq. ft., approx.)}$$

Example 4. A lawn measures 31 meters long and 22 meters wide. Find the number of square feet in the lawn.

Solution. First we find the number of square meters in the lawn. Then we use the information from Example 3: 1 sq. m. = 10.764 sq. ft. For the number of square meters, we have

$$(31)(22) = 682 \text{ (sq. m.)}$$

For the number of square feet, we have

$$(682)(10.764) = 7341 \text{ (approx.)}$$

Example 5. A hall measures 82 feet long and 50 feet wide. Find the number of square meters in the hall.

Solution. For the area in square feet, we have: $(82)(50) = 4100$. For the number of square meters, we divide by 10.764, the number of square feet in 1 square meter:

$$4100 \div 10.764 = 380.9 \text{ (sq. m., approx.)}$$

14.6. CUBIC MEASURE: MEASURES OF VOLUME

The relations between measures of volume are similar to the relations between measures of area. For volume, we take the *cubes of linear units.*

1 ft. = 12 in.	$1 \text{ ft}^3 = 12^3 \text{ in.}^3$	or	1 cu. ft. = 1728 cu. in.
1 yd. = 3 ft.	$1 \text{ yd.}^3 = 3^3 \text{ ft.}^3$	or	1 cu. yd = 27 cu. ft.
1 cm. = 10 mm	$1 \text{ cm.}^3 = 10^3 \text{ mm}^3$	or	1 cu. cm. = 1000 cu. mm.
1 dm. = 10 cm.	$1 \text{ dm.}^3 = 10^3 \text{ cm.}^3$	or	1 cu. dm. = 1000 cu. cm

Note. In chemistry, medicine, or other situations where liquid measurement is used, the abbreviation for cubic centimeter is "cc." *one thousand cubic centimeters (1000 cc) is called one* liter. *A liter has the volume of a little more than a liquid quart.*

204 GEOMETRY

The relations between units of volume in the English system and the metric system can be found by taking the *cubes* of the linear relations.

$$1 \text{ in.} = 2.54 \text{ cm.} \qquad 1 \text{ in.}^3 = (2.54)^3 \text{ cm.}^3 \quad \text{or} \quad 1 \text{ cu. in.} = 16.387 \text{ cu. cm.}$$

$$1 \text{ m.} = 3.28 \text{ ft.} \qquad 1 \text{ m.}^3 = (3.28)^3 \text{ ft.}^3 \quad \text{or} \quad 1 \text{ cu. m.} = 35.3 \text{ cu. ft.}$$

One cubic meter is called a *stere*. A stere is equal in volume to 1000 liters or 1 million cc.

Example 1. Find the number of cubic centimeters in a rectangular container that measures 8 inches long, 6.5 inches wide, and 5 inches deep.

Solution. We could first express each measurement in centimeters and then multiply the measurements together to get the volume. However, it is probably easier to find the volume in cubic inches first:

$$(8)(6.5)(5) = 260 \text{ (cu. in.)}$$

Now we use the relation: 1 cu. in. = 16.387 cu. cm. Then we have

$$(260)(16.387 \text{ cu. cm.}) = 4260.6 \text{ cu. cm.}$$

Exercise 14.2

1. A page of writing paper is 12 inches long and 8.5 inches wide. Find the number of square centimeters in the area.
2. A picture measures 62 cm. long and 44 cm. wide. Change these measurements to inches and then find the area of the picture in square inches. Check your answer by first finding the area in square centimeters.
3. A rectangular tablecloth is 80 inches long and 56 inches wide. How many square centimeters are in its area?
4. A rectangular lawn is 74 feet long and 50 feet wide. How many square meters are there in the area?
5. A building lot is 110 feet long and 62 feet wide. Find the number of square meters in the area. How long a fence will it take to enclose the lot on all sides?
6. One acre contains 160 square rods. One rod is equal to 16.5 feet. How many square feet are there in 1 acre? How many square meters?
7. One *hectare* is the name of an area equal to 10,000 square meters. How much more or less than 1 acre is one hectare?
8. A square table top measures 3.5 feet on a side. How many square centimeters are there in the area?
9. A chalk box has the following inside measurements: length, 6 in.; width, 4 in.; depth 3.6 in. Find its volume in cubic centimeters.
10. A suitcase has the following measurements: length, 70 cm.; width, 50 cm.; depth, 16.4 cm. What is its volume in cubic centimeters?

THE METRIC SYSTEM OF MEASUREMENT 205

11. One gallon holds 231 cubic inches. Find the number of liters in 20 gallons.
12. A room is 18 feet long, 15 feet wide, and 10 feet high. Find the volume of the room in *steres*.

14.7. METRIC SYSTEM OF WEIGHTS

In the English system, the primary unit of weight is the *pound*. From this primary unit, we get the *ounce*, which is $\frac{1}{16}$ of a pound. One *ton* is equal to 2000 pounds. Another unit of weight that is sometimes used is the *hundredweight*, which is equal to 100 pounds.

In the metric system, the primary unit of weight is the *gram*. The gram is a very small weight compared with the pound. It is equal approximately to $\frac{1}{30}$ of an ounce. For much ordinary measurement of weight, the gram is too small. Instead, a more convenient unit is the kilogram (kg), which is equal to 1000 grams. A kilogram is approximately equal to 2.2 pounds. Five kilograms of flour is approximately 11 pounds.

Other units in the metric system of weight are formed by multiplying or dividing the gram by 10 or some power of 10. We have seen that the prefix "kilo" means one thousand; therefore, 1 kilogram is 1000 grams. One important small unit of weight in chemistry, medicine, and other scientific studies is the *milligram*, which is equal to $\frac{1}{1000}$ of a gram.

The most common conversion factor between the English system and the metric system of weight is the relation:

$$1 \text{ kilogram} = 2.2 \text{ pounds (very nearly)}$$

To convert one system to the other, we have the rule:

RULE. To change kilograms to pounds, multiply the number of kilograms by 2.2. To change pounds to kilograms, divide the number of pounds by 2.2.

For greater accuracy, we might use the following approximate equivalents:

$$1 \text{ gram} = 0.03527 \text{ ounce}; \quad 1 \text{ kilogram} = 2.2046 \text{ pounds}$$

Exercise 14.3

1. A book page measures 9 inches by 7 inches. Find the perimeter and the area in convenient metric measurements.
2. An envelope measures 25 cm. by 12 cm. Find its area in square inches.
3. A room is 26 feet long and 18.5 feet wide. Find the number of meters in the perimeter, and the number of square meters in the area.

4. A flag pole is 62.5 feet high. How many meters is this?

Change the following metric measurements to some convenient English units:

5. 36 cm.
6. 15 m.
7. 40 km.
8. 8.2 dm.
9. 52 mm.
10. 600 m.
11. 500 km.
12. 400 cm.
13. 20 m^2.
14. 250 cm^2.
15. 20 dm^2.
16. 5 m^2.
17. 5000 mm^2.
18. Change 400 liters to quarts
19. Change 10 steres to cubic feet.

Change the following English measurements to some convenient metric units:

20. 45 feet
21. 15 inches
22. 700 feet
23. 0.375 inch
24. 5.2 miles
25. 65 yards
26. 20 rods
27. 0.625 miles
28. 15.2 miles
29. 34.5 rods
30. 5.5 yards
31. 5280 feet
32. 272.25 ft^2.
33. 160 rd^2.
34. 48 in^2.
35. 1 mile^2
36. "A 10-foot pole."
37. "A yard wide."
38. "An ounce of prevention."
39. "A pound of cure."
40. "A grain of salt." (1 pound = 7000 grains)
41. "A 10-gallon hat."
42. Sound travels about 1080 feet per second. How many meters is this?
43. Light and radio waves travel about 300,000,000 meters per second. Find the difference between this measurement and 186,000 miles per second.
44. A radio broadcasting station broadcasts on wave length of 480 meters. What is the wave length in feet?

III ALGEBRA

15 INTRODUCTION TO ALGEBRA

15.1. SYMBOLS IN ARITHMETIC

In arithmetic we use such numbers as 3, 4, 7, 10, 321, and so on. We also use fractions, both common and decimal, as $\frac{1}{2}$; $\frac{3}{4}$; 1.25. We shall call such numbers *arithmetic numbers*, because we use them in arithmetic.

In arithmetic, we also use the following symbols to show operations: To show addition, we use the symbol (+), which we call *plus*. To show subtraction, we use the symbol (−), which we call *minus*. To show multiplication, we use the symbol (×), which we call the *times* sign. Sometimes, to show multiplication, we use a raised dot between the two numbers. For example, to show 3×4, we can write: $3 \cdot 4$. We can also enclose the two numbers in parentheses without a sign between them, and write: (3)(4). When numbers are so enclosed in parentheses, they are always meant to be multiplied together. For example, $(6)(9) = 54$.

To indicate division, we use the symbol, (÷), to mean *divided by*. For example, $48 \div 3$ means 48 divided by 3. For division, we can also use a line between the two numbers, as in a fraction. For example, to show 36 divided by 4, we can write: $\frac{36}{4}$. In fact, the first symbol to indicate division was the horizontal line between the numbers. This symbol was used hundreds of years before the symbol (÷), which is now used.

We use the symbol (=) to indicate equality between two quantities. This symbol is read, *equals* or *is equal to*.

Example. To say "three times six and five more equals 23," we write

$$3 \cdot 6 + 5 = 23$$

Sometimes we wish to multiply a number by itself, such as $7 \cdot 7 = 49$. To show that two 7's are to be multiplied together, we use a small 2, placed just at the right and a little above the 7, as $7^2 = 49$. Then the small 2 is called an *exponent*. An *exponent* is a number that tells how many times a number is to be used as a factor in multiplication. The number used as the factor is called the *base*. In the expression, 7^2, the exponent is 2, and the base is 7.

If we wish to use the same number 3 times in multiplication, we use the exponent 3. For example, 5^3 means: $5 \cdot 5 \cdot 5 = 125$. In this example, the exponent is 3, and the base is 5. If we use the exponent 4, we mean that the number is to be taken 4 times as a factor: $3^4 = 81$.

When we take a number 2 times as a factor, the expression is called the *square* of the number, or the *second power*. If a number is used 3 times as a factor, the expression is called the *cube* of the number, or the *third power*. For higher powers, we simply name the power.

The expression, 7^2, is called "7 *squared*" or "7 to the *second power*."

The expression, 6^3, is called "6 *cubed*", or "6 to the *third power*."

The expression, 2^6, which equals 64, is called the *sixth power of 2*.

15.2. SYMBOLS IN ALGEBRA

In algebra, we use letters to represent numbers. We also use the same symbols as in arithmetic for addition, subtraction, multiplication, and division. We also use the equal sign (=).

By using a letter to represent a number, we can express a problem in a very short form. Then we shall find that it is often less difficult to find the answers to problems.

Suppose we ask: How do you find the cost of a number of books of the same kind if you know the number of books and the price of each? The answer: You multiply the price of one book by the number of books.

To state the rule in words, we can say

(number of books) × (price of each) = total cost

Actually, the expression is simply a rule for finding the total cost. Now, if we wish to state the rule in a still shorter form, we can use the letter N for the number of books, the letter P for the price of each, and the letter C for the total cost. Then the rule becomes

$N \cdot P = C$,   or simply $NP = C$

As another example, in geometry we learn that the area of a rectangle can be found by multiplying the length by the width. That is,

(length) × (width) = area

If we use the letter L for length, the letter W for width, and the letter A for area, we can indicate the rule by using letters:

$$L \cdot W = A, \quad \text{or simply} \quad LW = A$$

Notice how we can abbreviate ordinary language by using symbols. Then a statement can be expressed in a very short form.

15.3. LITERAL NUMBERS

When letters of the alphabet are used to represent quantities, as we have used the letters N, P, C, L, W, and A, in the previous section, such letters are called *literal numbers*. Numbers used only in arithmetic we shall call *arithmetic numbers*, such as 3, 5, 8, and so on. For a literal number we may use any letter of the alphabet. However, it is best to avoid using O for a literal number, since it resembles zero (0). The literal numbers often used are n, x, y, z, t, a, b, c, r, and some others. However, any letter may be used to represent an arithmetic number or some quantity. Literal numbers are often called *general* numbers.

Literal numbers cannot actually be added, subtracted, multiplied, or divided in the same way as arithmetic numbers. With literal numbers, we simply indicate the operation by symbols. For example, in arithmetic, the sum of 3 and 5 is 8. If we wish to show the sum of x and y we can only indicate the sum: $x+y$. The difference between x and y, is written: $x-y$. The product of x and y is written: xy. The quotient of x divided by y is written: $x \div y$; or $\frac{x}{y}$.

In algebra, the multiplication sign is usually omitted, except when we wish to emphasize the operation. Thus:

"3 times x" is written: $3x$; "4 times n" is written: $4n$

$5y$ means "5 times y"; $6hk$ means "6 times h times k"; st means "s times t."

Often the sum of two numbers, or their difference, is to be multiplied by a number. To indicate a quantity, we use various *signs of grouping*. The most common symbol for grouping is a set of parentheses, ( ). We can also use brackets, [ ], or braces, { }. For the sum of x and y, we can write: $x+y$. However, to write "3 times the *sum* of x and y," we write: $3(x+y)$. To express the quantity, "4 times the *difference* between b and c, we write $4(b-c)$. The expression, $5(2x+3y)$, means "5 times the sum of $2x$ and $3y$."

We show some examples using letters to represent numbers.

Example 1. Suppose we wish to say, "a number plus seven." Instead of writing the word *number* and other words, we can use symbols. We can use a letter such as *n* to represent the number. Then, instead of saying,

"a number plus seven"

we can say, in symbols: $n + 7$

Example 2. To say: "three less than a number,"

we can say, in symbols: $n - 3$

Notice that subtraction must be stated in the correct order.

Example 3. Write in symbols: "six less than 5 times a number." To express multiplication, we need not use the times sign. In multiplying a letter by an arithmetic number, we need not use a dot between the numbers. For example, if we represent the number by *t*, then we can indicate 5 times the number as: $5 \times t$, or $5 \cdot t$, or simply $5t$. Then, to write

"six less than 5 times a number"

we can write, in symbols: $5t - 6$

To show a number, such as *n*, multiplied by itself, we can write n^2, which we call the *square of n*.

Example 4. Express in symbols: "the square of a number increased by 4 times the number." For the square of the number, we use the exponent 2. In this example, let us use *x* to represent the number. Then we can write

$$x^2 + 4x$$

Example 5. Express in symbols: "a number plus 3 is multiplied by 5." In this case we first express the number plus 3. If we use *n* for the number, then we first write: $n + 3$. Now this quantity is multiplied by 5. To show this, we must use parentheses around the quantity, $n + 3$. Then the entire expression becomes

$5 \cdot (n+3)$;   or simply $5(n+3)$

Exercise 15.1

Write each of the following expressions in algebraic form, using symbols to show addition, subtraction, multiplication, division, and equals. Use some letter, such as *n*,

t, x, or y to represent the number. Do not use the same letter for all expressions. In some expressions, as in No. 19, you will need to use two letters. In No. 29 and 30, use parentheses to indicate the quantity.

1. A number plus two.
2. Three more than a number.
3. A number minus 6.
4. Five less than a number.
5. A certain number increased by 8.
6. A certain number decreased by 8.
7. Eighteen decreased by a number.
8. Ten increased by a certain number.
9. The product of 7 and a number.
10. A number multiplied by 6.
11. Eleven divided by a certain number.
12. A certain number divided by 4.
13. Eight times a number and 11 more.
14. Twice a certain number and 9 more.
15. Six less than five times a number.
16. Two less than 7 times a number.
17. Twice a number divided by 3.
18. Six times a number divided by 5.
19. The product of two different numbers.
20. The sum of two different numbers.
21. The quotient of one number divided by a different number.
22. Four more than the product of two different numbers.
23. A number multiplied by itself.
24. Six more than the square of a number.
25. The third power of some number.
26. The fourth power of some number.
27. The sum of two numbers divided by their product.
28. The product of two different numbers divided by their difference.
29. Three times the sum of some number and 6.
30. Four times the quantity: *a number minus 5*.

15.4. THE NUMERICAL VALUE OF AN ALGEBRAIC EXPRESSION

When we use a letter in an algebraic expression, we mean that it represents some arithmetic number. If we know what number the letter represents, then we can tell the numerical value of the expression.

Example 1. Find the value of the expression, $4n+7$, if $n=3$.

Solution. Since we know that n is equal to 3, we put 3 for n in the expression, and get

$$4 \cdot 3 + 7 = 12 + 7 = 19, \text{ the numerical value}$$

Example 2. Find the value of the expression, x^2+3x+7, if $x=4$.

Solution. The term, x^2, means: $x \cdot x$. Then, x^2 means: $4 \cdot 4$, which is 16. The term, $3x$, means: $3 \cdot 4$, which is equal to 12. Then, for the entire expression, we get

$$4^2 + 3 \cdot 4 + 7 = 35$$

Example 3. Find the numerical value of the expression, $5x^2 + 4xy - 3y^2$, if $x = 3$, and $y = 2$.

Solution. The term $5x^2$ means 5 times x^2. Then, since $x = 3$, the term $5x^2$, means: $5(x^2)$, or $5(3^2)$, which is equal to $5(9)$, or 45. The term, $4xy$, means: $4 \cdot x \cdot y$, or $4 \cdot 3 \cdot 2$, which is equal to 24. The third term, $3y^2$, means: $3(y^2)$, or $3(2^2)$, which is 3(4), or 12. Putting all the numerical values together, we have

$$5x^2 + 4xy - 3y^2$$

$45 + 24 - 12 = 57$, the numerical value

Exercise 15.2

Find the value of each of the following expressions for the given value of the letter in each.

1. $3n + 5$; $n = 8$
2. $4x - 3$; $x = 7$
3. $13 + 2t$; $t = 3$
4. $5x - 7$; $x = 8$
5. $20 - 3y$; $y = 4$
6. $4v + 5$; $v = 4.5$
7. $t^2 + 2t$; $t = 4$
8. $n^2 - 5n$; $n = 7$
9. $3x^2 + x$; $x = 3$
10. $4x^2 - 5x$; $x = 3$
11. $4t^2 - 3t$; $t = 1.5$
12. $5x - x^2$; $x = 3$
13. $\dfrac{12}{x} + 3x$; $x = 6$
14. $\dfrac{30}{n} - 5$; $n = 1.5$
15. $\dfrac{3t}{2} + 5$; $t = 7$
16. $5x - x^2$; $x = 5$
17. $x^3 - 4x$; $x = 2$
18. $16n - n^3$; $n = 4$
19. $3n + 2n + 4$; $n = 3$
20. $6x - 2x + 3x + 5$; $x = 4$
21. $3y + 7y - y - 2$; $y = 5$
22. $3t + 5t - t - 50$; $t = 10$
23. $\dfrac{20}{n} + 2n + n^2 + 5$; $n = 4$
24. $5t - \dfrac{24}{t} + 3t - 7$; $t = 6$
25. The product, xy; $x = 6$; $y = 3$
26. The product, $5n^2 t$; $n = 3$, $t = 2$
27. The fraction, $\dfrac{4a}{3b}$; $a = 12$; $b = 2$
28. $\dfrac{3n}{5x}$; $n = 5$; $x = 2$
29. $3x^2 - 5x + 2$; $x = 4$
30. $4x^2 - x - 7$; $x = 3$

31. $5n^2 + 2n - n - 9$; $n = 3$
32. $2y^2 - 6y + y^2 + 1$; $y = 2$
33. $2x^2 - 5x - 3$; $x = 3$
34. $4t^2 - 5t - 6$; $t = 2$
35. $3a^2 - 2ab + b^2$; $a = 2$; $b = 4$
36. $2a^2 + 3ab - b^2$; $a = 2$; $b = 3$
37. $n^3 - n^2 + 5n - 3$; $n = 2$
38. $2y^3 + y^2 - 5y + 3$; $y = 1$
39. LWH; $L = 12$; $W = 5.5$; $H = 3$
40. $\frac{1}{2}(H)(B + b)$; $H = 8$; $B = 13$; $b = 5$

16 THE EQUATION

16.1. DEFINITION OF THE EQUATION

In the preceding chapter (No. 15), we used letters to represent arithmetic numbers. We expressed certain quantities using letters. For example, to express the quantity, "a number plus 6," we can use a letter, such as n, to represent the number, and then write

$$n + 6$$

If we just say, "a number plus 6," this is not a complete statement. It does not say anything.

Now, suppose we say in words, "a certain number plus 6 equals 15." Now we have a complete statement. Then we might ask: "What is the number?" Up to this point, we do not know the number. It is called an *unknown*. An *unknown* is a number that we do not know, one which we try to determine. However, after a moment of thought, you answer: "The number is 9." Then we have the answer to the question.

If we use the letter n to represent the unknown number, we can make the same statement in symbols. Instead of saying

"a certain number plus 6 equals 15"

we write $$n + 6 = 15$$

The statement, $n + 6 = 15$, is called an *equation*. The equation says in symbols exactly what the problem says in words.

> *Definition.* An equation is an algebraic expression that makes a statement saying that two quantities are equal.

When we write the equation, $n + 6 = 15$, we are really asking: "What is the value of n?" After a moment of thought, you say, "n is 9." Every

218 ALGEBRA

equation really asks a question: "What is the value of the letter used?"

Notice a few facts about the equation. The equal sign (=) is the word that *makes the statement.* We have something on each side of the equal sign. On the left side, we have the quantity, $n+6$. On the right side, we have the quantity, 15. The equal sign says that the two sides of the equation are equal to each other.

The *equation* is probably the most important idea in all mathematics. In fact, algebra might be called the study of the equation. The equation enables us to solve practical problems in physics, chemistry, other sciences, as well as problems in everyday life. It is the equation that makes mathematics useful in finding answers to problems.

As another example, suppose we say, "If we subtract 5 from a certain number, we have 12 left. What is the number?" After a moment of thought, you say, "The number is 17." You can get the answer directly from the word statement.

Now, let us write the same statement using symbols. We can use any letter for the unknown number. This time, let us use x for the number. Then the statement can be written as an equation. To indicate subtraction, we must use the minus sign in the proper manner. Then, instead of saying

"A number minus 5 equals 12"

we write the equation $\qquad x - 5 = 12$

For the number, we see that $\qquad x = 17$

In this example, as in the previous example, we find the answer by *inspection.* By *inspection*, we mean we are able to tell the answer quickly without any systematic method.

You might think that we could always tell the answer from the word statement alone, and we would not need the equation. In simple problems this is true. However, in some problems the word statement can be long and complicated. Then the statement in the form of an equation using algebraic symbols becomes much simpler.

As an example of a complicated word statement, consider the following problem. An electric circuit has a current of a certain number of amperes. If we add 2 to this number, multiply the result by 3, subtract 8, add the number itself, and then divide the result by 2, the answer will be 24. What is the original number?

We shall see that it is much easier to find the answer to any problem if we write the statement in symbols. However, we shall need to know more about working with involved equations. At present, we shall work with very simple equations in which the answer can be determined by *inspection*. However, you should always write the statement in algebraic symbols, no matter how easy it is to tell the answer from the word statement alone.

Example 1. If a certain number is subtracted from 32, the remainder is 19. What is the number?

Solution

We tell what letter we want to use and what it is to represent. We write

Let x represent the number to be subtracted from 32

Now we write an equation using symbols and the equal sign, stating exactly what the problem says. The problem says

32 minus a number equals 19

In symbols, we write $\qquad 32 - x = 19$

As a third step, we state the answer. With a little thought, we see that

$$x = 13$$

Example 2. If we multiply a certain number by 3 and then subtract 5, the answer is 16. What is the number?

Solution

We first write: Let $x =$ the certain number.

Now we write the equation stating just what the problem says:

$$3x - 5 = 16$$

Then $3x$ must be 21, and x must be 7, the original number.

In working out the problem in the next exercise, show all of the three following steps;

1st *step:* Let some letter represent the unknown number.

2nd *step:* Write the equation that says what the problem says.

3rd *step:* Tell what the letter is equal to (that is, the unknown number).

Example 3. If 3 is added to 4 times a certain number, the result is 17. What is the number?

Solution. Here we show the three separate steps:

1st *step:* Let x represent the number (or, let $x =$ the number).

2nd *step:* $4x + 3 = 17$ (this equation tells what the problem says).

3rd *step:* $x = 3.5$ (this is the answer to the problem).

Exercise 16.1

In each of these problems follow the three steps: (1) Let a letter represent the unknown number; (2) Then write the equation that says the same thing that the problem says; (3) Finally, tell by inspection the value of the letter, which is the unknown number.

1. A number plus 3 equals 11. What is the number?
2. A certain number minus 9 equals 15. What is the number?
3. Five more than a number is 13.
4. Three less than a certain number is 17.
5. A certain number subtracted from 20 leaves 13.
6. Twice a number and 7 more makes 19.
7. Three times a certain number and 5 more makes 29.
8. Five times a number and 9 more makes 44.
9. Seven less than 3 times a number is 20.
10. Five less than 4 times a number is 43.
11. Six more than 5 times a number is 11.
12. Eight less than 7 times a number is 55.
13. Two less than 3 times a number is 25.
14. A certain number increased by 6 becomes 28.
15. If a certain number is decreased by 8, the result is 12.
16. If a number is subtracted from 20, the remainder is 16.5.
17. Four less than twice a number is equal to 13.
18. Two more than 3 times a number is equal to 18.
19. If 3 is subtracted from 5 times a number, the result is 15.
20. If 7 is added to twice a number, the result is 20.
21. Three less than 5 times a number is equal to 21 more than the number.
22. Six more than twice a number is equal to 30 minus 4 times the number.
23. Two less than 3 times a number is equal to the number subtracted from 8.
24. If twice a number is added to 5 times the same number, and then 4 is subtracted from the result, the answer is 42.
25. If 5 times a number is subtracted from 8 times the number, and then 7 is added, the result is equal to the number subtracted from 43.
26. If 3 is subtracted from 4 times a number, the remainder is 7 more than the number.
27. If 4 is subtracted from 5 times a number, the remainder is 10 more than the number.
28. A man says: "Take 3 times my age in years, and then subtract 10. The result is 65." What is his age?

16.2. THE SOLUTION OF AN EQUATION

In an equation such as the following, the letter n has a value that makes the equation true:

$$n + 6 = 15$$

THE EQUATION

By inspection, we see that n must have the value 9. Then 9 is called the *solution of the equation*.

> **Definition.** *The solution of an equation is a value of the letter that makes the equation true.*

If we put 9 for n in the equation, we get: $9+6=15$, which is true. The solution of an equation is sometimes called the *root* of the equation. In this equation, if we put another value, say, 7, for n, the statement becomes

$$7+6=15, \text{ which is not true}$$

Then 7 is *not* a solution of this equation.

To *solve* an equation means to find the root or solution. To solve an equation, we find a value of the letter that makes the equation true.

The foregoing equation was solved by *inspection*. By *inspection* we mean that we find the answer by a little simple reasoning. Later, we shall see that we need a more systematic method for solving most equations.

At present, we shall find the answers to some rather easy equations by inspection. However, a little study of the equation is often helpful.

Example 1. What is the value of n in the equation: $n+9=20$?

Solution. If we subtract 9 from 20, we get the answer: $n=11$.

Example 2. What is the value of x in the equation: $x-8=21$?

Solution. If we add 8 to the 21, we get the answer: $x=29$.

Example 3. What is the value of t in the equation: $3t-4=20$?

Solution. If we add 4 to the 20, we get the value of $3t$:

$$3t = 24$$

Then we see that

$$t = 8$$

Exercise 16.2

Solve each of the following equations by inspection:

1. $n+3=11$
2. $x+5=16$
3. $t-4=17$
4. $x-4=13$
5. $n-3=15$
6. $x-7=1$

7. $t-7=5.5$
8. $t+4=7.2$
9. $n+7=9.6$
10. $3n+2=23$
11. $15-2n=1$
12. $13-4x=5$
13. $20-3x=8$
14. $7-4x=7$
15. $16-5x=1$
16. $2y+3=17$
17. $4y-5=27$
18. $3x-2=28$
19. $4n+3=35$
20. $9+5x=64$
21. $3x+6=6$
22. $21-2x=11$
23. $40-3t=19$
24. $6y-5=49$
25. $9-2x=9$
26. $11+3x=11$
27. $2t+7=7$
28. $6n-7=13+n$
29. $19-3n=n-1$
30. $30-4n=15+2n$

Note. The purpose of this chapter is simply to show the meaning of the equation and its solution. In this chapter we have solved equations only by inspection. When we get to more difficult equations, we cannot always find the solution so easily. We shall need a systematic method consisting of certain definite steps. Then even more difficult equations are rather easily solved. The method is explained in Chapter 21.

However, before we can fully understand the steps in the method of solving equations, we must first understand by *positive* and *negative numbers.* Operations with such numbers are explained in Chapters 17–20.

17 SIGNED NUMBERS

17.1. DEFINITION

To understand the systematic solving of equations, we must understand what is meant by *positive* and *negative* numbers. In algebra, we make use of what we call *negative* numbers. Negative numbers are numbers that are *less than zero*. Then *positive* numbers are numbers that are *greater than zero*.

In arithmetic, we did not use negative numbers. Subtracting one number from a smaller number, such as $3-8$, is considered impossible in arithmetic. We cannot take 8 from 3. For that reason people once considered negative numbers to be impossible. Such numbers were called *absurd, ridiculous*, and they were called "fictitious numbers." Yet they forced themselves into mathematics. They have come to be very useful in many everyday situations, as well as in science.

As one example of the use of negative numbers, when the temperature drops to zero, it can drop still lower. Then numbers below zero can be called *negative numbers*. A temperature of 10 degrees below zero can be called a temperature of a *negative* 10 degrees. To indicate a number below zero, we place a minus sign (−) before the number. Then 10 degrees below zero can be called "−10" degrees. *This minus sign is not a sign of operation to show subtraction.* In the number, "−10," the minus sign tells only the *kind of number*. It shows a negative number, a number less than zero.

Numbers that are greater than zero are called *positive* numbers to distinguish them from negative numbers. A positive number can be indicated by a plus sign (+) before the number. Then a temperature of +40 degrees means a temperature of 40 degrees above zero.

When numbers are considered as positive or negative, they are called *signed* numbers. If a number has no sign before it, the number is considered to be positive. In arithmetic, we work with only positive numbers.

In arithmetic we use the signs, + and −, to indicate addition and subtraction, respectively. Then these signs are called signs of *operation*, because they show what operation is to be performed with the numbers.

However, when the signs are used to denote positive and negative numbers, they are called signs of *quality*. They indicate the kind of number.

There are many practical uses for negative numbers. We have seen how they can be used to show temperatures below zero. A golf score less than par can be shown by a negative number, such as "−3." If we represent elevation above sea level by a positive number, we can represent elevation below sea level by a negative number. For example, the elevation of Mt. Whitney in California is +14,495 feet. The elevation of the bottom of Death Valley in California is "−282" feet. A man's assets can be indicated by positive numbers, and his liabilities by negative numbers. If you have a debt of 10 dollars, we can say you have "−10" dollars. In some games you make a score that takes you backward. Such a score can be called a negative score, such as a score of "−20."

Positive and negative numbers can be used to show direction. If we indicate a distance of 3 miles east by +3, then we can indicate a distance of 3 miles west by −3. A flow of electricity can be indicated by signed numbers. If +5 amperes indicates a flow in one direction, then −5 amperes indicates a flow in the opposite direction.

Positive and negative numbers can be shown on a number scale represented by a horizontal line (Fig. 17.1). Positive numbers or distances are laid off at the right of a zero point; negative numbers or distances are laid off to the left of zero. Then zero is the dividing point between positive and negative numbers.

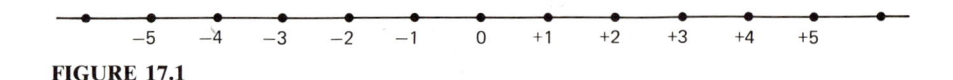

FIGURE 17.1

17.2. ABSOLUTE VALUE OF A NUMBER

The *absolute value* of a number is its *value without regard to the sign*. (The absolute value is sometimes called the *numerical* value.) The absolute value of +10 is the same as the absolute value of −10. We know that +10 is not the same as −10. A temperature of 10 degrees below is not the same as a temperature of 10 degrees above. However, their absolute values are the same. Actually, the absolute value refers to the distance from zero. The two numbers, +10 and −10, are the same distance from zero.

The absolute value of a number is usually indicated by a pair of vertical lines, as $|-10|$. Then we can say that $|+10|$ has the same value as $|-10|$. Notice these two statements:

This statement is not true: $10 = -10$ (not true)
This statement is true: $|10| = |-10|$ (true)

17.3. OPERATIONS WITH SIGNED NUMBERS: ADDITION

To work with positive and negative numbers, our first problem is to formulate rules for operating with such numbers. We must know how to add, subtract, multiply, and divide signed numbers so that we can always be sure to proceed correctly.

First, we consider addition. One of the simplest ways to understand the addition of signed numbers is to consider how we add scores in some games. In a game, you might make a score that increases your total score. On the other hand, you might make a score that decreases your total score. In the case that decreases your score, we sometimes say you "get set." If we use a positive score to indicate a score that increases your total, we might call that a *good* score. Then you can use a negative number to indicate a *bad* score that sets you back.

Suppose in a game you have a score of +20. Now make a good score of +5. Then your total score is +25. However, suppose you have a score of +20, and then make a *bad* score of −5. Then your total score is +15. We call this "adding" the scores.

Suppose, at the beginning of a game, you "get set" or go back 20 points. Your score is then actually 20 points less than nothing. In a game, this is sometimes called "going in the hole." Then we can call your score, −20. This simply means that, if someone else has a score of zero (0), then he is still 20 points ahead of you. It means also that you will have to make 20 points to make your score zero (0). If you have a score of −20, and then you make a score of +5, your score will remain at −15. Combining signed numbers in this way is called "adding" the numbers.

Examples. Suppose the following sets of scores are made. Find the total score in each set; in other words, *add* the scores. The sum of the scores is shown for each set. If no sign is shown, the number is positive.

+10	−8	+10	−12	9	3	−12	−14
+4	−4	−3	+3	−2	−8	18	6
+14	−12	+7	−9	+7	−5	+6	−8

From these examples we can formulate the rule for adding signed numbers.

RULE. To add two numbers with like signs, find the sum of their absolute values and prefix the common sign.

To add two numbers with unlike signs, find the difference between their absolute values and then prefix the sign of the number having the greater absolute value.

226 ALGEBRA

To add several signed numbers, we can first combine the positive numbers, then combine the negative numbers, and finally combine the two answers.

Examples. Add the scores in each of the following sets. Answers are shown.

+15	−10	+4	−6	2	−8	−2	7	−6
−4	+9	−7	−3	14	3	6	3	5
+8	−4	+5	+10	−3	−9	−12	9	−2
−6	+16	−8	−8	6	4	−8	−1	3
13	11	−6	−7	19	−10	−16	18	0

Exercise 17.1

Add the numbers in each set:

1.
+9	−8	−7	+10	+12	−15	+18
+4	−3	+2	−4	+18	−13	−11

2.
+8	−6	+9	−7	−23	−25	+18
−8	+6	0	0	+10	+24	−17

3.
−7	−4	0	0	−23	+30	−42
6	9	12	−10	10	−30	+41

4.
+8	−6	+7	8	−4	−5	−8
−3	+5	+9	1	−8	+9	−2
7	−9	−1	+9	+6	−4	4

5.
5	−7	+8	−6	−3	6	9
−4	1	+9	−8	−7	−9	+1
+9	−4	−2	5	−8	3	0

6.
20	−40	30
−14	21	15
10	−15	−10
13	+18	+12
−11	−10	−13

7.
−14	−27	−29
−6	−16	24
−13	16	−16
−20	−32	−11
16	14	32

8.
−328	493	−143	+48.3	−56.3	−23.4
−157	−218	+486	+15.6	+21.8	−17.6

9.
+585	−623	−173	−36.4	+64.9	−3.82
+158	+216	−246	73.8	−29.4	−1.56

17.4. SUBTRACTION OF SIGNED NUMBERS

Subtraction of signed numbers presents a little extra difficulty. It can best be understood if you remember that *subtraction* is the *inverse* or *opposite of addition*.

In subtraction we have three terms: the *minuend*, the *subtrahend*, and the *remainder*. The subtrahend is the number subtracted from the minuend. When we subtract one number from another, the result is called the difference or the *remainder*. Subtraction can always be checked by adding the remainder to the subtrahend. The result should be the minuend. Let us take a problem from arithmetic:

Example. Subtract 32 from 47. The work is shown.

$$\begin{array}{rr} \text{minuend} & 47 \\ \text{subtrahend} & 32 \\ \hline \text{remainder} & 15 \end{array}$$ add these to check

Now let us consider subtraction involving signed numbers. We shall compare it with addition. Suppose we have the two numbers: +10 and +3.

$$\text{Addition:} \quad \begin{array}{r} +10 \\ +3 \\ \hline +13 \end{array} \qquad \text{Subtraction:} \quad \begin{array}{r} +10 \\ +3 \\ \hline +7 \end{array} \quad \text{Check by adding}$$

Notice that the +3 increased the +10 by 3 points, which becomes 13. Since subtraction is the inverse of addition, then subtracting +3 from +10 will decrease the +10 by 3 points, which then becomes 7.

Notice especially:

If we subtract +3 from +10, we get +7.
If we add −3 to +10, we get 7.

Subtracting +3 from +10 is the same as adding −3 to +10.

Now, consider the two numbers, +20 and −4. Let us compare addition and subtraction:

$$\text{Addition:} \quad \begin{array}{r} +20 \\ -4 \\ \hline +16 \end{array} \qquad \text{Subtraction:} \quad \begin{array}{r} +20 \\ -4 \\ \hline +24 \end{array}$$

When we add −4 to +20, we get +16. Adding −4 to +20 decreases the +20 and we get +16. Since subtraction is the inverse of addition, then subtracting −4 from 20 increases the +20 by +4, and we get 24. That is, subtracting −4 from +20 is the same as adding +4 to +20.

The result is reasonable if we look upon the problem as follows. Suppose you have several scores, including −4, that add up to +20. Now, if you are

228 ALGEBRA

permitted to subtract (erase) a bad score of −4, your score will increase by 4 points. This is the meaning of subtraction. As another example, if a man is permitted to *erase* a debt of $4, he increases his worth by $4. That is, subtracting a −4 is the same as adding a +4.

The foregoing examples lead to the rule for subtraction involving signed numbers.

RULE. To subtract a signed number from another, mentally change the sign of the number subtracted and then proceed as in algebraic addition.

Note. Algebraic subtraction can be checked by addition in the same way as subtraction in arithmetic. The remainder added to the subtrahend should be equal to the minuend.

Examples showing subtraction of signed numbers. The remainder is shown. *Subtract* the bottom number from the top:

	+9	+10	−3	−8
−	+3	−6	−7	+4
	+6	+16	+4	−12

Check each of the examples by adding the remainder to the subtrahend. The result should be the minuend.

Exercise 17.2

In each set of numbers, subtract the bottom number from the top:

1. +9   +7   −8   −5   2. +8   +5   −7   −3
    +2   −3   −2   −9   +3   −4   −3   −9

3. +4   −8   +6   0   4. +7   −7   +9   0
    +7   +5   +6   +5   +4   −3   +9   −6

5. +13 +10 −12 −11   6. +9   −14 −15 −12
    +5   −4   −5   −16   +12 +6   −10 −23

7. −13 −16 −21 −13   8. −18 +23 −13   25
    52   −16 39   −12   29   23   −42 −11

9. 36   43   −26 −15 10. 48   55   −32 −21
    15   −14 −18 −28   16   −17 −15   49

11. 0   0   487   3.26 12. 0   0   468   4.17
    14   −23 153 −4.12   +16 34   213 −3.81

17.5. HORIZONTAL ADDITION AND SUBTRACTION

Sometimes we have signed numbers arranged horizontally. Then we must be careful to distinguish between the *signs of operation*, such as addition and subtraction, and the *signs of the numbers* themselves. Suppose we have the problem of adding the following numbers:

$$+9, -3, +6, \text{ and } -4.$$

```
 +9
 +3
 +6
 +4
 ---
 +8
```

Let us first place the numbers in vertical arrangement for addition as shown at the left. The sum is +8.

If we place these numbers in a horizontal line, we can indicate addition by placing the plus sign (+), between them to show the operation of addition. However, then we must enclose each of the numbers in parentheses so as not to confuse the signs. We get

$$(+9)+(-3)+(+6)+(-4)$$

The plus signs between the signed numbers are the signs of addition. The signs inside the parentheses are the signs of positive and negative numbers.

Now, if we omit the plus signs of operation and also omit the parentheses, we can write each number with its proper sign:

$$+9-3+6-4=8$$

Note that the numbers can be combined as if they were written in vertical form. For horizontal addition, we have this rule:

RULE. To add signed numbers horizontally, simply write each number with its proper sign and omit the signs of addition. Then combine the numbers.

Example. Combine $(+4)+(-7)+(+6)+(-8)$.
Writing only the signed numbers, we get $+4-7+6-8=-5$

To indicate subtraction in horizontal form, we enclose each signed number in parentheses, and then place the subtraction sign, (−), between them. For example, to subtract +4 from +11, we write

$$(+11)-(+4)$$

Now we recall that in subtraction, we change the sign of the subtrahend and then add. Then, instead of writing the problem as subtraction, we write it as addition and change the sign of the subtrahend. Then we get

$$(+11)+(-4)$$

Now we follow the rule concerning horizontal addition, and write

$$+11-4=7$$

RULE. In horizontal subtraction, omit the signs of subtraction and simply write each number to be subtracted with its sign changed. Then combine the terms as in horizontal addition.

Examples.

(a) $(-7)-(+3)=-7-3=-10$
(b) $(+4)-(-9)=4+9=13$
(c) $(+6)-(+5)=6-5=1$
(d) $(+3)-(+8)=3-8=-5$

When horizontal additions and subtractions occur in the same problem in a horizontal arrangement of signed numbers, the signs of operation, (+ and −), may be omitted provided the rules for horizontal addition and subtraction are observed.

Example. Combine

$$(+9)+(-2)-(+3)-(-4)+(+5)$$

Simplifying

$$+9-2-3+4+5=13$$

Example. Combine

$$(-3)+(-4)-(+5)-(-6)$$

Simplifying,

$$-3-4-5+6=-6$$

Exercise 17.3

Combine each of the following terms:

1. $(+6)+(-3)-(-4)-(+3)+(+2)$
2. $(+8)+(+7)-(-5)+(-2)-(+3)$
3. $(-9)+(+3)-(4)+(-6)-(-5)$
4. $(-7)-(2)+(-3)-(-8)+(4)$
5. $(3)-(-5)-(2)+(-6)-(7)$
6. $(-4)+(-8)-(3)-(-4)+(1)$
7. $(6)+(+7)+(-3)+(+4)+(-5)+(+9)$
8. $(+9)+(-7)+(+6)+(-3)+(+8)+(-5)$
9. $(+7)+(-2)+(+4)+(-5)+(+9)-(+6)$
10. $(3)+(-6)-(+9)+(-6)+(+8)-(+5)$

SIGNED NUMBERS 231

11. $(-5)+(-8)+(-1)-(-7)+(-3)-(-9)$
12. $(-7)+(+9)+(-2)+(+5)-(-6)-(+2)$
13. $(8)+(-7)+(+4)+(-6)-(+7)-(-3)-(-5)$
14. $(9)+(-6)+(+2)+(-8)-(-1)-(+5)-(-7)$
15. $(+10)-(+6)-(-5)+(-3)-(-12)-(+11)-(+7)$
16. $(12)-(+8)-(-7)+(-4)-(-11)-(13)+(-5)$
17. $(-15)-(-20)+(-8)-(+6)-(-10)-(+12)-(2)$
18. $(-18)-(15)-(-16)+(-13)-(-11)-(10)-(-1)$

17.6. MULTIPLICATION OF SIGNED NUMBERS

In multiplying signed numbers, we multiply the absolute values in the same way as in arithmetic. Our main problem then is to determine the sign of the product.

```
48
48
48
---
```

Multiplication can be considered as a shortened form of addition. In a problem in arithmetic, such as (3×48), the multiplier is 3. It shows how many times the multiplicand, 48, is to be added. As addition, the problem may be written as shown at the left.

Now, suppose we have the following problem in the multiplication of signed numbers: $(+3) \times (+4)$. Let us write one number above the other for multiplication:

$$\begin{array}{r} +4 \\ +3 \\ \hline \end{array}$$

The absolute value of the product is 12. Our problem then is to determine the sign of the product.

Suppose we consider the plus sign (+) before the 3 as the sign of the operation of addition. Then, since multiplication is a short form for addition, the problem means

add three "+4's,"   or $\begin{array}{r} +4 \\ +4 \\ +4 \\ \hline +12 \end{array}$

Then it can be seen that the product is a positive number, +12. Again, if we look upon the +4's as scores in a game, the problem means the following:

adding three +4's to your score changes your score by +12

As a second example, take the multiplication: $(+3) \times (-4)$. The absolute value of the product is 12. To determine the sign in this case, consider that the problem means: *add* three "−4's." Adding three −4's to your score changes your score by −12. Therefore,

$$(+3) \times (-4) = -12$$

Example 3. Multiply $(-3) \times (+4)$; or in column form:

$$\begin{array}{r} +4 \\ \underline{-3} \end{array}$$

Again, the absolute value of the product is 12. To determine the sign of the product, consider the minus sign (−) before the multiplier, 3, as the sign of subtraction. Then the problem means

subtract three "+4's"

Subtracting three +4's from your score changes your score by −12. Therefore,

$$(-3) \times (+4) = -12$$

Example 4. Multiply $(-3) \times (-4)$; or in column form:

$$\begin{array}{r} -4 \\ \underline{-3} \end{array}$$

Again, the absolute value of the product is 12. To determine the sign of the product, consider the minus sign (−) before the multiplier, 3, as a sign of subtraction. Then the problem means

subtract three "−4's"

Subtracting three −4's from your score changes your score by +12. Therefore,

$$(-3) \times (-4) = +12$$

From the foregoing examples we can formulate the rule for multiplying signed numbers.

RULE. The product of two numbers with like signs is a positive number. The product of two numbers with unlike signs is negative.

SIGNED NUMBERS

In symbols, the rule for the sign of a product can be stated in this way:

$$(+) \cdot (+) = + \qquad (+) \cdot (-) = -$$
$$(-) \cdot (-) = + \qquad (-) \cdot (+) = -$$

However, remember that this way of showing the rule for the multiplication of signed numbers is only symbolic, and the signs themselves are not to be considered as factors. The signs indicate only the kinds of numbers.

Examples.

$(+25)(-8) = -200$ $\qquad$ $(-12)(-9) = +108$

$(+8)(15) = +120$ $\qquad$ $(-7)(14) = -98$

$(-12)(-12) = +144$ $\qquad$ $(-9)^2 = +81$

$(+9)^2 = +81$ $\qquad$ $(-4)(-3)(+5) = +60$

$(-6)(-2)(-7) = -84$ $\qquad$ $(-\frac{1}{2})(-16)(+\frac{3}{4}) = +6$

Note. An odd number of negative factors will make the product negative, regardless of the number of positive factors. An even number of negative factors will make the product positive.

Exercise 17.4

Multiply the signed numbers in each of the following sets:

1. $(-3)(-2)(-4) =$
2. $(-2)(-1)(+5) =$
3. $(-6)(+3)(-5) =$
4. $(3)(-5)(-4) =$
5. $(-4)(-6)(+5) =$
6. $(-6)(+4)(-2) =$
7. $(+7)(-2)(-3)(-1) =$
8. $(-1)(+3)(-8)(+3) =$
9. $(-8)(-1)(+2)(+5) =$
10. $(-3)(-4)(-1)(-9) =$
11. $(-5)(+6)(-1)(-10) =$
12. $(-8)(+5)(-6)(-5) =$
13. $(+1.2)(+2.5)(-3.4) =$
14. $(-2.4)(+1.5)(-3.1) =$
15. $(-21)(-32)(-1)(-15) =$
16. $(18)(-25)(-16)(-25) =$
17. $(-\frac{3}{8})(-\frac{5}{6})(+\frac{4}{15}) =$
18. $(-\frac{2}{3})(-\frac{5}{8})(-\frac{3}{10}) =$
19. $(-3)(-2)(-1)(-5)(-4) =$
20. $(-5)(+3)(-1)(+1)(-6) =$
21. $(-2)(+6)(-1)(-1)(-6) =$
22. $(+6)(-3)(-2)(+5)(-1) =$

Multiply:

23. -32
    $\underline{-25}$

24. $+45$
    $\underline{-0.34}$

25. 6.5
    $\underline{-36}$

26. 0.24
    $\underline{2.1}$

17.7. DIVISION OF SIGNED NUMBERS

In division of signed numbers we have the same rules as in multiplication.

RULE. **In the division of numbers with like signs, the quotient is positive. If the signs are unlike, the quotient is negative.**

Examples.

$$(+12) \div (+4) = +3 \qquad (+12) \div (-4) = -3$$

$$(-12) \div (-4) = +3 \qquad (-12) \div (+4) = -3$$

Division can be checked by multiplication in the same way as in arithmetic; that is, if the divisor is multiplied by the quotient, the answer should be the dividend. This method of checking division applies also to the signs as well as to the numerical values.

In symbols, the rule for the sign of the quotient in division may be stated in this way:

$$(+) \div (+) = + \qquad (+) \div (-) = -$$

$$(-) \div (-) = + \qquad (-) \div (+) = -$$

Exercise 17.5

Divide as indicated:

1. $(+15) \div (+3) =$
2. $(-18) \div (-6) =$
3. $(-20) \div (+4) =$
4. $(-5) \div (-5) =$
5. $(+3) \div (+3) =$
6. $(+6) \div (-6) =$
7. $(-24) \div (+2) =$
8. $(36) \div (-9) =$
9. $(-42) \div (-7) =$
10. $(48) \div (-16) =$
11. $(-60) \div (+6) =$
12. $(+56) \div (+14) =$
13. $(120) \div (-8) =$
14. $(144) \div (-9) =$
15. $(184) \div 23 =$
16. $(-84) \div (-12) =$
17. $(-72) \div (-18) =$
18. $(-96) \div (-16) =$
19. $(+20) \div (-8) =$
20. $(-30) \div (+12) =$
21. $(15) \div (-7) =$
22. $(-\frac{5}{8}) \div (+\frac{3}{4}) =$
23. $(+\frac{2}{3}) \div (-\frac{5}{6}) =$
24. $(-\frac{3}{5}) \div (+\frac{4}{7}) =$
25. $\dfrac{-736}{-8}$
26. $\dfrac{+492}{+12}$
27. $\dfrac{+396}{-18}$
28. $\dfrac{-360}{+15}$

29. $\dfrac{-0.96}{+1.2}$

30. $\dfrac{-8.4}{-0.14}$

31. $\dfrac{+34.5}{+0.15}$

32. $\dfrac{+0.546}{-2.1}$

33. $\dfrac{+120}{-45}$

34. $\dfrac{-72}{+32}$

35. $\dfrac{-64}{-40}$

36. $\dfrac{-85}{-25}$

18 ALGEBRAIC EXPRESSIONS

18.1. DEFINITIONS

An *algebraic expression* is any expression that contains a literal number. The following are algebraic expressions:

$$3n; \quad 5x; \quad 4ab; \quad 2x+3y; \quad x^2+5x+1; \quad 3x^2-4xy+y^2-3x$$

Even an arithmetic number alone, such as 3, can be said to be an algebraic expression as used in algebra.

Now we must know the different kinds of expressions. We need a few specific definitions.

A *product* is the answer obtained by multiplying two or more quantities together. For example, in the multiplication, $3 \cdot 5 = 15$, the product is 15. The word *product* in mathematics always implies *multiplication*.

The quantities multiplied together are called the *factors* of the product. That is, the numbers 3 and 5 are the factors of 15. The factors of 30 are 2, 3, and 5. The factors of the product xy are x and y. The factors of $4n$ are 4 and n. The factors of $5xy$ are 5, x, and y.

A *coefficient* is one of *two factors* of a product. When two numbers are multiplied together, as $3 \cdot 7$, one factor is called the coefficient of the other. In this example, 3 is the coefficient of 7, and 7 is the coefficient of 3. In the product $5n$, 5 is the *numerical coefficient* of n, and n is the *literal coefficient* of 5. The word *co-efficient* means *working together*, or to be *efficient together*. The word *coefficient* alone usually refers to the *numerical* factor. The numerical coefficient of $3xyz$ is 3.

A product may have three or more factors, such as the product $5xy$. However, when we use the word *coefficient*, we think of the product separated into only *two* factors. Then either factor is the coefficient of the other. For example, in the product $5xy$,

5 is the coefficient of xy; $5x$ is the coefficient of y
$5y$ is the coefficient of x; x is the coefficient of $5y$
y is the coefficient of $5x$; xy is the coefficient of 5

238 ALGEBRA

If a literal number is shown without a numerical coefficient, its coefficient is always understood to be 1. The numerical coefficient of x^2 is 1. The numerical coefficient of $-xy$ is -1.

A *term* is any algebraic expression not separated within itself by a plus or minus sign. *A term consists of only factors.* The following expressions are single terms:

$$3xy; \quad 4x^2y^3; \quad +5abcd; \quad -2x; \quad y; \quad 6; \quad -20x^3y^4a^2b^3c$$

A term indicates a *product*, not a sum or difference. The *sign* of a term is the sign before it. If no sign is shown before the term, then the term is understood to be positive, and its sign is $+$. For example, the sign of $+3xy$ is $+$; the sign of $4ab$ is $+$; the sign of $-5xy$ is $-$.

Like terms (or *similar terms*) are terms that are *exactly alike in the letter part*. The following terms are *like* terms:

$$5x^2y; \quad -3x^2y; \quad +7x^2y; \quad -x^2y$$

Unlike terms (or *dissimilar terms*) are terms that are *not exactly alike in the letter part*. The following are unlike terms:

$$5x^2y; \quad -3xy^2; \quad -3axy; \quad 5x^3y; \quad -x; \quad 3y^2; \quad -3axy^2; \quad 7$$

A *monomial* is an algebraic expression consisting of *only one term*. The following expressions are *monomials*: (the prefix "mono" means *one*)

$$5x^2y; \quad -3xy^2; \quad -3x^2y^2; \quad -5x^2y; \quad 3xy^2$$

A *polynomial* is an algebraic expression of *more than one term*. (The prefix "poly" means many, or more than one). The following expressions are *polynomials*:

$$x-y; \quad 3x^2+4y^2; \quad x^2-3x+1; \quad 7x^3+4x^2y-5xy^2-6x+3$$

A polynomial of *two* terms is often called a *binomial*. A polynomial of *three* terms is called a *trinomial*.

binomials: $3x^2-4xy; \quad 5x-1; \quad 4x^2y^3-3xy^4; \quad x+1$

trinomials: $x^2-7x+1; \quad 4x-5xy+6y; \quad a+b+c; \quad r+s-t$

In Chapter 15, we defined an *exponent*. The meaning of an exponent must be thoroughly understood to avoid errors in mathematics. Perhaps more mistakes are made in mathematics because of a misunderstanding of the meaning of an exponent than for any other reason. Let us repeat the definition.

An exponent is a number placed at the right of and a little above another number, called the *base.* The exponent shows how many times the base is to be used as a factor. *The base is the number to be multiplied by itself.*

For example, the expression,

$$8^2 \quad \text{means} \quad 8 \cdot 8 = 64$$

The small 2 near the 8 means that two 8's are to be multiplied together. The expression is read, "8 squared," or "8 raised to the second power." The expression, x^3 means: $x \cdot x \cdot x$. It is read, "x cubed", or "x to the third power."

If any number is shown without an exponent, its exponent is understood to be 1. Thus the number 5 can be written: 5^1. Also, $x = x^1$.

18.2. ADDITION AND SUBTRACTION OF MONOMIALS

In adding monomials consisting of only like terms, we simply add their numerical coefficients. The letter part is not changed. We might say that the *name* of the term is the same. The following examples show *addition* of monomials consisting of like terms: *Add* the following:

+5 miles	$7x^2y$	$-7xy^2$	$+2abc$	$5xy$	$-4mn$	6 dollars
+3 miles	$2x^2y$	$3xy^2$	$-3abc$	$-xy$	mn	-2 dollars
+8 miles	$9x^2y$	$-4xy^2$	$-abc$	$4xy$	$-3mn$	4 dollars

In subtraction involving like terms, we follow the same rule as for signed numbers (Chapter 17). That is, we *mentally change the sign of the term subtracted,* and *then add algebraically.* In the following examples, the bottom monomial is subtracted from the top. *Subtract*:

$+8xy$	$+7x^2y$	$-4abc$	$-7x^3y$	$-5m$	$4x^2y^3$	xy^2
$+3xy$	$-2x^2y$	$-5abc$	$-6x^3y$	$+2m$	$5x^2y^3$	$-xy^2$
$5xy$	$+9x^2y$	abc	$-x^3y$	$-7m$	$-x^2y^3$	$2xy^2$

If two terms are *unlike,* their sum or difference can only be indicated. The following examples show *addition* of *unlike terms*: Add:

5 cents	$8x$	$-2a$	$6x^2$	$-4m$
3 yen	$5y$	$-3b$	$-2x$	$+3n$
5 cents + 3 yen	$8x + 5y$	$-2a - 3b$	$6x^2 - 2x$	$-4m + 3n$

In subtraction involving *unlike* terms, we use the rule for subtraction. In each of the following examples, the *bottom monomial* is *subtracted from the*

240 ALGEBRA

top. Subtract:

$$\begin{array}{cccccc} +8x & -3a & 3xy & -7x^2y & m & xy \\ +3y & +4b & -5x^2 & 2xy^2 & n & -x \\ \hline +8x-3y & -3a-4b & 3xy+5x^2 & -7x^2y-2xy^2 & m-n & xy+x \end{array}$$

Exercise 18.1

Add the sets of monomials in each of the following sets, No. 1–8:

1. $\begin{array}{c} 4x \\ 5x \end{array}$ $\begin{array}{c} -3x^2 \\ -6x^2 \end{array}$ $\begin{array}{c} -6xy \\ xy \end{array}$ $\begin{array}{c} +7n \\ -3n \end{array}$ $\begin{array}{c} 0 \\ -4ab \end{array}$ $\begin{array}{c} -4x^2y \\ 4x^2y \end{array}$ $\begin{array}{c} 6x^2y^2 \\ x^2y \end{array}$

2. $\begin{array}{c} 6y \\ 2y \end{array}$ $\begin{array}{c} -4n^2 \\ -8n^2 \end{array}$ $\begin{array}{c} -7st \\ st \end{array}$ $\begin{array}{c} +2a \\ -3a \end{array}$ $\begin{array}{c} 0 \\ -5x^2y \end{array}$ $\begin{array}{c} 4ab \\ -3a \end{array}$ $\begin{array}{c} -9y^2z \\ y^2z \end{array}$

3. $\begin{array}{c} -3a \\ 7a \end{array}$ $\begin{array}{c} -4n \\ 0 \end{array}$ $\begin{array}{c} 2x^3 \\ -8x^3 \end{array}$ $\begin{array}{c} -5xy^2 \\ -x^2y \end{array}$ $\begin{array}{c} -9x^3y \\ 2x^3y \end{array}$ $\begin{array}{c} -5xy^3 \\ -5xy^3 \end{array}$ $\begin{array}{c} -8b^2c \\ -2bc \end{array}$

4. $\begin{array}{c} -6c \\ 9c \end{array}$ $\begin{array}{c} -7n \\ 0 \end{array}$ $\begin{array}{c} 4a^2 \\ -5a^2 \end{array}$ $\begin{array}{c} -7xy^2 \\ -7xy^2 \end{array}$ $\begin{array}{c} -8x^3 \\ x^3 \end{array}$ $\begin{array}{c} 8xy^3 \\ 7xy^3 \end{array}$ $\begin{array}{c} -6b^2 \\ -3d \end{array}$

5. $\begin{array}{c} xy^2 \\ 6xy^2 \end{array}$ $\begin{array}{c} -2st^2 \\ -7st^2 \end{array}$ $\begin{array}{c} -9ab^3 \\ 8ab^3 \end{array}$ $\begin{array}{c} -2x^2y \\ 2x^2y \end{array}$ $\begin{array}{c} -rs \\ 3rs \end{array}$ $\begin{array}{c} -6x^2y^2 \\ -6x^2y^2 \end{array}$ $\begin{array}{c} 4xyz \\ 4xyz \end{array}$

6. $\begin{array}{c} an^3 \\ 7an^3 \end{array}$ $\begin{array}{c} -4xy \\ -9xy \end{array}$ $\begin{array}{c} -7x^2y \\ 6x^2y \end{array}$ $\begin{array}{c} -4rst \\ +4est \end{array}$ $\begin{array}{c} -4m^2n^3 \\ 3m^2n \end{array}$ $\begin{array}{c} -8x^3y \\ -8x^3y \end{array}$ $\begin{array}{c} -20x^2y \\ 20x^2y \end{array}$

7. $\begin{array}{c} 6x^3 \\ -5x^3 \end{array}$ $\begin{array}{c} -3ax^3 \\ 4ax^3 \end{array}$ $\begin{array}{c} -bx \\ 7bx \end{array}$ $\begin{array}{c} 8xy^2 \\ 8xy^2 \end{array}$ $\begin{array}{c} 7a^2b^2 \\ -5a^2b^2 \end{array}$ $\begin{array}{c} -4yz \\ -3x \end{array}$ $\begin{array}{c} -3a^2c^3d \\ -4a^2c^3d \end{array}$

8. $\begin{array}{c} -7ax \\ 6ax \end{array}$ $\begin{array}{c} 5m^2n \\ -6m^2n \end{array}$ $\begin{array}{c} -5x^2y \\ x^2y \end{array}$ $\begin{array}{c} -9xy \\ -9xy \end{array}$ $\begin{array}{c} -12ab^3 \\ 10ab^3 \end{array}$ $\begin{array}{c} 5x^2y^3 \\ 9xy \end{array}$ $\begin{array}{c} -6x^2y^4z \\ -5x^2y^4z \end{array}$

No. 9–16. Subtract the bottom term from the top in No. 1–8 above.

Add the monomials in each of the following exercises:

17. $\begin{array}{c} 6x^2y \\ -3x^2y \\ 5x^2y \\ -x^2y \end{array}$

18. $\begin{array}{c} xy^3 \\ -7xy^3 \\ 2xy^3 \\ 3xy^3 \end{array}$

19. $\begin{array}{c} -5x^2y^3 \\ x^2y^3 \\ -8x^2y^3 \\ 6x^2y^3 \end{array}$

20. $\begin{array}{c} -a^2b \\ 4a^2b \\ -6a^2b \\ 3a^2b \end{array}$

21. $\begin{array}{c} -x^2y \\ 4x^2y \\ -5x^2y \\ 3x^2y \end{array}$

22. $\begin{array}{c} 3mn^2 \\ -8mn^2 \\ 5mn^2 \\ -mn^2 \end{array}$

ALGEBRAIC EXPRESSIONS 241

18.3. HORIZONTAL ADDITION AND SUBTRACTION

If terms are arranged horizontally, they can be combined just as we combine any positive and negative numbers. For example, in the expression,

$$5x + 2x$$

the two terms can be combined and we get $7x$. The plus sign (+) between the two terms may be considered either as the sign of the $2x$, or as the sign of addition.

In the expression, $4n - n$, we can combine the two terms and get $3n$. Whatever number is represented by n, if we take the number from 4 times the number, we get 3 times the number. The minus sign (−) between the two terms can be considered either as the sign of the term n, or as the sign of subtraction. If we add $(+4n) + (-n)$, we get $3n$. If we take the minus sign as the sign of subtraction, we have: $(+4n) - (+n) = 3n$. To express the quantity, $5x$ minus $3x$, we write: $(5x) - (+3x) = 2x$. On the other hand, we could write: $(5x) + (-3x) = 2x$. The answer is the same.

It is sometimes important that we distinguish between the sign of a number and the sign of operation. Consider the expression

$$8x + (3x)$$

Here the plus sign (+) is the sign of addition. The sign of the $3x$ is not shown, so it is +. Consider another example:

$$6x - (2x)$$

Here, the minus sign between the terms is a sign of subtraction. The sign of the $2x$ is not shown; therefore, it is $+2x$. As another example:

$$7n + (-2n)$$

Here the plus sign is the sign of addition. The minus sign before the $2n$ is the sign of the term, which is $-2n$, which means a negative $2n$.

If several terms are arranged horizontally for addition or subtraction, or both, the like terms can be combined.

Example 1. Combine like terms in the following algebraic expression:

$$5x + 2y - 3n + x - 4y + 2n + 6 - n - 4x + y$$

Solution. Combining the x terms, we get: $5x + x - 4x = 2x$

Combining the y terms, we get: $2y - 4y + y = -y$

Combining the n terms, we get: $-3n + 2n - n = -2n$

The number 6 cannot be combined with any of the other terms. For the complete answer, we get

$$2x - y - 2n + 6$$

If some terms are enclosed in parentheses, the parentheses may be removed provided we follow the same rules as those for combining signed numbers. *For a term within parentheses, if the parentheses are preceded by a plus sign, omit the plus sign (+) before the parentheses and omit the parentheses. Then write each term with its proper sign and combine like terms.*

Example 2. Combine like terms in the following expression:

$$8x + (+3x) + (-4x) + (-x)$$

Solution. Removing parentheses, we get $8x + 3x - 4x - x = 6x$, answer.

If the parentheses are preceded by a minus sign, omit the minus sign of subtraction, and omit the parentheses. Then write the term with its sign changed.

Example 3. Combine like terms in the following expression:

$$6n - (+3x) - (-2n) - (4x) - (-y)$$

Solution. Removing parentheses, we get

$$6n - 3x + 2n - 4x + y = 8n - 7x + y$$

When both additions and subtractions of signed terms appear in the same problem in horizontal arrangement with the signs of operation (+ and −) between the terms, then we may omit these signs of operation and also omit the parenthesis, provided we observe the rules for removing parentheses.

Example 4. Remove the parentheses and combine like terms in the following algebraic expression:

$$(+7x) - (-8y) + (+6) - (+4x) + (-2x) - (3y) + (-2y) + (-4)$$

Removing parentheses,

$$7x + 8y + 6 - 4x - 2x - 3y - 2y - 4$$

Combining like terms,

$$x + 3y + 2$$

18.4. ADDITION AND SUBTRACTION OF POLYNOMIALS

To add or subtract polynomials, we write one polynomial below the other so that like terms fall in the same vertical column. Then we add or subtract in each column separately.

Example 1. Add the following polynomials:

$$3x^2 - 5xy + 7y^2; \quad a^2 + 3xy - x^2 - 4y^2; \quad 4x^2 + 3 - 2a^2 - 3y^2$$

Solution. We place one polynomial below the other with like terms in line:

$$\begin{array}{l} 3x^2 - 5xy + 7y^2 \\ -x^2 + 3xy - 4y^2 + a^2 \\ \underline{4x^2 - 3y^2 - 2a^2 + 3} \\ 6x^2 - 2xy - a^2 + 3 \end{array}$$

Adding each column, we get

Adding each column, we get

Note that the coefficients of y^2 add up to zero, which makes the sum zero.

Example 2. Subtract the polynomial, $4y^2 - 3x^2 - 2xy - 2y$, from the polynomial, $4x^2 - 2xy + 3y^2 - 5x$.

Solution. The polynomial to be subtracted is placed below the other.

Subtract
Subtracting in each column

$$\begin{array}{l} 4x^2 - 2xy + 3y^2 - 5x \\ \underline{-3x^2 - 2xy + 4y^2 - 2y} \\ 7x^2 - y^2 - 5x + 2y \end{array}$$

18.5. HORIZONTAL ADDITION AND SUBTRACTION OF POLYNOMIALS

The addition and subtraction of polynomials can be indicated horizontally by use of parentheses or other signs of grouping. The terms enclosed in parentheses must be considered as a single quantity. The addition of two polynomials, such as $3x^2 - 2xy - 4y^2$ and $5x^2 + 3xy - 4y^2$, can be shown by enclosing each in parentheses and placing the addition sign between them:

$$(3x^2 - 2xy - 4y^2) + (5x^2 + 3xy - 4y^2)$$

You recall that when we add like terms, we simply combine the terms according to the rules for algebraic addition. The plus sign (+) between the polynomials indicates addition. We have seen that none of the signs are changed in addition. Therefore, we can remove the parentheses without changing any signs of the terms, and then combine the like terms. Removing parentheses, we get

$$3x^2 - 2xy - 4y^2 + 5x^2 + 3xy - 4y^2 = 8x^2 + xy - 8y^2$$

Horizontal subtraction involving polynomials can be indicated by enclosing each polynomial in a set of parentheses and then placing the minus sign (−) for subtraction between the two polynomials. For example, to subtract the polynomial $3x^2 + 2x - 3$ from the polynomial, $5x^2 - x - 5$, we write

$$(5x^2 - x - 5) - (3x^2 + 2x - 3)$$

Before like terms can be combined, the parentheses must be removed. In doing so, we omit the minus sign of subtraction, and then change the sign of each term in the polynomial after the subtraction sign. Then we can combine the terms. We get

$$5x^2 - x - 5 - 3x^2 - 2x + 3 = 2x^2 - 3x - 2$$

Exercise 18.2

Horizontal addition and subtraction. Combine the following:

1. $3a + 5b - 3c - 2b + 4a - c + a - ab + 2c + 3b - 2a + 2ab - b$
2. $4x - (3y) - (-3x) + (-2z) - (2y) + (4z) - (-5x) - (z) + y - 5$
3. $5x + (-2y) - (-z) + (-2x) + (3y) - (+4z) - (-3x) + (-y) - (-2)$
4. $2ab + 3a^2 - (+5b^2) - (-ab) + (-a^2) - (-3b^2) - (-a) + (-4)$

Add the following polynomials by placing one below the other with like terms in line vertically:

5. Add: $4x^2 - 3xy + 5y^2$; $5xy - 4y^2 + x^2$; $-5x^2 - xy - 2y^2$
6. Add: $3x^3 - 4x^2y + 5y^3$; $3x^2y - x^3 - 2y^3$; $4y^2 - 4x^3 + 2x^2y - 3y^3$
7. Subtract $4x^2 + xy - y^2 + 2x$ from the polynomial, $5x^2 - 3xy + 4y^2$, by placing the polynomial to be subtracted below the other.
8. Subtract the quantity, $-4x^2 - xy - 3x - 2y$ from $-x^2 - 5xy + 2y^2 - 3x$, by placing the quantity to be subtracted below the other quantity.

Combine the terms in each of the following expressions by removing parentheses, and combining the terms horizontally:

9. $(4x^2 + 3xy - 2y^2) + (3y^2 - 2xy + x^2 + 5)$
10. $(5x^2 - 3xy - 6y^2 - 2x + y) + (2xy - 5y^2 - 2x^2 + 2x + y - 4)$

11. $(y+2xy-3x^2-y^2+6)+(2x^2+y^2-3x-2xy-5-2y)$
12. $(3nx-4x^2+5n^2-3n-x)+(5x^2-3n^2-nx+x-n)$
13. $(4x^2-5xy+y^2-3x+2y)-(3x^2-4xy-2y^2-2x-y)$
14. $(y^2+3y+x^2+2xy)-(4x^2+2x+4xy-y^2)$
15. $(x^2y-xy^2+5x^2y^2-6x+2y)-(3x^2y+4x^2y^2-2xy^2-5x+y)$
16. $(x^3-4x^2+4x+5)-(7x^3+4x^2+5x-3+2y)$

19 MONOMIALS: MULTIPLICATION AND DIVISION

19.1. EXPONENTS

We have already defined the meaning of an exponent. In multiplication and division it is essential that we understand clearly the meaning of the term. We repeat the definition of an exponent by examples.

In the expression, 5^2, the base is 5. The exponent, 2, indicates that the two 5's are to be multiplied together. Then

$$5^2 \text{ means } 5 \cdot 5, \text{ which is equal to } 25$$

In the expression, x^3, the base is x. The exponent, 3, indicates that three x's are to be multiplied together. Then

$$x^3 \text{ means } x \cdot x \cdot x; \qquad x^5 \text{ means } x \cdot x \cdot x \cdot x \cdot x; \qquad 10^4 = 10000$$

19.2. EXPONENTS IN MULTIPLICATION

Let us now see the result of multiplying two powers of the same base, such as $(x^2)(x^3)$. We write the factors of each expression.

$$x^2 \quad \text{means} \quad x \cdot x \quad \text{and} \quad x^3 \quad \text{means} \quad x \cdot x \cdot x$$

Then the product

$$(x^2) \cdot (x^3), \quad \text{means} \quad (x \cdot x)(x \cdot x \cdot x)$$

The product can be expressed by using another exponent:

$$(x^2) \cdot (x^3) = x^5$$

It appears that we can add the two exponents and get the exponent of the product.

247

If we analyze in the same way the multiplication of two other powers on the same base, such as, (x^5) (x^7), we can add the exponents on x in each factor and get the exponent of the product:

$$(x^5)(x^7) = x^{12}$$

From these examples we can formulate the rule for the multiplication of algebraic terms involving exponents.

RULE. In multiplying algebraic terms expressed as powers, we add exponents of the same base. (The base must be the same.) (We are not multiplying exponents; we are multiplying terms.)

The rule may be stated as a formula: if a and b are exponents on x,

$$(x^a) \cdot (x^b) = x^{a+b}$$

We illustrate the rule by use of arithmetic numbers:

$4^2 \cdot 4^3 = 4^5$;   that is, $4^2 = 16$; $4^3 = 64$; $4^5 = (16)(64) = 1024$

As another example, $(5^3) \cdot (5^4) = 5^7$; that is, $5^3 = 125$ and $5^4 = 625$. Then, $(5^3) \cdot (5^4)$ means $(125)(625) = 78,125$; and $5^7 = 78,125$.

The rule holds true for all kinds of exponents and for any base:

fractional exponents: $(x^{1/2}) \cdot (x^{1/3}) = x^{5/6}$
decimal exponents: $(10^{2.3}) \cdot (10^{3.5}) = 10^{5.8}$
negative exponents: $(x^7) \cdot (x^{-2}) = x^5$
zero exponent: $(x^3)(x^0) = x^3$
literal exponents: $(y^n) \cdot (y^c) \cdot (y) = y^{(n+c+1)}$
several factors: $(10^7) \cdot (10^{-3}) \cdot (10) \cdot (10^{-2}) = 10^3$

Note. We shall learn later the meaning of negative and fractional exponents. If no exponent is shown on a number, the exponent is understood to be 1. For example, $(x^3) \cdot (x) = x^4$; that is, $(x \cdot x \cdot x) \cdot (x) = x^4$.

19.3. MULTIPLICATION OF MONOMIALS

In the multiplication of two or more terms that include different bases, we add exponents of each base separately.

Example 1. Multiply $(x^3 y^2) \cdot (x^4 y^3)$.

Solution. For the product of the two monomials, we combine the exponents of x and then combine the exponents of y.

MONOMIALS: MULTIPLICATION AND DIVISION 249

For one factor of the product, we have

$$(x^3)(x^4) = x^7$$

For the other factor, we have

$$(y^2)(y^3) = y^5$$

Then the product of the two monomials becomes

$$(x^3 y^2)(x^4 y^3) = x^7 y^5$$

To multiply monomials consisting of powers of various bases, we can get the answer quickly by adding separately the exponents of each base. For example, the exponent on x in the product is equal to the sum of the exponents of x in the factors. The same is true with respect to the exponents on any other base.

Example 2. Multiply $(a^2 x^3 y) \cdot (ax^2 y^3) \cdot (axy^4)$.

Solution. The rule regarding exponents in multiplication of factors consisting of powers of various bases can be applied to any number of factors; then, adding exponents of like bases, we get

$$(a^2 x^3 y) \cdot (ax^2 y^3) \cdot (axy^4) = a^4 x^6 y^8$$

Remember, if no exponent is shown, it is understood to be 1.

Example 3. Multiply $(-2x^3 y^2)(-5axy^4)$.

Solution. Each monomial can be separated into factors as follows:

$$(-2)(x^3)(y^2)(-5)(a)(x)(y^4)$$

Since multiplication can be done in any order, we can rewrite the problem:

$$(-2)(-5)(a)(x^3)(x)(y^2)(y^4)$$

First multiplying, $(-2)(-5)$, we get the complete product: $+10ax^4 y^6$

In the multiplication of monomials, we have the following steps:

(1) *Determine the sign of the product by the rule for the multiplication of signed numbers.*

(2) *Multiply the numerical coefficients of the factors to get the numerical coefficient of the product.*

(3) *Multiply the literal factors by writing all literal factors and adding the exponents of like literal factors.*

Example 4. Multiply $(-3a^2bx^3)(-4abcx^2)(-5bcxy^2)$.

Solution. The product of three negative numbers is negative. That is, the product of the coefficients becomes $(-3)(-4)(-5) = -60$. Adding the exponents of like bases, we get

$$(a^2)(a) = a^3; \quad (b)(b)(b) = b^3; \quad (c)(c) = c^2; \quad (x^3)(x^2)(x) = x^6$$

The complete product is

$$-60a^3b^3c^2x^6y^2$$

Example 5. Multiply $(3)(4.2)(10^5)(10^{-3})(10^2)(10^{-1})$

Solution. The product of the first two factors is $(3)(4.2) = 12.6$. The product of the 10's is 10^3, which is equal to 1000. Then the entire product is

$$12.6(10^3) = (12.6)(1000) = 12,600$$

Exercise 19.1

Rewrite each of the following expressions by using exponents:

1. $2 \cdot 2 \cdot x \cdot x \cdot x \cdot x \cdot y \cdot y \cdot y$
2. $3 \cdot 3 \cdot 3 \cdot a \cdot a \cdot b \cdot b \cdot b \cdot c \cdot c \cdot c \cdot c \cdot c$
3. $5 \cdot 5 \cdot 5 \cdot 5 \cdot x \cdot x \cdot x \cdot y \cdot y \cdot z$
4. $7 \cdot 7 \cdot 7 \cdot n \cdot n \cdot n \cdot n \cdot n \cdot n \cdot t \cdot t \cdot t \cdot t$
5. $10 \cdot 10 \cdot 10 \cdot 10 \cdot 10 \cdot x \cdot x \cdot y \cdot z$
6. $3 \cdot 3 \cdot 3 \cdot 3 \cdot 5 \cdot m \cdot m \cdot m \cdot m \cdot n \cdot x \cdot x$

Multiply the following as indicated by the parentheses:

7. $(2x^4)(3x^3)$
8. $(12n^3)(5n^4)$
9. $(3n^3)(4n^4)$
10. $(6x^6)(3x^3)$
11. $(-4x^2y)(-3xy^4)$
12. $(-5ab^3)(-7a^2b)$
13. $(6ax^3y)(-5xy)$
14. $(-4a^2x)(5a^3bx^2)$
15. $(-7ab)(-4cde)$
16. $(-6mn)(-8xyz)$
17. $(3x^2y)(4xy^3)(2xy)$
18. $(4x^3y^2)(5xy)(3x^4y^2)$
19. $(+2x^3y)(-3x^4y^2z)(-xy)$
20. $(-4x^2y)(xy)(-7x^3y^4)$
21. $(-8xy^2z)(-3x^2z)(-x^3yz^3)$
22. $(-2x^3y)(-9ax^2y^2)(-axy)$
23. $(-ab^2)(-a^2b^2c)(c^2)(-cd^2)$
24. $(-x^3y)(-x^2y^2z)(-x)(-x^2z^3)$
25. $-(-3x)(-4x^2y)(-2y^2)(-xy)$
26. $-(-2x)(-5xy^3)(-y^3)(-xy)(-x^3)$
27. $(4)(10^5)(10^{-2})(10)(10^3)(10^{-1})$
28. $(5)(10^{-3})(10^6)(10^{-2})(10^4)(10^{-3})$
29. $(3.2)(10^{-2})(10^6)(10)(10^{-4})(10^3)$
30. $(5)(4.12)(10^2)(10^{-3})(10^5)(10^{-1})$

19.4. EXPONENTS IN DIVISION

Since division is the inverse of multiplication, then, in working with exponents, we should expect that in division we use a procedure opposite to that used in multiplication. In multiplication, we *add* exponents of the same base. In division, we *subtract* exponents of the same base. In division, *we do not divide the exponents.* We are dividing *quantities,* not exponents.

The rule for division of quantities expressed with exponents can be seen from an example.

Example 1. Divide as indicated: $x^6 \div x^2$.

Solution. Let us write the division as a fraction and then separate the numerator and denominator into prime factors:

$$\frac{x^6}{x^2} = \frac{x \cdot x \cdot x \cdot x \cdot x \cdot x}{x \cdot x} = x^4$$

Now we divide numerator and denominator by the quantity, $x \cdot x$, and get the quotient, $x \cdot x \cdot x \cdot x$, which can be written x^4.

If we were to analyze in the same way the division of two other powers, we should find that the quotient can be found by subtracting the exponents. For example,

$$x^9 \div x^4 = x^5; \qquad n^6 \div n^5 = n; \qquad y^8 \div y = y^7$$

Then we have this rule for division involving exponents.

RULE. In dividing algebraic quantities expressed as powers, we subtract the exponent of the divisor from the exponent of the dividend.

The rule may be stated as a formula:

$$x^a \div x^b = x^{a-b}$$

Division can be checked by multiplication:

$$x^6 \div x^2 = x^4; \qquad \text{check:} \qquad x^4 \cdot x^2 = x^6$$

The rule holds true for any base:

$$(10^5) \div (10^2) = 10^3; \qquad 4^8 \div 4^5 = 4^3; \qquad 2^{10} \div 2 = 2^9$$

The rule for subtracting exponents in division holds true for all kinds of exponents.

fractional exponents: $\quad x^{3/4} \div x^{1/2} = x^{1/4}$
decimal exponents: $\quad 10^{4.316} \div 10^{1.823} = 10^{2.493}$
negative exponents: $\quad n^4 \div n^{-2} = n^6; \qquad x^{-3} \div x^{-7} = x^4$
zero exponents: $\quad x^3 \div x^0 = x^3; \qquad x^0 \div x^5 = x^{-5}$
literal exponents: $\quad y^m \div y^n = y^{m-n}$

19.5. DIVISION OF MONOMIALS

In the division of two monomials that include powers on different bases, we subtract the exponents of each base separately.

Example 2. Divide as indicated: $(-12x^4y^3z) \div (-3x^3y^2)$.

Solution. To understand each step, let us first write the division as a fraction, and then separate the terms into prime factors:

$$\frac{-12x^4y^3z}{-3x^3y^2} = \frac{(-)(2)(2)(3) \cdot x \cdot x \cdot x \cdot x \cdot y \cdot y \cdot y \cdot z}{(-)(3) \cdot x \cdot x \cdot x \cdot y \cdot y} = +4xyz$$

Now the numerator and denominator can be divided by the factors: $3 \cdot x \cdot x \cdot x \cdot y \cdot y$. Remember, $(-) \div (-) = +$. Then we get the complete quotient: $+4xyz$.

In division of monomials, we have the following steps:

(1) *Determine the sign of the quotient by the rule for the division of signed numbers.*

(2) *Divide the numerical coefficients to get the numerical coefficient of the quotient.*

(3) *Divide the literal parts by writing all the literal factors of the dividend and subtracting the exponents of like factors in the divisor.*

Example 3. Divide as indicated: $(-28a^2bx^3y^5) \div (4abxy^5)$.

Solution. The sign of the quotient is negative. Subtracting exponents of like bases, we get the complete quotient: $-7ax^2$.

Exercise 19.2

Divide as indicated:

1. $(6x^2) \div (3x)$
2. $(8n^3) \div (4n)$
3. $(4n^2) \div (4n^2)$
4. $(-3x^4) \div (-3x^4)$
5. $(-20x^3) \div (-5x)$
6. $(24x^4) \div (6x)$
7. $(6x^6) \div (2x^2)$
8. $(8n^8) \div (2n^2)$
9. $(-60x^4y^3) \div (6xy)$
10. $(40x^3y^5) \div (-5x^2y)$
11. $(8a^3b) \div (-4ab)$
12. $(-18m^2n^4) \div (3m^2n)$
13. $(-12x^4y^2z^3) \div (3xy)$
14. $(10x^3yz^5) \div (-2x^2yz)$
15. $(-15a^3b^2c) \div (-5a^3bc)$
16. $(-20m^2n^3x^2) \div (-4m^2nx^2)$
17. $(6x^2y^3z) \div (6x^2y^3z)$
18. $(-8xy^2z^4) \div (8xy^2z^4)$
19. $(-24xy^2z) \div (6xy)$
20. $(30a^2bc^2) \div (-10abc)$
21. $(36a^4b^6c) \div (-9ab^2c)$
22. $(-48x^7y^3z) \div (-8xy^2z)$
23. $(-18x^3yz) \div (-18x^3yz)$
24. $(-16a^2b^3c^4) \div (16a^2b^3c^4)$

20 POLYNOMIALS: MULTIPLICATION AND DIVISION

20.1. MULTIPLICATION OF A POLYNOMIAL BY A MONOMIAL

A *polynomial* is an algebraic expression of more than one term. To indicate the multiplication of a polynomial by a monomial, we can place the expressions next to each other, but the polynomial must be enclosed in parentheses to denote a quantity.

Suppose we wish to take 3 times the quantity, $5x - 2y$. We can place the 3 before the binomial, $5x - 2y$, but the binomial must be enclosed in parentheses to indicate 3 times the entire quantity. We can write

$$3(5x - 2y); \quad \text{or} \quad (5x - 2y)(3)$$

Since multiplication is a short form for addition, then the problem can be written as the addition of 3 binomials:

$$\begin{array}{r} 5x - 2y \\ 5x - 2y \\ 5x - 2y \\ \hline 15x - 6y \end{array}$$

Adding the terms in each column, we get $15x - 6y$.

To get the answer for the problem, $3(5x - 2y)$, we can multiply each term of the binomial by the multiplier 3. Then we have the following rule:

RULE. To multiply a polynomial by a monomial, multiply each term of the polynomial by the monomial.

This rule is the application of the so-called *Distributive Law*, which states that *multiplication is distributive with respect to addition*. In general terms, the law states that

$$a(b + c) = ab + ac$$

254 ALGEBRA

Note. *Be careful to observe all the rules for sign, coefficients, and letters with their exponents.*

Example 1. Multiply the polynomial, $5x^2 - 4xy - 7$, by 4.

Solution. $4(5x^2 - 4xy - 7) = 20x^2 - 16xy - 28$

Example 2. Multiply as indicated: $2n(n^3 - 4n^2 + 5n + 1)$.

Solution. Applying the rule, we multiply each term of the polynomial by $2n$. We get the product:

$$2n^4 - 8n^3 + 10n^2 + 2n$$

Example 3. Multiply as indicated: $-3x^2y(5x^2 - 4xy - 2y^2 + 1)$

Solution. Applying the rule, we multiply each term of the polynomial by $(-3x^2y)$ and get

$$-15x^4y + 12x^3y^2 + 6x^2y^3 - 3x^2y$$

Example 4. Multiply as indicated: $3x(4x^2 - 5xy + 3y^2)(-2y)$

Solution. In this example, the polynomial is to be multiplied by two portions of a monomial; that is by $(3x)$ and by $(-2y)$. Notice that the $-2y$ must be placed in parentheses to avoid confusion.

In this example, the multiplication can be done in two different ways. Probably the best way is to multiply the monomials together first, and get the complete monomial multiplier: $(3x)(-2y) = -6xy$. Then we take $-6xy$ times the polynomial. Then we have the problem:

$$-6xy(4x^2 - 5xy + 3y^2) = -24x^3y + 30x^2y^2 - 18xy^3$$

The product could also be found by multiplying the polynomial first by $3x$, and then the resulting polynomial by $-2y$. However, the simplest way is to combine the monomials first into a single monomial.

Exercise 20.1

Multiply as indicated:

1. $3(2x + 5)$
2. $4(3x + 7)$
3. $-5(3n - 8)$
4. $-6(4n - 5)$
5. $4(2 - 7a)$
6. $5(7 - 8c)$
7. $5(7x - 3y)$
8. $4(3a + 7b)$

9. $3x(4x+5y)$
10. $5y(3x-7y)$
11. $2(4x^2-7x+1)$
12. $3(2x^2+8x-1)$
13. $-3(4x^2-5xy+6y^2)$
14. $-4(5n^2+4mn-3m^2)$
15. $4x(7x^2-3x+1)$
16. $-3x(6x^2+4x-1)$
17. $-2x^2(3x^2-4x-1)$
18. $-3n^3(4n^2+5n+1)$
19. $6n^3(3n^2+4nx-5x^2)$
20. $3x^2(4x^2-5xy-2y^2)$
21. $-5xy^2(3x^3-4x^2y+2xy^2-y^3)$
22. $-2x^2y(4xy^2+5x^2y-3x^2-x^3)$
23. $-3xy^3(4x^2-5xy-6y^2-1)$
24. $-4a^2b(3a^2-7ab+2b^2+1)$
25. $6x(3x^2-2xy-4y^2)(-y)$
26. $-3y^2(4x^2-3xy+2y^2)(-x)$
27. $-2c^2(5c^2+3cd-6d^2)(3d)$
28. $3h^2(4h^2-5hk+3k^2)(-k^2)$

20.2. DIVISION OF A POLYNOMIAL BY A MONOMIAL

Since division and multiplication are inverse processes, we have the following rule for the division of a polynomial by a monomial.

RULE. To divide a polynomial by a monomial, divide each term of the polynomial by the monomial. (Division can be checked by multiplication).

This rule is simply a restatement of the *Distributive Law*, this time in reverse. In general terms:

$$ab + ac = a(b+c)$$

As in the division of monomials, be careful to observe the rules for signs, numerical coefficients, and exponents.

Example 1. Divide as indicated: $(12x^4+8x^3) \div (2x)$.

Solution. We divide $2x$ into each term of the binomial and get:

$$6x^3 + 4x^2$$

Example. Divide as indicated: $(8n^2+4n) \div (4n)$.

Solution. Recall that $(4n) \div (4n) = 1$. Then the quotient is: $2n+1$.

Example 3. Divide as indicated: $(18x^4-12x^3+9x^2-3x) \div (3x)$.

Solution. A division problem may be written as a fraction, if desired.

This problem may be written:

$$\frac{18x^4 - 12x^3 + 9x^2 - 3x}{3x}$$

Dividing $3x$ into each term of the polynomial, we get: $6x^3 - 4x^2 + 3x - 1$.

Example 4. Divide as indicated:

$$\frac{15x^3y^2 - 12x^2y^3 + 6xy^4}{3xy}$$

Solution. Dividing $3xy$ into each term of the polynomial numerator, we get the quotient:

$$5x^2y - 4xy^2 + 2y^3$$

Example 5. Divide as indicated:

$$\frac{20x^3y^2 - 15x^2y^3 + 10xy^4}{-5xy^2}$$

Solution. Observing signs and exponents, we get the quotient:

$$-4x^2 + 3xy - 2y^2$$

Exercise 20.2

Divide as indicated:

1. $(6x + 8) \div 2$
2. $(8n - 12) \div 4$
3. $(9r - 12s) \div 3$
4. $(10x + 15y) \div 5$
5. $(4x + 3) \div 2$
6. $(6x + 4) \div 3$
7. $(8x - 3y) \div 4$
8. $(5x - 9y) \div 3$
9. $(10x^2 + 6x) \div (-2x)$
10. $(20n^2 - 15n) \div (-5n)$
11. $(16n^3 - 12n^2) \div (-4n)$
12. $(18x^4 + 15x^2) \div (-3x)$
13. $(12x^2 + 3x) \div 3x$
14. $(14x^2 - 2x) \div 2x$
15. $(18x^3 - 6x^2 + 4x) \div (-2x)$
16. $(12x^4 + 9x^3 - 6x^2) \div (-3x^2)$
17. $(20x^3 + 15x^2 - 5x) \div (5x)$
18. $(24x^3 - 12x^2 + 4x) \div (4x)$
19. $\dfrac{12n^3y^4 - 8n^2y^3 + 6ny^2}{-2ny}$
20. $\dfrac{30a^2c^3 + 18a^3c^2 - 12a^4c}{-6ac}$
21. $\dfrac{15a^2b + 12a^3b^2c - 3ab}{-3ab}$
22. $\dfrac{8x^2y^3z^5 - 4xy^2z^4 + 2yz^3}{-2yz^3}$
23. $\dfrac{10x^{10} - 8x^8 + 6x^6 - 4x^4 - 2x^2}{2x^2}$
24. $\dfrac{15x^{15} + 12x^{12} - 9x^9 + 6x^6 + 3x^3}{3x^3}$

20.3. MULTIPLICATION OF A POLYNOMIAL BY A POLYNOMIAL

The multiplication of a polynomial by another polynomial in algebra is somewhat similar in form to the multiplication by a two or three digit number in arithmetic. As an example, consider the following problem in arithmetic: (43)(32). For multiplication in arithmetic, we usually write one number below the other, as shown:

$$\begin{array}{r} 43 \\ \underline{32} \end{array} \qquad \begin{array}{cc} 4 & 3 \\ 3 & 2 \end{array}$$

At the right, we have separated the digits so that we might indicate the steps in multiplication by arrows. In this example, note that we must make *four* multiplications. Each digit in the multiplicand must be multiplied by both digits in the multiplier. The results of the multiplications are placed in proper position and then added.

In algebra, to multiply one polynomial by another polynomial, we place one below the other, just as in arithmetic. We show the method by examples.

$$\begin{array}{r} 4x - 5 \\ 3x + 2 \\ \hline 12x^2 \end{array}$$

$$\begin{array}{r} 4x - 5 \\ 3x + 2 \\ \hline 12x^2 - 15x \end{array}$$

$$\begin{array}{r} 4x - 5 \\ 3x + 2 \\ \hline 12x^2 - 15x \\ + 8x - 10 \\ \hline 12x^2 - 7x - 10 \end{array}$$

Example. Multiply $(4x-5)(3x+2)$.

Solution. We use parentheses around each binomial to indicate that it is to be considered as one quantity. For the multiplication, we place one binomial below the other, as shown at the left.

Here we shall have *four* separate multiplications, just as we have in the arithmetic example. The four multiplications are shown by arrows. Each term in the multiplicand must be multiplied by both terms of the multiplier.

In algebra we begin multiplication at the *left* instead of at the right as in arithmetic. The first multiplication is: $(3x)(4x)$; the next is $(3x)(-5)$. The third step is $(2)(4x) = 8x$, which is placed below the $-15x$, for convenience in adding the terms. The final multiplication is $(+2)(-5) = -10$. The partial products are then combined as shown in the complete product.

$$\begin{array}{r} 3x - 5y \\ 4x + 3y \\ \hline 12x^2 - 20xy \\ + 9xy - 15y^2 \\ \hline 12x^2 - 11xy - 15y^2 \end{array}$$

Example 2. Multiply $(3x-5y)(4x+3y)$.

Solution. The work is shown at the left. We place one binomial below the other, and then begin the multiplication at the *left*: $(4x)(3x) = 12x^2$. The next multiplication is $(4x)(-5y) = -20xy$. Then we take $3y$ times both terms of the multiplicand, and finally add the partial products to get the entire product.

Example 3. Multiply $(7x+5y)(4x-3y)$.

$$\begin{array}{r} 7x + 5y \\ 4x - 3y \\ \hline 28x^2 + 20xy \\ -21xy - 15y^2 \\ \hline 28x^2 - xy - 15y^2 \end{array}$$

Solution. The work is shown at the left. Begin multiplication at the *left* with $(4x)(7x) = 28x^2$. For the second multiplication, we get $(4x)(5y) = 20xy$. Third step: $(-3y)(7x) = -21xy$. For the fourth step we have: $(-3y)(5y) = -15y^2$. The two like terms are in the same column for easy adding.

Example 4. Multiply $(2+3x)(6-5x+4x^3-3x^2)$.

Solution. Before beginning the multiplication, we shall often find it desirable to rearrange the terms in multiplier and multiplicand. Although the multiplication could be done without this arrangement, it is usually convenient to arrange the terms in a *descending order of powers* of some letter. Here we arrange the terms so each starts with the *highest* power of x.

Rearranging terms: $4x^3 - 3x^2 - 5x + 6$
$3x + 2$
Multiplying by $3x$, $12x^4 - 9x^3 - 15x^2 + 18x$
Multiplying by 2, $+8x^3 - 6x^2 - 10x + 12$
The product is $12x^4 - x^3 - 21x^2 + 8x + 12$

Note that each term of the multiplicand is first multiplied by $3x$, then by $+2$. If like terms appear in the multiplications, they are placed in the same column and then added. If no like terms appear, they cannot be combined with other terms. Like terms are added *algebraically*.

Exercise 20.3

Multiply as indicated:

1. $(3x+2)(4x-5)$
2. $(5x-1)(2x-3)$
3. $(4x+3)(2x+5)$
4. $(3x+5)(4x+3)$
5. $(7+3x)(2x+5)$
6. $(8+5x)(2x-3)$
7. $(2x-5y)(4x+5y)$
8. $(4b+3a)(6b-5a)$
9. $(5n-4)(5n+4)$
10. $(7c+2)(7c-2)$
11. $(7xy-3)(3xy-2)$
12. $(5x^2+3)(7x^2-4)$
13. $(5ax+8)(5ax-8)$
14. $(4mn-11)(4mn+11)$
15. $(7n+3x)(7n-5x)$
16. $(8a-5b)(8a+7b)$
17. $(5x^2-6)(6+5x^2)$
18. $(9x^3-5)(5+9x^3)$
19. $(12nx-5)(12nx+5)$
20. $(8by-13)(13+8by)$
21. $(9mn+4)(7mn-3)$
22. $(4xy+7)(5xy-9)$
23. $(6x^2+3+4x)(3x-2)$
24. $(6x^2+11+8x)(3x-4)$
25. $(5x^2+3-7x)(4+3x)$
26. $(3x^2+4-7x)(5+2x)$

27. $(8x^2+6x^3-3-5x)(3x-4)$
28. $(4x^3+2-5x^2-4x)(5x-4)$
29. $(6x+4x^2+9)(2x-3)$
30. $(16+9x^2-12x)(3x+4)$
31. $(5x^3+3x^4-2x-4)(3x-5)$
32. $(6x^3+4x^4-5-3x)(2x-3)$
33. $(3x-5)(4x-3)(3x+5)$
34. $(4n+7)(3n-5)(4n-7)$
35. $(4n-3)(2n+1)(3n-2)(4n+3)$
36. $(2a-3)(3a-1)(2a+3)(4a-3)$
37. $(2x)(5x-2)(3x+4x^2-5)$
38. $-3n(2n+5)(3n^2-7-4n)$

20.4. DIVISION OF A POLYNOMIAL BY A POLYNOMIAL

In algebra, the division of one polynomial by another is similar in form to long division in arithmetic. If you will recall the separate steps in long division in arithmetic, you can follow the same general steps in long division in algebra; that is, division of one polynomial by another.

Suppose we have the problem: $(8x^3+22x^2+19x+14)\div(2x+3)$. Let us compare this problem with a problem in long division in arithmetic, for example, $(7439)\div(28)$. We write the two problems side by side and note the similar steps in each:

Arithmetic

$$\begin{array}{r} 2 \\ 28\overline{)7439} \\ \underline{56} \\ 183 \end{array}$$

Algebra

$$\begin{array}{r} 4x^2 \\ 2x+3\overline{)8x^3+22x^2+19x+14} \\ \underline{8x^3+12x^2} \\ 10x^2+19x \end{array}$$

Step 1. In arithmetic we first divide 28 into 74. We first try 2 into 7. The answer is 3. However, we find this is too large for the first digit of the quotient, so we write 2 as the first digit. In this step in arithmetic, there is a little bit of guesswork.

In algebra, we first divide $2x$ into $8x^3$, and we get $4x^2$. In algebra we do not need to guess at this answer. So we write $4x^2$ as the first term of the quotient.

Step 2. In arithmetic we multiply the whole divisor by the 2, the first digit of the quotient; that is, we "multiply back," taking 2 times the divisor 28, and get 56. We place the 56 below the 74 of the dividend.

In algebra, we follow the same pattern. We multiply the whole divisor by the first term of the quotient; that is, we "multiply back," taking the quotient term, $4x^2$ times the divisor, $2x+3$, and get $8x^3+12x^2$. We place this quantity below the first two terms of the dividend.

Step 3. In arithmetic, we subtract 56 from 74 of the dividend, and get a first remainder of 18.

In algebra, we also subtract the two terms, $8x^3 + 12x^2$, from the first two terms of the dividend, and get a first remainder of $+10x^2$.

Step 4. In arithmetic we bring down the next digit of the dividend, 3, and place it at the right of the 18, making 183, to be divided by 28.

In algebra, we bring down the next *term* and place it at the right of the first remainder; then we have $+10x^2 + 19x$ as the next dividend to be divided by the divisor, $2x + 3$.

Steps 5 to the end. From this point on, we continue to divide, repeating the same steps as in arithmetic. In algebra, we continue until we get a remainder of zero or a remainder whose highest power of x is lower than the highest power of x in the divisor. If there is a remainder other than zero, it may be placed over the divisor, as in arithmetic, forming a fraction, which is then added to the quotient as a fraction.

We summarize the steps in this problem. Follow the same in all division.

1. Divide the first term of the divisor into the first term of the dividend; we get $4x^2$.

2. Place the $4x^2$ directly above the $8x^3$ of the dividend. Then the $4x^2$ is the first term of the quotient.

3. "Multiply back," taking $4x^2$ of the quotient times the *entire* divisor. Place the result, $8x^3 + 12x^2$, below the like terms of the dividend.

4. Subtract. The remainder is $10x^2$.

5. Bring down the next term of the dividend, $19x$, and place it at the right of the first remainder, $10x^2$. Then we have the quantity, $10x^2 + 19x$, to be divided next by the divisor.

6. Divide $2x$, the first term of the divisor, into $10x^2$.

7. Place the result, $5x$, in the quotient just above the $22x^2$ of the dividend.

8. Multiply back, taking $5x$ times the *entire* divisor. Place the result, $10x^2 + 15x$, below the $10x^2 + 19x$.

9. Subtract, leaving a remainder of $4x$.

10. Bring down the 14 of the dividend, and place it at the right of the second remainder, $4x$; then the quantity $4x + 14$ is to be divided by the division $2x + 3$.

11. Divide $2x$, the first term of the divisor, into the $4x$, and place the result, 2, as the next term of the quotient.

12. Multiply back, 2 times the entire divisor, and get $4x + 6$.

13. Place the $4x + 6$ below the quantity, $4x + 14$, and subtract, leaving a remainder of 8.

14. The remainder, 8, may be placed above the divisor, $2x + 3$, and the result annexed to the quotient as a fraction.

POLYNOMIALS: MULTIPLICATION AND DIVISION

The problems in arithmetic and algebra are here shown in complete form:

$$265 + \tfrac{19}{28}$$
$$28\overline{)7439}$$
$$\underline{56}$$
$$183$$
$$\underline{168}$$
$$15$$
$$\underline{140}$$
$$19 = \text{remainder}$$

$$4x^2 + 5x + 2 + \frac{8}{2x+3}$$
$$2x+3\overline{)8x^3 + 22x^2 + 19x + 14}$$
$$\underline{8x^3 + 12x^2}$$
$$10x^2 + 19x$$
$$\underline{10x^2 + 15x}$$
$$4x + 14$$
$$\underline{4x + 6}$$
$$8 = \text{remainder}$$

Note 1. In division, it is necessary to arrange the terms of both dividend and divisor in a descending order of powers before beginning the division. Begin with the highest highest power of x (or whatever the letter). If the dividend contains two letters with various powers, then select a letter for the arrangement according to descending powers.

Note 2. If one of the powers is missing in the dividend, this means that the coefficient of the missing power is zero. Then a space must be allowed in the dividend for each missing power. This vacancy can be conveniently shown by placing +0 for each missing power.

Example 1. Divide as indicated: $(2x^2 + 8x^3 + 13 - 9x) \div (3 + 2x)$.

Solution. The work is shown at the right. First we arrange the terms of dividend and divisor in descending powers of x, beginning with the x^3 term. For the divisor we write: $2x+3$. Be especially careful about minus signs in the subtraction steps.

$$4x^2 - 5x + 3,\ \text{rem} = 4$$
$$2x+3\overline{)8x^3 + 2x^2 - 9x + 13}$$
$$\underline{8x^3 + 12x^2}$$
$$-10x^2 - 9x$$
$$\underline{-10x^2 - 15x}$$
$$+ 6x + 13$$
$$\underline{+ 6x + 9}$$
$$4 = \text{rem}$$

Example 3. Divide as indicated: $(18x^4 + 6 - 9x - 17x^2) \div (3x - 2)$.

Solution. The work is shown at the right. Note that the highest power of x is the 4th. Also note that the third power of x is missing. Then, in arranging the terms of the dividend in descending order, we place a zero in place of the missing term. The remainder in the division is -4, which may be written over the divisor. Then the quotient may

$$6x^3 + 4x^2 - 3x - 5,\ \text{rem} = -4$$
$$3x-2\overline{)18x^4 + 0 - 17x^2 - 9x + 6}$$
$$\underline{18x^4 - 12x^3}$$
$$+12x^3 - 17x^2$$
$$\underline{12x^3 - 8x^2}$$
$$-9x^2 - 9x$$
$$\underline{-9x^2 + 6x}$$
$$-15x + 6$$
$$\underline{-15x + 10}$$
$$-4$$

be written:

$$6x^3 + 4x^2 - 3x - 5 - \frac{4}{3x-2}$$

Example 3. Divide as indicated: $(12x^3 - 7x - 11x^2) \div (4x + 3)$.

Solution. The work is shown at the right. The terms are arranged in the descending order of powers. There is no constant term. A zero (0) may be attached to the dividend, but this is not necessary in this case.

$$\begin{array}{r}3x^2 - 5x + 2\\4x+3\overline{)12x^3 - 11x^2 - 7x + 0}\\12x^3 + 9x^2\\\hline -20x^2 - 7x\\-20x^2 - 15x\\\hline +8x\\+8x+6\\\hline -6 = \text{rem}\end{array}$$

Example 4. Divide as indicated: $(6x^4 + 4x - x^3 + 2) \div (2x^2 - 3x + 2)$.

Solution. The terms of dividend and divisor are arranged in descending order of powers of x. A zero (0) is placed in the dividend for the missing power of x. The dividend has no x^2 term. The work is shown below.

$$\begin{array}{r}3x^2 + 4x + 3\\2x^2-3x+2\overline{)6x^4 - x^3 + 0 + 4x + 2}\\6x^4 - 9x^3 + 6x^2\\\hline +8x^3 - 6x^2 + 4x\\+8x^3 - 12x^2 + 8x\\\hline +6x^2 - 4x + 2\\+6x^2 - 9x + 6\\\hline 5x - 4 = \text{remainder}\end{array}$$

The quotient may be written:

$$3x^2 + 4x + 3 + \frac{5x-4}{2x^2-3x+2}$$

Exercise 20.4

Divide as indicated (rearrange terms in descending order of powers):

1. $(6x^2 - 5x - 15) \div (2x + 3)$
2. $(12x^2 - 7x - 4) \div (3x + 2)$
3. $(x + 15x^2 - 16) \div (3x - 4)$
4. $(x + 12x^2 - 8) \div (4x - 5)$
5. $(8x^3 + 13 - 2x^2 - 21x) \div (2x - 3)$
6. $(8n^2 + 6n^3 - 23n + 18) \div (3n - 2)$
7. $(18x^3 + 7x - 6) \div (2 + 3x)$
8. $(18x^3 + 12 - 17x) \div (4 + 3x)$
9. $(n^2 + 12n^3 + 24) \div (4n - 5)$
10. $(30 + 20t^3 + t^2) \div (5t - 6)$
11. $(3x^4 - 7x^2 - 12x + 13) \div (x - 2)$
12. $(3x^4 + 8x^3 - 15x - 6) \div (x + 2)$

13. $(12-5x+2x^3) \div (2+x)$
14. $(5x^2-4x+x^3) \div (3+x)$
15. $(4x^2-21) \div (2x-5)$
16. $(18x^2-13) \div (3x-2)$
17. $(25x^2-36) \div (5x+6)$
18. $(4x^2-49) \div (2x+7)$
19. $(13x^2+12x^3-12x) \div (4x-1)$
20. $(14r^3-11r-r^2) \div (2r-1)$
21. $(8h^3+3-18h^2-9h) \div (2h-5)$
22. $(20n^3+5-13n-19n^2) \div (4n-3)$
23. $(v^2-9v+6v^4-8v^3) \div (3v+2)$
24. $(12y^4-y^2-22y^3-3y) \div (3y-4)$
25. $(81-72x^2+16x^4) \div (2x-3)$
26. $(32x^4-2x-58x^2) \div (4x-5)$
27. $(16x^4-81) \div (2x+3)$
28. $(32x^5+1) \div (2x+1)$
29. $(6a^4+22a^3b+6b^4+ab^3) \div (3a+2b)$
30. $(6x^3-3x^2y-12y^3-xy^2) \div (2x-3y)$
31. $(6x^4-8x^2-17+5x^3+18x) \div (2x^2+3x-4)$
32. $(25x+12x^4-x^3-20) \div (2x+3x^2-3)$
33. $(13c^2+3+7c^3+2c^4+11c) \div (3+c^2+2c)$
34. $(3y^5-2xy^4-3x^3y^2+6x^5) \div (y^2+2x^2+2xy)$

21 SOLVING EQUATIONS

21.1. DEFINITIONS

In Chapter 16, we defined the *equation*.

Definition. An equation is a statement saying that two quantities are equal.

As an example, consider the equation:

$$3n + 5 = 26$$

This statement says that the quantity $(3n + 5)$ is equal to the quantity 26.
  We also defined the *solution* or *root* of an equation.

Definition. The solution (or root) of an equation is a value of the letter that makes the statement true.

In the equation, $3n + 5 = 26$, we see that the value of n is 7 as the only value of n that makes the statement true. Then, 7 is a solution of the equation.
  In the foregoing equation, we found the solution by *inspection*. However, the solution of an equation is not always easy to find. We must now see how to solve an equation systematically. Then we shall be able to find the solution without guessing.
  There are certain specific rules and procedures by which even difficult equations can be solved rather easily. We shall now consider some of these procedures.

21.2. THE EQUATION AS A BALANCED SCALE

Let us first see how an equation resembles a balanced scale (Fig. 21.1). On the left pan of the scale is a bag of sand of an unknown weight. Let us call

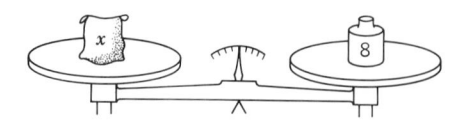

FIGURE 21.1

the weight x pounds. On the right pan is an 8-ounce weight. Suppose the scale is balanced. This means we have exactly the same weight on both sides of the scale.

An equation is like a balanced scale. The scale in Figure 21.1 can be represented by the equation
$$x = 8$$

The quantity on the left side of the equation is x. The quantity on the right side is 8. The equal sign ($=$) says that the quantities are equal. This is the same as saying that the scale is balanced.

If you will think of an equation at all times as a balanced scale, you will find that solving problems becomes very simple. An equation can always be treated as a balanced scale.

21.3. THE AXIOMS

To solve equations systematically, we make use of several important principles called *axioms*.

> *Definition.* An axiom is a general statement that we accept as true without proof.

To solve equations, we need to know the axioms and what they mean. They can be best understood if they are applied to a balanced scale.

Consider the scale in Figure 21.1. Suppose we add some weight, say, a 3-ounce weight to the left side of the scale, and a 3-ounce weight to the right side of the scale (Fig. 21.2). Do you think the scale will still balance? We feel sure that it will. We seem to know that the following statement is true:

If we add the same quantity to both sides of perfect scale that is perfectly balanced, the scale will still balance.

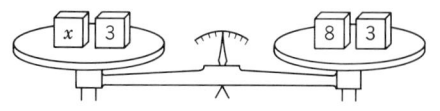

FIGURE 21.2

SOLVING EQUATION

The statement can be tested in a laboratory by using various weights. However, to verify the statement for *all* cases is impossible. Yet we feel that the general statement is true. *We accept it as true.*

The statement we have made concerning a balanced scale can also be applied to an equation. Suppose we have the equation

$$x = 8$$

If we add some quantity, say, 3, to both sides of the equation, the equation will still be true; that is,

$$x + 3 = 11$$

The principle we have just illustrated with a balanced scale and with an equation is called the *addition axiom*. It is often referred to as the *first axiom*. It is stated as follows:

> *Axiom 1. If the same quantity is added to both sides of an equation, the new equation will still be true.*

To see how this axiom is used in solving equations, consider this example.

$$x - 7 = 16$$

Let us add 7 to both sides of the equation

$$x - 7 + 7 = 16 + 7$$

Then we get $$x = 23$$

Now, let us consider a second possibility. Suppose we start with the balanced scale in Figure 21.1. If we *subtract* the same quantity from both sides of the scale, the scale will still balance. In the same manner, if we subtract the same number from both sides of an equation, the new equation will still be true. This is the basis of the *subtraction axiom*.

> *Axiom 2. If the same quantity is subtracted from both sides of an equation, the new equation will still be true.*

To see how this axiom is used in solving equations, consider the example:

$$x + 9 = 34$$

Let us subtract 9 from both sides of the equation

$$x + 9 - 9 = 34 - 9$$

Then we get $$x = 25$$

As a third possibility, if we start with the balanced scale in Figure 21.1 and multiply both sides of the scale by some number, say, 5, then each side of the scale will contain 5 times as much as before, but it will still balance. The same idea can be applied to an equation, and we have Axiom 3.

Axiom 3. If both sides of an equation are multiplied by the same number, the new equation will still be true.

To see how this axiom is used in solving equations, consider the example:

$$\frac{x}{3} = 6$$

On the left side of the equation, we have the fraction, $x/3$, which means that the number x has been divided by 3. If we multiply both sides of the equation by 3, we get $(3)(x/3)$, which is x, on the left; and 18 on the right. Then we get the solution,

$$x = 18$$

As a fourth possibility, if we divide both sides of the scale in Figure 21.1 by some number, say, 4, the scale will still balance. The same is true concerning an equation. Then we have the *division axiom.*

Axiom 4. If both sides of an equation are divided by the same quantity (not zero), the new equation will still be true.

Note. The one exception is that we cannot divide by zero.

To see how Axiom 4 is used in solving equations, consider this example:

$$5x = 60$$

We divide both sides of the equation by 5

$$\frac{5x}{5} = \frac{60}{5}; \quad \text{or} \quad x = 12$$

21.4. TRANSPOSING

In applying Axioms 1 and 2, in solving equations, we sometimes use a "trick" called *transposing.* By transposition, we mean that any term may be

SOLVING EQUATION 269

moved (or transposed) from one side of an equation to the other side, provided its sign is changed. Although transposing is not a mathematical process, it is a short cut that is simple, quick, and convenient if used properly. An *entire term* must be *transposed, not simply a factor*.

Let us see how transposing comes about through the use of Axioms 1 and 2. Suppose we begin with the following equation:

$$x + 3 = 12$$

By use of Axiom 2, we subtract 3 from both sides of the equation. To subtract 3 from the left side, we simply omit the 3. However, we show the subtraction on the right side. Then we get

$$x = 12 - 3$$

We have subtracted 3 from both sides of the equation. However, notice that the quantity 3 has disappeared from the left side but appears on the right side with its sign changed from + to −. Then $x = 9$.

Consider another example:

$$x - 5 = 14$$

By use of Axiom 1, we add 5 to both sides of the equation. To add 5 to the left side, we can simply omit the "−5." However, we show the addition of 5 to the right side. Then we get

$$x = 14 + 5$$

We have added 5 to both sides of the equation. Then $x = 19$.

In the two foregoing examples, notice that when a term is added to or subtracted from both sides of an equation, the term disappears from one side and then reappears on the other side with its sign changed. A simple way to apply Axioms 1 and 2 is to transfer or *transpose* any term we wish from one side of an equation to the other side (*across the equal sign*), and then change the sign of the term. This process is called *transposing*.

We shall now work several examples showing the application of the trick of transposing in applying Axioms 1 and 2, and also the application of Axiom 4, the division axiom.

Example 1. Solve the equation $\qquad x - 8 = 27$

Solution. Transposing the −8, we get $\qquad x = 27 + 8$
Combining the terms on the right, $\qquad x = 35$, the solution

Example 2. Solve the equation $\qquad n + 25 = 68$

Solution. Transposing the +25, $n = 68 - 25$
Combining the terms on the right, $n = 43$

Example 3. Solve the equation $5t = 80$

Solution. Applying Axiom 4, we divide both sides of the equation by 5

$$t = 16$$

Example 4. Solve the equation $4x - 3 = 27$

Solution. Transposing the -3, $4x = 27 + 3$
Combining the terms on the right, $4x = 30$
Dividing both sides by (Ax. 4), $x = 7.5$

Example 5. Solve the equation

$$5n - 7 = 2n + 18$$

Solution. Transposing to get the n terms together on one side,

$$5n - 2n = 18 + 7$$

Here we have moved the $2n$ to the left side of the equation, and the -7 to the right side of the equation. Now we combine the terms on the left side and on the right side, and get

$$3n = 25$$

Dividing both sides by 3, we get

$$n = \frac{25}{3},$$

or

$$n = 8\tfrac{1}{3}$$

Checking the answers. To check an answer, we substitute the value of the solution in the original equation to see if the equation is then true. We shall check the next problem.

Example 6. Solve the following equation and check the answer in the original equation:

$$5x - 13 = 2x + 11$$

Solution. We transpose the term $2x$ to the left side so it can be combined with $5x$. We move the -13 to the right side of the equation, so it can be combined with the 11. We get

$$5x - 2x = 11 + 13$$
Combining like terms, $\quad 3x = 24$
Dividing both sides by 3, $\quad x = 8$, the solution.

To check the answer, we put 8 for x in the original equation to see if the resulting statement is true. We ask

does $\quad 5(8) - 13 = 2(8) + 11$?
does $\quad 40 - 13 = 16 + 11$?
does $\quad 27 = 27$? Yes.

Example 7. Solve the equation $\quad 4x + 2 - x - 9 = 12 - 3x - 4$.

Solution. Transposing, $\quad 4x - x + 3x = 12 - 4 - 2 + 9$
Combining like terms, $\quad 6x = 15$
Dividing both sides by 6, $\quad x = 2.5$
(The check is left to the student.)

Example 8. Solve the equation: $\quad 5 + x = 4x - 13$

Solution. If we transpose the x terms to the left side, and others to the right side, we get,

$$-4x + x = -13 - 5$$
Combining, $\quad -3x = -18$
Dividing both sides by -3: $\quad x = +6$

Note. If we transpose the x terms to the right side and other terms to the left side, in this problem, we have positive terms, and get the same answer.

Example 9. Solve the equation $\quad 3x + 10 - x - 3 = 4 + 6x + 11 - 2x$

Solution. Transposing, $\quad 3x - x - 6x + 2x = 4 + 11 - 10 + 3$
Combining like terms on each side, $\quad -2x = 8$
Dividing both sides by -2, we get, $\quad x = -4$

Exercise 21.1

Solve the following equations and check the root in even-numbered problems.

1. $2x + 7 = 13$
2. $3x + 5 = 32$
3. $3x - 4 = 17$
4. $2x - 9 = 15$

5. $5y + 8 = 23$
6. $4n + 7 = 39$
7. $15 - 2x = 21$
8. $9 - 2y = 17$
9. $7 - 3t = 13$
10. $4x + 23 = 3$
11. $3n + 14 = 2$
12. $6y + 19 = 13$
13. $2y - 5 = 4$
14. $4n - 7 = 15$
15. $3x - 1 = 15$
16. $3t = 13 - t$
17. $5n = 14 + n$
18. $7y = 16 - 3y$
19. $5x - 3 = 2x + 15$
20. $6x + 2 = 2x - 10$
21. $2x + 5 = 7x - 15$
22. $7n + 2 = 9n - 16$
23. $3y - 5 = 5y + 13$
24. $7t + 4 = 4t - 13$
25. $10 - 3x = 4x - 22$
26. $9n + 5 = 7n - 12$
27. $1 - 5x = 15 - 2x$
28. $y - 28 = 6y - 5$
29. $16 - a = 8 + 4a$
30. $7 - 5x = x - 6$
31. $5w - 9 = 7w - 9$
32. $4a + 13 = 6a + 13$
33. $7c - 4 = 11c + 26$
34. $12 - 3v = 5v - 16$
35. $1 - b = 16 - 5b$
36. $8 + k = 9 + 5k$
37. $6x + 7 - x = 2x + 14$
38. $2y + 1 + y + 3 = 5y + 4 - 3y$
39. $21 - 3t = 5t + 6 - 3t$
40. $7 + 5R = 7R + 5 - R - 5$
41. $15 - 4b - 2 + 2 = 21 - 2b$
42. $4 - 4s + 5 = 5s - 7 - 3s + 4$
43. $4a - 6 + 13 = -5a - 2$
44. $2n - 2 = 7n + 13 - 10$
45. $5 + 6x + 3 = 9x + 25 - x$
46. $1 + 4x = 5 + x - 11 - 5x$
47. $x - 5 - 2x + 7 = 2x + 2 - 5x$
48. $7 - 3x - 7 - 2x = 5 + x - 5$

21.5. PARENTHESES IN EQUATIONS

Sometimes, in setting up an equation for a problem, we wish to enclose a quantity in parentheses. Then we may get an equation such as the following:

$$5x - 2(4x + 3) = 7 + (3x - 2)$$

To solve such an equation, the first step is to remove the parentheses. In doing so, we observe the following two-part rule.

RULE. **If a quantity within parentheses is preceded by a plus sign (+), this plus sign and the parentheses may be omitted without changing the sign of any term within the parentheses.**

  If a quantity within parentheses is preceded by a minus sign (−), this minus sign and the parentheses may be omitted, provided that the sign of each term within the parentheses is changed.

  Remember, if a coefficient appears before the parentheses, this coefficient is multiplied by each term within the parentheses. If there is no coefficient

SOLVING EQUATION

before the parentheses, it should be understood that the numerical coefficient is 1 (or −1 if preceded by a minus sign).

The foregoing rule, of course, holds true for any sign or symbol indicating a quantity, such as brackets, braces, or vinculum.

Example 1. Solve the equation $14-3(x-5)=5+(x-4)$.

Solution. In removing the first set of parentheses, we note the −3 before the parentheses. Then we have $-3(x-5)$, which becomes $-3x+15$. Note the plus sign before the second set of parentheses. Then we get,

Removing both sets of parentheses, $\quad 14-3x+15=5+x-4$
Transposing, $\quad -3x-x=5-4-14-15$
Combining, $\quad -4x=-28$
Dividing both sides by −4, we get $\quad x=7$

To check the answer, we put 7 for x in the original equation, and get

$$14-3(7-5)=5+(7-4)$$

which becomes $\quad 14-3(2)=5+3$

or $\quad 14-6=5+3$

Example 2. Solve the equation $\quad 4+2(3x-4)=5-(2x-3)$.

Solution. Removing parentheses, $\quad 4+6x-8=5-2x+3$
Transposing like terms, $\quad 6x+2x=5+3-4+8$
Combining like terms, $\quad 8x=12$
Dividing both sides by 8, $\quad x=1.5$

Example 3. Solve the equation $\quad 5-4(x-3)=8+(x-2)$

Solution. Removing parentheses, $\quad 5-4x+12=8+x-2$
Transposing terms, $\quad -4x-x=8-2-5-12$
Combining like terms, $\quad -5x=-11$
Dividing both sides by 5, $\quad x=\dfrac{11}{5}$, or 2.2

Example 4. Solve the equation

$$4+(x+3)(x+2)=20+(x+4)(x-1)$$

Solution. In solving this equation, we first multiply the binomials as indicated, before doing any adding or subtracting. That is, we first find,

$$(x+3)(x+2)=x^2+5x+6; \quad \text{and} \quad (x+4)(x-1)=x^2+3x-4$$

274 ALGEBRA

It is best to keep these products enclosed in brackets as a first step. Then we get

$$4+[x^2+5x+6]=20+[x^2+3x-4]$$

Now we remove the brackets: $\quad 4+x^2+5x+6=20+x^2+3x-4$
Transposing terms, $\quad\quad\quad\quad\quad x^2-x^2+5x-3x=20-4-4-6$
Combining like terms, $\quad\quad\quad\quad\quad\quad\quad 2x=6$
Dividing both sides by 2, $\quad\quad\quad\quad\quad\quad\quad x=3$

To check the answer, we put 3 for x in the original equation, and ask,

$$\begin{aligned}\text{does}\quad &4+(3+3)(3+2)=20+(3+4)(3-1)?\\ \text{does}\quad &4+\quad(6)\quad(5)\quad=20+\quad(7)\quad(2)\quad?\\ \text{does}\quad &4+\quad\quad 30\quad\quad=20+\quad\quad 14\quad\quad?\ \text{Yes}\end{aligned}$$

Example 5. Solve the equation $5x-(x-3)(x+2)=13-(x-3)(x-5)$.

Solution. In this equation, we must first perform the multiplication of the binomials as indicated by the parentheses. That is, we first find

$$(x-3)(x+2)=x^2-x-6;\quad\text{and}\quad(x-3)(x-5)=x^2-8x+15$$

Now we write the equation with these products enclosed in brackets:

$$5x-[x^2-x-6]=13-[x^2-8x+15]$$

Now we remove the brackets: $\quad 5x-x^2+x+6=13-x^2+8x-15$
Transposing terms, $\quad\quad\quad\quad\quad 5x-x^2+x+x^2-8x=13-15-6$
Combining like terms, $\quad\quad\quad\quad\quad\quad\quad -2x=-8$
Dividing both sides of the equation by -2, $\quad\quad x=4$

Check. To check the answer, we put 4 for x in the *original equation*, and ask

$$\begin{aligned}\text{does}\quad &5(4)-(4-3)(4+2)=13-(4-3)(4-5)?\\ \text{does}\quad &20-\quad(1)(6)\quad\quad=13-(1)(-1)?\\ \text{does}\quad &20-\quad\quad 6\quad\quad\quad=13-\quad(-1)?\\ \text{does}\quad &20-\quad\quad 6\quad\quad\quad=13+1?\ \text{Yes}\end{aligned}$$

Note. In examples 4 and 5, we get terms containing the square of a variable, such as x^2. Equations that contain such second degree terms are called quadratic equations. We are not ready at this point to solve general quadratic equations. They will be studied in a later chapter. However, at this time, it happens that such terms disappear in the process of solving. This happened in Examples 4 and 5, where the x^2 terms dropped out.

Exercise 21.2

Solve the following equations and check some answers:

1. $3x + 2(3x - 5) = 14 + 3(x - 4)$
2. $n + 4(2n - 3) = 19 + 2(n - 5)$
3. $4n - 3(n - 2) = 5 - (n + 3)$
4. $2x - 2(4x + 3) = 4 - (3x - 2)$
5. $x - (2x - 5) = 6 - (3x - 4)$
6. $3x - (4x - 5) = 7 - (5x - 4)$
7. $3y - 4(2y + 3) - 5 = 7 - 3(4 - 2y) - 5y$
8. $3y - 3(2y - 5) - 5 = 7 - 2(4y + 3) + 2y$
9. $4x + (x - 4)(x - 1) = 6 + (x - 5)(x + 2)$
10. $3n + (n + 4)(n - 1) = 2 + (n - 2)(n + 6)$
11. $5n - (n + 3)(n - 4) = 8 - (n - 2)(n + 4)$
12. $6x - (x - 2)(x - 3) = 9 - (x - 4)(x - 5)$
13. $(x + 2)(x - 5) - x(3x + 2) = 2 - 2x(3 + x) + 5x$
14. $3x - (x + 3)(x - 4) + 3 = 7 - x(x - 7) - 2x$
15. $3x - x(2x - 5) + 5 = 13 + 2x(5 - x)$
16. $4 + 3n(5 - n) - 2n(3 - n) = 7 - n(n - 3)$
17. $3c - (2c - 1)(3c + 2) + 5 = 2c - 2(c - 3)(3c + 1) + 5$
18. $6 - 2t(4 - 3t) - 3(t - 2)(t - 3) = 5 - 3(2 - t)(2 + t)$

22 STATED WORD PROBLEMS

22.1. THE IMPORTANCE OF STATED PROBLEMS

In this chapter, we show how equations can be used to solve the so-called "stated problems" or "word problems," By this, we mean the practical problems that are stated in words, problems that may arise in everyday life and in scientific study.

Equations enable us to find answers to problems concerning electric circuits, laws of motion, pressure of gases, speeds of electrons, and many other problems in science. Algebra can help us to find answers to problems such as the following.

1. How much water should be added to a gallon of a 5% solution of carbolic acid to make a new solution testing only 2%?

2. A messenger starts out by car to deliver a message at a distant point. He travels at an average of 40 miles per hour. Later it is discovered that the message must be changed. A second messenger starts out at 11 A.M. to overtake the first messenger who had started at 7 A.M. How fast must the second messenger travel to overtake the first by 5 P.M.?

3. A carpenter has a board 10 feet long to use for eight steps for a stepladder. Each step must be one-half inch shorter than the one below it. How long should the bottom step be if there is to be no waste?

4. A swimming pool can be filled through one opening in 9 hours. It can be filled by another opening alone in 12 hours. How long should it take to fill the pool if both openings are used?

5. A man has a motorboat that can travel 12 miles per hour in still water. How far can he travel downstream in a river and return so he can be back at his starting point in 6 hours if the current of the river is 4 miles per hour?

22.2. THE APPROACH TO PROBLEM SOLVING

In working word problems, your first concern should always be with the correct approach. Do not try to get the answer by guessing. Remember, at this time you are not simply looking for answers. Instead, you must try to learn how to use algebra to solve problems.

No matter how simple a problem might be, do not lose sight of your purpose. Your purpose now is to learn how to approach a problem in a systematic way through the use of algebra. That is the only way to acquire the practice necessary for working more difficult problems later.

In every problem, there are some quantities that are known. There is at least one quantity that we are asked to find. Such a quantity is called an *unknown*. There may be more than one unknown in a problem.

In solving a problem, we use some letter, say x, or n, or some other letter, to represent an unknown quantity; that is, a quantity we do not know, a quantity we are required to find. Then we work with this letter just as though we knew its value. As an example to show how we can use a letter for an unknown and express certain quantities, consider the following.

Suppose we do not known your age. We let x represent the number of years in your age now. Then we can express many things. Four years from now your age will be $(x+4)$ years. Five years ago, your age in years was $(x-5)$. If your father is twice your age, his age in years can be represented by $2x$. If your mother is a year younger than your father, her age in years can be represented by the quantity $(2x-1)$. If your uncle is three years older than your father, his age can be represented by the quantity $(2x+3)$.

22.3. THE "FIVE GOLDEN RULES" FOR SOLVING STATED PROBLEMS

Many people find it difficult to work stated problems. If you have trouble with such problems, it may be that you are uncertain as to just how to begin.

There is a definite procedure you can follow to get started right on the solution of a word problem. If you begin right, you will usually have little trouble in finding the answer.

There are five specific steps you can follow in solving any word problem. You may have some difficulty in following one or two of the steps, but you can always use these five as a guide.

Here are the "Five Golden Rules" for solving problems.

1. *Let some letter, such as x, represent one of the unknowns.* (This is usually, though not necessarily, the smallest unknown.)

2. *Then, try to express the other unknowns by using the same letter.*

3. *Write a true equation from the information given in the problem.* Make your equation say, in symbols, what the problem says in words.

4. *Solve the equation you have set up.*

5. *Check your answer to see whether it satisfies the conditions given in the problem.* Do *not* check in the equation you have set up, because you may have set up the wrong form of the equation.

If you can follow the first three steps carefully, you have a good start. Most people have trouble with the first three steps. In fact, many people have trouble with the first step. In the first step, let some letter represent one of the unknowns and tell definitely what the letter is to represent.

After the first two steps, you may have some difficulty in setting up the equation. In analyzing the problem, try to discover some relationship between the various quantities of the problem. Try to determine whether one quantity is equal to another, or whether the sum or the difference of two quantities is equal to a third quantity. Then make your equation state that relationship.

We shall work a few problems following the steps very carefully.

22.4. GENERAL PROBLEMS

Example 1. A student bought a chair and a desk for a total cost of $92. The desk cost $5 more than twice the cost of the chair. Find the cost of each.

Solution. Here you have two unknowns: the cost of the chair and the cost of the desk. Your thinking should be, "I am going to let the letter x represent the number of dollars the chair cost." Begin your statement with the word *let*. Now we shall follow each one of the separate steps.

Step 1. Let $x =$ the number of dollars the chair cost.

Step 2. Now we express the cost of the desk, using the same letter. Then $2x + 5 =$ the number of dollars the desk cost.

Step 3. In this step, we write an equation that states in symbols exactly what the problem says. The problem says that the total cost was $92. So we make the equation say exactly that:

$$\text{cost of chair} + \text{cost of desk} = 92$$

$$x \quad + \quad 2x + 5 \quad = 92$$

Step 4. In this step, we solve the equation:

$$3x = 92 - 5$$

then $$3x = 87$$

$$x = 29, \text{ number of dollars the chair cost}$$

and $$2x + 5 = 63, \text{ number of dollars the desk cost}$$

Step 5. To check the answer, we add the cost of the two. It should be $92, that is, $29 + 63 = 92$.

Example 2. A rectangular field has a perimeter of 124 rods. If the length is 10 rods less than 3 times the width, find the width and the length.

Solution. In this problem we have two unknowns, the width and the length.

Step 1. Let $x =$ the number of rods in the width.

Step 2. Then $3x - 10 =$ the number of rods in the length.

Now, to write the equation, we must remember that the perimeter of a rectangular means the distance around it; that is, the length of two sides plus two ends. Then we must take two times each of the quantities. To express the perimeter, we write: $2(x) + 2(3x - 10)$

$$\text{perimeter} = 2 \text{ ends} + 2 \text{ sides}$$

Step 3. The problem says that the perimeter is equal to 124 rods. Then we simply make that statement by an equation:

$$2x + 2(3x - 10) = 124$$

Step 4. We solve the equation:
$$2x + 6x - 20 = 124$$
$$8x = 124 + 20$$
$$8x = 144$$
$$x = 18, \text{ (rods in width)}$$

then, $3x - 10 = 44$, number of rods in length

Step 5. To check the solution, we add two ends and two lengths, and get

$$36 + 88 = 124$$

Example 3. A student has 560 pages of reading to be read in a total of four days. He wishes to arrange his reading so that he may reduce the number of pages read by 30 pages each day after the first. How many pages should he read the first day, and how many each day after that?

Solution. In this problem we have four unknowns, the number of pages to be read each day.

Step 1. Let $x =$ the number of pages to be read the first day.

Step 2. Then $x - 30 =$ the number of pages to be read the second day; and $x - 60 =$ the number of pages to be read the third day; and $x - 90 =$ the number of pages to be read the fourth day.

Step 3. Now we make the equation say that the total number of pages is 560:

$$x + x - 30 + x - 60 + x - 90 = 560$$
days: 1st   2nd   3rd   4th

STATED WORD PROBLEMS 281

Step 4. We solve the equation: $4x = 560 + 30 + 60 + 90$
$4x = 740$
then $x = 185$, pages for first day

Then, pages by days: 2nd day, 155; 3rd day, 125; 4th day, 95.

Step 5. We check by adding the number of pages for the four days: we get 560.

Exercise 22.1

1. A student bought a slide rule and a drawing set for a total of $43. The drawing set cost $8 more than the slide rule. Find the cost of each.
2. A fishing rod and reel together cost $31. If the reel cost $6 less than the rod, find the cost of each.
3. The length of a rectangle is 5 inches less than twice the width. If the perimeter of the rectangle is 83 inches, find the width and the length.
4. In a certain triangle, side b is twice as long as side a, and side c is 3 inches shorter than side b. If the perimeter of the triangle is 32 inches, find the length of each side.
5. In a certain triangle, angle A is 30° greater than angle B; angle C is twice as large as angle A. Find the size of each angle. (Recall that the sum of the degrees in the three angles of a triangle is 180°.)
6. A tract of land along a river is to be fenced along one side and the two ends. The length of the tract of land is 15 rods less than 3 times the width. If the total length of the fence is 90 rods, find width and length.
7. In a certain triangle, the first angle is 50° less than the second, and the third angle is twice the size of the first. Find the size of each angle.
8. A family on a tour drove 820 miles in 3 days. They drove twice as many miles the second day as the first. The third day they drove 60 miles less than on the second day. Find the distance driven each day.
9. An electric circuit has two branches. One branch carries a current that is 3 amperes more than the other. When the two currents combine, the total current is 20 amperes. Find the current in each branch.
10. On three tests, a student made a total combined score of 268 points. He made 15 points more on the second test than on the first; and on the third test he made 5 points less than on the second. How many points did he make on each test?

22.5. AGE PROBLEMS

In everyday life, we do not use algebra to find the ages of people. Why then do we work problems concerning ages? Such problems are useful at this point for at least two reasons.

In analyzing a problem about ages, we use the same kind of thinking that we use in analyzing a problem concerning electric circuits or any other scientific or strictly practical problem. Then, too, the subject matter is well known to everyone. We know what is meant by a person's age. It is better known at this point than the theory of electric circuits, Ohm's law, or Kirchhoff's laws. We can therefore concentrate on the thinking involved and the method of attacking such a problem.

We shall now work some problems involving ages.

Example 1. A man is 26 years older than his son. If the sum of their ages is 60 years how old is each?

Solution. Following the five basic steps in solving problems, we begin:
Let x = the son's age (in *years*).
Then $x + 26$ = the father's age (in *years*).
Equation: $x + (x + 26) = 60$ (this tells the sum of their ages, in years).
Then $x + x + 26 = 60$.
Transposing, $2x = 60 - 26$.
Combining, $2x = 34$,
Solving, $x = 17$, (number of years in son's age).
Then $17 + 26 = 43$, (number of years in the father's age).
Checking the answers, we add to see if the sum is 60, as it should be:

$$\text{son's age} + \text{father's age} = 60$$
$$17 + 43 = 60$$

Example 2. A man is 5 years older than twice his son's age. If the sum of their ages is 68 years, how old is each?

Solution. Let x = son's age (in years).
Then $2x + 5$ = father's age (in years).
Now we make the equation say that the sum of their ages is 68 (years).
Equation: $x + 2x + 5 = 68$
Transposing, $x + 2x = 68 - 5$
Combining, $3x = 63$
Dividing by 3, $x = 21$, son's age (in years)
Then the father's age is $(2x + 5)$, or 47 years.
Checking, we find that the sum of the ages is 68 years, as it should be.

Example 3. A man's age *now* is 2 years more than 3 times his son's age. In 5 years from now (*hence*, meaning in the future), the sum of their ages will be 62 years. How old is each one *now*?

Solution. Let x = the number of years in the son's age *now*.
Then $3x + 2$ = the father's age (in years) *now*.

Since the number 62 represents the sum of their ages 5 years hence, we first express the ages of both, 5 years from now.

5 years from now, the son's age will be $x+5$ (in years)
5 years from now, the father's will be $3x+2+5$ (in years)
Now, these are the two quantities whose sum is equal to 62. So we have

Equation: $\quad (x+5)+(3x+2+5)=62$ (the sum of the ages in 5 years)
Then $\quad\quad\quad\quad x+5+3x+2+5=62$
Transposing, $\quad\quad\quad x+3x=62-5-2-5$
Combining, $\quad\quad\quad\quad 4x=50$
Dividing by 4, $\quad\quad\quad x=12.5$ (son's age in years *now*)
Then the father's age is 39.5
In 5 years, the son will be 17.5 years, and the father will be 44.5 years. Then the sum of their ages will be 62, which checks with the problem.

Note. If you let x represent the son's age (in years) now, then whatever you do with the x, the answer will always be just what you let it represent.

Exercise 22.2

1. A father is 28 years older than his son. If the sum of their ages is 54 years, how old is each?

2. One man is 5 years older than another. If the sum of their ages is 62 years, how old is each?

3. A boy is 6 years younger than his sister. If the sum of their ages is 33 years, how old is each?

4. A mother is 3 times as old as her daughter. If the difference between their ages is 23 years, how old is each?

5. A man is now twice as old as his son. Three years hence (in the future), the sum of their ages will be 75 years. How old is each *now*?

6. A father is now 5 years older than 3 times his son's age. Four years ago, the sum of their ages was 45 years. How old is each one now?

7. A father is now twice as old as his son. Sixteen years ago, the father was 4 times as old as his son was then. How old is each *now*?

8. A woman is 6 times as old as her son. In 3 years, she will be 4 times as old as her son is then. How old is each *now*?

9. Of 3 brothers, the first is 4 years younger than the second, and the third is 3 years older than the second. If the sum of their ages is 80 years, how old is each now?

10. A man's age is now 6 years more than 3 times his son's age. Four years hence, the father will be 5 times as old as his son was 2 years ago. How old is each now?

22.6. CONSECUTIVE NUMBER PROBLEMS

An integer is a whole number, such as 2, 3, 5, 16, and so on. The integers also include the negative numbers, such as -5, -8, and so on. Consecutive integers are integers that have no other integer between them, such as the numbers, 5, 6, 7, 8, and so on. If we begin with an integer such as 12, then the next consecutive integer is $12+1$, or 13. The numbers, $-6, -5, -4$, are also consecutive integers.

If we let x represent the first of a series of three consecutive integers, then the second is $(x+1)$, the third is $(x+2)$.

Consecutive even integers are even integers that have no other even integer between then, such as the even numbers, 6, 8, 10. Notice that any two consecutive even integers have a difference of 2. In a series of three consecutive even integers, if we let x represent the first, then $(x+2)$ will represent the second, and $(x+4)$ will represent the third. The same is true if the integers are negative, such as $-8, -6, -4$.

Consecutive odd integers are odd integers (odd numbers) that have no other odd integer between them, such as the odd numbers, 5, 7, 9, and so on. Notice that any two consecutive odd integers have a difference of 2, the same as with even integers. In a series of three consecutive odd integers, if we let x represent the first, then $(x+2)$ will represent the second, and $(x+4)$ will represent the third. The expressions for odd integers are the same as we use for consecutive even integers.

In some problems, we have a series of numbers that differ by 3, or 5, or 10, or more. If the difference is uniform, then the problems are solved in the same way as consecutive number problems. In each case, we let x represent one of the numbers, and then express each of the other numbers using the same letter. We show the method by examples.

Example 1. Find four consecutive integers whose sum is 146.

Solution. Let $x =$ the first of the four consecutive integers.
Then $\qquad x+1 =$ the second
$\qquad x+2 =$ the third
$\qquad x+3 =$ the fourth

For the equation, we express the sum of the four integers and state the fact that the sum is equal to 146:

Equation: $\qquad x+(x+1)+(x+2)+(x+3) = 146$

Since all the signs are plus (+), we would not need parentheses; then we can write

$$x+x+1+x+2+x+3 = 146$$

Combining the x's, and transposing, we get:
$$4x = 146-1-2-3$$
Combining, $\qquad 4x = 140$

Dividing both sides by 4, $\quad x = 35$
Then the four integers are 35, 36, 37, 38.
To check the answer, we find that their sum is 146, as it should be.

Example 2. Find four consecutive odd integers whose sum is 192.

Solution. Let $x =$ the first of the consecutive odd integers.
Then $\quad x + 2 =$ the second,
$\quad x + 4 =$ the third,
$\quad x + 6 =$ the fourth

For the equation, we express the sum of the four integers, and say that the sum is equal to 192.

Equation: $\quad x + x + 2 + x + 4 + x + 6 = 192$
Then $\quad\quad\quad\quad\quad 4x = 192 - 2 - 4 - 6$
Combining like terms, $\quad 4x = 180$
Dividing both sides by 4, $\quad x = 45$
Then the four consecutive odd integers are 45, 47, 49, 51
To check the answer, we add the numbers and find their sum is 192.

Exercise 22.3

1. Find four consecutive integers whose sum is 158.
2. Find four consecutive even integers whose sum is −84.
3. Find five consecutive odd integers whose sum is 165.
4. A ladder is to have 5 steps, the longest at the bottom, and each step above the bottom is to be 1.5 inches shorter than the one immediately below it. If the steps are to be cut from a board 5 feet long without any portion of the board left over, find the length of each step.
5. On a trip of 1060 miles to be covered in four days, a family wishes to arrange their driving so that each day after the first they might drive 40 miles less than on the previous day. That is, they wish to reduce the driving by 40 miles each day. How many miles do they travel each day?
6. Find six consecutive multiplies of 5 whose sum is 255.
7. Find four consecutive multiplies of 8 whose sum is 144.
8. Each year, a man received a salary increase of $700 over the previous year. His total salary for a four-year period was $38,200. Find his salary for each of the four years.
9. Five shelves are to be made, one above the other, with the longest at the bottom. Each one above the bottom is to be 3 inches shorter than the one immediately below it, the shortest one at the top. If the five shelves are to be cut from a 10-foot board without any portion of the board left over as waste, find the length of each shelf.
10. A mechanic is asked to cut a strip of metal 4 feet long into four pieces so that each piece cut off is to be 2.5 inches longer than the previous piece. If there is to be no waste of the material, what should be the length of each piece?

22.7. COIN AND OTHER PROBLEMS ABOUT MONEY

In working coin problems, we have two things to consider: the number of coins (pieces of money), and the value of the coins (in cents). In any problem, for example, if we use x to represent the number of nickels, then we express the value of these nickels as $5x$ cents. Since each nickel is worth 5 cents, the value of x nickels is $5x$ cents. Since one dime is worth 10 cents, then the value of some dimes is equal to the number of dimes times 10, as the number of cents.

When you come to study electric circuits, for example, you will find that a problem in voltage involves the same kind of thinking as in a coin problem. For example, if a current of 5 amperes passes through a resistor, the voltage is equal to the resistance in ohms times 5. For a current of 10 amperes, the voltage through a resistance of 30 ohms is (30)(10) volts. A current of 10 amperes through a resistance of 80 ohms means that the voltage is (10)(80) volts. That is, the voltage is equal to the current times the resistance, using proper units for each. In the same way, the value of some coins of a particular denomination is equal to the number of coins times the value (in cents) of each coin.

Example 1. A collection of 53 coins consisting of nickels and dimes is worth $3.45. Find the number of coins of each denomination in the collection.

Solution. Let x = the number of nickels in this collection.
Then $53 - x$ = the number of dimes in the collection.

That is, to find the number of dimes, we subtract the number of nickels from 53. Now we express the value of the coins (in cents).

The value of x nickels is $5x$ cents.
The value of $(53-x)$ dimes is $10(53-x)$ cents.

Now we write the equation, stating that the total value is 345 cents ($3.45).

Equation: $$5x + 10(53 - x) = 345$$

In writing the equation, we do not use the word *cents*, because the numbers themselves refer to the number of cents.

Removing the parentheses, $5x + 530 - 10x = 345$
Transposing like terms, $530 - 345 = 10x - 5x$
Combining terms, $185 = 5x$
Dividing both sides by 5, $37 = x$

Since we let x represent the number of nickels, then there are 37 nickels. To get the number of dimes, we take $53 - 37 = 16$, number of dimes.

To check the solution, we now find the value of the coins:

37 nickels = $1.85;   16 dimes = $1.60;   total value = $3.45.

Example 2. A collection of 63 coins consisting of nickels and quarters is worth $6.95. Find the number of coins of each kind in the collection.

Solution. Here again we have two unknowns: the number of nickels and the number of quarters. For the first two steps we have the following:

Let x = the number of nickels in the collection.
Then $63 - x$ = the number of quarters in the collection.

The problem says that the total value is $6.95, which we call 695 cents. We state the value in cents because the value of the coins is to be stated in cents. Remember, the 695 is *not* the number of coins or pieces of money. Now we express the number of cents in the nickels and in the quarters.

The value of x nickels is $5x$ cents.
The value of $(63 - x)$ quarters is $25(63 - x)$ cents.

Now that we have the value stated in cents, we simply state that the total value of all the coins is equal to 695 cents; that is,

(value of nickels) + (value of quarters) = 695
     in cents   in cents

Equation: $5x + 25(63 - x) = 695$

Solving this equation, we get $x = 44$, number of nickels.
Then the number of quarters is $63 - 44 = 19$.
To check the work, we find that the total value is $6.95, as it should be.

Exercise 22.4

1. A collection of 14 coins consisting of nickels and dimes is worth 85 cents. Find the number of coins of each kind in the collection.
2. A collection of 72 coins consisting of nickels and dimes is worth $4.75. Find the number of coins of each kind in the collection.
3. A collection of 46 coins consisting of nickels and quarters is worth $5.30. Find the number of coins of each kind in the collection.
4. Find the number of dimes and quarters in a collection of a total of 54 coins worth $7.35.
5. In a collection of 40 coins, consisting of nickels, dimes, and quarters, there are 4 more dimes than quarters. The number of nickels is three times the number of quarters. If the total value of the coins in the entire collection is $6.90, find the number of coins of each kind.
6. At a school entertainment, children's tickets cost 25 cents each, and adults' tickets cost 60 cents each. If the total amount received was $37.30 from a total of 110 tickets, how many tickets of each kind were sold?

7. At a certain school entertainment, children's tickets cost 30 cents each, and adults' tickets cost 75 cents each. A total of 84 tickets were sold. If the same amount of money was received from all the children's tickets as from all the adults' tickets, how many tickets of each kind were sold?

8. At a ballgame, a total of 320 tickets were sold, some general admission and the rest as reserved seats. General admission tickets cost $1.50 each, and reserved tickets cost $2.50. If the total amount received from all the tickets was $560, how many tickets of each kind were sold?

22.8. PROBLEMS IN UNIFORM MOTION

When we use the words *uniform motion*, we mean that the rate of speed does not change. In most problems in uniform motion, the rate of speed varies from time to time, but we take the average rate and then assume that the average rate continues uniformly.

Suppose a car travels 120 miles in exactly 3 hours. To find the average speed, we divide the distance traveled by the number of units of time required. Then we have

$$120 \div 3 = 40$$

Then the average rate of speed is 40 miles per hour. By the average rate, we do not mean that the speed was exactly 40 miles per hour (mph) every minute of the time. It is impossible to drive at exactly 40 mph every second for 3 hours.

In working problems in uniform motion, we assume that the speed does not change. If there is any change or interruption in speed, then we must take such changes into account. We assume also that the speed is uniform even at the start, which is actually not possible. It is not correct to say "A car starts out at 30 mph." A car cannot start at 30 mph.

If we know the average rate of speed and the time of travel, we can find the distance traveled. For example, if a car travels at an average rate of speed of 30 mph for 5 hours, then distance traveled is 150 miles. In this example,

30 mph is called the rate (r)
5 hours is called the time (t)
150 miles is called the distance (d)

To find the total distance traveled in a given time at a given rate, we multiply the rate by the time. Of course, the factors must be stated in the proper units of measurement. Then the formula for the distance traveled is

$$(\text{rate}) \times (\text{time}) = \text{distance}, \quad \text{or} \quad rt = d$$

STATED WORD PROBLEMS 289

In beginning a problem in uniform motion we use a letter, such as x, to represent one of the unknowns. The unknown may be the rate, time, or distance. It is best to let the x represent the number of hours, or the rate of travel, rather than the distance traveled.

Example 1. A car starts on a trip at 10 A.M. and travels at an average rate of 35 mph. At 1 P.M., a second car starts out from the same point to overtake the first car and travels the same route at an average rate of 55 mph. How long will it take the second car to overtake the first? What time will it be then, and how far have the cars traveled?

Solution. Here we have several unknowns: the number of hours each car traveled; the number of miles they traveled; the time of day when the second car reached the first. We begin with the statement

Let $x = $ the number of hours the first car traveled.

Then $x - 3 = $ the number of hours the second car traveled.

(Since the second car started 3 hours later than the first, then the second car traveled 3 hours less than the first.)
Now we express the number of miles each car traveled:
  For the first car, we have the rate of 35 mph for x hours. Then the first car traveled a distance of: $35x$ miles
  For the second car, we have the rate of 55 mph for $(x-3)$ hours. Then the second car traveled a distance of: $55(x-3)$ miles
  It is sometimes convenient to set up a chart showing the *rate*, *time*, and *distance* for each car:

	Rate (mph)	Time (No. of hours traveled)	Distance (No. of miles traveled)
1st car	35	x	$35x$
2nd car	55	$x - 3$	$55(x - 3)$

To get the equation, we suddenly realize that the two cars traveled the same distance. Since they started at the same point, traveled the same route, and were together when the second reached the first, then they must have traveled the same distance. We make the equation say that the two distances were equal:

Equation: $\qquad 35x = 55(x-3)$

To solve the equation, we have:	$35x = 55x - 165$
Transposing,	$165 = 55x - 35x$
Combining like terms,	$165 = 20x$
Dividing both sides by 20,	$8.25 = x$

The result means that the first car traveled $8\frac{1}{4}$ hours. then the second car traveled $5\frac{1}{4}$ hours, since it started 3 hours later. To find the time of day when the second reached the first, we count $8\frac{1}{4}$ hours after the first car started, at 10 A.M., and get 6:15 P.M. If we count $5\frac{1}{4}$ hours after the start of the second car, 1 P.M., we also reach the time: 6:15 P.M.
To find the distance each traveled, we multiply:

1st car: $35(8.25) = 288.75$ miles;   2nd car: $55(5.25) = 288.75$ miles.

The answers check because they traveled the same distance.
    It took the second car 5.25 hours to overtake the first.

Example 2. A man taking a trip travels part of the way by car at an average of 45 mph. The rest of the way he travels by train at an average of 65 mph. If the entire trip was 405 miles and the total travel time was 7 hours, how far did he travel by car and how far by train?

Solution. In this problem, we could let x represent the number of miles traveled by car or by train. However, it is better to let x represent the time (in hours) for the car or for the train. We begin with the statement:
    Let $x =$ the number of hours traveled by car.
    Then $7 - x =$ the number of hours traveled by train.
Now we express the number of miles traveled by each mode of travel. The problem can be shown by a chart:

	Rate (mph)	Time of travel (hours)	Distance traveled (miles)
By car	45	x	$45x$
By train	65	$7 - x$	$65(7-x)$

Now the problem says that the total length of the trip was 405 miles. We use this information to write the equation; that is

(distance by car) + (distance by train) = 405

Equation: $\qquad\qquad 45x \;\; + \;\; 65(7-x) \;\; = 405$

To solve the equation, we first remove the parentheses, and get

$$45x + 455 - 65x = 405$$

Transposing like terms, $\qquad 455 - 405 = 65x - 45x$
Combining, $\qquad\qquad\qquad\quad 50 = 20x$
Dividing both sides by 20, $\qquad 2.5 = x$, number of hours by car
Then, $7 - 2.5 = 4.5$, the number of hours by train.
To get the distance traveled by each method of travel, we multiply the rate of each by the time in hours:

$$\begin{array}{lll} \text{by car:} & (2.5)(45) = & 112.5 \text{ miles} \\ \text{by train,} & (4.5)(65) = & \underline{292.5 \text{ miles}} \\ \text{total distance} & & 405.0 \text{ miles} \end{array}$$

Exercise 22.5

1. One car starts out from a town at 9 A.M. and travels at an average rate of 40 mph. Two hours later a second car starts out from the same town to overtake the first. If the second car travels at an average rate of 60 mph, how long will it take the second car to overtake the first? What time of day will it be then?

2. One car starts out on a trip at 8 A.M. with a message and travels at an average rate of 45 mph. At 11 A.M. a second car starts from the same point with a changed message to overtake the first car. If the second car travels at an average rate of 60 mph, how long will it take the second car to overtake the first? What time of day will it be then?

3. Two cars start from the same point at the same time and travel in opposite directions, the first at an average of 55 mph, the second at an average of 45 mph. How long will it take until they are 525 miles apart?

4. Two towns, A and B, are 440 miles apart. At 8 A.M., a car starts from A traveling toward B at an average of 50 mph. At 10 A.M., a second car starts from B and travels toward A at an average of 30 mph. How long will it be until they meet? What time of day will it be then?

5. A car starts out at 9 A.M. with a message and travels at an average of 42 mph. Later it is discovered the message must be changed. A second car starts at 11 A.M. from the same point with a changed message. How fast must the second car travel to overtake the first by 6 P.M.?

6. A man starts walking on a hike at an average of 2.5 mph. Two hours later another man starts on the same route and walks at an average of 4.5 mph. If the first man started at 7 A.M. when will the second overtake the first?

7. A boy starts on a hike at 8 A.M. walking at an average rate of 3 mph. At noon, another boy on a bicycle travels the same route at an average of 9 mph. When will the boy on the bicycle overtake the other?

8. On a trip of 450 miles, a man traveled part of the way by car at an average of 40 mph, and the rest of the way by train at an average of 60 mph. If his total travel time was 8 hours, how far did he travel by car and how far by train?

9. On a flight between two cities, one plane averages 640 mph. Another plane traveling the same route averages 480 mph. but requires 1.5 hrs longer than the first plane. Find the distance between the cities and the time required for each plane.

23 SPECIAL PRODUCTS AND FACTORING

23.1. DEFINITIONS

In arithmetic, we learn the multiplication table so we can multiply two numbers quickly without hesitation. In algebra, also, multiplication occurs so often that we should be able to do it quickly; that is, by *inspection*. That is to say, as we look at a problem in multiplication, we are able to tell the answer at once just as easily as we give the answer to 7 times 9, which is 63.

When two or more numbers are multiplied to gether, the answer is called a *product*. A product that is found quickly by inspection is called a *special product*. A special product does not give us a different answer. It is simply a product that is found quickly. It can be called a "short cut."

23.2. MULTIPLICATION OF MONOMIALS BY INSPECTION

We have already seen in Chapter 19 how to multiply two or more monomials. To multiply monomials, we take note of three things:

(1). *The sign of the product is determined by the rule for the signs in multiplication of signed numbers.*

(2). *The numerical coefficient of the product is found by multiplying the numerical coefficients of the monomials.*

(3). *For the literal numbers, we add exponents of like bases.*

Example 1. Multiply by inspection: $(-5x^2y^3)(-6x^3y^4)$.

Solution. For the numerical coefficient, we have: $(-5)(-6) = 30$.
For the letter part, we get: $(x^2)(x^3) = x^5$; $(y^3)(y^4) = y^7$.

Then we get the complete product: $+30x^5y^7$. The product can be found quickly.

Example 2. Multiply by inspection: $(-2a^3b)(-3ab^4c)(-7bc^4)$.

Solution. For the numerical coefficient, we have: $(-2)(-3)(-7) = -42$. For the letter part, we get: $(a^3)(a) = a^4$; $(b)(b^4)(b) = b^6$; $(c)(c^4) = c^5$. Then we get the complete product: $-42a^4b^6c^5$.

Exercise 23.1

Multiply the following by inspection:

1. $(3x)(4x)(5x)(2x)$
2. $(-2x^2)(-4x)(3x^3)(x)$
3. $(5n^3)(2n)(-3n^2)(n)$
4. $(4x^2y)(3xy^3)(-x^3y)$
5. $(-2a^2b^2)(-5ab^3)(a^2b^2)$
6. $(-xy)(-3x^3y^2)(-7x^2y)$
7. $(3x^2y)(-4x^4y^2)(-xy^3)$
8. $(-2rs)(-2r^3)(-4r^2s^3)$
9. $(-4rs^2)(-3r^2st)(-r^3t)$
10. $(-a^3b)(4a^2c)(-3abc^2)$
11. $(2ab^3)(-4a^3b^2)(3a^2b^4)$
12. $(4x^3y^2)(-2xy^2z)(yz^3)$
13. $(-3x^2y)(-2y^3)(-4x)(-x^4y^2)$
14. $(-a^2b^3c)(-2a^3b)(-4ac)(-2ab)$
15. $(-x^2y)(-xy^3)(-x^4y)(-x)(-y)$
16. $(-r^3s)(-r^2s^3)(-rs)(-r)(-s^2)$
17. $(-2x^3)(-5xy^2)(-3y^4)(-3x^3)^2$
18. $(-4xy^3)(-3y^4)(-2xy)(-2y^3)^2$

23.3. FACTORING MONOMIALS

Factoring is the reverse of multiplication. When two or more quantities are multiplied together, the answer is called the product. Each of the quantities multiplied is called a *factor* of the product. For example, the product of 3 and 7 is 21. Then the *factors* of 21 are 3 and 7.

It is often necessary and useful to find the factors of a product. Finding the factors is called *factoring*. To factor a number, such as 15, means to find the factors of 15, which are 3 and 5.

There are many uses for factoring. In many practical problems, we can actually find the answers by factoring, as we shall see later. One of the chief uses of factoring is in reducing fractions. For example, suppose we have the fraction

$$\frac{21}{35}$$

SPECIAL PRODUCTS AND FACTORING

In most cases, we reduce such a fraction to lower terms. The fraction, $\frac{21}{35}$, can be reduced by dividing numerator and denominator by 7. Then

$$\frac{21}{35} = \frac{3}{5}$$

Now our question is: How can we tell when a fraction can be reduced? In a fraction such as $\frac{21}{35}$, we can tell at a glance that 7 is a factor of 21 and of 35. Actually, what we do is to factor numerator and denominator; that is,

$$\frac{21}{35} = \frac{3 \cdot \cancel{7}}{5 \cdot \cancel{7}} = \frac{3}{5}$$

When we have factored numerator and denominator, we see that both can be divided by 7, and we get the reduced fraction, $\frac{3}{5}$.

We usually reduce fractions to lowest terms because it is more convenient to work with smaller numbers. For example, if we have the fraction, $\frac{24}{36}$, we change it to $\frac{2}{3}$ by dividing numerator and denominator by 12.

To reduce a fraction in algebra, we must first find the factors of the numerator and the denominator of the fraction. Actually, this is not so difficult as one might think. In many ways, factoring is easier in algebra than in arithmetic. Of course, in arithmetic we can often tell the factors of a number at sight. For example, the factors of 55 are 5 and 11. However, in arithmetic it is not always easy to find the factors. Suppose we wish to reduce the following fraction:

$$\frac{221}{493}$$

To reduce the fraction we must find a common factor of 221 and 493. After some study we finally discover that the fraction can be written:

$$\frac{221}{493} = \frac{13 \cdot \cancel{17}}{\cancel{17} \cdot 29}$$

Then we know the numerator and the denominator can be divided by 17, and we get the reduced fraction: 13/29.

Remember: To reduce a fraction to lower terms, in algebra as in arithmetic, we must know some common factor of numerator and denominator. That means factoring is necessary. When we have studied a few special products, we shall find that factoring in algebra is not difficult.

Some numbers in arithmetic cannot be factored, such as the number 17. The number 17 cannot be separated into any factors other than 1 and the number 17 itself. Such a number is said to be *prime*. A prime number is a number that cannot be separated into any integral factors other than 1 and the number itself. Just so, we have some expressions in algebra that are prime; that is, they cannot be factored.

23.4. PRIME FACTORS

A prime factor is a factor that cannot be further separated into any factors other than itself and 1. For example, we can say the factors of 60 are 6 and 10, because 6 times 10 equals 60. However, the factors, 6 and 10, are not prime factors because they can be split into other simpler factors. The factors of 6 are 2 and 3; the factors of 10 are 2 and 5. Then the number 60 can be separated into the prime factors: 2, 2, 3, and 5.

In algebra, a monomial can be separated into its prime factors simply by writing all the factors separately. As an example,

$$12x^2y^3z = 2 \cdot 2 \cdot 3 \cdot x \cdot x \cdot y \cdot y \cdot y \cdot z$$

However, in most work in algebra, a monomial is not usually separated into its prime factors in this manner. Sometimes this is done to emphasize the separate factors.

To see how factoring is used in reducing fractions in algebra, consider the following fraction:

$$\frac{15x^3y}{21x^2y^2}$$

In this example, we shall show all the separate factors of numerator and denominator:

$$\frac{15x^3y}{21x^2y^2} = \frac{3 \cdot 5 \cdot x \cdot x \cdot x \cdot y}{3 \cdot 7 \cdot x \cdot x \cdot y \cdot y}$$

Now we see that we can divide numerator and denominator by the following factors: $3 \cdot x \cdot x \cdot y$. Then we get the reduced fraction:

$$\frac{5x}{7y}$$

SPECIAL PRODUCTS AND FACTORING 297

However, in a fraction of this kind in which numerator and denominator are monomials consisting of only factors, we need not show the separate factors as we have done in this example. We can at once divide numerator and denominator by any factors contained in both. In the foregoing fraction, we can at once divide numerator and denominator by 3, by x^2, and by y, showing the result by crossing out these factors:

$$\frac{\cancel{15}^{5x}\cancel{x^3}\cancel{y}}{\cancel{21}_{7}\cancel{x^2}\cancel{y^2}_{y}} = \frac{5x}{7y}$$

Exercise 23.2

Separate each of the following monomials into prime factors, (No. 1–20):

1. $4x^2$
2. $8xy$
3. $6y^2$
4. $10abc$
5. $9x^2y$
6. $12xy^3$
7. $15a^2b^2$
8. $14m^2n^3$
9. $21x^3y^2$
10. $18rs^4$
11. $20x^3y^3$
12. $16a^2b^3c$
13. $24a^3b^4$
14. $27x^3y^3z^2$
15. $25m^3n^4t$
16. $22x^4y^2t^3$
17. $36x^3y^5z^2$
18. $48a^4b^2c^3$
19. $60r^3s^2t^6$
20. $80a^4s^5t^2$

Reduce the following fractions to lowest terms:

21. $\dfrac{6x^2}{8xy}$
22. $\dfrac{4x^2y}{6xy^2}$
23. $\dfrac{6ab}{10b^2}$
24. $\dfrac{16a^2b^2}{20a^4b}$
25. $\dfrac{4x^3y}{8x^2y}$
26. $\dfrac{6x^2y^2z}{3xyz}$
27. $\dfrac{24ax^3y^2}{40axy^3}$
28. $\dfrac{40rs^2t^3}{50r^2st^2}$

23.5. A SPECIAL PRODUCT: A MONOMIAL TIMES A POLYNOMIAL

We have seen (Chapter 20) how a polynomial is multiplied by a monomial. For example, $5(3x-7) = 15x - 35$. A product of this kind can be written quickly. For this type of product, we have the following rule.

RULE. To multiply a polynomial by a monomial, multiply each term of the polynomial by the monomial.

This rule is called the *distributive law for multiplication.* It may be stated as a formula. If a, b, and c represent three numbers, respectively, then

$$a(b+c) = ab + ac$$

If such a product is indicated horizontally, then the polynomial must be enclosed in parentheses. By following this rule, we can write such a product at once by inspection.

Examples

(1) $3(2n-5) = 6n - 15$
(2) $4x(5x^2 - 3x + 7) = 20x^3 - 12x^2 + 28x$
(3) $3x^2y(4x^2 - 2xy + 3y^2) = 12x^4y - 6x^3y^2 + 9x^2y^3$
(4) $(6x^2 - 5xy + 4y^2)(2xy) = 12x^3y - 10x^2y^2 + 8xy^3$

Exercise 23.3

Multiply by inspection. Try to work the entire set in 5 minutes.

1. $3(2x-7)$
2. $5(3x-8)$
3. $7n(1-3n)$
4. $4a(2-5a)$
5. $12(3x-5)$
6. $15(2t-3)$
7. $6ax(8ax^2 - 3x)$
8. $3mn(6mn^3 - 5n)$
9. $20rst(4st - 1)$
10. $30a^2bc(3ab - 1)$
11. $2(5x - 3y + 4)$
12. $4(7x - 3xy + 5y)$
13. $-7n(3 - 5n - n^2)$
14. $-5n(6 + 2n - 3n^2)$
15. $5x^2(4x^2 - 7xy + 1)$
16. $6xy(3x^2 + 5xy - 1)$
17. $-3xy^2(4x - 5y - 1)$
18. $-7x^2y(2x^2 - 3xy + 1)$
19. $4ab^2(3a^2 - 5ab + 7b^2 + 1)$
20. $3m^3n^2(5m^3 - 6m^2n - 3mn^2 - 1)$
21. $3x^2y^3(1 - 4x^2y - 2x^4y^2 - 5x)$
22. $-4xy^3(5 - 3x^3y - 4x^5y^3 + 2y)$
23. $-6x^4(3x^4 - 4x^3 - x^2 - x + 2)$
24. $-3x^5(5x^4 + 2x^3 + 3x^2 - x - 1)$

23.6. FACTORING BY TAKING OUT A COMMON FACTOR

Let us look at the multiplication: $3(2x - 5) = 6x - 15$. In this example, we have the product of two factors. One factor is 3; the other factor is the binomial, $2x - 5$. The product is $6x - 15$.

SPECIAL PRODUCTS AND FACTORING

Now suppose we know the product of two factors is $6x-15$, but we do not know the factors. Then our question is this: since we know that the quantity, $6x-15$, is the product of two factors, how can we find the factors?

As we look at the product, $6x-15$, we try to determine what number has been used as a multiplier. To get the two terms in the product, $6x$ and 15, we might guess that we used a "3" times a binomial, because 6 and 15 can both be divided by 3. Then we know that one factor is 3.

That is, we know that we have multiplied 3 times some binomial. To get the binomial factor, we divide 3 into both terms, $6x$ and -15, and we get the result, $2x-5$. We can then write the given quantity as the product of two factors:

$$6x - 15 = 3(2x - 5)$$

The answer of course can be checked by multiplying the two factors together.

Now let us look at some examples.

Example 1. Find the factors of the binomial: $16x^2 - 24y$.

Solution. Now we assume that we have multiplied some binomial by a number to get this product. We first try to determine what number must have been used as a multiplier to get the numbers, 16 and 24, in the product. It must be some number that can be divided into 16 and 24. The largest common factor of 16 and 24 is 8. Therefore, we know that some binomial must have been multiplied by 8. Then 8 is one of the factors of the product. To find the binomial that was multiplied by 8, we divide 8 into both terms of the product, $16x^2 - 24y$. Dividing, we get the other factor; $(2x - 3y)$. Then the product can be written as the product of two factors:

$$16x^2 - 24y = 8(2x^2 - 3y)$$

To check the answer, multiply the two factors together.

Example 2. Find the factors of the polynomial: $12x^2 - 20x + 32$.

Solution. First we try to determine what number has been used to multiply a polynomial to get the product: $12x^2 - 20x + 32$. It must be a factor of all the terms of the polynomial. The largest common factor of the three terms is 4. Then one of the factors of the polynomial is 4. Now we divide 4 into each term and get: $3x^2 - 5x + 8$. This is the other factor.

Then the given polynomial can be written as the product of two factors:

$$12x^2 - 20x + 32 = 4(3x^2 - 5x + 8)$$

One factor is 4; the other factor is the trinomial: $3x^2 - 5x + 8$.

Note. *In the foregoing example, we divide 4 into each term of the given polynomial. This process is called "taking out a common factor."*

Example 3. Factor the following polynomial by taking out the largest common factor of each term:

$$12n^3 - 15n^2 + 6n$$

Solution. We look for the largest factor that can be divided into all the terms of the given polynomial. In this problem the largest common factor of the three terms is $3n$. We can divide the factor $3n$ into each term of the polynomial.

Now we take the factor $3n$ out of each term. Dividing each term by $3n$, we get $4n^2 - 5n + 2$. The result is the other factor. Then we get

$$12n^3 - 15n^2 + 6n = (3n)(4n^2 - 5n + 2)$$

The result can be checked by multiplying the two factors together. We should get the original polynomial, which is the product.

For factoring any polynomial by taking out a common factor, we have the following rule.

RULE. Inspect the terms of the given polynomial to determine the greatest common factor that is contained in each term of the polynomial. This common factor is one of the factors of the given expression.
Divide the common factor into each term of the polynomial. The result is the other factor.

Example 4. Factor the trinomial, $2x^3 - 4x^2 + x$.

Solution. We can divide each term by x. Taking out the x, we have

$$2x^3 - 4x^2 + x = x(2x^2 - 4x + 1)$$

The result can be checked by multiplication.

SPECIAL PRODUCTS AND FACTORING

Example 5. Find the factors of the following polynomial by taking out a common factor:

$$6x^3y - 8x^2y^2 - 10xy^3 + 4xy$$

Solution. The largest quantity that can be divided into each term of the given polynomial is $2xy$. This is the common factor. Dividing the common factor, $2xy$, into each term of the polynomial, we get the other factor. It is

$$3x^2 - 4xy - 5y^2 + 2$$

Then we can write the original expression as the product of two factors:

$$6x^3y - 8x^2y^2 - 10xy^3 + 4xy = 2xy(3x^2 - 4xy - 5y^2 + 2)$$

Exercise 23.4

Separate the following expressions into factors by taking out the largest common factor:

1. $5x + 15$
2. $18y + 12$
3. $14x + 35$
4. $20n + 4$
5. $15t - 5$
6. $24y - 6$
7. $9x^2 + 6x$
8. $10x^2 - 15x$
9. $40n^2 - 25n$
10. $15t^3 - 40t^2$
11. $24n^3 + 36n$
12. $50x^2 - 30x^4$
13. $18xy - 27x^2y$
14. $12x^5y^2 - 8x^3y^3$
15. $33x^2y^3 - 44xy^4$
16. $9R^2h - 4r^2h$
17. $\pi R^2 h - \pi r^2 h$
18. $2\pi r^2 + 2\pi rh$
19. $6x^6 - 2x^3$
20. $5x^2 + 10x^5$
21. $6x^3 - 18x^4$
22. $8n^4 + 24n^5$
23. $10n^3 - 30n^6$
24. $3x^3y - 12x^4y^3$
25. $8x^2y^3 + 12x^3y^2 - 6xy$
26. $10ab^3 - 15a^2b^2 + 20ab$
27. $12a^3b - 18a^2b + 3ab$
28. $25r^3s^2 - 15r^2s^2 - 5rs$
29. $3x^3 + 15x^2y - 6xy^2$
30. $2x^2y - 6xy^2 + 8y^3$
31. $8x^3y - 16x^2y^2 - 12xy^3$
32. $5x^4y - 15x^3y + 20x^2y$
33. $12a^3b^2 + 30a^2b^3 - 18ab^4$
34. $24h^2k^4 - 36h^3k^3 + 12h^2k^2$
35. $10x^4 + 8x^3 - 6x^2 + 2x$
36. $3x - 6x^2 + 9x^3 - 12x^4$
37. $x^5 + x^4 - x^3 + x^2 - x$
38. $y^2 - y^3 + y^4 - y^5 + y^6$

23.7. A SPECIAL PRODUCT: THE SUM OF TWO NUMBERS TIMES THEIR DIFFERENCE

Suppose we multiply two binomials that are exactly alike except for one sign, such as $(3x + 5y)(3x - 5y)$. We shall first multiply these by the usual

$$\begin{array}{r}3x + 5y \\ 3x - 5y \\ \hline 9x^2 + 15xy \\ -15xy - 25y^2 \\ \hline 9x^2 - 25y^2\end{array}$$

method of long multiplication as shown at the left. Notice that the two binomials are exactly alike except for one sign. If we assume that $3x$ represents one number, and that $5y$ represents another number, then the first binomial can be called the *sum* of two numbers, and the second binomial can be called their *difference*. That is, $3x+5y$ represents the sum of two numbers, and $3x-5y$ represents their difference. Then the problem means that we multiply *the sum of two numbers by their difference.*

Notice that the product contains only two terms. The first term of the product, $9x^2$, is the square of the first number, $3x$. The second term, $25y^2$, of the product is the square of the second term, $5y$. Notice there is no middle term in the product because we get a $+15xy$ and a $-15xy$, and these two terms cancel each other. The product, then, contains two terms which are the squares of the two numbers, with a minus sign between them. That is, the product is the *difference between the squares* of the two numbers.

Example 1. Multiply the two binomials: $(4y-7)(4y+7)$, and try to determine the rule for such a product.

$$\begin{array}{r}4y - 7 \\ 4y + 7 \\ \hline 16y^2 - 28y \\ +28y - 49 \\ \hline 16y^2 - 49\end{array}$$

Solution. We first multiply the two binomials by the usual method of long multiplication, as shown at the left. Again, we note that the product contains only two terms with a minus sign between them. The binomial, $4y+7$, represents the sum of two numbers, and the binomial, $4y-7$, represents their difference. The product again becomes the square of the first number minus the square of the second number.

Let us see how we can formulate a rule for finding a product of this kind by inspection. If we represent one number by x and another number by y, then $x+y$ represents the sum of the two numbers, and $x-y$ represents their difference. If we multiply the sum, $(x+y)$, by the difference, $(x-y)$, we shall always get a product that is the difference between the two squares; that is, x^2-y^2. For this kind of product we have the following rule.

RULE. The product of the sum of two numbers times their difference is equal to the difference between their squares.

The rule may be stated as a formula:

$$(a+b)(a-b) = a^2 - b^2$$

Note. To apply this rule, the two binomials must be exactly the same except that in one case we have a *plus* sign and in the other binomial we have a *minus* sign between the two numbers.

Example 2. Multiply by inspection: $(7a-11b)(7a+11b)$.

Solution. Applying the rule we get the product: $49a^2 - 121b^2$.

Example 3. $(8c+1)(8c-1) = 64c^2 - 1$.

Example 4. $(6x-0.4)(6x+0.4) = 36x^2 - 0.16$.

Example 5. $(5t-3x)(3x+5t) = 25t^2 - 9x^2$.

Example 6. Multiply: $(72)(68)$, using the foregoing principle.

Solution. Let us write the two numbers: $72 = 70 + 2$; and $68 = 70 - 2$. Then we can write the problem: $(70+2)(70-2) = 4900 - 4 = 4896$.

Exercise 23.5

Multiply by inspection. (Time: 6 minutes.)

1. $(x-5)(x+5)$
2. $(4-n)(4+n)$
3. $(x+7)(x-7)$
4. $(y+1)(y-1)$
5. $(3x-1)(3x+1)$
6. $(1-5x)(5x+1)$
7. $(5n-2)(5n+2)$
8. $(3x+4)(3x-4)$
9. $(9+4t)(4t-9)$
10. $(5t-7)(7+5t)$
11. $(8+3y)(3y-8)$
12. $(6y-11)(11+6y)$
13. $(7n+2a)(2a-7n)$
14. $(8b-5a)(5a+8b)$
15. $(20x+17y)(20x-17y)$
16. $(30a-23b)(30a+23b)$
17. $(40t-19x)(40t+19x)$
18. $(35k-21h)(21h+35k)$
19. $(3n+\frac{1}{2})(3n-\frac{1}{2})$
20. $(6n-\frac{1}{3})(6n+\frac{1}{3})$
21. $(8x-\frac{1}{4})(8x+\frac{1}{4})$
22. $(6y-0.5)(6y+0.5)$
23. $(n^3+7)(n^3-7)$
24. $(2a^3+5b)(2a^3-5b)$
25. $(4r^4-9s)(4r^4+9s)$
26. $(12xy-7z)(12xy+7z)$
27. $(10ab+11c)(10ab-11c)$
28. $(15x^2y-8ab)(15x^2y+8ab)$
29. $(80+1)(80-1)$
30. $(50-3)(50+3)$
31. $(300+2)(300-2)$
32. $(82)(78)$
33. $(603)(597)$
34. $(504)(496)$

23.8. FACTORING THE DIFFERENCE BETWEEN TWO SQUARES

In the preceding section, we multiplied two binomial factors, such as

$$(3x+5y)(3x-5y)$$

One factor, $3x+5y$, is called the *sum* of two numbers. The other factor, $3x-5y$, is called the *difference* between the two numbers. We get the

product
$$9x^2 - 25y^2$$

This product is the *difference between two squares*.

Now, suppose we have the expression, $9x^2 - 25y^2$, but we do not know the factors. Our problem then is to find the factors of the expression. To do so, we reverse the process of the preceding section. We notice that the expression contains two terms that are squares with a minus sign (−) between them. That is, we begin with the *difference between two squares*.

From the preceding section, we can assume that the expression, $9x^2 - 25y^2$, came about by multiplying two binomials; one binomial was the sum of two numbers, and the other binomial was their difference. To get the two original numbers, we take the square root of each of the terms, $9x^2$ and $25y^2$. Then we show the sum and the difference between the two numbers.

To find the factors of the difference between two squares, we follow a definite procedure.

Example 1. Find the factors of the expression $9x^2 - 25y^2$.

Solution. We recognize the expression as the difference between two squares. First, we indicate the two factors by two sets of parentheses:

$$(\quad)(\quad)$$

Now we take the square root of each term of the given binomial. The square root of $9x^2$ is $3x$, which becomes the first term of each factor:

$$(3x\quad)(3x\quad)$$

The square root of the second term, $25y^2$, is $5y$, which becomes the second term of each binomial:

$$(3x\quad 5y)(3x\quad 5y)$$

Now we connect the terms in one factor with a plus sign (+), and the terms in the other factor with a minus sign (−). Then we have the two factors:

$$(3x + 5y)(3x - 5y)$$

The given expression can now be expressed as the product of two factors:

$$9x^2 - 25y^2 = (3x + 5y)(3x - 5y)$$

We can always check the result by multiplication of the two factors.

Note. *Either factor may be written first.*

SPECIAL PRODUCTS AND FACTORING 305

In general, if we have given an expression that represents the difference between two squares, we can find the factors by the following rule.

RULE. To factor the difference between two squares, find the square root of each quantity. Then connect the square roots with a plus sign for one factor and with a minus sign for the other factor.

Factoring the difference between two squares is one of the most useful types of factoring. It should be completely understood and memorized. The rule can be represented in symbols. If one quantity is represented by x, and another quantity by y, then we have the formula

$$x^2 - y^2 \text{ can be factored as } (x+y)(x-y)$$

Example 2. Factor the expression $3x^3 - 48x$ into prime factors.

Solution. In all cases of factoring, any common factor should be taken out of the term *first*. As a first step in this example, we get

$$3x^3 - 48x = 3x(x^2 - 16)$$

Now the second factor, $x^2 - 16$, can be factored as the difference between two squares, and we get

$$3x^3 - 48x = 3x(x^2 - 16)$$
$$= 3x(x-4)(x+4)$$

Example 3. Factor the expression $16x^2 - 49$.

Solution. We recognize the expression as the difference between two squares. We first set down the two sets of parentheses:

$$(\quad)(\quad)$$

Now we find the square root of each term of the given expression. The square root of $16x^2$ is $4x$, which becomes the first term of each factor:

$$(4x\quad)(4x\quad)$$

The square of 49 is 7, which becomes the second term of each factor:

$$(4x\quad 7)(4x\quad 7)$$

We connect the terms in one factor with a plus sign (+), and in the other factor with a minus sign (−). Then we have the two factors:

$$16x^2 - 49 = (4x+7)(4x-7)$$

Note. To be sure of the correct factors, the following procedure should be followed: First set down the sets of parentheses; then find the square root of each term and place the square roots as the first and second terms of the factors; finally, insert the plus sign in one factor and the minus sign in the other.

Example 4. Find the factors of the expression $36n^2 - \frac{1}{9}$.

Solution. Following the suggested steps, we get the factors:

$$36n^2 - \frac{1}{9} = \left(6n + \frac{1}{3}\right)\left(6n - \frac{1}{3}\right)$$

Example 5. Find the prime factors of the following expression $25x^2 - 100$.

Solution. We recognize the expression as the difference between two squares. We could take the square root of each term as it stands. However, we should always take out any common factor first. Then we have

$$25x^2 - 100 = 25(x^2 - 4)$$
$$= 25(x-2)(x+2)$$

If we were to factor the expression in its original form as the difference between two squares, we should get the factors: $(5x-10)(5x+10)$. However, these factors are not prime, since they contain common factors.

Example 6. Factor the expression $(x-3)^2 - (y+5)^2$.

Solution. Here we have the squares of entire quantities. The rule for factoring the difference between two squares applies also when we have the squares of quantities. Applying the rule, we take the square root of each quantity and then connect the square roots, first with a plus sign, and then with a minus sign. It is well to keep each square root in parentheses for the first step. Then the factors can be indicated by pairs of brackets.

The square root of $(x-3)^2$ is $(x-3)$.
The square root of $(y+5)^2$ is $(y+5)$.
We first indicate the factors by brackets:

[   ][   ]

Now we insert the square root of each quantity, keeping each square root

in parentheses:
$$[(x-3) \quad (y+5)][(x-3) \quad (y+5)]$$

For the next step, we use a plus sign in one factor and a minus sign in the other factor:
$$[(x-3)+(y+5)][(x-3)-(y+5)]$$

Remove inner parentheses:
$$[x-3+y+5][x-3-y-5]$$
$$=[x+y+2][x-y-8]$$

Example 7. Factor $144x^6 - 169y^8$.

Solution. The square root of $144x^6$ is $12x^3$. The square root of $169y^8$ is $13y^4$. Then we have the factors:
$$144x^6 - 169y^8 = (12x^3 - 13y^4)(12x^3 + 13y^4)$$

Example 8. In geometry we had the problem of finding the area of a ring; that is, the space between two concentric circles. To find the area of a ring, we took the area of the outer large circle minus the area of the smaller inner circle. If we use R for the radius of the large outer circle, and r for the radius of the small inner circle, then

πR^2 is the area of the large outer circle, A_L;
πr^2 is the area of the small inner circle, A_S.

For the area of the ring, we subtract as follows:
$$A_L - A_S = \pi R^2 - \pi r^2 = \text{area of the ring}$$

As an example, suppose we have a circular mirror having a diameter of 35 inches surrounded by a 3-inch ring. What is the area of the ring? The outside diameter is 41 inches. Then we have the following radii of the two circles: $R = 20.5$ inches; $r = 17.5$ inches.

For the area of the ring, let us use the formula: $A = \pi R^2 - \pi r^2$. We shall first work the problem using this formula, and then see how the problem can be simplified by use of factoring.

Inserting the numerical values, we get
$$A = \pi(20.5)^2 - \pi(17.5)^2$$

Now we must find the following squares: $(20.5)^2 = 420.25$; $(17.5)^2 = 306.25$. For the area, we get:
$$A = 420.25\pi - 306.25\pi = 114\pi.$$

Now, let us see how we might simplify the problem somewhat by factoring. We factor the formula

$$A = \pi R^2 - \pi r^2$$
$$= \pi(R^2 - r^2) = \pi(R+r)(R-r)$$

Now we insert the numerical values:

$$A = \pi(20.5 + 17.5)(20.5 - 17.5)$$
$$= \pi(38)(3) = 114\pi$$

The use of factoring avoids the necessity of squaring large numbers.

Exercise 23.6

Find prime factors:

1. $4x^2 - 9$
2. $9n^2 - 25$
3. $25y^2 - 16$
4. $R^2 - r^2$
5. $s^2 - t^2$
6. $u^2 - v^2$
7. $9x^2 - 1$
8. $1 - 4y^2$
9. $64x^2 - 1$
10. $c^2 - 16d^2$
11. $9a^2 - b^2$
12. $y^2 - 25x^2$
13. $36x^2 - 25$
14. $16n^2 - 49$
15. $9t^2 - 64$
16. $25u^2 - 16v^2$
17. $16s^2 - 9t^2$
18. $36x^2 - 121y^2$
19. $49s^2 - 144t^2$
20. $169x^2 - 225$
21. $196 - 25n^2$
22. $4x^2 - \dfrac{1}{4}$
23. $9n^2 - \dfrac{4}{9}$
24. $25x^2 - \dfrac{9}{16}$
25. $3n^3 - 27n$
26. $5xy - 80x^3y$
27. $28t^3 - 7x^2t$
28. $36x^2 - 9y^2$
29. $4n^2 - 100t^2$
30. $16a^2 - 64b^2$
31. $25n^2 - 0.16$
32. $0.36 - 49y^2$
33. $9x^2 - 0.09$
34. $x^8 - 1$
35. $y^4 - 16x^4$
36. $n^4 - 81$
37. $(x+y)^2 - n^2$
38. $(a-b)^2 - 25$
39. $x^2 - (y+2)^2$
40. $(m-n)^2 - (x-y)^2$
41. $(x-3)^2 - (y-2)^2$
42. $(2x+3)^2 - (3y-5)^2$

23.9. THE SQUARE OF A BINOMIAL: A SPECIAL PRODUCT

If we multiply two binomials that are exactly alike, we call the product the *square of a binomial*. We can indicate such multiplication by the use of the exponent 2 on the one binomial. For example,

$$(3x+4)(3x+4) \text{ is usually written as } (3x+4)^2$$

The expression indicates a perfect square. By a perfect square, we mean a product of two factors that are exactly alike. The number 49 is a perfect square, because it is the product of the two identical factors, 7 and 7.

In the expression, $(3x+4)^2$, if we multiply the two factors by the usual long method, we get

$$(3x+4)^2 = 9x^2 + 24x + 16$$

$$
\begin{array}{r}
3x + 4 \\
3x + 4 \\
\hline
9x^2 + 12x \\
+ 12x + 16 \\
\hline
9x^2 + 24x + 16
\end{array}
$$

as shown at the left. The product, $9x^2+24x+16$, is called the *square of a binomial*.

Let us see how we might get the product by inspection. Notice that the product has *three* terms. The first and third terms of the product are simply the *squares* of the two terms of the binomial, $3x+4$. The middle term of the product is *twice the product of the two terms of the binomial; that is,* $(2)(3x)(4)$.

For the square of a binomial of this kind, we have the following rule.

RULE 1. The square of the sum of two quantities, such as $(a+b)^2$, is equal to the square of the first term, plus twice the product of the two terms, plus the square of the second term.

Stated as a formula, we have

$$(a+b)^2 = a^2 + 2ab + b^2$$

Example 1. Find the following indicated product by inspection.

$$(5x+7)^2$$

Solution. The product has three terms. The first term is $(5x)^2$, or $25x^2$. The middle term is twice the product of $(5x)$ and (7): $(2)(5x)(7) = 70x$. The third term is $(7)^2$, which is 49. Then the product is: $25x^2 + 70x + 49$.

If a binomial indicates the *difference* between two numbers, as in the binomial, $3x-4$, then we can get the square of such a binomial also by using a similar rule. By actual long multiplication, we shall find that

$$(3x-4)^2 = 9x^2 - 24x + 16$$

$$\begin{array}{r} 3x - 4 \\ 3x - 4 \\ \hline 9x^2 - 12x \\ -12x + 16 \\ \hline 9x^2 - 24x + 16 \end{array}$$

as shown at the left. Notice that the product is exactly the same as for the sum of the two numbers except for the sign of the second term. The expression $(3x-4)^2$ indicates the square of the difference between two numbers. For the product of this kind, we have the following rule.

RULE 2. **The square of the difference between two quantities, such as $(a-b)^2$, is equal to the square of the first term, minus twice the product of the two terms, plus the square of the second term.**

Stated as a formula, we have

$$(a-b)^2 = a^2 - 2ab + b^2$$

Example 2. Find the following indicated product by inspection.

$$(7x - 4y)^2$$

Solution. The product has three terms. The first term is $(7x)^2$, or $49x^2$. The middle term is twice the product of $(7x)$ and $(-4y)$: $(2)(7x)(-4y) = -56xy$. The third term is $(-4y)^2$, which is $+16y^2$. Then the entire product is

$$49x^2 - 56xy + 16y^2$$

Example 3. Find by inspection $(4n + 1.5)^2$.

Solution. Applying the formula, we get the product: $16n^2 + 12n + 2.25$.

Example 4. Find the following product by inspection:

$$[(x+y)+5]^2$$

Solution. We apply the rule for the binomial $(x+y)$ the same as for any other term. We get

$$(x+y)^2 + (2)(5)(x+y) + 25$$

Expanding further,

$$= x^2 + 2xy + y^2 + 10x + 10y + 25$$

Exercise 23.7

Find the following products by inspection:

1. $(m+n)^2$
2. $(r+s)^2$
3. $(c-d)^2$
4. $(h-k)^2$
5. $(R-r)^2$
6. $(u+v)^2$
7. $(x-8)^2$
8. $(n+11)^2$
9. $(x-10)^2$
10. $(12+n)^2$
11. $(13-y)^2$
12. $(15+y)^2$
13. $(2x-5)^2$
14. $(7+3x)^2$
15. $(4x-7)^2$
16. $(6n+7)^2$
17. $(8x-5)^2$
18. $(5n+12)^2$
19. $(3x+8y)^2$
20. $(4a-9b)^2$
21. $(8n-7m)^2$
22. $(4a-11b)^2$
23. $(20x+5y)^2$
24. $(30c+11d)^2$
25. $(x^2+5y)^2$
26. $(3y^2+4b)^2$
27. $(5n^3-7x)^2$
28. $(2ab-7cd)^2$
29. $(7rst+9)^2$
30. $(11xy+z^3)^2$
31. $(12x+11y^2)^2$
32. $(10a^2b-c)^2$
33. $(15mn^2-t)^2$
34. $(20rs+3uv)^2$
35. $(5x^2y+11)^2$
36. $(8x^2-15y)^2$
37. $\left(x+\dfrac{1}{2}\right)^2$
38. $\left(6n+\dfrac{1}{3}\right)^2$
39. $\left(8y-\dfrac{1}{4}\right)^2$
40. $(n-0.4)^2$
41. $(y-0.6)^2$
42. $(t+1.4)^2$
43. $(16x+0.75)^2$
44. $\left(6t-\dfrac{2}{3}\right)^2$
45. $\left(2x-\dfrac{3}{4}\right)^2$
46. $(40+5)^2$
47. $(60+5)^2$
48. $(80+5)^2$
49. $[(x-2)-y]^2$
50. $[(a+b)+c]^2$
51. $[x-(y-3)]^2$
52. $[m-(n+4)]^2$

Write the square of each of these binomials:

53. $5x+9$
54. $4n-7$
55. $8y-3$
56. $6-5t$
57. $4n+\dfrac{1}{4}$
58. $3x-\dfrac{1}{6}$
59. $2x+\dfrac{1}{4}$
60. $4t+\dfrac{1}{8}$

Square both sides of each of these equations:

61. $x = 5a + 3$ **62.** $6x - 5 = y$
63. $\sqrt{x - 2} = 5$ **64.** $2x = \sqrt{3 - x}$

23.10. FACTORING TRINOMIALS THAT ARE PERFECT SQUARES

We have seen in the preceding section that the square of a binomial contains three terms. For example, we have seen that

$$(3x + 4)^2 = 9x^2 + 24x + 16$$

Then we know that the trinomial, $9x^2 + 24x + 16$, can be factored into two equal factors, and the trinomial is a perfect square.

Now, if we have any given trinomial, we may be able to recognize it as a perfect square. If so, then we can tell the factors at once. Let us see then how we can recognize a perfect square.

First, the expression must contain three terms. The first and the third terms are perfect squares. In the trinomial, $9x^2 + 24x + 16$, the first term, $9x^2$ is the square of $3x$. The third term is the square of 4. These conditions are necessary if the trinomial is a perfect square.

However, now we must look at the middle term. Before any trinomial can be called a perfect square, the middle term must also be correct for a square. The middle term, $24x$, must be twice the product of the square roots, $3x$ and 4. That is, the middle term must be $(2)(3x)(4)$, or $24x$. Then the trinomial is a perfect square.

Example 1. Is the trinomial, $16x^2 + 40x + 25$, a perfect square? If so, tell the factors.

Solution. The first term, $16x^2$, is a perfect square of $4x$. The third term, 25, is the square of 5. Then the trinomial may be a perfect square. However, before we can be sure, we must check the middle term. The middle term must be twice the product of the square roots, $4x$ and 5. That is, the middle term must be: $(2)(4x)(5) = 40x$. Since this is the middle term, the given trinomial is a perfect square, and it can be factored as such. Since the $40x$ is positive, the given expression is the square of the sum of two numbers. Then we have

$$16x^2 + 40x + 25 = (4x + 5)(4x + 5) \quad \text{or} \quad (4x + 5)^2$$

Example 2. Is the following trinomial a perfect square? If so, find the factors and write them as the square of a binomial: $16x^2 - 40x + 25$.

Solution. The first and the third terms are perfect squares. However, we recall that the square root of the third term, 25, can be (-5). That is, $(-5)^2 = 25$. Since the middle term is negative, we can take $4x$ as the square root of the first term, and (-5) as the square root of the third term. Then for the middle term we must have $(2)(4x)(-5)$, which is equal to $-40x$. Since this is the middle term, the trinomial is a perfect square. In this case we have the square of the difference between two quantities, and we get

$$16x^2 - 40x + 25 = (4x - 5)(4x - 5) \quad \text{or} \quad (4x - 5)^2$$

Note. Although the first and the third terms of a trinomial are perfect squares, the entire expression is not always a perfect square of a binomial. Before we can be sure that the expression is a perfect square, the middle term must be checked.

Example 3. Is the following expression a perfect square: $4x^2 + 30x + 25$?

Solution. The first and third terms are perfect squares. That is, $4x^2$ is the square of $2x$; 25 is the square of 5. However, if the expression is a perfect square, the middle term must be: $(2)(2x)(5) = 20x$. For a perfect square, the middle term must be $20x$. Since it is $30x$, the given trinomial is not a perfect square. The expression cannot be factored as a perfect square of a binomial. However, in some trinomials, there are other factors, as we shall see later.

Exercise 23.8

Determine which of the following trinomials are perfect squares. Then factor the perfect squares and write the two factors as the square of a binomial.

1. $y^2 + 10y + 25$
2. $x^2 - 16x + 64$
3. $x^2 - 12x + 16$
4. $y^2 + 15y + 36$
5. $9x^2 - 48x + 64$
6. $4y^2 - 12y + 9$
7. $4x^2 - 28x + 49$
8. $36n^2 + 60n + 25$
9. $t^2 + 25t + 100$
10. $25n^4 + 30n^2 + 9$
11. $49y^2 + 126y + 81$
12. $25n^2 - 80n + 64$
13. $16t^2 - 72t + 81$
14. $16x^2 - 20x + 9$
15. $36 - 84x + 49x^2$
16. $144x^2 + 36x + 1$
17. $169y^2 + 26y + 1$
18. $225t^2 + 40t + 1$
19. $9n^4 - 24n + 16$
20. $1 + 18x^2 + 81x^4$
21. $25x^2 + 100x + 100$
22. $4x^2 - 2x + 1$
23. $t^2 + t + \dfrac{1}{4}$
24. $144x^2 + 8x + \dfrac{1}{9}$

23.11. FINDING THE UNKNOWN MIDDLE TERM OF A PERFECT SQUARE TRINOMIAL

In some problems, it may be that we have given the first and third terms of a perfect square trinomial, such as the following:

$$9x^2 \cdots + 16y^2$$

Our problem in such expressions may be to determine the proper middle term for a perfect square trinomial.

To find the necessary middle term, we first take the square root of each of the end terms; the square root of $9x^2$ is $3x$; the square root of $16y^2$ is $4y$. We know that the middle term must be twice the product of these two square roots; that is, it must be $(2)(3x)(4y)$, or $24xy$. This middle term may be either positive or negative. If we call the middle term positive, then we have, for the perfect square (which can be factored):

$$9x^2 + 24xy + 16y^2 = (3x + 4y)(3x + 4y) \quad \text{or} \quad (3x + 4y)^2$$

If we choose to call the middle term negative, we have

$$9x^2 - 24xy + 16y^2 = (3x - 4y)^2$$

Exercise 23.9

Supply a proper middle term to make each of the following expressions a perfect square. Then factor each expression and write the factors as the square of a binomial.

1. $x^2 + \cdots + 16$
2. $y^2 + \cdots + 36$
3. $n^2 - \cdots + 49$
4. $9a^2 - \cdots + 64$
5. $4x^2 - \cdots + 25$
6. $9x^2 - \cdots + 16$
7. $25n^2 + \cdots + 1$
8. $49y^2 + \cdots 1$
9. $36y^2 + \cdots 1$
10. $81x^2 - \cdots 25y^2$
11. $121t^2 - \cdots 9x^2$
12. $25 - \cdots + 9R^2$
13. $9x^2 + \cdots + \dfrac{1}{9}$
14. $4r^2 + \cdots + \dfrac{1}{16}$
15. $x^2 - \cdots + \dfrac{1}{4}$
16. $16x^2 + \cdots 0.04$
17. $9x^2 + \cdots 0.09$
18. $81n^2 + \cdots + \dfrac{4}{9}$
19. $F^2 - \cdots 400$
20. $I^2 - \cdots + 361$
21. $144 - \cdots + t^2$
22. $a^2b^2 + \cdots 9c^2$
23. $x^2y^6 + \cdots 49n^4$
24. $169x^2 + \cdots + 4y^2z^2$

23.12. COMPLETING A SQUARE BY ADDING A THIRD TERM

In some expressions, it happens that we have the first two terms of a perfect square. Our problem in some instances is to determine what number should be added to the given expression to produce a perfect square. This process is called *completing a square*. The device of *completing a square* is very useful in much work in practical mathematics.

For example, suppose we have the expression: $x^2 + 8x$. Now we may find it useful to add some quantity to the expression so that the result will be a perfect square. We try to determine the proper number to be added. Let us see how this can be done.

We recall that the middle term is *twice the product of the square roots of the end terms*. Then we see that this product must be $8x$. Therefore, the product of the square roots must be $4x$. Now, the square root of the first term is x. Then the square root of the third term must be 4. The third term itself must then be $(4)^2$, or 16. This is the number that must be added to make the expression a perfect square. We get

$$x^2 + 8x + 16 = (x+4)^2$$

If the middle term is negative, we use the same procedure.

Example 1. Complete the square on the following expression. Then factor the expression and write the factors as a square: $x^2 - 12x$.

Solution. We know that the $-12x$ must be twice the product of the square roots of the end terms. Then their product itself must be $-6x$. Since the first term is x^2, the square root of the first term is x. Then the square root of the third term must be -6. The third term itself must be $(-6)^2$, or 36. That is the number to be added to make the expression a perfect square. We get

$$x^2 - 12x + 36 = (x-6)^2$$

For completing a square we have the following first rule.

RULE 1. If the coefficient of x^2 is 1, take one-half of the coefficient of x and square this number. The result is the number to be added.

Example 2. Complete the square on the expression: $x^2 + 7x \cdots$

Solution. We take one-half of the coefficient, $(+7)$, which is $(+\frac{7}{2})$. Then the number to be added is $(+\frac{7}{2})^2$, or 49/4. We get

$$x^2 + 7x + \frac{49}{4}$$

Factoring, we get $(x+\frac{7}{2})(x+\frac{7}{2})$, or $(x+\frac{7}{2})^2$

To complete a square, in some cases it may be a little more difficult to determine the number to be added. This is true *if the coefficient of the squared term is something other than* 1. Consider the following example:

$$9x^2 + 30x + \cdots$$

In this example, the term, $30x$, is twice the product of the square roots. Then the product itself is $15x$. Now, the square root of the first term is $3x$. Therefore, the square root of the third term must be $(15x) \div (3x)$, which is equal to 5. This number must be squared for the term to be added. Adding 25, we have

$$9x^2 + 30x + 25 = (3x + 5)^2$$

As a second rule for completing a square we have the following:

RULE 2. **If the coefficient of x^2, (or first term), is something other than 1, take one-half the coefficient of x in the middle term. Divide this by the square root of the coefficient of x^2, and square the result. This is the number to be added.**

Example 3. Complete the square on the following expression. Then factor the result and write the factors as a square: $36n^2 - 60n$.

Solution. First step: take $(\frac{1}{2})(60) = 30$. Now we divide the 30 by 6, which is the square root of 36. The result is 5. We square the 5, and get 25, which is the number to be added. We get

$$36n^2 - 60n + 25 = (6n - 5)^2$$

Example 4. Complete the square on the expression: $25x^2 - 40xy$.

Solution. For the first step, we have: $(\frac{1}{2})(40y) = 20y$. The coefficient of x^2 is 25. We take the square root of 25 and get 5. Then we divide: $(20y) \div (5) = 4y$. We square the $4y$, and get $16y^2$, which is the quantity to be added for a perfect square. Then we have

$$25x^2 - 40xy + 16y^2 = (5x - 4y)^2$$

SPECIAL PRODUCTS AND FACTORING 317

Exercise 23.10

Complete the square on each of the following expressions. Then factor the resulting trinomial and write the factors as the square of a binomial.

1. $x^2 + 10x \cdots$
2. $y^2 - 14y \cdots$
3. $t^2 - 16t \cdots$
4. $n^2 - 20n \cdots$
5. $x^2 + 30x \cdots$
6. $x^2 + 18x \cdots$
7. $y^2 + 22y \cdots$
8. $t^2 - 40t \cdots$
9. $m^2 - 24m \cdots$
10. $x^2 - x \cdots$
11. $y^2 + y \cdots$
12. $r^2 + 3r \cdots$
13. $n^2 + 5n \cdots$
14. $x^2 - 7x \cdots$
15. $y^2 - 9y \cdots$
16. $9x^2 - 42x \cdots$
17. $16n^2 + 40n \cdots$
18. $4x^2 + 28x \cdots$
19. $x^2 + 0.8x \cdots$
20. $y^2 - 1.2y \cdots$
21. $t^2 - 0.2t \cdots$
22. $25t^2 - 10t \cdots$
23. $49x^2 + 14x \cdots$
24. $16n^2 + 8n \cdots$
25. $49x^2 + 42xy \cdots$
26. $25a^2 - 30ab \cdots$
27. $36c^2 - 60cd \cdots$
28. $4x^2 - 3x \cdots$
29. $9n^2 + 5n \cdots$
30. $16x^2 + 7x \cdots$
31. $16m^2 + 6mn \cdots$
32. $9s^2 - 4st \cdots$
33. $4x^2 - 9xy \cdots$
34. $4x^2 + 5x \cdots$
35. $9x^2 + x \cdots$
36. $4x^2 - x \cdots$

23.13. A SPECIAL PRODUCT OF THE FORM $(x+a)(x+b)$

As another type of special product, consider the following problem:

$$(x+5)(x+2)$$

This problem is different from those we have already studied. Notice that the binomials have the same first term, which is simply x with a coefficient of 1. The second terms of the binomials are different. Let us first multiply the factors by the usual long method, as shown at the left. Then we shall try to determine how the product can be found quickly.

$$\begin{array}{r} x + 5 \\ x + 2 \\ \hline x^2 + 5x \\ + 2x + 10 \\ \hline x^2 + 7x + 10 \end{array}$$

Notice that the first term of the product is simply x^2, the square of the first term of each binomial. The third term is the product of the last two terms of the binomials; that is, $(2)(5)$. The coefficient of x in the middle term is equal to the sum of the two terms, $+2$ and $+5$, which is equal to 7. If we represent the product by the two binomials in general $(x+a)(x+b)$, then we can say:

the first term of the product is x^2; the third term is $(a)(b)$;
the coefficient of x in the middle term is equal to $(a+b)$.

Now, let us find a product of this kind by using these ideas.

Example 1. Find the product $(x+5)(x+8)$.

Solution. The first term of the product is simply x^2. The third term is the product $(5)(8)$, which is equal to 40. The middle term has the letter x with a coefficient equal to $(5+8)$, which is equal to 13. Then we can write the product without the usual long multiplication; we get

$$(x+5)(x+8) = x^2 + 13x + 40$$

Example 2. Find the following product by inspection: $(x-7)(x+3)$.

Solution. The first term of the product is simply x^2. The third term is the product of the second terms; that is, $(-7)(3)$, which is -21. Note the negative sign. To get the middle term, we take the sum of (-7) and $(+3)$, which is (-4), as the coefficient of x. Then we can write the product at once:

$$(x-7)(x+3) = x^2 - 4x - 21$$

Notice especially, in the product, the *third term*, -21, is the *product of the second terms* of the factors; the coefficient of x in the middle term is the *algebraic sum of the second terms*.

For a product of two factors of this type, we state the rule concisely.

RULE. **The first term of the product is x^2. The second term is a term containing x with a coefficient equal to the algebraic sum of the second terms of the factors. The third term is the product of the second terms.**

Examples:

(a) $(x-6)(x-9) = x^2 - 15x + 54$,
(b) $(n+6)(n-5) = n^2 + n - 30$;
(c) $(y-8)(y+7) = y^2 - y - 56$.

Exercise 23.11

Multiply by inspection:

1. $(x+4)(x+7)$
2. $(x-5)(x-4)$
3. $(x+6)(x+8)$
4. $(n+7)(n+5)$
5. $(n-6)(n-2)$
6. $(y-3)(y-8)$
7. $(y-8)(y-4)$
8. $(t+5)(t+9)$
9. $(n+5)(n+12)$
10. $(x+4)(x-6)$
11. $(x+5)(x-10)$
12. $(y+9)(y-7)$

13. $(x-8)(x+7)$
14. $(y-6)(y+7)$
15. $(t-9)(t+8)$
16. $(n+12)(n-1)$
17. $(x-1)(x+16)$
18. $(x-10)(x+1)$
19. $(y+3)(y-10)$
20. $(y-4)(y+12)$
21. $(y-15)(y+4)$
22. $(t-8)(t+8)$
23. $(t+6)(t-6)$
24. $(t-9)(t+9)$
25. $(n+30)(n-4)$
26. $(n-18)(n+5)$
27. $(n+48)(n-3)$
28. $(8-x)(10+x)$
29. $(7+y)(9-y)$
30. $(12-t)(4-t)$
31. $(x^2-5)(x^2+8)$
32. $(x^3+9)(x^3-10)$
33. $(x^4-2)(x^4-8)$
34. $(n-7)(13+n)$
35. $(x+5)(10-x)$
36. $(t-8)(9+t)$

23.14. FACTORING TRINOMIALS OF THE TYPE: $x^2 + px + q$

This is the type of trinomial we get for products like those of the previous section. The first term is x^2, whose coefficient is 1. The second term contains the first power of x, with some coefficient which is represented by p. The third term is a constant represented by q.

As an example, consider again the following problem:

$$(x+2)(x+6)$$

For the product we get

$$x^2 + 8x + 12$$

The first term of the product is x^2. In the second term, the coefficient of x is 8, which is the sum of +2 and +6. The third term, +12, is the product of +2 and +6; that is, $(+2)(+6) = +12$.

Now suppose we are given the trinomial, $x^2 + 8x + 12$, and we do not know the factors. Our problem is to find the factors. As a first step, we set down two sets of parentheses to indicate the factors:

$$(\quad)(\quad)$$

Now we write x for the first term of each factor, since the product must be x^2:

$$(x\quad)(x\quad)$$

As a next step, we look for two numbers whose product is +12. The sum of these two numbers must be +8, the coefficient of x in the middle term. By inspection we see that the two numbers must be +2 and +6. Their product is +12, and their sum is +8. Therefore, these two numbers are the second terms of the factors. Then we have

$$x^2 + 8x + 12 = (x+2)(x+6)$$

The result can be checked by multiplication.

Example 1. Find the factors of the trinomial: $x^2 - 8x + 15$.

Solution. Since the third term of the trinomial is positive (+), then the second terms of the factors must have the same sign; both may be positive, or both may be negative. For the factors of +15, we can use (+3) and (+5), or (−3) and (−5). The sum of these two numbers must be −8, the coefficient of x in the middle term. Then we take the negative numbers. For the factors, we have

$$x^2 - 8x + 15 = (x-3)(x-5)$$

Example 2. Find the factors of the trinomial, $n^2 - 12n + 20$.

Solution. To indicate the factors, we first set down the two sets of parentheses, with n as the first term in each factor:

$$(n \quad)(n \quad)$$

Now we look for two factors of +20 whose sum is −12, the coefficient of n. By inspection, we see that the numbers are (−2) and (−10). These numbers are the second terms of the factors, and we have

$$n^2 - 12n + 20 = (n-2)(n-10)$$

Example 3. Factor the trinomial, $n^2 - n - 30$.

Solution. In this example, we must find two factors whose product is −30. The sum of the two factors must be −1, the coefficient of n in the middle term. By inspection, we find that the numbers are (−6) and (+5). Then these numbers are the second terms of the factors, and we have

$$n^2 - n - 30 = (n-6)(n+5)$$

Example 4. Factor the trinomial, $2x^3 - 10x^2 - 48x$.

Solution. First we take out the common factor: $2x(x^2 - 5x - 24)$. For the factors of the trinomial, $x^2 - 5x - 24$, we set down the parentheses:

$$(x \quad)(x \quad)$$

SPECIAL PRODUCTS AND FACTORING

Now we look for two factors of (−24) whose sum is (−5), the coefficient of x. The two factors of (−24) must have opposite signs because the product is negative. By inspection, we see that the numbers are (−8) and (+3). The product of these numbers is (−24), and their sum is (−5), the coefficient of x. Therefore, these numbers are the second terms of the factors. For the complete factoring, we must show the *monomial factor* as well as the *two binomial factors*. Then we have

$$2x^3 - 10x^2 - 48x = 2x(x-8)(x+3)$$

In general terms, to factor a trinomial of the type, $x^2 + px + q$, we look for two factors of the constant term q, such that their *algebraic sum* is equal to the value of p, the coefficient of x in the middle term. Usually the numbers can be found by inspection.

Note. *If there are no factors of the third term, q, whose sum is equal to the coefficient in the middle term, then the trinomial cannot be factored in rational terms. If the two required factors of the third term are exactly alike, then the two factors are identical, and the trinomial is a perfect square.*

Example 5. $x^2 + 9x + 12$. This trinomial cannot be factored because there are no factors of 12 such that their sum is 9.

Example 6. $x^2 + 8x + 16$. In this example, we look for two factors of +16 such that their sum is 8. The two numbers are (+4) and (+4). Then we have

$$x^2 + 8x + 16 = (x+4)(x+4)$$

Therefore, the trinomial is a perfect square because the two factors are equal.

Exercise 23.12

Factor the following if possible. Tell which are perfect squares. Take out common factors first.

1. $x^2 + 5x + 6$
2. $x^2 + 7x + 10$
3. $x^2 - 8x + 15$
4. $n^2 + 7n + 12$
5. $n^2 - 9n + 20$
6. $n^2 - 9n + 18$
7. $y^2 - 11y + 28$
8. $y^2 - 11y + 24$
9. $y^2 + 13y + 36$
10. $14x + x^2 + 24$
11. $40 + x^2 - 22x$
12. $30 + x^2 - 13x$
13. $t^2 - 20 - 8t$
14. $t^2 - 30 + 7t$
15. $t^2 - 40 - 18t$
16. $3x^3 + 18x^2 - 48x$

17. $2x^4 - 10x^3 - 28x^2$
18. $4x^5 + 40x^4 - 96x^3$
19. $n^2 + n - 30$
20. $n^2 + n - 56$
21. $n^2 - n - 72$
22. $xy^2 - 8xy - 9x$
23. $2xy^2 + 22xy - 24x$
24. $5xy^2 + 75xy - 80x$
25. $x^2 - 10x + 25$
26. $x^2 - 12x + 36$
27. $n^2 + 14n + 49$
28. $t^2 + 10t + 12$
29. $t^2 - 7t - 24$
30. $t^2 - 12t + 18$
31. $3x^2 + 57x + 54$
32. $5x^2 + 75x + 70$
33. $2x^4 - 46x^3 - 48x^2$
34. $n^2 - 21n - 72$
35. $n^2 - 14n - 72$
36. $n^2 + 34n - 72$
37. $y^2 + 40y - 84$
38. $y^2 - 8y - 84$
39. $y^2 - 17y - 84$
40. $x^2 + 17x - 60$
41. $x^2 - 11x - 60$
42. $x^2 + 28x - 60$
43. $x^2y^2 + 13xy^2 - 48y^2$
44. $x^4 - 22x^3 - 48x^2$
45. $n^3x^2 - 8n^3x - 48n^3$
46. $n^2x^2 + 37xn^2 + 36n^2$
47. $x^4 - 24x^2 + 144$
48. $x^6 + 18x^3 + 81$

23.15. A MORE DIFFICULT SPECIAL PRODUCT: $(ax+b)(cx+d)$

We now come to a special product of two binomials that is more difficult than those we have studied so far. Consider the product

$$(4x+5)(3x+2)$$

First, we shall multiply the two binomials by the usual long method of multiplication, as shown at the left. We get the product

$$12x^2 + 23x + 10$$

$$
\begin{array}{r}
4x + 5 \\
3x + 2 \\
\hline
12x^2 + 15x \\
+ 8x + 10 \\
\hline
12x^2 + 23x + 10
\end{array}
$$

Now, let us see how we can get the product quickly by inspection. Notice the following facts about the product.

1. The first term of the product, $12x^2$, is obtained by multiplying the first terms of the binomials: $(3x)(4x) = 12x^2$.

2. The third term of the product, 10, comes from multiplying the second terms of the binomials: $(5)(2) = 10$.

3. The middle term of the product, $23x$, is the *algebraic sum* of the *cross products* shown by arrows: $(3x)(5) + (4x)(2)$; that is, $15x + 8x = 23x$.

If we write the two factors in a horizontal form, we can show the cross products by arrows:

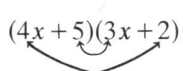

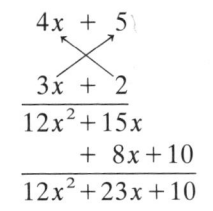

$$(4x+5)(3x+2)$$

SPECIAL PRODUCTS AND FACTORING

The middle term is the algebraic sum of these cross products. If some terms are negative and a product is negative, the cross products must be added algebraically.

Example 1. Find the following product by inspection: $(5x-2)(3x+4)$.

Solution. For the *first term* of the product, we take the product of the first terms of the binomials:

$$(5x)(3x) = 15x^2.$$

For the *third term* of the product, we multiply the second terms of the factors:

$$(-2)(+4) = -8$$

For the *middle term* of the answer, we find the algebraic sum of the cross products;

$$(5x)(4) + (-2)(3x) = 20x - 6x = 14x$$

Then we can write the entire product:

$$(5x-2)(3x+4) = 15x^2 + 14x - 8$$

For the multiplication of two binomials of the type $(ax+b)(cx+d)$, we have the following rule.

RULE. **The first term of the answer is the product of the first terms of the factors: $(ax)(bx) = abx^2$.**
 The middle term of the answer is the algebraic sum of the cross products: $(adx)+(bcx)$, which is equal to $(ad+bc)x$.
 The third term of the answer is the product of the second terms of the factors: $(b)(d)$.

Example 2. Multiply by inspection: $(4x-7)(3x+5)$.

Solution. For the first term of the answer, we have: $(4x)(3x) = 12x^2$.
For the middle term, we have $(4x)(5) + (-7)(3x) = 20x - 21x = -x$.
The third term is the product of the second terms: $(-7)(5) = -35$.
Now we can write the entire product:

$$(4x-7)(3x+5) = 12x^2 - x - 35$$

Exercise 23.13

Multiply the following by inspection:

1. $(2x+3)(3x+5)$
2. $(3x+2)(2x+5)$
3. $(4x-3)(3x-4)$
4. $(3x-4)(5x+2)$
5. $(7x+2)(4x+3)$
6. $(6x-5)(5x+3)$
7. $(4x+5)(5x-4)$
8. $(3x-5)(5x+3)$
9. $(5x-7)(4x+5)$
10. $(7t-4)(5t+3)$
11. $(4t+3)(5t-4)$
12. $(3t+5)(5t-8)$
13. $(n+3)(8n+1)$
14. $(6x-1)(x-5)$
15. $(4n-1)(n-6)$
16. $(9a+2)(2a-3)$
17. $(8c-3)(3c+2)$
18. $(7x-5)(5x-2)$
19. $(6-7x)(3+2x)$
20. $(5+4n)(7-2n)$
21. $(4-5y)(6+5y)$
22. $(3xy+8)(2xy-1)$
23. $(4ab-7)(ab+2)$
24. $(9cd-7)(4cd+3)$
25. $(x^2-5)(3x^2+7)$
26. $(n^3+3)(4n^3-3)$
27. $(2y^2-5)(y^2+2)$
28. $(2x+3y)(5x-8y)$
29. $(4a-5b)(7a+9b)$
30. $(6c+7d)(5c-6d)$
31. $(4n-1)(n-25)$
32. $(9n+1)(n+16)$
33. $(25n-1)(n-49)$
34. $(6x-y)(y+7x)$
35. $(8a-b)(b+9a)$
36. $(10x-3y)(2y+7x)$

23.16. FACTORING A TRINOMIAL OF THE TYPE: ax^2+bx+c

This kind of trinomial is one of the most difficult to factor. Finding the factors is largely a matter of trial and error. At first you might have some difficulty in finding the correct set of factors. However, there is a definite approach that should be helpful. We show the method by examples.

Example 1. Find the factors of the following expression: $3x^2+7x+2$.

Solution. First we indicate the factors by pairs of parentheses:

$$(\quad)(\quad)$$

Now we notice that the first terms of the factors must be $3x$ in one factor and x in the other, because their product must be $3x^2$. So we show these terms in the parentheses as the first terms:

$$(3x\quad)(x\quad)$$

SPECIAL PRODUCTS AND FACTORING 325

The only possible factors of the last term are 2 and 1. Since the original trinomial has a positive 2, (+2) for the third term, then the last terms of the factors must have the same sign, either + or −. However, the sum of the cross products must be +7x, so both signs must be positive.

Now, our problem is to pair up the last terms, 2 and 1, with the first terms so that the cross products will be 7x. By inspection, we see that we must have

$$(3x+1)(x+2)$$

These two factors have cross products whose sum is 7x. Therefore, we have the correct factors:

$$3x^2 + 7x + 2 = (3x+1)(x+2)$$

Example 2. Factor the expression $6x^2 - 11x - 10$.

Solution. This problem is more difficult than Example 1 because the first term and the last term may have more than one set of factors.

First we set up the pairs of parentheses to indicate the factors:

$$(\quad)(\quad)$$

We know the first terms of the factors must have a product of $6x^2$. Then the first terms might be either set shown here:

$$(3x\quad)(2x\quad), \quad \text{or} \quad (6x\quad)(x\quad)$$

Let us list all the factors of the first term, $6x^2$, in column form, and the factors of the last term, -10, in another column:

	factors of $6x^2$	factors of -10	
one set	3x 2x	2 5	one of these must be negative
or the set	6x x	10 1	one negative

Now we try to pair up a set in the first column with a set in the second column so that the sum of the cross products will equal $(-11x)$.

Notice that if we take the pairs: $(3x)(5)$ and $(2x)(2)$, we shall have the sum of these products equal to $-11x$, if one term is negative. Then we place the terms, $3x$ and 5, in the proper positions in parentheses so that the sum of the cross products will be $-11x$:

$$(3x\quad 2)(2x\quad 5)$$

Now, if we use a minus sign before the 5, we shall have the proper factors:

$$(3x+2)(2x-5)$$

The sum of the cross products is $-11x$, and the last term becomes -10, as it should be.

Exercise 23.14

Factor the following. Identify perfect squares.

1. $3x^2 - 17x + 10$
2. $2x^2 - 9x + 10$
3. $5x^2 - 17x + 6$
4. $2n^2 + 15n + 18$
5. $3n^2 + 17n + 20$
6. $3n^2 - 10n + 8$
7. $4x^2 - 11x + 6$
8. $6x^2 + 17x + 12$
9. $8x^2 + 26x + 15$
10. $6n^2 - 19n + 10$
11. $4n^2 + 11n + 6$
12. $10x^2 - 19x + 6$
13. $8x^2 + 10x - 3$
14. $6x^2 - 5x - 6$
15. $5x^2 - 11x + 6$
16. $9t^2 - 24t + 16$
17. $9y^2 - 48y + 64$
18. $4x^2 + 20x + 25$
19. $10n^2 - n - 21$
20. $15n^2 + n - 40$
21. $30n^2 - n - 42$
22. $16x^2 + 30x + 9$
23. $9x^2 + 26x + 16$
24. $25t^2 - 61t + 36$
25. $36n^2 - 75n + 25$
26. $25n^2 - 50n + 9$
27. $4n^2 + 35n + 49$
28. $6x^2 - 35x - 6$
29. $8x^2 + 47x - 6$
30. $8x^2 - 33x - 4$

23.17. FACTORING THE SUM AND THE DIFFERENCE OF TWO CUBES

If we divide the expression $x^3 + y^3$ by the quantity $x + y$, we get the quotient, $x^2 - xy + y^2$. Then we can write the factors:

$$x^3 + y^3 = (x+y)(x^2 - xy + y^2)$$

By division, we should also find that

$$x^3 - y^3 = (x-y)(x^2 + xy + y^2)$$

The best way to learn to factor the sum or difference of two cubes is to become familiar with the pattern of the factors. In either case, one factor contains two terms, the other factor contains three terms. Notice that the

SPECIAL PRODUCTS AND FACTORING 327

binomial factor is made up of the cube roots of the two cubes. The sign between them is the same as the sign between the cubes.

The second factor, the trinomial, in each case is made up of the squares of the cube roots, with the product of the cube roots as the middle term. However, the middle term has a sign opposite to that between the cubes. The results can be verified by long division, and by multiplication.

Example 1. Factor $m^3 - n^3 = (m-n)(m^2 + mn + n^2)$.

Example 2. Factor $8a^3 - 27$.

Solution. The two cube roots are: $2a$ and 3. Then we write the factors:

$$8a^3 - 27 = (2a - 3)(4a^2 + 6a + 9)$$

Exercise 23.15

Factor the following expressions:

1. $R^3 - r^3$
2. $a^3 - 8b^3$
3. $x^3 + 27$
4. $8x^3 + 125$
5. $64x^3 + 1$
6. $125 - n^3$
7. $T^3 - 1000$
8. $216y^3 + 27$
9. $t^3 - 343$
10. $27y^3 - 1$
11. $x^6 + y^{12}$
12. $8y^3 + 1$
13. $64x^3 + \dfrac{8}{27}$
14. $27x^3 - \dfrac{1}{8}$
15. $64x^3 - \dfrac{1}{27}$
16. $64x^6 - \dfrac{1}{64}$
17. $64x^6 - y^6$
18. $n^6 - 729$

23.18. FACTORING BY GROUPING TERMS

Sometimes a polynomial can be factored by grouping certain terms together. We show the method by examples.

Example 1. Factor the expression: $3x + ax - 3y - ay$.

Solution. We could group the terms as they stand, but let us first rearrange them:

$$3x - 3y + ax - ay$$

Grouping terms, we get

$$(3x-3y)+(ax-ay)$$

Now we take out a common factor of each group:

$$3(x-y)+a(x+y)$$

Taking out the factor, $x-y$,

$$(x-y)(3+a)$$

If we multiply the two factors, we shall find that we get the original expression.

Example 2. Factor the expression x^3+x^2-4x-4.

Solution. Grouping the terms,

$$(x^3+x^2)-(4x+4)$$

Factoring each part,

$$x^2(x+1)-4(x+1)$$

Taking out the factor, $x+1$,

$$(x+1)(x^2-4)$$

Also factoring, x^2-4, we get

$$(x+1)(x-2)(x+2)$$

To check the answer, we can multiply the three factors. We should get the original polynomial.

Example 3. Factor the expression: x^2-y^2-6x+9.

Solution. Rearranging terms so as to show a square,

$$x^2-6x+9-y^2$$

The first three terms can be now written as a square of a binomial:

$$(x-3)^2-y^2$$

SPECIAL PRODUCTS AND FACTORING 329

Now we have the difference between two squares. Factoring, we get

$$= [(x-3)-y][(x-3)+y]$$

$$= [x-3-y][x-3+y], \quad \text{the two factors}$$

Exercise 23.16   Review

Find the prime factors of the following expressions. Always take out a common factor first. If any factor can be factored, find the prime factors.

1. $16x^2 - 36$
2. $48n^2 - 75$
3. $5a^2 - 180b^2$
4. $9t^3 - 49t$
5. $9x^4 - 81x^2y^2$
6. $16x^3y - 100xy^3$
7. $n^4 - 81$
8. $m^8 - n^8$
9. $n^{10} - n^6$
10. $x^4 - 27x$
11. $8x^4 + x^7$
12. $64x^2 - x^5$
13. $n^5 - n^4 - 42n^3$
14. $3n^3 + 3n^2 - 60n$
15. $6n^4 - 6n^3 - 72n^2$
16. $9x^3 + 63x^2 + 90x$
17. $6x^4 - 24x^3 - 72x^2$
18. $4x^5 + 36x^4 + 72x^3$
19. $18x^4 - 33x^3y - 30x^2y^2$
20. $28a^2y^3 + 35a^3y^2 - 42a^4y$

Factor the following expressions by grouping terms:

21. $a^2 - 8a + 16 - x^2$
22. $x^5 - x^3 + 8x^2 - 8$
23. $y^3 + 3y^2 - 9y - 27$
24. $16y^2 + x^2 - 16 - 8xy$
25. $6x + 4y^2 - 9 - x^2$
26. $n^2 - 36 - 12x - x^2$
27. $x^3 - 9x + 2x^2 - 18$
28. $x^3 - 5x^2 - 4x + 20$
29. $x^2 - y^2 + 2x + 6y - 8$
30. $x^2 - y^2 - 10x + 4y + 21$

24. FRACTIONS

24.1. DEFINITION

In algebra, just as in arithmetic, we often have fractions. In order to understand clearly how to work with fractions in algebra, let us recall some of the facts about fractions in arithmetic.

In arithmetic a *fraction* consists of two numbers, one written above the other with a horizontal line between them, as $\frac{3}{4}$. The number above the line is called the *numerator*. The number below the line is called the *denominator*. The numerator and denominator are called the *terms* of the fraction.

In algebra, a fraction has the same form as in arithmetic. The numerator and the denominator may be algebraic expressions. The following are algebraic fractions:

$$\frac{3x}{5y}; \quad \frac{2x-3}{x+4}; \quad \frac{x^2-5x+6}{x-2}; \quad \frac{x^3+5x}{x^2+7x+4}$$

A fraction in arithmetic or in algebra may be considered as an indicated division. For example,

$$\frac{3}{4} \quad \text{can be taken to mean} \quad 3 \div 4$$

$$\frac{150}{37} \quad \text{can be taken to mean} \quad 150 \div 37$$

$$\frac{3x}{5y} \quad \text{can be taken to mean} \quad 3x \div 5y$$

$$\frac{x^2-5x+6}{x-2} \quad \text{means} \quad (x^2-5x+6) \div (x-2)$$

In algebra, as in arithmetic, we must learn how to work with fractions in several ways. In arithmetic, we learn how to reduce fractions to lower terms. We must do the same with algebraic fractions. We have already mentioned how to reduce some fractions in algebra (Chapter 23). In arithmetic, we learn how to multiply and divide fractions. We must know how to add and subtract fractions. In algebra we must learn how to do the same with algebraic fractions.

24.2. FUNDAMENTAL PRINCIPLE OF FRACTIONS

Before we consider the various operations with fractions, it is important that we understand the following principle.

FUNDAMENTAL PRINCIPLE OF FRACTIONS. **If the numerator and the denominator of any fraction are multiplied or divided by the same quantity (other than zero), the value of the fraction will not be changed.**

We have already seen how this principle applies to a fraction in arithmetic. For example, we use it in reducing a fraction to lower terms. As an example, consider the fraction:

$$\frac{15}{21}$$

To reduce the fraction, we divide numerator and denominator by some number that is a factor of both. Sometimes this is easy. In this example, we can tell at a glance that both terms of the fraction can be divided by 3. Then the fraction is reduced to the form: $\frac{5}{7}$.

Sometimes we need to multiply both terms of the fraction by some number. For example, we may want to change the fraction, $\frac{3}{4}$, to higher terms. Then we make use of the first part of the *Fundamental Principle*. In adding and subtracting fractions, we must often change the form of the fractions so that they have the same denominator. Suppose we want to change the fraction, $\frac{3}{4}$, to a form in which the denominator is 24. Then we multiply both terms of the fraction by 6, and get the new form of the same fraction $\frac{18}{24}$.

24.3. REDUCING FRACTIONS

In reducing fractions to lower terms, we must be careful not to cross out anything but factors. In any problem in reducing fractions, first find the factors of numerator and denominator. Then you can tell at a glance what number can be divided into both terms of the fraction.

Example 1. Reduce the fraction $\frac{21}{91}$ to lower terms.

Solution. First we write the fraction with the factors of numerator and denominator:

$$\frac{21}{91} = \frac{3 \cdot 7}{7 \cdot 13}$$

Now we can tell at once that both terms of the fraction can be divided by 7. Then we get the reduced fraction: $\frac{3}{13}$. This fraction is equal to the original fraction but in a different form.

Be especially careful not to cross out terms connected to other terms by a plus or minus sign. If you will always remember that you can divide out only factors, you will avoid many errors in working with fractions.

Example 2. Can the following fraction be reduced:

$$\frac{5+2}{5+6}?$$

Solution. You may be tempted to cross out the "5's" in this example, and also to divide the "2" into the "6." This would be wrong. The fraction can be written as

$$\frac{5+2}{5+6} = \frac{7}{11}$$

Therefore, the fraction cannot be reduced to lower terms.

Example 3. Can the following fraction be reduced:

$$\frac{28}{84}?$$

Solution. Some students might go so far as to cross out the "8's," and then divide the "2" into the "4", which, of course, would be wrong. Instead, let us factor numerator and denominator:

$$\frac{28}{84} = \frac{2 \cdot 2 \cdot 7}{2 \cdot 2 \cdot 3 \cdot 7} = \frac{1}{3}$$

When the terms of the fraction are in factored form, we can at once see that numerator and denominator can be divided by 2, 2, and 7.

In algebra also, we must be careful about reducing fractions. Divide numerator and denominator by any factor found in both. We can show this division by crossing out such common factors, *but do not cross out separate terms connected to other terms by a plus or a minus sign.* To be sure of the right procedure, *first factor numerator and denominator into prime factors.*

A monomial indicates only factors. Therefore, in a monomial, the separate factors need not be written out, as we have already shown by an example in Chapter 23. Here we show another example in reducing an algebraic fraction in which the numerator and the denominator are monomials.

Example 4. Reduce the fraction to lower terms:

$$\frac{21a^3b^2}{28ab^3}$$

Solution. In this example, the numerator and the denominator have only one term each. Since monomials indicate only multiplication of factors, then the numerator and denominator already show the factors. Both terms can be divided by 7, by a, and by b^2. We get the reduced fraction:

$$\frac{3a^2}{4b}$$

Example 5. Reduce the fraction:

$$\frac{x^2-9}{x^2-5x+6}$$

Solution. Do not be tempted to cross out the "x^2's" in numerator and denominator. This error happens often when we use the word "cancel." Some people cross out any two things that look alike. Crossing out such terms is like crossing out the "5's" in Example 2. If you will first think of factoring numerator and denominator, you will avoid such mistakes. Factoring, we get

$$\frac{x^2-9}{x^2-5x+6} = \frac{(x-3)(x+3)}{(x-2)(x-3)} = \frac{x+3}{x-2}$$

After factoring, we see that numerator and denominator can be divided by the factor $(x-3)$.

FRACTIONS

To avoid mistakes in reducing fractions, follow these steps:

(1) *Factor numerator and denominator into prime factors.*
(2) *Divide numerator and denominator by any factor or factors found in both. These can be crossed out to show the division.*

Example 6. Reduce the fraction:

$$\frac{6x^4 - 6x^3 - 12x^2}{8x^3 - 24x^2 - 32x}$$

Solution. First, we take out any common factor in numerator and denominator. We get

$$\frac{6x^2(x^2 - x - 2)}{8x(x^2 - 3x - 4)}$$

Now we find the factors of the trinomials and get

$$\frac{6x^2(x-2)\cancel{(x+1)}}{8x\cancel{(x+1)}(x-4)}$$

Now we can divide numerator and denominator by $2x$, and by the factor $(x+1)$: and get

$$\frac{3x(x-2)}{4(x-4)}$$

The result *cannot* be further reduced.

Example 7. Reduce the fraction:

$$\frac{x^4 + 4x^3 + 4x^2}{x^4 + 6x^3 + 8x^2}$$

Solution. Factoring into prime factors:

$$\frac{x^2(x+2)\cancel{(x+2)}}{x^2\cancel{(x+2)}(x+4)}$$

Now we can divide both terms of the fraction by x^2, and by $(x+2)$, and get the reduced fraction

$$\frac{x+2}{x+4}, \quad \text{which cannot be further reduced.}$$

Example 8. Reduce the fraction:

$$\frac{x^2-16}{x-4}$$

Solution. Factoring,

$$\frac{(x-4)(x+4)}{x-4} = x+4$$

Now we can divide numerator and denominator by the factor $(x-4)$, and the denominator becomes 1, which can be omitted.

Exercise 24.1

Reduce the following fractions to lowest terms:

1. $\dfrac{4xy}{6x^2}$
2. $\dfrac{16x^2y}{12xy^2}$
3. $\dfrac{9x^3y^2}{21xy^3}$
4. $\dfrac{14a^3b}{35ab^2}$
5. $\dfrac{20r^2s^4}{36r^3s}$
6. $\dfrac{16m^3n}{6mn^2}$
7. $\dfrac{30x^3yz}{24xyz^2}$
8. $\dfrac{25xy^2z^3}{15xy^3}$
9. $\dfrac{15y^3z}{18xyz}$
10. $\dfrac{15r^2s^2}{24r^2st}$
11. $\dfrac{16m^3n^2x}{28m^4nx}$
12. $\dfrac{27a^3bc}{12a^2bc}$
13. $\dfrac{8x^2y}{4x^2y}$
14. $\dfrac{2R^2h}{6r^2h}$
15. $\dfrac{15x^3y}{10x^3y}$
16. $\dfrac{5a^2b^3}{5a^2b^3}$
17. $\dfrac{-4xy^2}{-4xy^2}$
18. $\dfrac{6m^3n}{-6m^3n}$

19. $\dfrac{x^2-3x-10}{x^2-4}$

20. $\dfrac{x^2+7x+12}{x^2-9}$

21. $\dfrac{x^2-25}{x^2+8x+15}$

22. $\dfrac{n^2-n-20}{n^2+3n-40}$

23. $\dfrac{n^2+3n-10}{n^2+8n-20}$

24. $\dfrac{n^2-n-30}{n^2+2n-15}$

25. $\dfrac{x^2-6x-16}{x^2-2x-48}$

26. $\dfrac{x^2-36}{x^2-4x-12}$

27. $\dfrac{x^2-11x+24}{x^2-7x+12}$

28. $\dfrac{6x^3-12x^2-48x}{3x^4-48x^2}$

29. $\dfrac{2x^4+6x^3+4x^2}{6x^3-6x}$

30. $\dfrac{8x^5+8x^4-16x^3}{2x^4-18x^3+16x^2}$

31. $\dfrac{2x^5-72x^3}{6x^3-48x^2+72x}$

32. $\dfrac{4x^4-64x^2}{2x^5+12x^4+16x^3}$

33. $\dfrac{3x^5-75x^3}{9x^3-63x^2+90x}$

34. $\dfrac{2x^3-162x}{6x^4-48x^3-54x^2}$

24.4. MULTIPLICATION OF FRACTIONS

In arithmetic, we use the following rule for multiplication of fractions.

RULE. To multiply fractions, multiply the numerators together for the numerator of the answer. Multiply the denominators together for the denominator of the answer.

The same rule is used in algebra.

Example 1. Multiply

$$\frac{3x}{4y} \cdot \frac{5x}{7y} = \frac{15x^2}{28y^2}$$

In arithmetic we can often divide numerator and denominator by some common factor before multiplying. This will simplify the work.

Example 2. Multiply

$$\frac{8}{9} \cdot \frac{15}{16}$$

Solution. If we multiply the numerator together and the denominators together as they appear, we get the fraction:

$$\frac{120}{144}$$

Now this fraction can be reduced to lower terms by dividing the numerator and the denominator by 24. Then we get the reduced fraction: $\frac{5}{6}$. However, in this example, we can divide a numerator and a denominator by 8, and by 3 before doing the multiplication. Here, again, we often use the word "cancel." Then we have

$$\frac{\overset{1}{\cancel{8}}}{\underset{3}{\cancel{9}}} \cdot \frac{\overset{5}{\cancel{15}}}{\underset{2}{\cancel{16}}} = \frac{5}{6}$$

Example 3. Multiply

$$\frac{5x^3}{8y^2} \cdot \frac{16y}{25x}$$

Solution. Before multiplying, we look for common factors in numerators and denominators. Both terms of the fraction can be divided by 5, by 8, by x, and by y. Dividing out these factors, we get

$$\frac{\overset{x^2}{\cancel{5x^3}}}{\underset{y}{\cancel{8y^2}}} \cdot \frac{\overset{2}{\cancel{16y}}}{\underset{5}{\cancel{25x}}} = \frac{2x^2}{5y}$$

Example 4. Multiply

$$\frac{5}{6} \cdot \frac{15}{16} \cdot \frac{12}{25}$$

Solution. Before multiplying, we can first divide numerators and denominators by the following factors: 3, 4, 5, and 5. We show this division by crossing out common factors:

$$\frac{\overset{1}{\cancel{5}}}{\underset{1}{\cancel{6}}} \cdot \frac{\overset{3}{\cancel{15}}}{\underset{}{\cancel{16}}} \cdot \frac{\overset{2}{\cancel{12}}}{\underset{8}{\cancel{25}}} = \frac{3}{8}$$

FRACTIONS

Example 5. Multiply

$$\frac{8a^2b}{9xy^3} \cdot \frac{5x^2y}{6b^2} \cdot \frac{3byz}{10a^3} = \frac{2xz}{9ay}$$

Solution. We first look for factors that can be divided into some numerator and some denominator. We can divide out the following factors in both terms of the fractions: 2, 2, 3, 5, a^2, b, b, x, y, and y.

To the student: Cross out each of these factors by dividing each one into a numerator and a denominator as we do in arithmetic. Answer:

$$\frac{2xz}{9ay}$$

Example 6. Multiply

$$10 \cdot \frac{3}{8} = \frac{10}{1} \cdot \frac{3}{8} = \frac{15}{4}$$

Solution. In multiplying a whole number times a fraction, we multiply the whole number times the numerator only. If we wish, we can place the whole number over the denominator 1, and then multiply.

Example 7. Multiply

$$(ax^2) \cdot \frac{3y}{2x}$$

Solution. We write the whole number over the denominator 1, and have

$$\frac{ax^2}{1} \cdot \frac{3y}{2x}$$

Now we divide x into a numerator and a denominator and get the answer:

$$\frac{3axy}{2}$$

Example 8. Multiply

$$\frac{x^2-16}{3x^2-15x} \cdot \frac{6x^3-30x^2}{x^2-2x-8}$$

Solution. First we factor all numerators and denominators into prime factors; then look for any factor or factors found in a numerator and in a denominator:

$$\frac{\cancel{(x-4)}(x+4)}{\cancel{3x}\cancel{(x-5)}} \cdot \frac{\overset{2x}{\cancel{6x^2}}\cancel{(x-5)}}{\cancel{(x-4)}(x+2)}$$

Now we see that a numerator and some denominator can be divided by the following factors: $3x$, $(x-4)$, and $(x-5)$. Crossing out these common factors, we get the answer:

$$\frac{2x(x+4)}{x+2}$$

The answer cannot be further reduced because the numerator and the denominator are not divisible by any quantity.

Example 9. Multiply

$$\frac{x^2-4x+3}{x^2-4} \cdot \frac{x^2-6x+8}{x^2+2x-3} \cdot \frac{x^2+7x+12}{x^2-7x+12}$$

Solution. We factor all numerators and denominators into prime factors:

$$\frac{\cancel{(x-1)}\cancel{(x-3)}}{\cancel{(x-2)}(x+2)} \cdot \frac{\cancel{(x-2)}\cancel{(x-4)}}{\cancel{(x-1)}\cancel{(x+3)}} \cdot \frac{\cancel{(x+3)}(x+4)}{\cancel{(x-3)}\cancel{(x-4)}} = \frac{x+4}{x+2}$$

We cross out factors found in numerators and denominators, and get the answer.

Example 10. Multiply

$$\frac{4x^3-36x}{5x^2+5x-30} \cdot \frac{15x^2+60x-180}{8x^4+8x^3-96x^2}$$

Solution. We first find the prime factors of numerators and denominators:

$$\frac{4x\cancel{(x-3)}\cancel{(x+3)}}{\cancel{5}\cancel{(x-2)}\cancel{(x+3)}} \cdot \frac{\overset{3}{\cancel{15}}(x+6)\cancel{(x-2)}}{\underset{2x}{\cancel{8x^2}}(x+4)\cancel{(x-3)}} = \frac{3(x+6)}{2x(x+4)}$$

FRACTIONS 341

Dividing numerators and denominators by all common factors, we get the answer.

Exercise 24.2

Multiply as indicated:

1. $\dfrac{3}{5} \cdot \dfrac{4}{7}$

2. $\dfrac{15}{23} \cdot \dfrac{17}{32}$

3. $\dfrac{a}{b} \cdot \dfrac{x}{y}$

4. $\dfrac{3x}{5y} \cdot \dfrac{4x}{7y}$

5. $\dfrac{7rs}{8xy} \cdot \dfrac{5m}{6t}$

6. $\dfrac{3}{5} \cdot \dfrac{x-2}{x+4}$

7. $\dfrac{15}{32} \cdot \dfrac{16}{25}$

8. $\dfrac{35}{48} \cdot \dfrac{16}{21}$

9. $\dfrac{45}{56} \cdot \dfrac{8}{15}$

10. $\dfrac{3x}{4y} \cdot \dfrac{8ay^2}{9x}$

11. $\dfrac{5a^3}{8b^2} \cdot \dfrac{16bc}{25a^2}$

12. $\dfrac{14c^2}{15d} \cdot \dfrac{20d^2x}{21cx}$

13. $\dfrac{5a^3}{12bc^2} \cdot \dfrac{16bc}{15a^2}$

14. $\dfrac{35xy^3}{24ab^2} \cdot \dfrac{8ab}{15xy^2}$

15. $\dfrac{20x^3y}{27ab^2} \cdot \dfrac{18a^2b}{25xy^3}$

16. $\dfrac{18xy^3}{35mn} \cdot \dfrac{14m^2n}{27x^2y}$

17. $\dfrac{24xy}{25a^2b^2} \cdot \dfrac{15ab}{8x^3y^2}$

18. $\dfrac{30r^2s^3}{49a^3b^2} \cdot \dfrac{14ab}{45rs}$

19. $\dfrac{26x^3y}{30a^2b} \cdot \dfrac{18a^2b}{39x^3y}$

20. $\dfrac{8r^2}{27xy} \cdot \dfrac{9xy}{4r^2}$

21. $\dfrac{21xy^2}{8m^2n} \cdot \dfrac{24m^2n}{35xy^2}$

22. $\dfrac{21x^4y}{32r^3} \cdot \dfrac{16r^2}{35xy}$

23. $\dfrac{28ab^3}{27r^2s} \cdot \dfrac{9rs}{14ab}$

24. $\dfrac{34r^2s}{45ab} \cdot \dfrac{25a^2b^2}{51rs^2}$

25. $\dfrac{39ab^2}{44xy^4} \cdot \dfrac{33y^4}{26b^4}$

26. $\dfrac{12x^3y}{35ab} \cdot \dfrac{25ab^2}{8x^2y}$

27. $\dfrac{32a^2b}{15c} \cdot \dfrac{35c^2d}{16a^2b}$

28. $\dfrac{36a^2b}{25cd} \cdot \dfrac{15c^4d^2}{16abx}$

29. $\dfrac{42r^3s}{9xyz} \cdot \dfrac{36x^2y}{49rs}$

30. $\dfrac{15xy^2}{26ab} \cdot \dfrac{39a^2b^3}{20xy^5}$

31. $\dfrac{80x^3y^4}{33a^2b^3} \cdot \dfrac{11ab^3}{40x^3}$

32. $\dfrac{40uv}{21x^3y} \cdot \dfrac{42x^3}{65u^2v^2}$

33. $\dfrac{35x^4y}{24a^3b^4} \cdot \dfrac{8a^4b^3}{15x^4y^2}$

34. $\dfrac{4x^2y^3}{3a^3c} \cdot \dfrac{7ax^2}{8bc} \cdot \dfrac{9ac^2}{14x^3y}$

342 ALGEBRA

35. $\dfrac{5xy^2}{6ac^2} \cdot \dfrac{9c^3}{10xy} \cdot \dfrac{4a}{3cy}$

36. $\dfrac{14by^2}{15a^2x} \cdot \dfrac{10a}{21b^2y} \cdot \dfrac{9abx}{4y}$

37. $\dfrac{8x^2}{15a^3} \cdot \dfrac{5b}{6c} \cdot \dfrac{4bx}{3a^2}$

38. $(3x^2) \cdot \dfrac{5y^3}{6ax}$

39. $(7a^2bc) \cdot \dfrac{13xy^2}{21ab^2c^3}$

40. $\dfrac{15a^3b}{16x^2y} \cdot (8bxy^2)$

41. $\dfrac{x^2-x-6}{x^2+2x-24} \cdot \dfrac{x^2-5x+4}{x^2-4x+3}$

42. $\dfrac{x^2+8x+15}{x^2-25} \cdot \dfrac{x^2-6x+5}{x^2+2x-3}$

43. $\dfrac{2x^3+18x^2+40x}{3x^2+6x-45} \cdot \dfrac{6x^2-30x+36}{4x^4-12x^3-16x^2}$

44. $\dfrac{2n^3+8n^2-24n}{3n^2+15n+18} \cdot \dfrac{6n^2+60n+96}{n^4+2n^3-8n^2}$

45. $\dfrac{2x^4-18x^2}{3x^3-75x} \cdot \dfrac{5x^2-25x}{4x^5-12x^4}$

46. $\dfrac{4x^4-144x^2}{3x^4-12x^2} \cdot \dfrac{x^3-2x^2}{6x^3-36x^2}$

47. $\dfrac{x^3-6x^2+9x}{3x^2+6x-9} \cdot \dfrac{6x^2-12x+6}{x^4-2x^3-3x^2} \cdot \dfrac{x^2-1}{4x^2-36}$

48. $\dfrac{2t^4-20t^3+18t^2}{t^3+t^2-20t} \cdot \dfrac{2t^2-2t-24}{3t^3-18t^2+15t} \cdot \dfrac{6t^4-150t^2}{4t^2-36}$

24.5. DIVISION OF FRACTIONS

In arithmetic, when we wish to divide a fraction by a fraction or by an integer, *we invert the divisor and then multiply*. That is, we *multiply the dividend by the reciprocal of the divisor*. Remember, the divisor is the number that follows the division sign (÷). In algebra we do the same in the division of fractions.

Example 1. Divide as indicated

$$\dfrac{5}{8} \div \dfrac{3}{4}$$

Solution. Inverting the divisor, we have

$$\dfrac{5}{\underset{2}{\cancel{8}}} \cdot \dfrac{\cancel{4}}{3} = \dfrac{5}{6}$$

Example 2. Divide

$$\dfrac{4}{5} \div 3 = \dfrac{4}{5} \cdot \dfrac{1}{3} = \dfrac{4}{15}$$

FRACTIONS

Example 3. Divide

$$\frac{4x^3}{5y^2} \div \frac{3x^2}{10y} = \frac{4x^3}{5y^2} \cdot \frac{10y}{3x^2} = \frac{8x}{3y}$$

When we have multiplications and divisions of fractions in the same problem, these operations are performed *in the order in which they occur.* Then we invert only the fraction or fractions immediately after the division sign.

Example 4. Simplify

$$\frac{8}{15} \div \frac{7}{10} \cdot \frac{3}{4}$$

Solution. Here we invert only the second fraction, and get

$$\frac{8}{15} \cdot \frac{10}{7} \cdot \frac{3}{4} = \frac{4}{7}$$

Example 5. Simplify

$$\frac{8x^3}{15y^2} \cdot \frac{5cy}{6ax} \div \frac{4bx}{3ay}$$

Solution. Here we invert only the third fraction:

$$\frac{8x^3}{15y^2} \cdot \frac{5cy}{6ax} \cdot \frac{3ay}{4bx} = \frac{cx}{3b}$$

Example 6. Simplify

$$\frac{15}{16} \div \frac{9}{8} \div \frac{5}{12}$$

Solution. Here we invert the second and the third fractions, since both are indicated divisors. Then we have

$$\frac{15}{16} \cdot \frac{8}{9} \cdot \frac{12}{5} = 2$$

Example 7. Simplify

$$\frac{3x^4}{4y^3} \div \frac{9b^2x}{16ay} \div \frac{5ax}{6by}$$

Solution. In this example, we invert the second and third fractions:

$$\frac{3x^4}{4y^3} \cdot \frac{16ay}{9b^2x} \cdot \frac{6by}{5ax} = \frac{8x^2}{5by}$$

Sometimes we wish to indicate that a certain operation is to be performed first. If an operation is indicated by a quantity within parentheses (including brackets or braces), then that part of the work must be done first. This type of problem is shown in the next example.

Example 8. Simplify

$$\frac{10}{21} \div \left(\frac{2}{3} \div \frac{4}{5}\right)$$

Solution. In this example, we first work out the portion enclosed in parentheses. We get

$$\frac{2}{3} \div \frac{4}{5} = \frac{2}{3} \cdot \frac{5}{4} = \frac{5}{6}$$

Then the problem becomes

$$\frac{10}{21} \div \frac{5}{6} = \frac{10}{21} \cdot \frac{6}{5} = \frac{4}{7}$$

Exercise 24.3

Perform the indicated operations:

1. $\dfrac{27a^2x}{20by^2} \div \dfrac{18ax^2}{25by^3}$

2. $\dfrac{9x^3}{10y^2} \div \dfrac{12bx^2}{5ay}$

3. $\dfrac{5a^3b}{12mn^2} \div \dfrac{20ab}{9m^2}$

4. $\dfrac{33abc}{56x^2y} \div \dfrac{11abc^2}{28x^2y^2}$

5. $\dfrac{32m^2n}{15xy} \div \dfrac{16m^3n}{25x^2y}$

6. $\dfrac{25ab^2}{32x^3y} \div \dfrac{35ab}{16x^2y}$

7. $\dfrac{22xy}{21abc^3} \div \dfrac{8y}{9bc} \cdot \dfrac{14a^2bc}{33x^3}$

8. $\dfrac{18r^2s}{35x^2y} \div \dfrac{9rs}{8ax^2} \cdot \dfrac{16s^2}{15y^2}$

9. $\dfrac{5ac^2}{16xy^3} \div \left(\dfrac{3bc}{8x^2y} \cdot \dfrac{25ac}{6by^2}\right)$

10. $\dfrac{9ab}{2xy} \div \left(\dfrac{15x^2}{32y} \div \dfrac{5ax^2}{8by}\right)$

11. $\dfrac{x^5 - x^3y^2}{4x^2 - 9y^2} \div \dfrac{3x^3 - 3x^2y}{8x - 12y}$

12. $\dfrac{x^3y - 25xy}{5x^2 + 10x} \div \dfrac{x^2 - 6x + 5}{15x^3 + 30x^2}$

FRACTIONS

13. $\dfrac{x^2-6x-16}{x^2-8x+16} \div \dfrac{6x+12}{4x-16}$

14. $\dfrac{x^2-9}{3x^3-6x^2} \div \dfrac{x^2+3x-18}{4x^2-8x}$

15. $(3x-6) \div \dfrac{x^2+6x-16}{x+4}$

16. $\dfrac{a^2+8a-9}{a+3} \div (5a-5)$

17. $\dfrac{9x^2-6x+1}{6x} \div (18x^3-2x)$

18. $(8x^3-2x) \div \dfrac{4x^4-4x^3+x^2}{4x}$

19. $\dfrac{16x^4-100x^2}{x+2} \div (8x^2-20x)$

20. $(15c+20) \div \dfrac{45c^2-80}{c+4}$

21. $\dfrac{x^3-6x^2+9x}{6x^2+30x+36} \cdot \dfrac{x^4+8x^3+16x^2}{x^5-6x^4+5x^3} \div \dfrac{x^4-16x^2}{3x^2-6x-45}$

22. $\dfrac{6t^3-36t^2+30t}{20t^2-45} \div \dfrac{3t^4-9t^3-30t^2}{t^2-t-12} \cdot \dfrac{8t^3+4t^2-24t}{2t^2-10t+8}$

23. $\dfrac{x^5+5x^4-14x^3}{8x^2-8x-48} \div \dfrac{x^3+8x^2+7x}{2x^2-18x+36} \div \dfrac{x^4-8x^3+12x^2}{3x^2+7x+2}$

24. $\dfrac{3y^3+27y^2+24y}{8y^2-24y-80} \div \dfrac{3y^4-3y^3-6y^2}{2y^3-18y^2+40y} \div \dfrac{10y^2-160}{9y^4-36y^2}$

24.6. ADDITION AND SUBTRACTION OF FRACTIONS

In arithmetic, we have seen that if fractions are to be added or subtracted, they must have the same denominators. When two or more fractions have the same denominators, the fractions can be added or subtracted by simply combining the numerators, placing the result over the common denominator.

Example 1. Combine into a single fraction

$$\frac{5}{9}+\frac{8}{9}-\frac{7}{9}$$

Solution. All these fractions have the same denominators. Then we combine them by placing all numerators over the same denominator. The result should always be reduced to lowest terms:

$$\frac{5}{9}+\frac{8}{9}-\frac{7}{9} = \frac{5+8-7}{9} = \frac{6}{9} \quad \text{which reduces to} \quad \frac{2}{3}$$

Example 2. Combine the fractions

$$\frac{3x}{5}+\frac{7x}{5}-\frac{6x}{5}$$

Solution. Algebraic fractions are added or subtracted in the same way as fractions in arithmetic. Since all denominators are the same in these fractions, we place all numerators over the common denominator:

$$\frac{3x}{5} + \frac{7x}{5} - \frac{6x}{5} = \frac{3x + 7x - 6x}{5} = \frac{4x}{5}$$

Example 3. Combine the fractions:

$$\frac{10}{3x} + \frac{16}{3x} - \frac{14}{3x}$$

Solution. If the denominators contain a term with an unknown, such as x, we follow the same procedure. Here all the denominators are the same. Then

$$\frac{10}{3x} + \frac{16}{3x} - \frac{14}{3x} = \frac{10 + 16 - 14}{3x} = \frac{12}{3x}, \text{ which reduces to } \frac{4}{x}$$

Example 4. Combine the fractions

$$\frac{3x+1}{5x} + \frac{4x-3}{5x}$$

Solution. When we place the numerators over the common denominator, it is well to *enclose each numerator in parentheses* to indicate a quantity as a first step; then we get

$$\frac{(3x+1) + (4x-3)}{5x}$$

Removing parentheses and combining

$$\frac{3x + 1 + 4x - 3}{5x} = \frac{7x - 2}{5x}$$

Example 5. Combine the fractions

$$\frac{5x - 3}{7x} - \frac{2x - 5}{7x}$$

Solution. A fraction preceded by a minus sign should be treated in the same way as any quantity preceded by a minus sign. Notice the minus sign before the second fraction. This sign means that the entire numerator is to be considered negative. Therefore, when we place each numerator over the common denominator, it is important that we *first place each numerator in parentheses*. This will avoid many mistakes in connection with signs.

As a first step we have

$$\frac{(5x-3)-(2x-5)}{7x}$$

Now we remove parentheses and combine

$$\frac{5x-3-2x+5}{7x} = \frac{3x+2}{7x}$$

Example 6. Combine

$$\frac{6x-5}{4x^2} - \frac{2x-1}{4x^2} - \frac{x+3}{4x^2}$$

Solution. All these fractions have the same denominator. Then we place all the numerators over the common denominator. In doing so, we first enclose each numerator in parentheses to indicate a quantity; we get

$$\frac{(6x-5)-(2x-1)-(x+3)}{4x^2}$$

We remove parentheses and combine

$$\frac{6x-5-2x+1-x-3}{4x^2} = \frac{3x-7}{4x^2}$$

Example 7. Combine

$$\frac{5x-2y}{6xy} - \frac{4x-5y}{6xy} - \frac{3y-2x}{6xy}$$

Solution. All these fractions have the same denominator. Then we place all the numerators over the common denominator, keeping each numerator in a set of parentheses to indicate the quantity. We get

$$\frac{(5x-2y)-(4x-5y)-(3y-2x)}{6xy}$$

Now we remove the parentheses and combine like terms. When the terms of the numerator are combined, the answer can be reduced:

$$\frac{5x-2y-4x+5y-3y+2x}{6xy} = \frac{3x}{6xy} = \frac{1}{2y}$$

In arithmetic, if fractions have different denominators, then before the fractions can be added or subtracted, they must be changed in form so that all have the same denominator.

Example 8. Add the fractions

$$\frac{3}{5}+\frac{4}{7}$$

Solution. These fractions cannot be added until the form is changed. To change the form of the fractions, we first look for a common denominator that will contain the denominators, 5 and 7. The smallest number that will contain the 5 and 7 as factors is 35. This is called the *lowest common denominator* (*LCD*). Then we change each fraction to a new form with 35 as the denominator. Remember, when we change the form of the fraction, *we do not change its value; we change only the form.*

To change the first fraction, $\frac{3}{5}$, to a new fraction with a denominator of 35, we multiply numerator and denominator of the original fraction by 7. Then we get

$$\frac{3}{5} = \frac{(7)(3)}{(7)(5)} = \frac{21}{35}$$

For the second fraction, $\frac{4}{7}$, we multiply numerator and denominator by 5:

$$\frac{4}{7} = \frac{(5)(4)}{(5)(7)} = \frac{20}{35}$$

Now that the fractions have the same denominator, we can add them:

$$\frac{3}{5}+\frac{4}{7} = \frac{21}{35}+\frac{20}{35} = \frac{41}{35}$$

Notice that when we change a fraction to higher terms, *we do not change the value of the fraction; we change only the form.*

In adding or subtracting fractions in algebra, we follow the same procedure as in arithmetic. If fractions have different denominators, the fractions must be changed in form so that they have the same denominators.

Note. At this point, you should review the method of finding the lowest common denominator (LCD), as explained in Sec. 2.7 (*in the section on arithmetic*).

In combining fractions having different denominators, observe these steps:

(1) *Find the lowest common denominator (LCD) of all the fractions; that is, the smallest number that can be divided by all the denominators.*

(2) *Rewrite each fraction in a new form so that each fraction has the lowest common denominator of all the fractions.*

(3) *Combine the numerators and place the result over the common denominator. Simplify the result as much as possible.*

Example 9. Combine the fractions

$$\frac{3x}{4} + \frac{5x}{12} - \frac{2x}{9}$$

Solution. The lowest common denominator (LCD) is 36. That is the smallest number that can be divided by the denominators, 4, 12, and 9. Now we change the form of each fraction so that each has 36 as its denominator. The best way to begin is to set down the form of each fraction first:

$$\frac{}{36} + \frac{}{36} - \frac{}{36}$$

To give the first fraction a denominator of 36, we must multiply its denominator by 9. Then we must also multiply the numerator by 9. For the first fraction, we get

$$\frac{(9)(3x)}{(9)(4)} = \frac{27x}{36}$$

It is important to understand at this point, that we have not multiplied the fraction by 9; we have simply changed the form of the fraction. The new fraction, $27x/36$, has exactly the same value as the original fraction.

The second fraction is changed:

$$\frac{(3)(5x)}{(3)(12)} = \frac{15x}{36}$$

The third fraction: $\quad \dfrac{(4)(2x)}{(4)(9)} = \dfrac{8x}{36}$

Now the problem, $$\frac{3x}{4}+\frac{5x}{12}-\frac{2x}{9}$$

becomes $$\frac{27x}{36}+\frac{15x}{36}-\frac{8x}{36}$$

The value of each fraction is not changed. Now we combine:

$$\frac{27x}{36}+\frac{15x}{36}-\frac{8x}{36}=\frac{34x}{36}=\frac{17x}{18}$$

Example 10. Combine the fractions

$$\frac{2x-3}{4}-\frac{3x-1}{10}-\frac{5-4x}{20}$$

Solution. The LCD is 20. Then we first set up the form of the fractions:

$$\frac{\quad}{20}-\frac{\quad}{20}-\frac{\quad}{20}$$

To find the multiplier for the numerator and denominator of the first fraction, we divide: $20 \div 4 = 5$. Then we must multiply numerator and denominator of the first fraction by 5; we get

$$\frac{2x-3}{4}=\frac{(5)(2x-3)}{(5)(4)}=\frac{(5)(2x-3)}{20}$$

In the numerators, keep the multiplier separate from the numerator. Then the original fractions are changed as follows:
  The original fractions

$$\frac{2x-3}{4}-\frac{3x-1}{10}-\frac{5-4x}{20}$$

are changed to $$\frac{5(2x-3)}{20}-\frac{2(3x-1)}{20}-\frac{1(5-4x)}{20}$$

Now we write all the numerators over the common denominator, 20, keeping each numerator in parentheses; we get

$$\frac{5(2x-3)-2(3x-1)-1(5-4x)}{20}$$

Removing parentheses and combining, we get

$$\frac{10x-15-6x+2-5+4x}{20} = \frac{8x-18}{20} = \frac{4x-9}{10}$$

Example 11. Combine the fractions

$$\frac{x-1}{4x} - \frac{3}{2x^2} - \frac{2-x}{3x}$$

Solution. The LCD is $12x^2$. Now change each fraction to a new form having the denominator: $12x^2$; we enclose binomial numerators in parentheses:

$$\frac{3x(x-1)}{12x^2} - \frac{18}{12x^2} - \frac{4x(2-x)}{12x^2}$$

We write the numerators over the common denominator, combine terms, and simplify:

$$\frac{3x(x-1)-18-4x(2-x)}{12x^2} = \frac{3x^2-3x-18-8x+4x^2}{12x^2} = \frac{7x^2-11x-18}{12x^2}$$

Exercise 24.4

Combine the following fractions. Reduce if possible.

1. $\dfrac{3x}{4} + \dfrac{7x}{4} - \dfrac{5x}{4}$

2. $\dfrac{2y}{5} + \dfrac{4y}{5} - \dfrac{3y}{5}$

3. $\dfrac{7t}{12} + \dfrac{t}{12} - \dfrac{5t}{12}$

4. $\dfrac{8x}{15} - \dfrac{2x}{15} + \dfrac{4x}{15}$

5. $\dfrac{7n}{20} + \dfrac{11n}{20} - \dfrac{3n}{20}$

6. $\dfrac{13x}{18} - \dfrac{3x}{18} - \dfrac{7x}{18}$

7. $\dfrac{5}{3x} + \dfrac{2}{3x} - \dfrac{4}{3x}$

8. $\dfrac{7}{5n} - \dfrac{4}{5n} + \dfrac{2}{5n}$

9. $\dfrac{8}{9x} + \dfrac{2}{9x} + \dfrac{5}{9x}$

10. $\dfrac{5a}{9x^2} - \dfrac{7a}{9x^2} + \dfrac{8a}{9x^2}$

11. $\dfrac{7x}{st} + \dfrac{2x}{st} - \dfrac{4x}{st}$

12. $\dfrac{4n}{5y} + \dfrac{7n}{5y} - \dfrac{6n}{5y}$

13. $\dfrac{3n}{4} + \dfrac{4n}{9}$

14. $\dfrac{7x}{8} + \dfrac{5x}{6}$

15. $\dfrac{9n}{10} + \dfrac{4n}{15}$

16. $\dfrac{x+2}{3} + \dfrac{x-5}{4}$

17. $\dfrac{x-3}{6}+\dfrac{2x-1}{9}$
18. $\dfrac{5x-2}{8}+\dfrac{2x-3}{12}$
19. $\dfrac{5x-1}{12}-\dfrac{2x-3}{9}$
20. $\dfrac{3x-2}{10}-\dfrac{2x+1}{15}$
21. $\dfrac{4x-5}{6}-\dfrac{2x-3}{3}$
22. $\dfrac{2x+5}{4}-\dfrac{3x+2}{6}$
23. $\dfrac{3x+4}{9}-\dfrac{4x-1}{12}$
24. $\dfrac{2x+7}{12}-\dfrac{3x-4}{18}$
25. $\dfrac{7n}{3}-\dfrac{4n}{5}+\dfrac{2n}{15}$
26. $\dfrac{5n}{6}+\dfrac{2n}{9}-\dfrac{7n}{18}$
27. $\dfrac{3n}{8}+\dfrac{5n}{12}-\dfrac{7n}{24}$
28. $\dfrac{3x}{4}+\dfrac{5x}{8}-\dfrac{7x}{12}$
29. $\dfrac{2r}{5}-\dfrac{4r}{15}+\dfrac{3r}{10}$
30. $\dfrac{4t}{9}+\dfrac{5t}{6}-\dfrac{2t}{3}$
31. $\dfrac{3n}{x^2}+\dfrac{4n}{3x}-\dfrac{5n}{2x}$
32. $\dfrac{4x}{5t}-\dfrac{3x}{2t}+\dfrac{5x}{t^2}$
33. $\dfrac{5t}{4x}-\dfrac{3t}{x^2}-\dfrac{2t}{3x^2}$
34. $\dfrac{t+2}{4}-\dfrac{t-3}{6}+5$
35. $\dfrac{n+5}{3}-\dfrac{n-3}{5}-4$
36. $\dfrac{n+4}{2}-\dfrac{n-3}{3}-2$
37. $\dfrac{5x-2}{2x}-\dfrac{2y-3}{3y}$
38. $\dfrac{3a+2}{6a}-\dfrac{2b-5}{4b}$
39. $\dfrac{4m-3}{6m}-\dfrac{2n+5}{3n}$
40. $\dfrac{6x-5}{8x}-\dfrac{3y-2}{4y}$

24.7. FRACTIONS WITH POLYNOMIAL DENOMINATORS

In the addition or subtraction of fractions, if the fractions have polynomial denominators, we must be especially careful in stating the lowest common denominator. However, if all the fractions have the same denominator, we simply combine the numerators as usual and place the result over the given denominator.

Example 1. Combine the fractions

$$\dfrac{5x-3}{x-4}+\dfrac{2x-1}{x-4}-\dfrac{3x-2}{x-4}$$

Solution. All these fractions have the same denominator. Then we place all the numerators over the common denominator. In doing so, we first

FRACTIONS **353**

enclose each numerator in parentheses to indicate a quantity. We get

$$\frac{(5x-3)+(2x-1)-(3x-2)}{x-4}$$

Now we remove the parentheses and combine terms in the numerator:

$$\frac{5x-3+2x-1-3x+2}{x-4}=\frac{4x-2}{x-4}=\frac{2(2x-1)}{x-4}$$

In the answer, the numerator is sometimes factored, if possible.

Example 2. Combine the fractions

$$\frac{7x+5}{3x+2}+\frac{4x+3}{3x+2}-\frac{5x+4}{3x+2}$$

Solution. We place all numerators over the common denominator. As a first step, we enclose each numerator in parentheses to indicate a quantity:

$$\frac{(7x+5)+(4x+3)-(5x+4)}{3x+2}$$

Removing parentheses and combining terms of the numerator, we get

$$\frac{7x+5+4x+3-5x-4}{3x+2}=\frac{6x+4}{3x+2}$$

which reduces to 2.

Example 3. Combine the fractions

$$\frac{3}{x}-\frac{2x-1}{x-3}$$

Solution. These fractions have different denominators. Neither denominator contains the other as a factor. The two denominators are as different as the arithmetic numbers 7 and 4. Therefore, the lowest common denominator must be the product of the two denominators, $x(x-3)$. We write each fraction with the denominator, $x(x-3)$:

$$\frac{3(x-3)}{x(x-3)}-\frac{x(2x-1)}{x(x-3)}$$

Note especially that each fraction has the same value as in the original fractions. We have not changed the value of either fraction; we have changed only its form. Now we write the numerators over the LCD, keeping the numerators in parentheses at first; we get

$$\frac{3(x-3)-x(2x-1)}{x(x-3)}$$

We combine the numerators and simplify:

$$\frac{3x-9-2x^2+x}{x(x-3)} = \frac{4x-2x^2-9}{x(x-3)}$$

Note. *We usually leave the denominator in factored form. In some problems the final numerator can be factored and the fraction may be reduced to lower terms.*

Example 4. Combine the fractions

$$\frac{7}{x+4} - \frac{5}{x+2}$$

Solution. The LCD is the product of the denominators: $(x+4)(x+2)$. Changing the fractions, we get

$$\frac{7(x+2)}{(x+4)(x+2)} - \frac{5(x+4)}{(x+4)(x+2)}$$

Again, we have changed the form of each fraction but not its value. Simplifying, we get

$$\frac{7(x+2)-5(x+4)}{(x+4)(x+2)} = \frac{7x+14-5x-20}{(x+4)(x+2)} = \frac{2x-6}{(x+4)(x+2)} = \frac{2(x-3)}{(x+4)(x+2)}$$

In the answer the numerator is factored. This is often done if possible because it may be possible to reduce the fraction. The fraction could be reduced if the numerator happened to contain one of the factors of the denominator.

Example 5. Combine the fractions

$$\frac{4}{x-5} - \frac{3}{5-x}$$

Solution. At first glance, the fractions appear to have different denominators. However, the denominators can be made identical. If we multiply the numerator and denominator of the second fraction by -1, we get

$$\frac{4}{x-5} - \frac{-3}{x-5}$$

Now the fractions have the same denominator, and the answer is

$$\frac{7}{x-5}$$

Note. In working with fractions, as in Example 5, it is well to understand the following facts about the signs in a fraction. Any fraction has three basic signs associated with it:

1. The sign of the numerator.
2. The sign of the denominator.
3. The sign of the fraction itself.

PRINCIPLE. If any two of the signs mentioned above are changed, then the value of the fraction will not be changed.

As an illustration of this principle, consider the fraction:

$$+\frac{+6}{+3} = +2$$

The value of the fraction is $+2$.
Changing the signs of numerator and denominator, we get

$$+\frac{-6}{-3} = +2$$

Changing the signs of numerator and fraction, we get

$$-\frac{-6}{+3} = +2$$

Changing signs of denominator and fraction, we get

$$-\frac{+6}{-3} = +2$$

As another example, $\dfrac{5}{x-2} = \dfrac{-5}{2-x} = -\dfrac{-5}{x-2} = -\dfrac{5}{2-x}$

Example 6. Combine the fractions

$$\dfrac{3}{x-4} - \dfrac{2x-5}{x^2-6x+8}$$

Solution. Factoring the second denominator:

$$\dfrac{3}{x-4} - \dfrac{2x-5}{(x-2)(x-4)}$$

The second denominator contains the first denominator as a factor. Therefore, it is the LCD. This example is like combining the fractions: $\tfrac{2}{3} - \tfrac{4}{15}$. Since the LCD is the quantity, $(x-2)(x-4)$, we change the form of only the first fraction:

$$\dfrac{3(x-2)}{(x-2)(x-4)} - \dfrac{2x-5}{(x-2)(x-4)}$$

Now we write both numerators over the LCD, keeping the numerators in parentheses to indicate quantities; then we remove parentheses and combine numerators:

$$\dfrac{3(x-2)-(2x-5)}{(x-2)(x-4)} = \dfrac{3x-6-2x+5}{(x-2)(x-4)} = \dfrac{x-1}{(x-2)(x-4)}$$

Example 7. Combine the fractions

$$\dfrac{7}{x+2} - \dfrac{2x-1}{x^2-4} - \dfrac{3}{x-2}$$

Solution. If we factor the second denominator, we get $(x-2)(x+2)$. Then the second denominator contains the other denominators as factors. Therefore, the LCD is the quantity $(x-2)(x+2)$. We change the form of the first and third fractions but leave the second unchanged:

$$\dfrac{7(x-2)}{(x-2)(x+2)} - \dfrac{2x-1}{(x-2)(x+2)} - \dfrac{3(x+2)}{(x-2)(x+2)}$$

FRACTIONS 357

Note that each fraction has the same value as in the original fractions. Combining,

$$\frac{7(x-2)-(2x-1)-3(x+2)}{(x-2)(x+2)} = \frac{7x-14-2x+1-3x-6}{(x-2)(x+2)} = \frac{2x-19}{(x-2)(x+2)}$$

Example 8. Combine the fractions

$$\frac{4}{x-2} + \frac{5}{x-3} - \frac{3x-4}{x^2-5x+6}$$

Solution. The third denominator can be factored into $(x-2)(x-3)$. Since the third denominator contains the other two denominators as factors, the LCD is the third denominator: $(x-2)(x-3)$. We change the form of the first and second fractions so that each has the denominator: $(x-2)(x-3)$:

$$\frac{4(x-3)}{(x-2)(x-3)} + \frac{5(x-2)}{(x-2)(x-3)} - \frac{3x-4}{(x-2)(x-3)}$$

Now we write the numerators over the common denominator, keeping each numerator in parentheses to indicate a quantity:

$$\frac{4(x-3)+5(x-2)-(3x-4)}{(x-2)(x-3)} = \frac{4x-12+5x-10-3x+4}{(x-2)(x-3)}$$

$$= \frac{6x-18}{(x-2)(x-3)} = \frac{6(x-3)}{(x-2)(x-3)} = \frac{6}{x-2}$$

Note that when the fractions are combined, the numerator can be factored and the fraction can be reduced by dividing numerator and denominator by $(x-3)$.

Example 9. Combine the fractions

$$\frac{4}{x-3} - \frac{x+2}{x^2-9} - \frac{3x-2}{x^2-6x+9}$$

Solution. In this example, the LCD is $(x-3)(x+3)(x-3)$, which may be written: $(x+3)(x-3)^2$. We must use the factor $(x-3)$ *twice because it is contained twice in one of the given denominators.* The fractions are

358 ALGEBRA

changed to new forms having the LCD as the denominators; we get

$$\frac{4(x-3)(x+3)}{(x+3)(x-3)^2} - \frac{(x+2)(x-3)}{(x+3)(x-3)^2} - \frac{(3x-2)(x+3)}{(x+3)(x-3)^2}$$

$$= \frac{4(x-3)(x+3) - (x+2)(x-3) - (3x-2)(x+3)}{(x+3)(x-3)^2}$$

$$= \frac{4(x^2-9) - (x^2-x-6) - (3x^2+7x-6)}{(x+3)(x-3)^2}$$

$$= \frac{4x^2 - 36 - x^2 + x + 6 - 3x^2 - 7x + 6}{(x+3)(x-3)^2} = \frac{-6x - 24}{(x+3)(x-3)^2} = \frac{-6(x+4)}{(x+3)(x-3)^2}$$

Exercise 24.5

Combine the terms into a single fraction and simplify:

1. $\dfrac{5}{x-2} - \dfrac{4}{x+3}$
2. $\dfrac{2}{x-4} - \dfrac{5}{x-1}$
3. $\dfrac{7}{x-3} + \dfrac{1}{x+4}$
4. $\dfrac{3x}{x+5} - \dfrac{2x}{x-2}$
5. $\dfrac{4x}{x-2} + \dfrac{2}{x+2}$
6. $\dfrac{6x}{x-5} - \dfrac{3}{x+1}$
7. $\dfrac{4}{x-3} - \dfrac{3}{3-x}$
8. $\dfrac{7}{x-5} + \dfrac{2}{5-x}$
9. $\dfrac{9}{x-4} - \dfrac{5}{4-x}$
10. $\dfrac{n+3}{n-2} + \dfrac{n+1}{n-1}$
11. $\dfrac{y+2}{y-3} - \dfrac{y-1}{y-4}$
12. $\dfrac{t-2}{t+3} - \dfrac{t+2}{t-3}$
13. $\dfrac{n-2}{n^2-9} - \dfrac{n-1}{n-3}$
14. $\dfrac{n+3}{n+4} + \dfrac{n-2}{n^2-16}$
15. $\dfrac{n-4}{n-2} - \dfrac{n-3}{n^2-4}$
16. $\dfrac{3}{x-3y} - \dfrac{4}{x+2y}$
17. $\dfrac{5}{3x-y} - \dfrac{2}{3x+2y}$
18. $\dfrac{2}{2x-y} + \dfrac{4}{3x-2y}$
19. $\dfrac{3n-2}{n+3} - \dfrac{2n+3}{n^2+n-6}$
20. $\dfrac{4n+3}{n+2} - \dfrac{2n-5}{n^2-2n-8}$
21. $\dfrac{2x-5}{x^2-5x+6} - \dfrac{x+3}{x-2}$
22. $\dfrac{3x-1}{x^2-3x-4} - \dfrac{x+4}{x+1}$

23. $\dfrac{4}{n+3} - \dfrac{3n-1}{n^2+3n} - \dfrac{2}{n}$

24. $\dfrac{3}{n-2} - \dfrac{2n-1}{n^2-2n} - \dfrac{4}{n}$

25. $\dfrac{2x}{x-1} - \dfrac{x+1}{x-2} - \dfrac{4x-5}{x^2-3x+2}$

26. $\dfrac{n+1}{n-2} - \dfrac{n+2}{n+3} - \dfrac{3n-4}{n^2+n-6}$

27. $\dfrac{2x-7}{x-5} - \dfrac{3}{x} - 4x - 2$

28. $\dfrac{3x-5}{x+3} - \dfrac{4}{x} - 2x - 3$

29. $\dfrac{2x-3}{x-2} - \dfrac{4}{x} - 3x - 4$

30. $\dfrac{4-3x}{x-3} - \dfrac{5}{x} - 3x - 2$

25 FRACTIONAL EQUATIONS

25.1. DEFINITION

A fractional equation is an equation containing a fraction. The fraction may be a common fraction or a decimal fraction. Recall the definition of an equation: *An equation is a statement of equality between two equal quantities.* The expression must contain the equal sign (=), with some quantity on each side of the equal sign. One side may be only zero.

The following expression is not an equation:

$$\frac{x+2}{4} + \frac{x-5}{3}$$

All we can do with these fractions is to combine them into a single fraction. We cannot solve for x because the expression is not an equation. On the other hand, the following expression is an equation; it can be solved for x:

$$\frac{x+2}{4} + \frac{x-5}{3} = 0$$

Here are some examples of fractional equations:

(a) $\dfrac{2x}{3} = 8$ (b) $\dfrac{3x}{5} + 6 = 4x$ (c) $0.25x = 1.8x + 3.6$

(d) $\dfrac{3}{x-2} - \dfrac{4x+5}{x-3} = 2$ (e) $\dfrac{2}{x} = \dfrac{4}{3x-5}$

The expression "*fractional equation*" is sometimes restricted to refer only to such equations as (d) and (e), in which the unknown x appears in the denominator. However, the same procedure is used in solving all types of equations in which any fraction of any kind appears, So we call all of these *fractional equations*.

362 ALGEBRA

It often happens that in setting up the equation for a practical problem, a fraction appears in the equation. For example, suppose that a current of 10.5 amperes in an electric circuit is to be separated into two branches, and it is necessary that one current be $\frac{2}{3}$ of the other. Our problem then is to determine the current in each branch, such that the total current will be 10.5 amperes, and one current $\frac{2}{3}$ of the other. In working such a problem, we get a fractional equation.

As another example involving a fractional equation, suppose that we have a wire 100 inches in length. The wire is to be used as the perimeter of a rectangle, and it is necessary that the width of the rectangle be exactly $\frac{3}{5}$ of the length. We might begin the problem:

Let x = the number of inches in the length of the rectangle;
then $\frac{3}{5}x$ = the number of inches in the width; which we write $3x/5$.

Then, taking twice the width and twice the length, we get the equation:

$$2\left(\frac{3x}{5}\right) + 2x = 100$$

Notice that the equation contains a fraction. Therefore, we need to know how to solve such equations. Many problems in electronics and other scientific studies involve fractional equations.

25.2. ELIMINATING THE DENOMINATOR OF A FRACTION

To solve a fractional equation, we must first understand an important point concerning fractions. *If any fraction is multiplied by its denominator, then the denominator disappears.* The denominator will also disappear if the fraction is multiplied by a larger number that contains the denominator as a factor.

Study the following examples to see what happens to the denominator of each fraction:

$$4\left(\frac{3x}{4}\right) = 3x; \qquad 3\left(\frac{2x}{3}\right) = 2x; \qquad 5\left(\frac{4x}{5}\right) = 4x; \qquad 7\left(\frac{5x}{7}\right) = 5x$$

$$8\left(\frac{5x}{2}\right) = 20x; \qquad 12\left(\frac{3x}{2}\right) = 18x; \qquad 20\left(\frac{7x}{4}\right) = 35x; \qquad 18\left(\frac{4n}{3}\right) = 24n$$

$$10x\left(\frac{7}{2x}\right) = 35; \qquad 12t\left(\frac{9}{2t}\right) = 54; \qquad 18x\left(\frac{5}{3x}\right) = 30; \qquad 8n\left(\frac{5t}{2n}\right) = 20t$$

$$6\left(\frac{5x-3}{2}\right) = 3(5x-3); \qquad 20\left(\frac{3n+7}{4}\right) = 5(3n+7)$$

FRACTIONAL EQUATIONS

Exercise 25.1

Write the answers for each of the following.

1. $6\left(\dfrac{5x}{6}\right)$
2. $5\left(\dfrac{2x}{5}\right)$
3. $7\left(\dfrac{4x}{7}\right)$
4. $4\left(\dfrac{5n}{4}\right)$
5. $8\left(\dfrac{3n}{4}\right)$
6. $6\left(\dfrac{3n}{2}\right)$
7. $10\left(\dfrac{7t}{2}\right)$
8. $12\left(\dfrac{5t}{3}\right)$
9. $24\left(\dfrac{5t}{3}\right)$
10. $36\left(\dfrac{2t}{9}\right)$
11. $2x\left(\dfrac{3t}{2x}\right)$
12. $6x\left(\dfrac{5n}{3x}\right)$
13. $3\left(\dfrac{2x-5}{3}\right)$
14. $4\left(\dfrac{3x+2}{4}\right)$
15. $5\left(\dfrac{4-3x}{5}\right)$
16. $6\left(\dfrac{4x-1}{3}\right)$
17. $6\left(\dfrac{5x-3}{2}\right)$
18. $12\left(\dfrac{2n-7}{4}\right)$
19. $20\left(\dfrac{3x+2}{5}\right)$
20. $18\left(\dfrac{4t-3}{6}\right)$
21. $30\left(\dfrac{2x-7}{2}\right)$
22. $12\left(\dfrac{5c+2}{3}\right)$
23. $24\left(\dfrac{2a-1}{4}\right)$
24. $36\left(\dfrac{5t+1}{4}\right)$

25.3. SOLVING A FRACTIONAL EQUATION

In solving a fractional equation, the first step is to get rid of the denominator. To do so, we multiply both sides of the equation by the *lowest common denominator* (LCD). The LCD may be called the "*multiplying operator.*" This first step is simply the application of the multiplication axiom.

Axiom. If both sides of an equation are multiplied by the same quantity, the equation is still true.

In applying this rule, we must be sure to multiply each fraction and each term by the lowest common denominator. After the denominators have been eliminated, the rest of the problem is usually not difficult.

Example 1. Solve this equation

$$\frac{2x}{3} = 4.$$

Solution. To eliminate the denominator, we multiply both sides of the equation by 3. The axiom says that we have a right to do this. Then we write

$$3\left(\frac{2x}{3}\right) = 3(4)$$

We have multiplied both sides of the equation by 3. Now we see that the left side becomes simply $2x$; the right side becomes 12. Then

the original equation $\quad \dfrac{2x}{3} = 4$

becomes $\quad 2x = 12$

Now we can solve the equation as we have done before. Dividing both sides by 2, we get

$$x = 6$$

To check the answer, we put 6 for x in the original equation to see if the result is true:

$$\frac{(2)(6)}{3} = 4$$

which we see is true.

Example 2. Solve the equation

$$\frac{3x}{5} + x = 16.$$

Solution. Here we multiply both sides of the equation by 5. This means we must multiply the fraction and both the other terms by 5. We get

$$5\left(\frac{3x}{5}\right) + (5)(x) = (5)(16)$$

FRACTIONAL EQUATIONS

The equation reduces to $\qquad 3x + 5x = 80$

Combining terms, $\qquad 8x = 80$

Dividing both sides by 8, $\qquad x = 10$

The answer, $x = 10$, means that 10 is the root of the equation. If we put the number 10 for x in the equation, then the equation should be true. So we ask:

$$\text{does} \quad \frac{3(10)}{5} + 10 = 16?$$

$$\text{does} \quad 6 + 10 = 16? \quad \text{Yes}$$

Example 3. Solve the equation:

$$\frac{3x}{4} - 2 = \frac{2x}{3}.$$

Solution. We multiply both sides of the entire equation by 12, the LCD.

$$(12)\left(\frac{3x}{4}\right) - (12)(2) = (12)\left(\frac{2x}{3}\right)$$

Now we see that the denominators disappear. In the first term, we divide 4 into 12. In the third term, we divide 3 into 12. Then

the original equation $\qquad \dfrac{3x}{4} - 2 = \dfrac{2x}{3}$

becomes $\qquad 9x - 24 = 8x$

Solving this equation, we get $\qquad x = 24$

To check the answer, we put 24 for x in the *original equation*. The check should always be done by using the *original equation*. Substituting 24 for x in the original equation, we ask:

$$\text{does} \quad \frac{3(24)}{4} - 2 = \frac{2(24)}{3}?$$

Reducing the fractions, we ask:

$$\text{does} \quad 18-2=16? \quad \text{Yes.}$$

Combining:

$$\text{does} \quad 16=16? \quad \text{Yes.}$$

Example 4. Solve the fractional equation

$$\frac{3x-2}{5} = \frac{2x-3}{4}.$$

Solution. We multiply both sides of the equation by 20, the LCD. Then we write

$$20\left(\frac{3x-2}{5}\right) = 20\left(\frac{2x-3}{4}\right).$$

In the first fraction, we divide 5 into 20. On the right, 4 into 20. Then the original equation

$$\frac{3x-2}{5} = \frac{2x-3}{4}$$

has the new form, $\quad 4(3x-2) = 5(2x-3)$

Removing parentheses, $\quad 12x - 8 = 10x - 15$

Transposing, $\quad 12x - 10x = 8 - 15$

Combining like terms, $\quad 2x = -7$

Dividing both sides by 2, $\quad x = -3.5$

The answer should check in the original equation.

Example 5. Solve the fractional equation:

$$\frac{3n}{4} + \frac{2n}{9} - 2 = \frac{5n}{6}.$$

Solution. The LCD is 36. Then we multiply each fraction and term by 36:

$$36\left(\frac{3n}{4}\right) + 36\left(\frac{2n}{9}\right) - 36(2) = 36\left(\frac{5n}{6}\right)$$

Simplifying the terms, we get $\quad 27n + 8n - 72 = 30n$

Transposing, $\quad\quad\quad\quad\quad\quad 27n + 8n - 30n = 72$

Combining like terms, $\quad\quad\quad\quad\quad 5n = 72$

Dividing both sides by 5, $\quad\quad\quad\quad n = 14.4$

Note. *When you have worked a few problems in fractional equations, it is not necessary to write the multiplier before each fraction and term. You can tell at a glance the result of multiplying both sides of the equation by the LCD. You just think of the LCD before each term, and then at once write the result.*

Exercise 25.2

Solve the following equations. Check every fourth problem.

1. $\dfrac{5x}{3} = 35$

2. $\dfrac{4x}{5} = 24$

3. $\dfrac{3x}{7} = 18$

4. $\dfrac{4x}{3} = 7$

5. $\dfrac{5x}{6} = 2$

6. $\dfrac{2x}{9} = 5$

7. $\dfrac{3n}{5} = n - 4$

8. $\dfrac{7n}{3} - n = 8$

9. $\dfrac{n}{4} + 6 = n$

10. $x + 6 = \dfrac{2x}{5}$

11. $6 + x = \dfrac{x}{4}$

12. $\dfrac{5x}{7} = 4 + x$

13. $x - 5 = \dfrac{3x}{7}$

14. $2x = \dfrac{3x}{4} + 6$

15. $x - 4 = \dfrac{5x}{8}$

16. $\dfrac{3t}{5} - 2 = \dfrac{t}{3}$

17. $\dfrac{4t}{5} - 2 = \dfrac{t}{2}$

18. $\dfrac{2t}{5} = \dfrac{3t}{4} - 1$

19. $6 - n = \dfrac{3 - 2n}{5}$

20. $n - 3 = \dfrac{3n - 5}{7}$

368 ALGEBRA

21. $2n - 5 = \dfrac{n+3}{6}$ **22.** $\dfrac{3x-5}{4} = \dfrac{4x-3}{6}$

23. $\dfrac{x+3}{2} = \dfrac{x-7}{3}$ **24.** $\dfrac{3x+2}{8} = \dfrac{2x-5}{6}$

25. $\dfrac{2x}{5} + 2 = x - \dfrac{x}{3}$ **26.** $\dfrac{3x}{5} + 2 = x - \dfrac{x}{4}$

27. $\dfrac{3x}{4} - x = 1 - \dfrac{2x}{3}$ **28.** $\dfrac{3x}{8} - \dfrac{x}{4} + \dfrac{x}{6} = 70$

29. $\dfrac{x}{5} + \dfrac{3x}{10} - \dfrac{x}{4} = 40$ **30.** $\dfrac{2x}{3} - \dfrac{x}{2} + \dfrac{3x}{4} = 110$

31. The width of a rectangle is $\tfrac{3}{5}$ of the length. If the perimeter of the rectangle is 100 inches, find the length and the width.

32. A man has $\tfrac{3}{8}$ of his savings invested in bonds, $\tfrac{1}{3}$ in stocks, and the remainder, $7000, in Savings and Loan. Find the total amount of his savings.

33. In one branch of an electric circuit, the current is $\tfrac{3}{7}$ as much as in another branch. When the two branches join, the total current is 25 amperes. Find the current in each branch.

34. A certain street was to be paved. The first week, $\tfrac{1}{12}$ of the street was paved; the second week, $\tfrac{3}{20}$ of the street was completed. Then the length remaining was 3680 feet. Find the total length of the street.

25.4. FRACTIONS AND FRACTIONAL EQUATIONS: THE DIFFERENCE

We have combined fractions and we have solved fractional equations. There is a difference between the two operations. It is important to understand this difference. Consider the following expressions:

Example 1.

(a) $\dfrac{x-5}{3} + \dfrac{3x-2}{4}$ (b) $\dfrac{x-5}{3} + \dfrac{3x-2}{4} = 0$

The first expression (a) *is not an equation.* An equation must have something on each side of the equal sign, even though one side may be only zero. In (a), we can combine the fractions, but we cannot solve for *x*.

The second expression (b) *is an equation.* It has something on each side of the equal sign. The equation can be solved for *x*; that is, we can find the value for *x* that will make the equation true.

Now, notice what happens to the fractions in the first expression (a). To combine the fractions, we must have a common denominator, which is 12. We change each fraction to a new form that has the denominator 12, the lowest common denominator. Then the original expression

$$\frac{x-5}{3} + \frac{3x-2}{4}$$

becomes the new form $\quad \dfrac{4(x-5)}{12} + \dfrac{3(3x-2)}{12}$

Note especially that each fraction has exactly the same value as in the original expression. We have not changed the value of either fraction; we have changed only its form. Now we can add the fractions:

$$\frac{4(x-5)+3(3x-2)}{12} = \frac{4x-20+9x-6}{12} = \frac{13x-26}{12}$$

We do not lose the denominator.

Now let us consider the second expression (b). This is an equation. We can find the value of x that makes the equation true. We first multiply both sides of the equation by 12, the LCD, and get

$$12\left(\frac{x-5}{3}\right) + 12\left(\frac{3x-2}{4}\right) = 12(0)$$

Dividing 12 by the denominators eliminates the denominators, and we get

$$4(x-5)+3(3x-2)=0$$

The denominators have disappeared!!!

Removing parentheses, we get $\quad 4x-20+9x-6=0$

Transposing, $\quad\quad\quad\quad\quad\quad\quad\quad\quad 4x+9x=20+6$

Combining like terms, $\quad\quad\quad\quad\quad\quad 13x=26$

Dividing both sides by 13, $\quad\quad\quad\quad\quad x=2$

Let us summarize the differences between the two types of expressions.

370 ALGEBRA

For the first expression, (a), we can say:

1. *It is not an equation.*
2. *It cannot be solved for x; that is, we cannot find the value of x.*
3. *When the fractions are combined, the denominator (LCD) remains.*

The second expression is different in three ways:

1. *It is an equation.*
2. *It can be solved for x.*
3. *In the first step in solving the equation, the denominators disappear.*

Example 2. Compare the following expressions. Then combine the fractions in the first; solve the equation in the second.

$$\text{(a)} \quad \frac{3x+2}{5} + \frac{x+4}{3} \qquad \text{(b)} \quad \frac{3x+2}{5} + \frac{x+4}{3} = 4$$

Compare the two expressions in the three ways mentioned.

(a) To combine the fractions in (a), we change the form of each fraction so that each has the common denominator 15:

the original expression (a)

$$\frac{3x+2}{5} + \frac{x+4}{3}$$

is changed to

$$\frac{3(3x+2)}{15} + \frac{5(x+4)}{15}$$

Now we place the numerators over the common denominator, keeping the numerators in parentheses:

$$\frac{3(3x+2) + 5(x+4)}{15}$$

Removing parentheses,

$$\frac{9x+6+5x+20}{15} = \frac{14x+26}{15}, \text{ the answer}$$

(b) To solve the equation in (b), we multiply both sides of the equation by

15, the LCD, and get, as a second step:

$$3(3x+2)+5(x+4)=60$$

Removing parentheses, $\qquad 9x+6+5x+20=60$

Transposing, $\qquad 9x+5x=60-6-20$

Combining like terms, $\qquad 14x=34$

Dividing both sides by 14, $\qquad x=\dfrac{34}{14}$, or $2\tfrac{3}{7}$

Example 3. Solve the equation $0.24x+4=0.3x+2.5$

Solution. If we write the decimal fractions as common fractions, we shall see that the lowest common denominator is 100. Then we multiply both sides of the equation by 100, by simply moving the decimal point two places.

We get $\qquad 24x+400=30x+250$

Transposing $\qquad 400-250=30x-24x$

Combining like terms, $\qquad 150=6x$

Solving for x, $\qquad 25=x$

The check is left for the student. Check in the *original equation*.

Exercise 25.3

Solve the following equations. Check every fourth equation.

1. $\dfrac{5x-3}{7}=\dfrac{3x-1}{5}$

2. $\dfrac{4x-5}{9}=\dfrac{2x-1}{5}$

3. $\dfrac{3x-1}{5}=\dfrac{4x+3}{6}$

4. $\dfrac{4x+1}{3}=\dfrac{2x-7}{4}$

5. $\dfrac{3x-2}{5}=\dfrac{2x-7}{8}$

6. $\dfrac{2x-1}{3}=\dfrac{7x+2}{8}$

7. $\dfrac{3x-5}{4}+\dfrac{x-1}{3}=6$

8. $\dfrac{5x-2}{6}+\dfrac{7-3x}{5}=2$

9. $\dfrac{x-3}{6}+\dfrac{3x-1}{8}=1$

10. $\dfrac{6x-5}{3}+\dfrac{4-3x}{6}=5$

11. $\dfrac{4n-3}{5}+\dfrac{3-2n}{4}=3$

12. $\dfrac{2n-3}{6}+\dfrac{5-4n}{9}=2$

13. $\dfrac{2t-1}{4}+\dfrac{5t-3}{6}=4$

14. $\dfrac{7t-3}{6}+\dfrac{5-4t}{2}=5$

15. $\dfrac{2x+1}{5}-x=\dfrac{4-3x}{4}-2$

16. $\dfrac{3x+5}{4}+1=\dfrac{x-2}{3}+x$

17. $\dfrac{2t+1}{5}-2+\dfrac{4t-3}{15}=t$

18. $\dfrac{3t-5}{10}+3=\dfrac{4t-7}{5}-t$

19. $\dfrac{4x-5}{7}+2=\dfrac{x-3}{4}+x$

20. $\dfrac{5x-4}{9}+3=\dfrac{x+1}{18}+x$

21. $\dfrac{6x-7}{8}+1=\dfrac{2x-5}{12}+x$

22. $\dfrac{4x-3}{15}+1=\dfrac{3x-1}{5}-x$

23. $\dfrac{4t-1}{7}+1=\dfrac{t-3}{2}-t$

24. $\dfrac{4t+5}{9}-2=\dfrac{2-5t}{12}+t$

25. $0.7x-8.4=0.2x+31.6$

26. $0.05x+6.2=0.13x+3$

27. $0.4x-7=0.36x-4.6$

28. $0.05x+0.07(4000-x)=234$

29. $0.06x+0.08(5600-x)=380$

30. $0.6x-10.8=0.16x+9$

25.5. A FRACTION AS AN INDICATED QUANTITY

A fraction should always be considered as a quantity. If a fraction is preceded by a minus sign (−), this sign makes the entire fraction negative. The minus sign before the fraction should be treated in the same way as a minus sign before parentheses. As an example, consider the following equation.

Example 1. Solve for x: $\quad \dfrac{4x-5}{3}-\dfrac{x-1}{2}=8$

Solution. Notice the minus sign before the second fraction. When we multiply both sides of the equation by 6, the LCD, it is well to keep each binomial numerator in parentheses to indicate a quantity. Multiplying both sides by 6, we get

$$6\left(\dfrac{4x-5}{3}\right)-6\left(\dfrac{x-1}{2}\right)=6(8)$$

This first step can be omitted if we remember that the denominators disappear. Then we get

$$2(4x-5)-3(x-1)=48$$

Removing parentheses, $\quad 8x-10-3x+3=48$
Transposing, $\quad 8x-3x=48+10-3$
Combining like terms, $\quad 5x=55$
Dividing both sides by 5, $\quad x=11$

Example 2. Solve the equation

$$\frac{3x-1}{4}-\frac{2x-3}{5}-\frac{4x-5}{20}=2.$$

Solution. Note the minus signs before the second and third fractions. The third fraction is most likely to cause an error, because its denominator is itself the lowest common denominator. Then be especially careful. We multiply both sides of the equation, and omit the first step. We get, multiplying both sides by 20,

$$5(3x-1)-4(2x-3)-(4x-5)=40$$

Removing parentheses, $\quad 15x-5-8x+12-4x+5=40$
Transposing and combining
  like terms, $\quad 3x=28$
Dividing both sides of
  the equation by 3, $\quad x=\dfrac{28}{3}$, or $9\frac{1}{3}$.

Example 3. Solve the equation

$$\frac{5x-3}{4}-\frac{3x-7}{5}-\frac{4x-5}{20}=0$$

Solution. We need not show the multiplier, 20, if we remember what happens to the denominators. Omitting the first step, we get

$$5(5x-3)-4(3x-7)-(4x-5)=0$$

Removing parentheses, $\quad 25x-15-12x+28-4x+5=0$
Transposing, $\quad 25x-12x-4x=15-28-5$
Combining like terms, $\quad 9x=-18$
Dividing both sides by 9, $\quad x=-2$

The check is left for the student.

Example 4. Solve the equation:

$$\frac{2x+5}{3} - \frac{4x+3}{5} - \frac{3x-2}{15} = 3$$

Solution. Multiplying both sides of the equation by 15, the LCD, we get

$$5(2x+5) - 3(4x+3) - (3x-2) = 45$$

Removing parentheses, $\quad 10x + 25 - 12x - 9 - 3x + 2 = 45$

Transposing and combining like terms, $\quad -5x = 27$

Dividing both sides by -5, we get the root, $\quad x = -5.4$

Exercise 25.4

Watch the minus signs before fractions in solving these equations.

1. $\dfrac{3x+4}{5} - \dfrac{2x-5}{3} = 2$
2. $\dfrac{4x+3}{5} - \dfrac{2x-1}{3} = 4$
3. $\dfrac{5x-2}{6} - \dfrac{4x-7}{5} = 1$
4. $\dfrac{6x+5}{7} - \dfrac{3x-2}{4} = 1$
5. $\dfrac{2t+3}{4} - \dfrac{3t-1}{5} = 1$
6. $\dfrac{2t+5}{4} - \dfrac{3t-1}{5} = 2$
7. $\dfrac{4n+3}{2} - \dfrac{5n-1}{3} = 3$
8. $\dfrac{2n+5}{3} - \dfrac{4n-1}{7} = 1$
9. $\dfrac{3x-2}{6} - \dfrac{2x-1}{9} = 2$
10. $\dfrac{5x-4}{10} - \dfrac{4x-1}{15} = 2$
11. $\dfrac{x-3}{4} - \dfrac{2x-1}{3} - \dfrac{5x-4}{6} = 4$
12. $\dfrac{3x-4}{6} - \dfrac{2x-3}{3} - \dfrac{2x-1}{9} = 2$
13. $\dfrac{5n-1}{9} - \dfrac{2n-5}{3} - \dfrac{3n-4}{18} = 4$
14. $\dfrac{2n-3}{3} - \dfrac{4n-1}{5} - \dfrac{2n-7}{15} = 5$
15. $\dfrac{3x-2}{4} - \dfrac{2x-1}{3} - \dfrac{4x-5}{12} = 2$
16. $\dfrac{4x-1}{5} - \dfrac{2x-3}{2} - \dfrac{x-2}{10} = 3$
17. $\dfrac{3x-4}{4} - \dfrac{2x-7}{5} - \dfrac{5x-4}{20} = 0$
18. $\dfrac{5x-2}{6} - \dfrac{2x-3}{4} - \dfrac{3x-5}{24} = 0$

25.6. FRACTIONAL EQUATIONS WITH VARIABLE DENOMINATORS

A fractional equation may have the variable x in the denominator. Such equations are solved in the same way as other fractional equations. We multiply both sides of the equation by the lowest common denominator (LCD).

Example 1. Solve the equation

$$\frac{4}{x} - \frac{3}{x^2} = \frac{5}{2x}$$

Solution. In this example, the lowest common denominator is $2x^2$. Multiplying both sides by $2x^2$, we get

$$2x^2\left(\frac{4}{x}\right) - 2x^2\left(\frac{3}{x^2}\right) = 2x^2\left(\frac{5}{2x}\right)$$

The equation reduces to $\qquad 8x - 6 = 5x$
Transposing and combining like terms, $\qquad 3x = 6$
Then $\qquad x = 2$

To check the answer, we substitute 2 for x in the *original fractional equation*:

$$\frac{4}{2} - \frac{3}{4} = \frac{5}{4}, \qquad \text{which is true.}$$

Example 2. Solve the equation

$$\frac{3}{x} = \frac{7}{x+8}$$

Solution. The LCD is $x(x+8)$.
Multiplying both sides by $x(x+8)$, we get $\qquad 3(x+8) = 7x$
Removing parentheses and transposing, $\qquad 24 = 4x$
Dividing both sides by 4, $\qquad 6 = x$

Example 3. Solve the equation

$$\frac{2x}{x-3} - \frac{4}{x} = 2$$

Solution. Multiplying both sides by $x(x-3)$, we get

$$x(x-3)\left(\frac{2x}{x-3}\right) - x(x-3)\frac{4}{x} = x(x-3)(2)$$

376 ALGEBRA

The equation reduces to $\qquad 2x^2 - 4(x-3) = 2x(x-3)$
Removing parentheses, $\qquad 2x^2 - 4x + 12 = 2x^2 - 6x$
Solving the equation, we get $\qquad x = -6$

Note. *In this example, we first get a term containing x^2. However, this term drops out in the solution.*

Example 4. Solve the equation

$$\frac{4}{x} - \frac{2}{x-1} = \frac{5}{x^2 - x}$$

Solution. Notice that the third denominator is the product of the other denominators. Therefore, the LCD is the quantity $x(x-1)$ or $x^2 - x$.

Multiplying by the LCD, we get $\qquad 4(x-1) - 2x = 5$
Removing parentheses, $\qquad 4x - 4 - 2x = 5$
Transposing and combining like terms, $\qquad 2x = 9$
Dividing both sides by 2, we get $\qquad x = 4.5$

Example 5. Solve the equation

$$\frac{5x}{x+3} - \frac{2x}{x-2} = 3$$

Solution. Multiplying both sides by $(x+3)(x-2)$,

$$5x(x-2) - 2x(x+3) = 3(x+3)(x-2)$$

Removing parentheses, $\qquad 5x^2 - 10x - 2x^2 - 6x = 3x^2 + 3x - 18$
Transposing and combining, $\qquad -19x = -18$
Dividing both sides by -19, $\qquad x = \dfrac{18}{19}$

The best way to check the answer is to go over the work again.

Example 6. Solve the equation

$$\frac{3x}{x-3} - \frac{x+1}{x-4} - \frac{x+6}{x^2 - 7x + 12} = 2$$

Solution. Note that the third denominator is the product of the other denominators. Then the LCD is the quantity $(x-3)(x-4)$, or $x^2 - 7x + 12$.

Multiplying by the LCD,
$$3x(x-4)-(x+1)(x-3)-(x+6)=2(x^2-7x+12)$$
Removing parentheses, $\quad 3x^2-12x-x^2+2x+3-x-6=2x^2-14x+24$
Transposing and combining, $\qquad\qquad\qquad\qquad\qquad 3x=27$
Dividing both sides by 3, $\qquad\qquad\qquad\qquad\qquad\quad x=9$

Example 7. Solve the equation

$$\frac{3}{x+2}=\frac{1}{x-2}-\frac{4}{x^2-4}$$

Solution. Multiplying both sides of the equation by the LCD, x^2-4, we get

$$3(x-2)=(x+2)-4$$

Removing parentheses, $\quad 3x-6=x-2$
Transposing and combining, $\quad 2x=4$
Dividing both sides by 2, $\qquad x=2$

The answer, $x=2$, seems like a very logical answer to the problem. The procedure is entirely correct. Yet, when we check the answer in the original equation, we get a zero (0) in the denominator. In fact, we get a zero in two denominators. Therefore, *the equation has no solution.*

Note. Whenever the check of a solution produces a denominator of zero (0), that solution must be discarded.

The equation, therefore, has no solution. If we carelessly say that a number divided by zero is equal to infinity, we are no better off, because we cannot check an answer with infinity.

Exercise 25.5

Solve the following fractional equations:

1. $\dfrac{x+1}{x-3}=3$
2. $\dfrac{y-2}{y-6}=5$
3. $\dfrac{n+4}{n-1}=2$
4. $\dfrac{2n+1}{4n-3}=\dfrac{3}{5}$
5. $\dfrac{2x-3}{4x-5}=\dfrac{2}{5}$
6. $\dfrac{2t+3}{3t+2}=\dfrac{5}{6}$
7. $\dfrac{5}{x-4}=\dfrac{3}{x-2}$
8. $\dfrac{5}{x-2}=\dfrac{7}{x-4}$

ALGEBRA

9. $\dfrac{4}{3x+2} = \dfrac{3}{2x+1}$

10. $\dfrac{2}{x-4} = \dfrac{4}{x-3}$

11. $\dfrac{5}{4x+1} = \dfrac{2}{3x-2}$

12. $\dfrac{6}{2x-3} = \dfrac{5}{3x+1}$

13. $\dfrac{4}{x-3} = \dfrac{6}{x-2}$

14. $\dfrac{3}{x-5} = \dfrac{7}{x+3}$

15. $\dfrac{2}{x-6} = \dfrac{5}{x+1}$

16. $\dfrac{x-6}{x-2} = \dfrac{7}{3}$

17. $\dfrac{2t+3}{3t-2} = \dfrac{5}{6}$

18. $\dfrac{3x-2}{4x+1} = \dfrac{2}{7}$

19. $\dfrac{2}{x} + \dfrac{3}{x+2} = \dfrac{10}{3x}$

20. $\dfrac{5}{2x} - \dfrac{4}{3x} = \dfrac{2}{x-5}$

21. $\dfrac{2x}{x-3} - \dfrac{5}{x} = 2$

22. $\dfrac{2x-1}{x-2} = 3 - \dfrac{x-2}{x-3}$

23. $\dfrac{2x-3}{x+2} = 3 - \dfrac{x-2}{x-1}$

24. $\dfrac{x-3}{x+1} = 4 - \dfrac{3x-2}{x+3}$

25. $\dfrac{6}{x^2-9} + \dfrac{2x-1}{x-3} = 2 + \dfrac{2}{x+3}$

26. $\dfrac{3n-2}{n-3} = \dfrac{3n+1}{n+3} + \dfrac{5n-8}{n^2-9}$

27. $\dfrac{4n}{n+2} - \dfrac{2n-1}{n-3} - \dfrac{n-7}{n^2-n-6} = 2$

28. $\dfrac{5x}{x-4} - \dfrac{2x-1}{x-2} - \dfrac{x+8}{x^2-6x+8} = 3$

25.7. LITERAL EQUATIONS

A *literal equation* is an equation in which constants are represented by letters; that is, letters appear where we would expect arithmetic numbers. For example, this is a literal equation: $ax = b$. In this equation, we assume that x is the unknown, and we take the a and b to represent constants.

A literal equation is solved in the same way as any other equation. Let us first review a simple equation as those we have already solved.

Example 1. Solve for x: $3x = 17$.

Solution. To solve for x, we divide both sides of the equation by 3, the coefficient of x. We get

$$x = \dfrac{17}{3}$$

Now, suppose we have a literal equation in which we find letters in place of 3 and 17.

Example 2. Solve for x: $ax = b$.

Solution. Again, to solve for x, we divide both sides of the equation by a, the coefficient of x. We get

$$x = \frac{b}{a}$$

The equation is now solved for x. The root of the equation is b/a.

If an equation has several terms containing the unknown, x, or whatever unknown we wish to find, then all the terms containing the unknown are first isolated just as in any other equation. To review the procedure, let us first solve an equation such as those we have already solved. Note carefully the steps in solving this equation.

Example 3. Solve for x: $\quad 5x + 3 = 15 - 2x$.

Solution. We first transpose to isolate the terms containing x:

$$5x + 2x = 15 - 3$$

Combining like terms, $\quad 7x = 12$

Dividing both sides by 7, $\quad x = \dfrac{12}{7}$, the root of the equation.

Example 4. Solve this literal equation for x: $\quad ax - 3 = b - 6x$

Solution. Transposing to isolate x-terms: $\quad ax + 6x = b + 3$
We cannot combine the x-terms into a single term. However, by factoring the expression, $ax + 6x$, into two factors, $x(a+6)$, we see that the coefficient of x is the binomial $(a+6)$. Therefore, we divide both sides of the equation by the quantity $(a+6)$, and get the value of x, which is the root of the equation:

$$x = \frac{b+3}{a+6}$$

Example 5. Solve the following formula for n:

$$I = \frac{nE}{R + nr}$$

Solution. As in any fractional equation, we multiply both sides of the equation by the LCD, in this case $(R + nr)$. This clears the equation of fractions:

$$I(R + nr) = nE$$

Removing parentheses, $\quad IR + nrI = nE$
Transposing to isolate n-terms, $\quad IR = nE - nrI$
Factoring to show the coefficient of n, $\quad IR = n(E - rI)$
Dividing both sides by $(E - rI)$, $\quad \dfrac{IR}{E - rI} = n;\quad$ or $\quad n = \dfrac{IR}{E - rI}$

Note. Any literal equation may be solved for any letter we wish by isolating the terms containing that particular letter and then dividing both sides of the equation by its coefficient. This process is called "changing the subject of the formula," and is important in much work in connection with formulas.

Exercise 25.6

Solve each formula for the letter indicated at the right of each:

1. $V = LWH \quad (W)$
2. $V = \pi r^2 h \quad (h)$
3. $A = 2\pi rh \quad (r)$
4. $t = \dfrac{d}{r} \quad (r)$
5. $A = \dfrac{ab}{2} \quad (b)$
6. $I = \dfrac{E}{R + r} \quad (R)$
7. $\dfrac{1}{f} = \dfrac{1}{a} + \dfrac{1}{b} \quad (a)$
8. $\dfrac{W_1}{W_2} = \dfrac{L_2}{L_1} \quad (W_2)$
9. $F = \dfrac{9}{5}C + 32 \quad (C)$
10. $Z_t = \dfrac{Z_1 Z_2}{Z_1 + Z_2} \quad (Z_2)$
11. $S = \dfrac{a - rl}{1 - r} \quad (r)$
12. $\dfrac{1}{t} = \dfrac{1}{a} + \dfrac{1}{b} + \dfrac{1}{c} \quad (c)$
13. $A = \dfrac{h}{2}(B + b) \quad (B)$
14. $S = \dfrac{N}{2}(A + L) \quad (A)$
15. $C = \dfrac{5}{9}(F - 32) \quad (F)$
16. $V = \dfrac{1}{3}\pi r^2 h \quad (h)$
17. $\dfrac{E}{e} = \dfrac{R + r}{r} \quad (r)$
18. $F = k\dfrac{M_1 M_2}{d^2} \quad (M_1)$
19. $l = a + (n - 1)d \quad (n)$
20. $A = 2\pi r(h + r) \quad (h)$
21. $V = \pi R^2 h - \pi r^2 h \quad (h)$
22. $S = vt - \dfrac{1}{2}gt^2 \quad (v)$

23. Formula 1 is used to find the volume V of a rectangular solid, given the length L, the width W, and the height H. After solving the formula for W, find W when $H = 15$ in., $L = 24$ in., and $V = 3024$ cu. in.

FRACTIONAL EQUATIONS 381

24. Formula 2 is used to find the volume of a right circular cylinder, given the radius r, and the altitude h. Use the formula solved for h to find h when $r = 6$ in., and $V = 900\pi$ cu. in.

25. Formula 3 is used to find the lateral area of a right circular cylinder, given the radius r and the altitude h. Use the formula solved for h to find h when $r = 8$ in. and $A = 96\pi$ sq. in.

26. Formula 4 is used to find the time t when the distance d and the rate r are known. After solving for r, find r when $d = 280$ miles and $t = 5.5$ hours.

27. Formula 5 is used to find the area of a triangle when the base b and the altitude a are known. After solving for a, find a when $A = 85.8$ sq. in. and $b = 5.2$ in.

28. Formula 6 is used to find the current I in an electric circuit when the voltage E, the external resistance R, and the internal resistance r, are known. After solving for r, find r when $E = 39$ volts, $I = 1.2$ amperes, and $R = 32$ ohms.

29. Formula 7 is used to express the relation between the focal length f of a lens, the distance a to an object, and the distance b to the image. Use the formula solved for b to find b when $a = 60$ in. and $f = 10$ in.

30. Formula 8 expresses the relation between the two weights and the two arm lengths of a balanced lever. After solving for W_2, find W_2 when $L_1 = 18$ in., $L_2 = 12$ in., and $W_1 = 40$ pounds.

31. Formula 9 is used to change a Celsius (Centigrade) temperature reading to a Fahrenheit reading. After solving for C, find C when $F = 68°$.

32. Formula 11 is used to find the sum S of a geometric series, where a is the first term, l is the last term; and r is the common ratio. After solving for r, find r when $a = 2$, $l = 486$, and $S = 728$.

33. Formula 13 is used to find the area A of a trapezoid when the altitude h, and the bases, B and b, are known. After solving for b, find b when $B = 17.5$ cm, $h = 8$ cm, and $A = 104$ sq. cm.

34. Formula 16, is used to find the volume V of a circular cone, when the radius r and altitude h are known. After solving for h, find h when $r = 6$ in., and $V = 240\pi$ cu. in.

35. Formula 19 is used to find the last term l of an arithmetic progression in which a is the first term, n is the number of terms, and d is the common difference. Using the formula solved for n, find n when $a = 5$, $d = 1.5$, and the last term is 41.

36. Formula 20 is used to find the total area of a right circular cylinder, in which r is the radius of the circular end and h is the altitude. After solving for h, find h when $r = 6$ in. and $A = 312\pi$ sq. in.

37. Formula 21 is used to find the volume of material in a hollow right circular cylinder, in which R is the radius of the outside circle, r is the radius of the inside circle, and h is the altitude. After solving for h, find h when $R = 8$ in., $r = 6$ in., and volume $V = 490\pi$ cu. in.

25.8. STATED PROBLEMS INVOLVING FRACTIONAL EQUATIONS

In solving practical problems, we often get fractional equations, as shown in the following examples. Such equations are then solved by the methods explained in this chapter.

Example 1. A wire 72 inches long is to be bent into the shape of a rectangle. If the width of the rectangle is to be $\frac{3}{5}$ of the length, find the width and the length of the rectangle.

Solution. We usually let x represent the smaller number. But in this case, it is better to let it represent the length.
  Let $x =$ the number of inches in the length.
  Then $\frac{3x}{5} =$ the number of inches in the width.

For the equation, we have	$2x + 2\left(\frac{3x}{5}\right) = 72$ (2 ends plus 2 sides)
Multiplying both sides by 5,	$10x + 6x = 360$
Combining like terms:	$16x = 360$
Dividing both sides by 16,	$x = 22.5$ (inches in length)
For the width, we have,	$\frac{3}{5}(22.5) = 13.5$ (inches in width)

Example 2. A farmer has 80 acres that he wishes to divide into two fields so that one field shall be $\frac{2}{3}$ as large as the other. Find the size of each field.

Solution. Let $x =$ the number of acres in the larger field.
  Then $\frac{2x}{3} =$ the number of acres in the smaller field.

For the equation, we have	$x + \frac{2x}{3} = 80$
Multiplying both sides by 3,	$3x + 2x = 240$
Combining like terms,	$5x = 240$
Dividing both sides by 5,	$x = 48$ (acres in larger field)
For the smaller field, we have	$\frac{2}{3}(48) = 32$ (acres in smaller field)

Example 3. A man has $6000 invested, part at a rate of return of 6% a year, the rest at 8% a year. If his total income from these two investments is $394 a year, how much has he invested at each rate? What is the return on each?

Solution. Let $x =$ amount ($) invested at 6%.
  Then $6000 - x =$ amount ($) invested at 8%.
The income on an investment is found by multiplying the rate times the money invested, that is,

$$\text{income} = (\text{rate}) \times (\text{money invested})$$

The income on x dollars at 6% is $0.06x$; on $(6000 - x)$ dollars at 8%, the

FRACTIONAL EQUATIONS

income is $0.08(6000-x)$. The total income is $394. Then we have the equation

$$0.06x + 0.08(6000-x) = 394$$

Multiplying both sides by 100, $\quad 6x + 8(6000-x) = 39400$
Removing parentheses, $\quad 6x + 48000 - 8x = 39400$
Transposing and combining
like terms, $\quad -2x = -8600$
Dividing both sides by -2, $\quad x = 4300 \text{ (dollars at 6\%)}$
For the amount invested at 8%,
we have $\quad 6000 - 4300 = 1700$
For the return on each investment,
we have $\quad$ 6% of $4300 = $258
$\quad$ 8% of $1700 = $136
$\quad$ Total yearly income $= $394

Example 4. The denominator of a fraction is 3 more than the numerator. If the numerator is increased by 2 and the denominator increased by 5, the new fraction is equal to $\frac{3}{5}$. Find the numerator and denominator of the original fraction.

Solution. Let $x =$ the numerator of the original fraction.

Then $x + 3 =$ the denominator. Then the fraction is shown by $\dfrac{x}{x+3}$.

When the numerator is increased by 2 and the denominator is increased by 5, the new fraction becomes

$$\frac{x+2}{x+8}$$

We set the new fraction equal in value to $\frac{3}{5}$:

$$\frac{x+2}{x+8} = \frac{3}{5}$$

Clearing the equation of fractions, we get

$$5(x+2) = 3(x+8)$$

Solving the equation, we get $x = 7$; then the fraction is $\frac{7}{10}$.

Example 5. Tom can mow a certain lawn in 6 hours working alone. Dick can mow the same lawn in 4 hours working alone. If each gets a mower and both work at the same time, how long should it take them to mow the lawn?

Solution. First we consider how much of the lawn can be mowed in 1 hour.

In 1 hour, Tom can do $\frac{1}{6}$ of the work.
In 1 hour, Dick can do $\frac{1}{4}$ of the work.

Let $x=$ the number of hours it should take them working together.

Then $1/x =$ the part of the work both can do in 1 hour. That is, the part done in 1 hour is simply the reciprocal of the number of hours. For the equation, we have

$$\frac{1}{6}+\frac{1}{4}=\frac{1}{x}$$

The equation shows the total amount of work done in 1 hour by both.
Multiplying both sides of the equation by $12x$, $\quad 2x+3x=12$
Combining like terms, $\quad 5x=12$
Dividing both sides of the equation by 5, $\quad x=\dfrac{12}{5}$, or 2.4

Then the number of hours required when both work is 2.4.

Note. *In solving a work problem of this kind, we must assume that the two work at the same rate as when each is working alone. We cannot make allowance for their getting into each other's way, which actually might happen. We must assume that the work goes along steadily.*

Example 6. The difference between two numbers is 5. If 2 times the smaller is divided by the larger, the quotient has the value $\frac{6}{5}$. Find each number.

Solution. Let $x=$ the smaller number.
Then $\quad x+5=$ the larger number.
For the equation, we have

$$\frac{2x}{x+5}=\frac{6}{5}$$

Multiplying both sides by $5(x+5)$, $\quad 10x=6(x+5)$
Removing parentheses, $\quad 10x=6x+30$
Transposing and combining, $\quad 4x=30$
Dividing both sides by 4, $\quad x=7.5$, the smaller number
For the larger number, we have $\quad 7.5+5=12.5$, the larger number.

Example 7. A solution of 30 gallons of alcohol and water contains 8% alcohol. How much pure alcohol should be added to the 30 gallons to produce a new solution testing 25% alcohol? (Disregard the very slight solubility of alcohol in water.)

Solution. In a problem of this kind, keep in mind an important principle: The total amount of alcohol in the final solution will be equal to the amount of alcohol in the original solution plus the amount of alcohol added. That is,

$$\left\{\begin{matrix}\text{alcohol in}\\\text{original solution}\end{matrix}\right\} + \left\{\begin{matrix}\text{amount of}\\\text{alcohol added}\end{matrix}\right\} = \left\{\begin{matrix}\text{total alcohol in}\\\text{final solution.}\end{matrix}\right\}$$

The original solution contains 8% of 30 gallons or 2.4 gallons of alcohol.
  Let x = the number of gallons of pure alcohol to be added.
Then $30 + x$ = the total number of gallons in the final solution.
  Now, 2.4 = number of gallons of alcohol in the original solution.
  Then $2.4 + x$ = number of gallons of alcohol in the final solution.
  The $(2.4 + x)$ gallons of alcohol must be 25% of the final solution. That is,

$$2.4 + x = 0.25(30 + x)$$

Multiplying both sides by 100, $240 + 100x = 25(30 + x)$
or, $240 + 100x = 750 + 25x$
Transposing and combining, $75x = 510$
Dividing both sides by 75, $x = 6\frac{4}{5}$, or 6.8

If we add 6.8 gallons of pure alcohol, we shall find that the final solution tests 25% alcohol. The new solution will contain 9.2 gallons of alcohol, which is 25% of the total solution, 36.8 gallons.

Example 8. How much of an alloy of silver containing 15% silver and how much of an alloy of 90% silver should be mixed together to produce 24 ounces of a new alloy testing 35% silver?

Solution. Let x = number of ounces of the 15% alloy.
  Then $24 - x$ = number of ounces of the 90% alloy.
  The amount of actual silver in each alloy is as follows:
In x ounces of the 15% alloy, there are $0.15x$ ounces of silver
In $(24 - x)$ ounces of the 90% alloy, there are $0.90(24 - x)$ ounces of silver. The total amount of silver in the new alloy must be $0.35(24) = 8.4$ ounces.
Then we have the equation: $0.15x + 0.90(24 - x) = 8.4$
Multiplying both sides by 100, $15x + 90(24 - x) = 840$
Removing parentheses, $15x + 2160 - 90x = 840$
Transposing and combining like terms, $1320 = 75x$
Dividing both sides by 75, $17.6 = x$

Then we have 17.6 ounces of the 15% alloy; 6.4 ounces of the 90% alloy.

Example 9. A man goes into a store with some money. He spends $\frac{3}{4}$ of his money for a suit, $\frac{1}{10}$ for a hat, and has $24 left. How much did he have at first, and how much was spent for each item?

Solution. Let $x =$ the amount of money ($) he had at first. The amount spent for a suit was $(\frac{3}{4})x$; for the hat he spent $(\frac{1}{10})x$. Now we add the amounts spent and the amount left. The total is x dollars.

Equation: $$\frac{3x}{4} + \frac{x}{10} + 24 = x$$

Multiplying both sides by 20, $\quad 15x + 2x + 480 = 20x$
Transposing and combining, $\quad -3x = -480$
Dividing both sides by -3, $\quad x = 160$ (dollars he had)

The suit cost $(\frac{3}{4})(160) = \$120$; the hat cost $(\frac{1}{10})(\$160) = \$16$. To check the solution, we add: $\$120 + \$16 + \$24 = \$160$.

Exercise 25.7

1. A man went into a store with a certain amount of money. He spent $\frac{5}{8}$ of his money for a suit, $\frac{1}{6}$ of his money for a hat, and $\frac{1}{16}$ of his money for gloves. If he had $17.50 left, how much had he at first?

2. A lady went to a store with a certain amount of money. She spent $\frac{5}{9}$ of her money for a dress, $\frac{1}{6}$ of her money for a hat, $\frac{1}{12}$ of her money for gloves, and had $26.25 left. How much had she at first?

3. The width of a certain rectangle is $\frac{2}{3}$ of the length. If the perimeter of the rectangle is 75 inches, find the length, the width, and the area.

4. The width of a rectangle is $\frac{3}{5}$ of the length. If the perimeter is 76 inches, find the length, the width, and the area.

5. A certain rectangle is $\frac{3}{5}$ as wide as it is long. If the width is increased by 2 inches and the length decreased by 3 inches, the area is the same as before. Find the dimensions of the original rectangle. What change takes place in the perimeter with the changes in width and length?

6. The width of a certain rectangle is $\frac{2}{3}$ of the length. If the width is decreased by 3 and the length increased by 3, the area of the new rectangle is 36 square inches less that the original rectangle. Find the dimensions of the original rectangle. What change takes place in the perimeter?

7. The denominator of a certain fraction is 6 more than the numerator. If the numerator is increased by 7 and the denominator decreased by 3, the value of the new fraction is $\frac{5}{4}$. Find the original fraction.

8. The denominator of a certain fraction is 4 more than the numerator. If 3 is subtracted from the numerator and added to the denominator, the value of the new fraction is $\frac{4}{9}$. Find the original fraction.

9. What number can be subtracted from numerator and denominator of the fraction $\frac{8}{11}$ so that the value of the new fraction will be $\frac{3}{5}$?

10. If a certain number is subtracted from the numerator, and twice the number subtracted from the denominator of the fraction $\frac{19}{30}$, the value of the new fraction is $\frac{2}{3}$. What is the certain number?

11. What number must be subtracted from the numerator and the denominator of the fraction $\frac{12}{7}$ so that the value of the new fraction will be $\frac{3}{4}$?

12. What number must be added to the numerator and the denominator of the fraction $\frac{5}{7}$ so that the value of the new fraction will be $\frac{5}{3}$?

13. A man has $8000 invested, part of it at an annual return of 6%, and the rest at 8%. If his total income from these two investments is $544 a year, find the amount invested at each rate, and the income from each.

14. A man has $8400 invested, part at 5% and the rest at 7% return per year. If the income from both investments is the same, find the amount invested at each rate, and the return on each investment.

15. Frank can mow a lawn in 8 hours working alone. Joe can mow the same lawn alone in 6 hours. If each gets a lawn mower and both work at the rate each works when working alone, how long will it take both to mow the lawn?

16. A water tank can be filled through one opening in 8 hours. The same tank can be filled through a second opening alone in 12 hours. If both openings are used, how long will it take to fill the tank?

17. A swimming pool can be filled through one opening alone in 12 hours. It can be filled by a second opening alone in 9 hours. A third opening can fill the same pool in 15 hours. How long will it take to fill the pool if all three openings are used?

18. A farmer has an orchard containing plum, apple, and cherry trees. The number of plum trees is $\frac{1}{3}$ of the total number, $\frac{2}{5}$ of the total are apple trees, and the remainder, which is 48 trees, are cherry. Find the total number of trees in the orchard.

19. A farmer has a tract of his land planted to various crops. Of the total number of acres, $\frac{4}{15}$ is planted in corn, $\frac{1}{6}$ is in wheat, $\frac{1}{3}$ is planted in oats, and the remainder, which is 28 acres, is in pasture. Find the total number of acres and the number used for each crop.

20. How much pure alcohol must be added to 20 gallons of an 8% solution to make the new solution test 25% alcohol?

21. How much water must be added to 12 quarts of a 40% solution to make the new solution test only 15%?

22. How much of a 5% solution of some concentrate and how much of a 80% solution must be used to make 24 gallons of a solution testing 30%?

23. How much of a 12% solution and how much of a 60% solution must be mixed to make a new solution of 45 quarts testing 20%?

24. How much of a 10% alloy of silver and how much of a 90% alloy of silver must be mixed together for 48 ounces of an alloy containing 35% silver?

26 SYSTEMS OF EQUATIONS

26.1. INDETERMINATE EQUATIONS

Let us review a kind of equation we have already solved. Up to this point, we have solved equations that contain only one letter, or unknown, such as the following:

$$2x + 5 = 17$$

Transposing the +5, we get $\quad 2x = 17 - 5$

Combining the numbers on the right, $\quad 2x = 12$

Dividing both sides by 2, $\quad x = 6$

This equation can be solved, and we get the solution: $x = 6$.

Now, it sometimes happens that we find it convenient to set up an equation that contains two letters, or unknowns. As an example, consider the following problem:

Suppose you buy 2 apples and 3 oranges, and the cost is 77 cents. How can you tell the price of each apple and of each orange? In this problem we cannot be sure of the price of each. If an apple costs 7 cents, then we find that an orange costs 21 cents. But the price might be different. If one apple costs 10 cents, then the price of an orange is 19 cents. If an apple costs 16 cents, then one orange costs 15 cents. From the one statement alone, we cannot tell definitely the price of each.

Let us represent the cost by an equation. We use one letter to represent the price of 1 apple (in cents); we use another letter to represent the price of 1 orange (in cents). First we write:

let x = the number of cents one apple costs
let y = the number of cents one orange costs.

Then 2 apples will cost $2x$ cents; and 3 oranges will cost $3y$ cents. Now we

389

can write an equation stating that 2 apples and 3 oranges cost 77 cents:

$$2x + 3y = 77$$

In this equation, how can we tell the value of *x* and the value of *y*? There may be several possible answers:

if $x = 7$, $y = 21$;   if $x = 10$, $y = 19$;   if $x = 16$, $y = 15$

This problem and the equation show that when we have two letters (unknowns) in an equation, we cannot tell what value each letter must have. There may be many sets of answers.

Definition. An equation, such as $2x + 3y = 77$, is called an indeterminate equation because the value of one letter cannot be determined independently of the other letter.

In an indeterminate equation, containing two letters, the letters may have various values. The letters are called *variables*.

26.2. A SYSTEM OF EQUATIONS

Consider again the problem in the foregoing section. The cost of 2 apples and 3 oranges is 77 cents. We express the cost by an equation:

$$2x + 3y = 77$$

Now, suppose you go to the same store and buy 1 apple and 3 oranges and the cost is 64 cents. We can write an equation for this purchase:

$$x + 3y = 64$$

The equation is almost the same as the first. The only difference is that in the first purchase you bought one more apple. Then that extra apple must have cost the extra money; that is, from 64 cents to 77 cents. From this information we know that the price of 1 apple is 13 cents. Then we can find the price of 1 orange; it is 17 cents.

If we consider the two equations together, then both equations must be true for the same values of the letters. Then we have a *system of equations*.

Definition. A system of equations is a set of two or more equations that contain two or more letters, such that the two or more equations must be true for the same values of the letters in all the equations.

A system of equations is sometimes called a set of *simultaneous equations*.

26.3. SOLVING A SYSTEM OF EQUATIONS BY ADDING OR SUBTRACTION

In the foregoing section, we were able to tell the values of x and y by inspection. However, we need a systematic method for solving such equations.

There are several methods for solving systems of equations. One of the most useful and convenient is the method called *addition or subtraction*. Consider again the two equations:

$$2x + 3y = 77$$
$$x + 3y = 64$$

In a system of equations, one equation may be subtracted from the other. Subtracting the bottom equation from the top, we get

$$x = 13$$

The result immediately gives us the value of one of the letters. Notice that the *y-terms* have been *eliminated*. When we have found the value of one letter, we can use this value to find the value of the other. When we find that $x = 13$, we substitute 13 for x in one of the equations to find the value of y. Substituting 13 for x in the second equation, we get

$$13 + 3y = 64$$

Transposing, $\qquad\qquad\qquad 3y = 64 - 13$

Combining like terms, $\qquad\quad 3y = 51$

Dividing both sides by 3, $\qquad y = 17$

To check the answers, we put 13 for x, and 17 for y, in both equations. Then we find that both equations are true. Then we have the solution of the system: $x = 13$; $y = 17$.

Sometimes one letter can be eliminated by adding two equations. Suppose we have the system:

$$x + y = 13$$
$$x - y = 2$$

Adding the two equations, we get $\;\; 2x = 15$
Dividing both sides by 2, $\qquad\qquad\;\; x = 7.5$

Substituting 7.5 for x in the first equation, we get

$$7.5 + y = 13$$

Then

$$y = 5.5$$

In solving a system of equations by the method of addition or subtraction, *we seek to eliminate one of the letters.* This can happen only if the coefficients of one letter are numerically equal. If the coefficients are equal and have the same sign, then we *subtract* one equation from the other.

Example 1. Solve the system

$$3x + 7y = 27$$
$$3x + 5y = 21$$

Solution. Notice that the coefficients of x are numerically equal and have the same sign. Then, when we subtract one equation from the other, the x-*terms* will be eliminated. Subtracting the bottom equation from the top, we get

$$2y = 6$$

Dividing both sides of this equation by 2, $\quad y = 3$

Now we can find the value of x by putting 3 for y in one of the equations. Then we get the value of x: $x = 2$. The solution is: $x = 2$; $y = 3$.

Example 2. Solve the system

$$7x - 3y = 29$$
$$4x - 3y = 14$$

Solution. Subtracting the bottom equation from the top, we get

$$3x = 15$$

Dividing both sides by 3, $\quad x = 5$

Substituting 5 for x in either equation we find that $y = 2$.

Example 3. Solve the system

$$7x + 2y = 22$$
$$5x - 2y = 26$$

SYSTEMS OF EQUATIONS 393

Solution. If the coefficients of one letter are numerically equal but *opposite in sign*, we *add the two equations* to eliminate one letter. Adding the two equations, we get

$$12x = 48$$

Dividing both sides of the equation by 12, $\quad x = 4$

Substituting 4 for x in either equation, we find that $y = -3$.

In some systems, the coefficients of a letter are not numerically equal, as in the next example. However, they can be made equal.

Example 4. Solve the system

$$5x + 3y = 21$$
$$2x - y = 4$$

Solution. Adding, we get

$$7x + 2y = 25$$

If we add these two equations, no letter is eliminated. We get the new equation, $7x + 2y = 25$. This operation of adding is perfectly legal, but it does not lead to a solution. If we subtract one equation from the other, we are no better off. The new equation will still contain both letters. In adding or subtracting equations, we seek to eliminate one letter. This can happen only if the *coefficients of one letter are equal numerically*.

In this system, we can make the y-coefficients numerically equal by multiplying the second equation by 3. The multiplication axiom says that we can multiply both sides of any equation by the same quantity. Then the second equation becomes

$$6x - 3y = 12$$

Writing the first equation, $\quad 5x + 3y = 21$

Adding the two equations, $\quad 11x = 33$
Dividing both sides by 11, $\quad x = 3;$ and $y = 2$

Example 5. Solve the system

$$4x - 3y = 34$$
$$x + 2y = 3$$

Solution.
Multiplying the second equation by 4, $\quad 4x + 8y = 12$
Rewriting the first equation, $\quad\quad\quad 4x - 3y = 34$

Subtracting, $\quad\quad\quad\quad\quad\quad\quad\quad\quad\quad 11y = -22$
Dividing both sides by 11, $\quad\quad\quad\quad\; y = -2;\quad$ then $x = 7$.

In some systems of equations, we must *multiply both equations*, each by a different number, so that the coefficients of one letter are equal numerically as in the following example.

Example 6. Solve the system

$$5x - 3y = 22$$
$$2x - 4y = 13$$

Solution. Now we seek to make the coefficients of one letter numerically equal. To do so, we can multiply the first equation by 4, and the second equation by 3. We get

$$\begin{aligned}4 \text{ times the first equation:} &\quad 20x - 12y = 88\\ 3 \text{ times the second equation:} &\quad\; 6x - 12y = 39\end{aligned}$$

Subtracting the equations, $\quad\quad\quad\quad 14x \quad\quad = 49$
Dividing both sides by 14, $\quad\quad\quad\quad\;\; x \quad\quad\quad = 3.5$

Now, instead of substituting 3.5 for *x*, to find the value of *y*, we can equalize the *x*-coefficients:

$$\begin{aligned}2 \text{ times the first equation:} &\quad 10x - \;\;6y = 44\\ 5 \text{ times the second equation:} &\quad 10x - 20y = 65\end{aligned}$$

Subtracting the equations, $\quad\quad\quad\quad +14y = -21$
Dividing both sides by 14, $\quad\quad\quad\quad\quad\;\; y = -1.5$

Summary of steps in solving systems of equations by addition or subtraction:

1. Multiply one or both of the given equations, if necessary, by some factor or factors, that will make the coefficients of one letter numerically equal.
2. Eliminate this variable (letter) by addition or subtraction.
3. Solve the resulting equation for one letter.
4. Find the value of the other letter by substituting the known value of the one letter in one of the equations. (The value of the second letter may be found in the same way as the first.)

Exercise 26.1

Solve the following systems by the method of addition or subtraction:

1. $x + y = 20$
   $x - y = 4$

2. $x + y = 12$
   $x - y = 2$

3. $a + b = 15$
   $a - b = 1$

4. $a - b = 13$
   $a + b = 3$

5. $m - n = 9$
   $m + n = 5$

6. $x - y = 7$
   $x + y = 40$

7. $2x + y = 12$
   $2x - y = 8$

8. $4x - y = 9$
   $4x + y = 15$

9. $3s - t = 14$
   $3s + t = 16$

10. $x + 3y = 15$
    $5x - 3y = 12$

11. $2a - 5y = 13$
    $3a + 5y = 7$

12. $m - 4n = 11$
    $3m + 4n = 1$

13. $3y + 5x = 18$
    $2x - y = 8$

14. $y + 2x = 9$
    $3x - 4y = 8$

15. $5x + y = 14$
    $4x - 3y = 15$

16. $7a - 5b = 23$
    $2a - b = 7$

17. $5c - 6d = 3$
    $3c - d = 7$

18. $5k + 8h = 19$
    $3h + 5k = 4$

19. $2y + x = 5$
    $4x + 5y = 23$

20. $3x - 5y = 9$
    $x - 4y = 10$

21. $5x - 3y = 12$
    $x - 2y = 7$

22. $4x - 3y = 26$
    $2y + 5x = 21$

23. $5x - 4y = 14$
    $2x - 3y = 14$

24. $x - 3y = 2$
    $7x - 5y = 22$

25. $7y + 9x = 17$
    $5x - 3y = 6$

26. $7x - 3y = 19$
    $5y + 3x = 5$

27. $5x - 3y = 6$
    $7x - 5y = 12$

28. $4y = 30 - 3x$
    $2x + 3 = 5y$

29. $4a = 23 - 5b$
    $3a + 6 = 4b$

30. $3y = 21 - 2x$
    $7x + 4 = 5y$

31. $3b = 10 - 5a$
    $4a = 12 - 5b$

32. $3c = 13 - 7d$
    $3d = 6 - 2c$

33. $3y = 9 - 5x$
    $6x = 5 - 7y$

34. $3x = 2y - 7$
    $4 + 6y = 5x$

35. $4y = 3x + 3$
    $8y + 9 = 5x$

36. $7 + 3y = 4x$
    $2 + 7x = 9y$

26.4. SOLVING SYSTEMS OF EQUATIONS BY SUBSTITUTION

Substitution is another method for solving systems of equations. This method can sometimes be used to advantage, although it is perhaps not so practical or convenient as the method of addition or subtraction. One advantage of the method of substitution is that it can be useful in the solution of higher degree equations.

In using the method of substitution, we follow these steps:

1. *Solve one of the equations for one letter in terms of the other.*
2. *Substitute the resulting expression for its equal in the other equation.*
3. *Solve the new equation for the value of one letter.*
4. *Find the value of the other letter by the usual method.*

Example 1. Solve this system by substitution:
$$5x + 3y = 13 \quad \text{(A)}$$
$$4x + y = 9 \quad \text{(B)}$$

Solution. Solving the second equation for y, we get $y = 9 - 4x$ (C)
Now we substitute the value, $9 - 4x$, for y in the *other* equation; we get

$$5x + 3(9 - 4x) = 13$$

Notice that one letter has been eliminated. Now we solve this equation:

$$5x + 27 - 12x = 13$$

Transposing and combining like terms, $\quad -7x = -14$
Dividing both sides of the equation by -7, $\quad x = 2$

To find the value of y, we can substitute the value, $x = 2$, in any one of the equations (A) or (B). We can also use equation (C). Then

$$y = 9 - 4(2) = 1$$

Example 2. Solve this system by substitution:
$$3x - 4y = 7$$
$$2x + y = 12$$

Solution.
Solving the second equation for y, we get $\quad y = 12 - 2x \quad$ (C)
Substituting $(12 - 2x)$ for y in the *other* equation, we get

$$3x - 4(12 - 2x) = 7$$

Now we solve this equation: $\quad 3x - 48 + 8x = 7$
Transposing and combining like terms, $\quad 11x = 55$
Dividing both sides by 11, $\quad x = 5.$

SYSTEMS OF EQUATIONS

Now we find the value of y by putting 5 for x in the equation (C); we get

$$y = 12 - 2(5) = 12 - 10 = 2$$

Example 3. Solve this system by substitution: $\begin{aligned} 4x - 3y &= 23 \\ x + 2y &= 3 \end{aligned}$

Solution.
Solving the second equation for x, we get $x = 3 - 2y$ (C)
Substituting $(3 - 2y)$ for x in the *other* equation, we get

$$4(3 - 2y) - 3y = 23$$

Now we solve this equation: $12 - 8y - 3y = 23$
Transposing and combining like terms, $-11y = 11$
Dividing both sides of the equation by -11, $y = -1$

Now we find the value of x by putting (-1) for y in equation (C); we get

$$x = 3 - 2(-1) = 3 + 2 = 5$$

Example 4. Solve this system by substitution: $\begin{aligned} 3x - 2y &= 7 \\ 4x - y &= 11 \end{aligned}$

Solution.
From the second equation, we get $-y = 11 - 4x$
However, we must first find the value
of $+y$, *not* $-y$: $+y = 4x - 11$ (C)
Now we substitute the value, $4x - 11$,
for y in the *other* equation; we get $3x - 2(4x - 11) = 7$
Now we solve this equation: $3x - 8x + 22 = 7$
Transposing and combining like terms, $-5x = -15$
Dividing both sides of the equation by -5, $x = 3$

To find the value of y, we put (3) for x in equation (C); we get

$$y = 4(3) - 11 = 1$$

Example 5. Solve this system by substitution: $\begin{aligned} 2x + y &= -3 \\ 3x - 2y &= 13 \end{aligned}$

Solution.
Solving the first equation for y, we get: $y = -2x - 3$ (C)

Now we substitute the value $(-2x-3)$ for y in the *other* equation; we get

$$3x - 2(-2x - 3) = 13$$

Solving this equation: $\qquad\qquad\qquad\qquad 3x + 4x + 6 = 13$
Transposing and combining like terms, $\qquad\qquad 7x = 7$
Dividing both sides of the equation by 7, $\qquad\qquad x = 1$

To find the value of y, we substitute (1) for x in equation (C); we get

$$y = -2 - 3 = -5$$

Example 6. Solve this system by substitution: $\quad\begin{array}{r} 4x - y = 9 \\ 3x - 2y = 5 \end{array}$

Solution.
Solving the first equation for y, we have $\qquad\qquad y = 4x - 9 \qquad$ (C)
Substituting in the other equation, $\qquad\qquad 3x - 2(4x - 9) = 5$
Now we solve this equation, $\qquad\qquad 3x - 8x + 18 = 5$
Transposing and combining like terms, $\qquad\qquad -5x = -13$
Dividing both sides of the equation by -5, we get $\quad x = 13/5$

To find the value of y, we use equation (C):

$$y = 4\left(\frac{13}{5}\right) - 9 = \frac{7}{5}$$

Example 7. Solve this system by substitution: $\quad\begin{array}{r} 2x - 3y = 14 \\ 5x + 4y = 12 \end{array}$

Solution. In this system, no letter has a coefficient of 1; therefore, we shall get fractions. We solve the first equation for x:

$$2x = 3y + 14$$

Dividing both sides of the equation by 2, $\qquad\qquad x = \dfrac{3y + 14}{2} \qquad$ (C)

Now we substitute the fraction for x in the other equation; we get

$$5\left(\frac{3y + 14}{2}\right) + 4y = 12$$

Multiplying both sides by 2, $\qquad\qquad 5(3y + 14) + 8y = 24$
Solving this equation, $\qquad\qquad y = -2$
Substituting (-2) for y in equation (C), $\qquad\qquad x = 4$

Exercise 26.2

Solve the following systems by substitution:

1. $3x + 2y = 12$
   $4x + y = 11$

2. $4x + 3y = 20$
   $5x + y = 14$

3. $5x + 4y = -2$
   $y + 3x = 4$

4. $3a + b = 7$
   $4a - 3b = 18$

5. $4c + d = 14$
   $3c - 5d = 22$

6. $4s + t = 22$
   $5s - 4t = 17$

7. $5x - 4y = 7$
   $2x - y = 3$

8. $4m - 5n = 8$
   $3m - n = 2$

9. $5a - 3b = 6$
   $3a - b = 4$

10. $3x - 5y = 9$
    $x + 2y = 2$

11. $4x + 5y = 8$
    $x + 2y = 3$

12. $3y - 5x = 4$
    $x - 2y = 3$

13. $4x - 5y = 18$
    $3x - 2y = 10$

14. $6x - 5y = 13$
    $2x - 3y = 11$

15. $2x - 5y = 16$
    $3x - 4y = 4$

16. $4x + 3y = 5$
    $5x + 2y = 2$

17. $3x - 4y = -1$
    $4x + 3y = 1$

18. $6x - 5y = 3$
    $5x - 2y = -1$

26.5. SOLVING A SYSTEM OF EQUATIONS BY COMPARISON

A system of two equations can also be solved by a method called *comparison*. In the method of comparison, we *solve both equations* for the same letter. Then we set the two resulting expressions equal to each other.

Example 1. Solve the following system by comparison: $\quad 2x - 3y = 9$
$\qquad 3x + 4y = 5$

Solution. Let us solve the two equations for x.
From the first equation, we get

$$x = \frac{9 + 3y}{2}$$

From the second equation, we get

$$x = \frac{5 - 4y}{3}$$

Since x must have the same value in both equations, we set one fraction equal to the other:

$$\frac{9+3y}{2} = \frac{5-4y}{3}$$

Multiplying both sides of the equation by 6

$$3(9+3y) = 2(5-4y)$$

Solving this equation for y, we get

$$y = -1; \quad \text{then } x = 3$$

Example 2. Solve this system by comparison: $\quad 3x - 2y = 4$
$\phantom{\textbf{Example 2.} \text{Solve this system by comparison:} \quad\ } 4x - 5y = 7$

Solution. We can solve both equations for either x or y. Again we solve each for x:

From the first equation, we get $\quad x = \dfrac{4+2y}{3}$

From the second equation, we get $\quad x = \dfrac{7+5y}{4}$

Setting the expressions equal to each other,

$$\frac{4+2y}{3} = \frac{7+5y}{4}$$

We multiply both sides of the equation by 12,

$$4(4+2y) = 3(7+5y)$$

Solving the equation for y, we get

$$16 + 8y = 21 + 15y$$

$$-7y = 5$$

$$y = -\frac{5}{7}$$

To find the value of x, we put $-\frac{5}{7}$ for y in one of the fractions:

$$x = \frac{4 + 2\left(-\dfrac{5}{7}\right)}{3} = \frac{4 - \dfrac{10}{7}}{3} = \frac{6}{7}$$

Exercise 26.3

Solve the following systems by comparison:

1. $2x + 5y = 2$
   $3x - 2y = 6$

2. $5x - 3y = 7$
   $3x - 4y = 2$

3. $2x - 3y = 2$
   $5x - 4y = 7$

4. $4x - 5y = 3$
   $3x - 6y = 7$

5. $7x - 4y = 6$
   $4x + 3y = 2$

6. $3x + 2y = 7$
   $2x - 3y = 5$

Solve the following systems by any convenient method:

7. $\dfrac{x-3}{2} - \dfrac{y-2}{3} = 5$
   $\dfrac{x-5}{5} - \dfrac{y+3}{4} = 0$

8. $\dfrac{x+2}{3} + \dfrac{y-4}{5} = -1$
   $\dfrac{5-x}{2} + \dfrac{2-y}{7} = 0$

9. $\dfrac{x+3}{3} + \dfrac{2y-1}{5} = 1$
   $\dfrac{5-2x}{4} + \dfrac{y+7}{6} = 3$

10. $\dfrac{2x-3}{4} - \dfrac{y-4}{5} = 1$
    $\dfrac{x+2}{3} - \dfrac{2y-5}{7} = 3$

11. $\dfrac{2x-1}{4} - \dfrac{5-2y}{3} = 0$
    $\dfrac{7-2x}{6} + \dfrac{2y+5}{9} = 2$

12. $\dfrac{5-x}{2} + \dfrac{3+2y}{3} = 4$
    $\dfrac{3x+7}{5} + \dfrac{2-y}{4} = 1$

26.6. SYSTEMS OF EQUATIONS IN THREE OR MORE UNKNOWNS

Three equations containing three unknowns (letters) can be solved by methods similar to those used for solving a system of two unknowns. This is also true for equations containing more than three letters. One of the best methods to use for such systems is the method of *addition or subtraction.*

If we have three unknowns (letters), then we must have three independent equations. In general, the number of equations must be the same as the number of unknowns, and the equations must be independent; that is, one equation must not be dependent upon another.

In solving a system of equations containing three letters, we try to eliminate one letter by addition or subtraction. This can be done if the coefficients of one letter are the same numerically in two of the equations. After one letter has been eliminated by using a set of two equations, then the same letter must be eliminated by using a new set of equations. That is, the equation not used the first time must be used with one of those already used. We show the method by examples.

Example 1. Solve the system:

$$\begin{aligned} 3x+2y-z &= 12 \quad &(A) \\ 4x+3y+2z &= 9 \quad &(B) \\ 2x-4y+z &= 10 \quad &(C) \end{aligned}$$

Solution. The equations are lettered (A), (B), and (C), for convenience. Notice that in equations (A) and (C), the coefficients of z are -1 and $+1$. Then the letter z can be eliminated immediately by adding equations (A) and (C):

Adding the first and third equations, we get

$$5x - 2y = 22 \quad (D)$$

Now, we must eliminate the same letter z by using equation (B) with one of the others. We write equation (B) as it is:

$$4x+3y+2z = 9$$

Next, we multiply equation (A) by 2:

$$6x+4y-2z = 24$$

Adding these two equations, we get $\quad 10x+7y = 33 \quad (E)$

When we added equations (A) and (C),

we got $\quad 5x-2y = 22 \quad (D)$

Solving equations (E) and (D) as we solve any system of two equations, we get the values for x and y:

$$x = 4 \quad \text{and} \quad y = -1$$

After we have found the values of x and y, we go back to one of the original equations and substitute the values of x and y to find the value of z. Then we find that

$$z = -2$$

We now have the values of all three unknowns. Checking should always be done in the three original equations to see if the answers are correct.

Example 2. Solve the following system:

$$\begin{aligned} 2x+5y+3z &= 7 \quad &(A) \\ x-2y+5z &= 15 \quad &(B) \\ 3x-y-z &= 8 \quad &(C) \end{aligned}$$

Solution. We letter the equations, (A), (B), and (C) for convenience.

First we decide upon the letter to be eliminated. Note that the y and z letters have each a coefficient of −1 in equation C. If we multiply equation (C) by 5 first and then by 3, we can add the results to equations (B) and (A). Try to select a set of two equations in which the letter can be eliminated by *addition rather than by subtraction.*

Multiplying equation (C) by 5 $15x - 5y - 5z = 40$
Now we write equation (B) as it is: $x - 2y + 5z = 15$

Adding, we get the new equation (D): $16x - 7y = 55$ (D)

Now we eliminate the same letter z, using *another combination of two equations.* We shall use equation (C) with equation (A).

Multiplying equation (C) by 3 $9x - 3y - 3z = 24$
Now we write equation (A) as it is: $2x + 5y + 3z = 7$

Adding, we get the new equation (E): $11x + 2y = 31$ (E)

We now have two equations, (D) and (E), from which the letter z has been eliminated:

$$16x - 7y = 55 \quad \text{(D)}$$
$$11x + 2y = 31 \quad \text{(E)}$$

Solving this system of two equations as we solve any such system we get the values of x and y:

$$x = 3 \quad \text{and} \quad y = -1$$

Now we substitute these values for x and y in one of the original equations to find z. Then we find that

$$z = 2$$

Example 3.   Solve the system of equations: $3x - 2y - 3z = 9$ (A)
$5x + 3y - 2z = 7$ (B)
$4x + y - 7z = 5$ (C)

Solution.   To eliminate the first letter, it is well to look for some letter that has a coefficient of 1 or −1. Note that in equation (C), the letter y has a coefficient of 1. Then we shall use this equation first with equation (A) and then with equation (B).

Multiplying equation (C) by 2, $8x + 2y - 14z = 10$
We write equation (A) just as it is: $3x - 2y - 3z = 9$

Adding these two equations we get $11x \quad - 17z = 19$ (D)

We have eliminated y, using equations (A) and (C). Now we eliminate the same letter y by using equation (C) with equation (B).

Multiplying equation (C) by 3, we get $12x + 3y - 21z = 15$
We write equation (B) just as it is: $5x + 3y - 2z = 7$

Subtracting, we get a new equation (E): $7x \quad - 19z = 8$ (E)

Now we have a new system of two equations, (D) and (E), containing the two letters, x and z. Solving this system of two equations, we get

$$x = 2.5 \quad \text{and} \quad z = 0.5$$

Substituting these values in one of the original equations, we find the value of y, which is

$$y = -1.5$$

Example 4. Solve the system of three equations:

$$\begin{aligned} 2x + 3y - 4z &= 15 & \text{(A)} \\ 3x + 7z &= -6 & \text{(B)} \\ 5x - y + 3z &= 1 & \text{(C)} \end{aligned}$$

Solution. Since the second equation (B) has no y-term, we eliminate y, using equations (A) and (C):

Multiplying equation (C) by 3,
we get $15x - 3y + 9z = 3$
We write equation (A) just as it is: $2x + 3y - 4z = 15$

Adding these two equations we get $17x \quad + 5z = 18$ (D)
Now we solve this equation with
    equation (B): $3x \quad + 7z = -6$ (B)

The complete solution is left to the student. The answers are 2; $-\frac{3}{2}$; $\frac{3}{2}$, although these answers are not necessarily in the same order as the letters.

Exercise 26.4

Solve the following systems of equations. The answers are given for some, though not necessarily in the same order as the unknowns.

1. $3x + 2y - z = 12$
   $4x - 3y - 2z = 4$
   $5x - 4y + z = 8$
   (Answers: 1, 2, 3)

2. $5x - 2y + z = 11$
   $4x + 5y - 3z = 4$
   $2x + 4y - z = 5$
   (Answers: 1, 2, 3)

3. $3x - y + 2z = 4$
   $2x + 3y - 3z = 5$
   $4x + y - z = 5$
   (Answers: 1, 2, 3)

4. $2x + 2y + 3z = 7$
   $3x - y - 3z = 11$
   $4x + 4y - 5z = 3$
   (Answers: 1, −2, 4)

5. $5x - 3y + z = 2$
   $2x + 5y - 3z = 3$
   $3x + 4y - z = 8$
   (Two ans: 2, 3)

6. $3x + 2y - z = 5$
   $4x - 3y + 2z = 11$
   $5x - 4y + z = 9$
   (Two ans: 1, 2)

7. $4a - b + 3c = 13$
   $2b + 3a - c = 12$
   $2a - 4c + b = 4$
   (Two ans: 2, 1)

8. $2b + 5a - 3c = 4$
   $4a - 5c + 3b = 3$
   $3a + c - 2b = 6$
   (Two ans: −1, 1)

9. $4y + 5x - 2z = 5$
   $2z + 4x + 3y = 8$
   $2x - 4z - y = 4$

10. $5r + 3s - 2t = 5$
    $t + 3r - 2s = 4$
    $2s + 4r - t = 3$

11. $4r + s + 3t = 6$
    $2r + 3s - 2t = 6$
    $3r + 4t = 5$

12. $4x + 2y - 3z = 8$
    $3x + 4y = 18$
    $2x + 3y - 2z = 9$

13. $2z + 7x - 5y = 3$
    $2y + 3x = 7$
    $3z + 5x = 14$

14. $3x - 2z + 4y = 2$
    $2x - 3y = 14$
    $2y + 7z = 3$

15. $3y + 5x - 2z = 7$
    $4x - 2y + 5z = 10$
    $3y - 3z + 2x = 0$

16. $4x - 5y + 3z = 11$
    $2x + 3y + 4z = -5$
    $5z + 3x - y = 0$

17. $5x - 3y - 4z = 6$
    $3x - 5y - 3z = 4$
    $4x - 5z + 2y = -1$
    (Ans: fractions)

18. $3x + 4y - 2z = -2$
    $x + 3y - 4z = -4$
    $x + 3z - 2y = 5$
    (Ans: fractions)

19. $3a + 4b - 5c = 3$
    $4a - 3b + 2c = 18$
    $5a - 3c + 2b = 12$
    (Ans: fractions)

20. $5a - 3b - 5c = 7$
    $3a - 4b + 8c = 3$
    $5b + 2a - 3c = 12$
    (Ans: fractions)

21. $c + 3d - 4e = 6$
    $2d - e - 5c = -9$
    $4e - 3c - 3d = -8$

22. $4z + 5x - 3y = 10$
    $2y - 4x - 5z = 10$
    $3z - 2x - 4y = 0$

23. $2w-3x+4y-2z= 5$
    $3w+ x-3y+4z= 8$
    $w-2x+ y-3z= 7$
    $4w+3x-2y+5z=11$
    (Two ans: $-1, -2$)

24. $3a+2b+4c-5d= 8$
    $2a-3b-2c-3d= 4$
    $4a-5b+3c-2d= 22$
    $5a+4b+5c+4d=-6$
    (Two ans: $-1, -3$)

26.7. EQUATIONS: DEPENDENT; DERIVED; INCONSISTENT; INDEPENDENT

We have mentioned the fact that the equations of a system must be *independent equations* if the system is to have a solution. That is, one equation must not be dependent upon or derived from another. We must now explain what is meant by a *dependent or derived equation*.

Consider the equation in two unknowns:

$$4x-3y=9$$

If we multiply this equation by 2, we get the new equation

$$8x-6y=18$$

The new equation is called a *derived* equation because it is derived from the first by multiplication. If the second equation is first given, it can be reduced to the first by dividing both sides by 2. In either case, *one of these equations can be derived from the other*. Then the two equations are said to be *dependent equations*.

Two dependent equations can be reduced to one. Therefore, any set of values for x and y that satisfies one equation will also satisfy the other. For example, the values, $x=6$ and $y=5$, will satisfy both equations. The same is true with any set of values we choose. Both equations are satisfied by the same set of values.

Let us see what happens when we try to solve a set of two dependent equations. Try to solve the set:

$$3x-2y=5$$
$$7x-15=6y-2x$$

Transposing in the second equation, we get $\quad 7x+2x-6y=15$
Combining like terms, we get $\quad 9x-6y=15$
Multiplying the first equation by 3, we get $\quad 9x-6y=15$

Subtracting $\quad 0=0$

Such an answer indicates a set of dependent equations. One can be derived from the other. The result (0 = 0) is true but it does not help in finding the answer to a practical problem.

If two dependent equations appear in a practical problem, then the problem must be more carefully analyzed to find a set of equations that are not dependent. To get the answer to a problem, the two equations must be independent. That is, one equation must not be the same as the other through any process of transposing, multiplication, or division.

Now, let us consider another set of equations:

$$2x + 3y = 4$$
$$4x + 6y = 15$$

Multiplying the first equation by 2, $\quad 4x + 6y = 8$

Subtracting from the second, $\quad 0 = 7$

The answer shows an impossible situation. The two equations in this set are called *inconsistent* equations. They cannot both be true in the same problem. This means that there is no set of values for x and y that will satisfy both equations. There is no answer. If inconsistent equations appear in the same problem, then the problem must be further analyzed to discover the error in setting up the equations.

To summarize:

1. In a set of dependent equations, any pair of values that will satisfy one of the equations will also satisfy the other. Therefore, such a set will not help in solving a practical problem.

2. Two inconsistent equations have no solution. Therefore, they cannot both be true in the same problem.

3. Independent equations have a limited number of pairs of values that satisfy the equations. Therefore, independent equations are useful in finding the answers to practical problems.

Exercise 26.5

Try to solve the following systems of equations. Tell whether each set is dependent, independent, or inconsistent. Solve those that are independent.

1. $3x - 5y = 9$
   $6x - 12 = 4y$

2. $2x - 7 = 3y$
   $6y - 10 = 4x$

3. $3y + 5 = 4x$
   $8x - 6y = 10$

4. $2x - 4 = 5y$
   $15y + 12 = 4x$

5. $4x + 5y = 3$
   $2x + 7 = -2y$

6. $3x + 4y = -6$
   $8y = 9 - 6x$

7. $5x - 3 = 2y$
   $4y + 5 = 10x$

8. $y + 5 = 4x$
   $8x - 2y = 10$

9. $3x - 4 = y$
   $9 - 2y = 6x$

10. $4y + 5x = 7$
    $3x = 4 + 2y$

11. $4y + 2x = 9$
    $x = 3 - 2y$

12. $3x = y - 2$
    $4y - 12x = 8$

13. $x + 3y = 5$
    $9y = 8 - 3x$

14. $3x = 8 + 9y$
    $7 - 3y = x$

15. $2x = 3y - 4$
    $6y = 4x + 2$

16. $4x + 5y - 5 = 3z$
    $4y - 2z = 4 - 3x$
    $z - 9 = 3y - 2x$

17. $x - 11 + 3y = 4z$
    $6y - 5z + 2x = 7$
    $z + 6 + 9y = -3x$

18. $4x - 3z = 2y - 5$
    $2x + 5y = 2z - 7$
    $y + 2x = 3z$

26.8. SOLVING WORD PROBLEMS BY USE OF TWO OR MORE UNKNOWNS

In solving word problems, we have previously used a letter to represent an unknown quantity. Then we represent any other unknowns in terms of the same letter. You recall the first two steps in solving word problems by using only one letter:

Step 1. Use some letter to represent one of the unknown quantities.

Step 2. Then express the other unknown using the same letter.

Sometimes it is more convenient to use a different letter for each unknown quantity in a problem. Many problems can be solved by using either one letter alone, or by using two or more letters in the same problem. However, some problems are more easily solved by using only one letter; others are more easily solved by using two or more letters. You should know how to use both methods. Then you can always choose the method that seems most convenient in a particular problem.

We shall now work a problem using both methods; first, by using only one letter; then, by using two letters.

Example. A collection of 27 coins, consisting of nickels and dimes, is worth $1.75. How many coins of each kind are in the collection?

Solution. Using only one letter, we begin as usual:
Let x = the number of nickels.
Then $(27 - x)$ = the number of dimes.

For the value of each, stated in cents, we have the following:

the value of x nickels is $5x$ cents;
the value of $(27-x)$ dimes is $10(27-x)$ cents.

The total value of the entire collection is $1.75, or 175 cents. Then we have the equation, showing the total value:

$$5x + 10(27-x) = 175 \text{ (value in cents)}$$

Removing parentheses	$5x + 270 - 10x = 175$
Transposing and combining like terms,	$-5x = -95$
Dividing both sides of the equation by -5,	$x = 19$ (nickels)
For the number of dimes, we have	$27 - 19 = 8$ (dimes)

Therefore, the collection contains 19 nickels and 8 dimes. These answers check for a total value of $1.75.

Now we shall work the same problem by using two *letters.*

Let $x =$ the number of nickels.
Let $y =$ the number of dimes.

When we use *two letters* in the same problem, we must write *two independent equations*. To write the two equations, we use the information given in the problem. *To be sure the equations are independent, we use different information for each equation.* For one equation, we use the information that the total number of coins is 27. Then we have the equation:

$$x + y = 27 \text{ (total number of coins)}$$

For the other equation, we use the information that the total value is 175 cents. First we express the value of all nickels; then the value of all the dimes. For the values, we have the following:

the value of x nickels is $5x$ cents;

the value of y dimes is $10y$ cents.

Then, for the total value we have the equation:

$$5x + 10y = 175$$

Now we have the two independent equations, or a system of equations:

$$x + y = 27$$

$$5x + 10y = 175$$

To solve the system, we could multiply the first equation by 5. However, let us divide the second equation by 5; we get

$$x + 2y = 35$$

Now we write the first equation: $x + y = 27$

Subtracting, $y = 8$ (number of dimes)
Then $x = 19$ (number of nickels)

Steps in solving word problems by use of two unknowns:

1. *Let some letter such as x represent one of the unknowns.*
2. *Let some other letter represent the other unknown.*
3. *Write two independent equations from the given information. Use different information for each equation.*
4. *Solve the system of two equations for the values of both letters.*
5. *Check the answers in the original word problem, not in your equations.*

For three or more unknowns, we follow the same procedure. Let a different letter represent each of the unknowns. Then write three or more independent equations. *You must have as many independent equations as the number of letters used to represent the unknowns.*

Example 2. At an entertainment, a total of 180 tickets were sold, children's tickets for 40 cents each, and adults' tickets for 85 cents each. If the total amount received was $91.80, how many tickets of each kind were sold?

Solution. Let x = the number of children's tickets at 40 cents each.
Let y = the number of adults' tickets at 85 cents each.
For the total number of tickets, we have the equation

$$x + y = 180 \quad \text{(A)}$$

Now we express the number of cents received for each kind of tickets:

for x children's tickets, the amount received was $40x$ cents.
for y adults' tickets, the amount received was $85y$ cents.

Now we write another equation, showing the total value (in cents) received:

$$40x + 85y = 9180 \text{ (cents received)}$$

We multiply equation (A) by 40:
$$40x + 40y = 7200$$

Subtracting,
$$45y = 1980$$

Dividing both sides by 45,
$$y = 44, \text{ number of } 85¢ \text{ tickets.}$$

Then the number of 40¢ tickets is $180 - 44 = 136$

The total amount received: $40(136) + 85(44) = 9180$; or $91.80.

Example 3. At a certain refreshment stand, a group of 15 people buy 6 hamburgers and 9 hot dogs, for a total cost of $5.58. Another group of 12 people buy 5 hamburgers and 7 hot dogs at the same stand for a total cost of $4.49. Find the price of a hamburger and of a hot dog.

Solution. Let $x =$ the number of cents one hamburger costs.
let $y =$ the number of cents one hot dog costs.

Then, 6 hamburgers cost $6x$ cents; and 9 hot dogs cost $9y$ cents. For the total cost for the first group we have the equation, showing the total cost:

$$6x + 9y = 558 \text{ (cents)} \quad \text{(A)}$$

For the second group, we find the cost in the same way:

5 hamburgers cost $5x$ cents;   7 hot dogs cost $7y$ cents.

For the total cost for the second group, we have the equation:

$$5x + 7y = 499 \text{ (cents)} \quad \text{(B)}$$

Now we have the system consisting of equation (A) and equation (B):

$$6x + 9y = 558$$
$$5x + 7y = 499$$

To equalize the x-coefficients, we multiply the first equation by 5, and the second equation by 6:

$$30x + 45y = 2790$$
$$30x + 42y = 2694$$

Subtracting, $3y = 96$

Dividing both sides by 3, $y = 32$ (cents, cost of hot dog)

412 ALGEBRA

To find the value of x, we substitute the value, 32, for y in the second equation:

$$5x + 7(32) = 449$$
or, $\quad 5x + 224 = 449$
Transposing, $\quad 5x = 449 - 224$
Combining like terms, $\quad 5x = 225$
Dividing both sides by 5, $\quad x = 45$ (cents, cost of hamburger)

Example 4. Find two complementary angles that differ by 20 degrees.

Solution. Complementary angles are two angles whose sum is 90 degrees.

Let $x =$ the number of degrees in the larger angle.
Let $y =$ the number of degrees in the smaller angle.

For the sum, we have the equation: $\quad x + y = 90$
For the difference, we have the equation: $\quad x - y = 20$

Adding the two equations, $\quad 2x = 110$
Dividing both sides by 2, $\quad x = 55$ (deg., large angle)
For the smaller angle, we have $\quad 90 - 55 = 35$ (deg., small angle)

Example 5. A man takes a trip of 510 miles, part of the way by car at 40 miles per hour, and the rest of the way by train at 64 miles per hour. If his total travel time is 9 hours, how many hours and how far does he travel by each method of travel?

Solution. Let $x =$ the number of hours by car at 40 mph.
Let $y =$ the number of hours by train at 64 mph.

For the total travel time, we have the equation: $x + y = 9$ (hours).
Now we express the distance traveled by each method:

the number of miles by car at 40 mph is: $\quad 40x$ (miles),
the number of miles by train at 64 mph is: $\quad 64y$ (miles).

For the total distance of travel, we have the equation: $40x + 64y = 510$.

SYSTEMS OF EQUATIONS

Now we have the system of equations:

$$x + y = 9 \quad \text{(A)}$$
$$40x + 64y = 510 \quad \text{(B)}$$

Multiplying the first equation by 40,
$$40x + 40y = 360$$

Subtracting this new equation from (B),
$$24y = 150$$

Dividing both sides by 24,
$$y = 6.25 \text{ (hours by train)}$$

For the number of hours by car, we have
$$x = 9 - 6.25 = 2.75 \text{ (hours by car)}$$

Distance by car: $(2.75)(40) = 110$ (miles).
Distance by train: $(6.25)(64) = 400$ (miles).

For the total distance, we have
$$110 + 400 = 510 \text{ (miles)}$$

Example 6. At 10 A.M. a car starts out on a trip and averages a speed of 35 miles per hour. Two hours later, a second car starts out from the same point to overtake the first car. The second car averages a speed of 55 miles per hour. How long will it take the second car to overtake the first? What time will it be then?

Solution. Let x = the number of hours the first car travels.
Let y = the number of hours the second car travels.

Since the second car started 2 hours later, its travel time was 2 hours less. Then we have the equation for the time

$$x - y = 2 \text{ (difference in time)} \quad \text{(A)}$$

Now we express the distance each car travels:

the first car travels x hours at 35 mph, or $35x$ miles;
the second car travels y hours at 55 mph, or $55y$ miles.

By a study of the conditions of the problem, we realize the two cars must have traveled the same distance. Then we write the second equation:

$$35x = 55y$$

This equation can be rewritten:
$$35x - 55y = 0$$

Multiplying equation (A) by 35,
$$35x - 35y = 70$$

Subtracting,
$$-20y = -70$$

Dividing both sides by -20,
$$y = 3.5 \text{ (hours, 2nd car)}$$

The number of hours for the first car:
$$x = 3.5 + 2 = 5.5 \text{ hours.}$$

To find the time when the second car overtakes the first, we use the fact that the second car started at noon and overtook the first car in 3.5 hours, which makes the time at 3:30 P.M.

Example 7. A man has a total of $9600 invested, part of it at an annual rate of 6% and the rest at an annual rate of 8%. If his total annual income from these two investiments is $632, how much has he invested at each rate?

Solution. Let x = the number of dollars invested at 6%.
Let y = the number of dollars invested at 8%.

For the total investment, we have the first equation:

$$x + y = 9600 \quad (A)$$

The income for a year on any investment is found by taking the rate times the amount invested. A rate of 6% is called 0.06; 8% is called 0.08.
The income on x dollars at 6% is $\quad 0.06x$;
The income on y dollars at 8% is $\quad 0.08y$.

Now we can write the equation showing the total income from both investments:

$$0.06x + 0.08y = 632 \text{ (income)} \quad (B)$$

We can multiply this equation by 100: $\quad 6x + 8y = 63200$
Now we write the equation (A): $\quad x + y = 9600$
We multiply this equation (A) by 6: $\quad 6x + 6y = 57600 \quad (C)$
Subtracting equation (C) from equation (B), $\quad 2y = 5600$
Dividing both sides of the equation by 2, $\quad y = 2800$
Now we see that $2800 is invested at 8%. Then $\quad 9600 - 2800 = 6800$

Then the amount invested at 6% is $6800.
To check the answers, we find the income on each investment:

$0.06(\$6800) = \$408; \quad 0.08(\$2800) = \$224. \quad$ Total income = $632.

Example 8. How much of a 12% solution of alcohol and how much of a 60% solution should be mixed together for 30 quarts of a new solution testing 24%?

Solution. Let x = the number of quarts of the 12% solution
Let y = the number of quarts of the 60% solution.

For the first equation we use the fact that the entire new solution is 30 quarts: that is,

$$x + y = 30 \text{ (number of quarts)} \quad \text{(A)}$$

Now we consider the amount of the alcohol in the new solution:

In x quarts of the 12% solution, there are $0.12x$ quarts of alcohol.
In y quarts of the 60% solution, there are $0.60y$ quarts of alcohol.

If the new solution of 30 quarts is to test 24%, then the new solution must contain the following amount of alcohol:

$$0.24(30) = 7.2 \text{ (quarts of alcohol)}$$

Now we write an equation showing the amount of alcohol in the two solutions is equal to the total amount in the new solution; that is

$$\begin{pmatrix} \text{alcohol in one} \\ \text{solution} \end{pmatrix} + \begin{pmatrix} \text{alcohol in second} \\ \text{solution} \end{pmatrix} = \begin{pmatrix} \text{total alcohol} \\ \text{in final solution} \end{pmatrix}$$

The equation: $\quad 0.12x + \quad 0.60y \quad = \quad 7.2$

Multiply this equation by 100, $\quad 12x + 60y = 720$
Now we multiply equation (A)
above, by 12: $\quad 12x + 12y = 360$
Subtracting, $\quad\quad\quad\quad\quad\quad\quad 48y = 360$
Dividing both sides by 48, $\quad\quad\quad y = 7.5$ (quarts of 60% sol.)

For the number of quarts of the 12% solution, we have

$$30 - 7.5 = 22.5$$

The problem can be checked by finding the amount of alcohol in each of the original solutions. The total amount of alcohol in both should be 7.2 quarts.

Example 9. A candy merchant has one kind of candy selling at 35 cents a pound, and another kind selling at 85 cents a pound. To increase sales, he wishes to make a mixture of the two kinds to sell at 50 cents a pound. How many pounds of each kind should he use for 60 pounds of the mixture so that he will receive the same amount as if the two kinds were sold separately?

Solution. Let $x =$ the number of pounds of the 35¢ candy.
Let $y =$ the number of pounds of the 85¢ candy.

For one equation, we state that the total number of pounds is 60. Then we write

$$x + y = 60 \quad \text{(A)}$$

If he is to sell 60 pounds at 50 cents a pound, he will receive $30.00 for the entire mixture of 60 pounds. We write $30.00 as 3000 cents. Now we express the value (in cents) of each kind:

the value of x pounds of 35¢ candy is $35x$ cents.
the value of y pounds of 85¢ candy is $85y$ cents.

Now we write the second equation which expresses the *total value:*

$$35x + 85y = 3000 \text{ (cents)} \quad \text{(B)}$$

We multiply equation (A) by 35:
$$35x + 35y = 2100$$
Subtracting this equation from (B),
$$50y = 900$$
Dividing both sides of the equation by 50,
$$y = 18 \text{ (pounds of 85¢ candy)}$$
For the number of pounds of the 35¢ candy:
$$x = 60 - 18 = 42$$

To check the answers, we find the total value of each kind of candy in the mixture:

$$42(35¢) = 1470 \text{ cents, or } \$14.70$$
$$18(85¢) = 1530 \text{ cents, or } \$15.30$$
Total value $30.00, which is 60(50¢)

Example 10. A student has a grade-point average of 2.1, which he has acquired in two semesters of 15 hours each (30 hours in all). The grade points are as follows: A, 4 points; B, 3 points; C, 2 points; D, 1 point. During the next two semesters covering 30 hours of course work, the student wishes to bring his average up to 2.5. He is sure of a grade A in two 3-hour courses. In the remaining 24 hours of course work, how many B's and how many C's must he make to bring his grade-point average up to 2.5?

Solution. The student now has a grade-point average of 2.1 in 30 hours of course work, which means a total of 63 grade points. If he is to have an average of 2.5 for 60 hours of course work, he must have a total of 150 grade points. He now has 63 grade points, and he is sure of 24 more for a grade of A in 6 hours of work. Then he can count on having 87 points. Then he needs to make up the difference, $150-87=63$, in B's and C's in the remaining 24 hours of course work in the next two semesters. We omit the 6 hours of A's of which he is sure. Then we

Let $x =$ the number of hours of B's to be made;
Let $y =$ the number of hours of C's to be made.

Then we get the equation for the total number of hours of B's and C's:

$$x + y = 24$$

For the second equation, we consider the number of grade points:

On x hours of B, he gets $3x$ grade points,
On y hours of C, he gets $2y$ grade points.

The sum of these grade points must make up the total extra points required:

$$3x + 2y = 63$$

Now we have a system of two independent equations:

$$x + y = 24$$
$$3x + 2y = 63$$

Multiplying the first equation by 2, $\quad 2x + 2y = 48$
Subtracting from the second equation, $\quad x \quad\quad = 15$ (hours of B)

For the value of y, we subtract: $\quad 24 - 15 = 9$ (number of hours of C)

The result means, that, for 3-credit courses, he must make a B in 5 courses, and a C in 3 courses.

Exercise 26.6 (use two unknowns)

1. Find two complementary angles whose difference is 22.8°.
2. Find two supplementary angles whose difference is 27°.
3. In an electric circuit consisting of two branches, the current in one branch is 5 amperes more than in the other branch. When the currents join, the total current is 18 amperes. Find the current in each branch.

4. The sum of two voltages in an electric circuit is 1200 volts. If the polarity of one is reversed, the total voltage is 36 volts. Find each voltage.

5. A collection of 83 coins consisting of nickels and dimes is worth $5.45. Find the number of coins of each kind in the collection.

6. A collection of 67 coins, consisting of nickels and quarters, is worth $7.75. Find the number of coins of each kind in the collection.

7. A collection of 53 coins, consisting of dimes and quarters, is worth $7.55. Find the number of coins of each kind in the collection.

8. A man buys some 8-cents stamps and some 11-cent stamps for a total cost of $4.42. If the total number of stamps is 50, how many of each kind does he buy?

9. At an entertainment, children's tickets cost 20 cents each and adults' tickets cost 50 cents each. If a total of 425 tickets were sold and the total amount received was $118, how many of each kind were sold?

10. At an entertainment, children's tickets cost 25 cents each and adults' tickets cost 60 cents each. If the total amount received was $164.90 for a total of 500 tickets, how many of each kind were sold?

11. At a ball game, general admission tickets cost $1.20 each and grandstand tickets cost $2.50 each. If a total of $2886 was received for a total of 1820 tickets, how many of each kind were sold?

12. A car starts on a trip at 9 A.M. and averages 45 miles per hour. At 11 A.M. the same day, a second car starts from the same point and sets out to overtake the first. If the second car averages 60 miles per hour, how long will it take the second car to overtake the first? What time of day will it be then? How far has each car traveled?

13. One car starts on a trip at 7 A.M. and travels at an average of 35 miles per hour. At 10 A.M. the same day, a second car starts from the same point and sets out to overtake the first. If the second car travels at an average of 55 miles per hour, what time will the second car overtake the first? How far has each car traveled then?

14. A man takes a trip of a total of 670 miles, part of the way by car at an average of 40 miles per hour, and the rest of the way by train at an average of 76 miles per hour. If the total travel time was 10 hours, how far does he travel by each method of travel? (First find number of hours traveled by each method of travel.)

15. A man takes a trip of a total of 560 miles, part of the way by car at an average of 40 miles per hour, and the rest of the way by train at an average of 72 miles per hour. If the total travel time is 9 hours, how far does he travel by each method of travel?

16. A man has $7200 invested in savings, part of it at an annual rate of return of 5% and the rest at 7%. If the total income from these two investments is $406 a year, find the amount invested at each rate.

17. A man has $12,600 invested, part of it at an annual return of 5% and the rest at 7%. If he receives the same annual income from both investments, how much has he invested at each rate?

18. How much of a 10% solution of alcohol and how much of a 60% solution should be mixed together for 30 quarts of a solution testing 25%?

19. How much of an 8% solution and how much of a 40% solution must be mixed together for 20 quarts of a new solution testing 15%?

20. How much of an 80% alloy of silver and how much of a 5% alloy should be mixed together for 10 ounces of an alloy that is 20% silver?

21. A tea merchant sells one kind of tea for 60 cents a pound and another kind for $1.10 a pound. He wishes to make a mixture of the two kinds and sell the mixture at 75 cents a pound. How many pounds of each kind should he use for 50 pounds of the mixture so that he will receive the same amount as if the two kinds were sold separately?

22. A grain feed dealer sells one kind of grain for $1.40 a bushel and another kind of grain for $2.20 a bushel. He wishes to make a mixture of the two kinds to sell at $1.70 a bushel. How many bushels of each kind should he use for 100 bushels of the mixture so that he receives the same amount as if the two kinds were sold separately?

23. A student hopes for a grade-point average of 2.8 for 30 hours of course work, which is ten 3-credit courses. If he is sure of an A in two of the 3-credit courses (6 hours of course work), how many B's and how many C's must he make in the remaining 24 hours of course work in order to get a grade-point average of 2.8 for the 30 hours of course work? (A's count 4 points; B's, 3 points; C's, 2 points; D's, 1 point.)

24. A student has a grade-point average of 2.8 for the first semester in which he took five 3-credit courses, or 15 hours of course work. During the next semester he takes 15 hours of course work and hopes to end up with a 3.1 average. How many A's and how many B's must he earn to get a 3.1 average for the first two semesters of 30 hours of course work? (Grade points count the same as in problem 23.)

27 GRAPHING

27.1. IMPORTANCE OF GRAPHS

Hardly a day goes by that we do not see a graph of some kind. A graph is a pictorial representation of certain number facts. We often see graphs in books, magazines, and newspapers. We see graphs of such things as temperature changes, business losses and gains, economic changes and so on. Many important facts in engineering and other sciences are represented by graphs, such as the relations between current, voltage, and power in electrical engineering.

The advantage of a graph is that it shows important relations at a glance. Looking at a business graph, we can tell immediately the high and the low points of business activity. A temperature graph shows instantly the high and low points of temperature. Many ideas in mathematics are more easily understood when they are represented graphically. For these and many other reasons, graphs are extremely important in mathematics and science.

27.2. THE RECTANGULAR COORDINATE SYSTEM

To understand the meaning and construction of graphs, we must understand what is meant by a coordinate system for locating points in a plane. In most cities, at least part of the street system is laid out so that some streets have an east-west direction, and others have a north-south direction.

Let us suppose a city has one main street, which we shall call street X, extending in an east-west direction. It has another main street, which we shall call street Y, extending in a north-south direction (Fig. 27.1). All other streets are parallel to one or the other of these two main streets. All points in the city can be located with reference to these two main streets. All distances are measured from the two main streets.

If we wish to direct someone to a particular location in the city, we might say, starting at the point of intersection of the two main streets, "Go 5

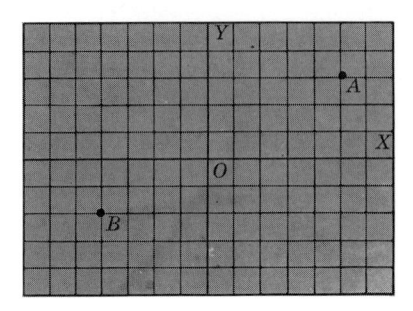

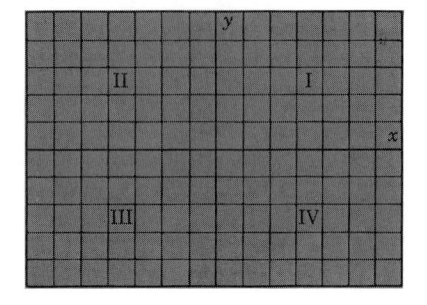

FIGURE 27.1

FIGURE 27.2

blocks east and 3 blocks north." The location is indicated by point *A* in Figure 27.1. To another, we might say, "Go 4 blocks west and 2 blocks south." This location is shown by point *B* in the figure.

If we wish to locate a point in a plane, we use a system like the system of streets. We set up two perpendicular lines of reference, which correspond to the two main streets of a city. Each reference line is called an *axis*. We call the horizontal line the *x-axis*, and vertical line the *y-axis*. The point of intersection of the two axes is called the *origin*, denoted by *O* (Fig. 27.2).

This system for locating points is called the rectangular coordinate system. The two axes divide the entire plane into four quarters called *quadrants*. The quadrants are numbered I, II, III, IV, starting at the upper right-hand quadrant and going around in a counterclockwise direction. By means of this system, we can tell the location of any point with reference to the two *axes*.

Suppose a point is located 8 units to the right of the *y*-axis, and 5 units upward from the *x*-axis (Fig. 27.3). Then these two numbers, 8 and 5, taken in that order, will definitely tell the location of the points. The *distance, 8, in the x-direction* is called the *abscissa* of the point. The *distance, 5, in the y-direction*, is called the *ordinate* of the point. The two numbers, *abscissa*

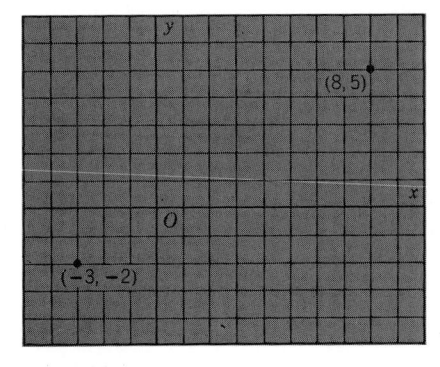

FIGURE 27.3

GRAPHING 423

and *ordinate*, are called the *coordinates* of the point. They are like house numbers in a city.

The coordinates of a point are enclosed in parentheses and separated by a comma, thus: (8, 5). The coordinates are usually written near the point on the graph. The abscissa is always written first. With this understanding, the pair of numbers (8, 5) will definitely determine one and only one point on the graph. For this particular point, *the two numbers cannot be reversed.* Since the two numbers must be taken in proper order, they are called an *ordered pair.*

In a city, we speak of distances east or west and north or south. In the rectangular coordinate system, distances to the right are called *positive* (+); distances to the left are called *negative* (−); distances upward are called *positive*; distances downward are called *negative*.

If we wish to indicate a point 3 units to the left of the y-axis, we say the abscissa, or x-distance, is −3. If the point is 2 units downward from the x-axis, we say the y-distance, or ordinate, is −2. Then the two numbers, −3 and −2, are the coordinates of the point; written: (−3, −2). This point is shown in Figure 27.3.

The locating of points on a graph is most conveniently done on paper called "rectangular coordinate" graph paper. On this kind of paper, lines are evenly spaced horizontally and vertically. To locate points on a graph, we first draw the two axes, x and y, at right angles so that the origin is approximately in the center of the portion of the paper to be used.

In measuring the distances indicated by the coordinates of a point, the

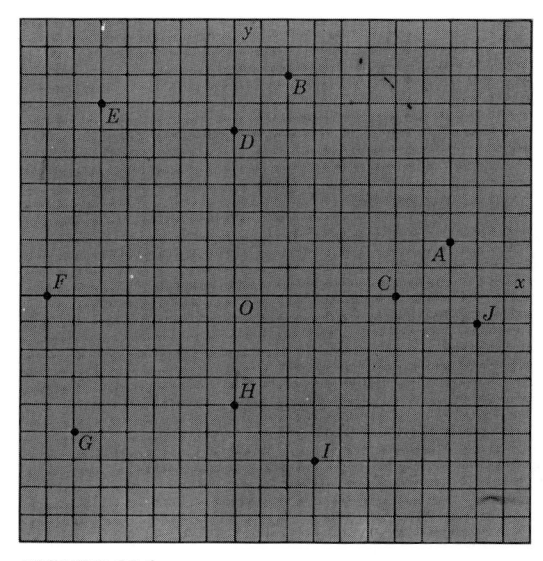

FIGURE 27.4

units may be of any convenient length. In most cases, the units should have the same length on both axes. To *plot* points means to locate points on the graph. The following points are shown in Figure 27.4: $A(8, 2)$; $B(2, 8)$; $C(6, 0)$; $D(0, 6)$; $E(-5, 7)$; $F(-7, 0)$; $G(-6, -5)$; $H(0, -4)$; $I(3, -6)$; $J(9, -1)$, and the origin $(0, 0)$.

Exercise 27.1

1. Plot these points on one graph: $A(7, 3)$; $B(1, 6)$; $C(0, 3)$; $D(-8, 2)$; $E(-5, -2)$; $F(0, 0)$; $G(3, -5)$; $H(0, -7)$; $I(6, -6)$; $J(9, 0)$.
2. Plot the following points on the same graph, and then with straight line segments connect the points in the order in which they are given: $A(8, 0)$; $B(0, 8)$; $C(-8, 0)$; $D(0, -8)$; $A(8, 0)$; $E(4, 0)$; $F(0, 4)$; $G(-4, 0)$; $H(0, -4)$; $E(4, 0)$.

Plot the points in each of the following sets (no. 3–10) on a graph, and then connect the points in order by straight line segments to form polygons by connecting the last point with the first as a final step. Use a different portion of graph paper for each set:

3. $A(8, 7)$; $B(-5, 3)$; $C(2, -6)$
4. $A(7, 1)$; $B(-2, 8)$; $C(4, -3)$
5. $A(3, 0)$; $B(-3, 7)$; $C(-8, -5)$
6. $A(-2, 8)$; $B(-6, 0)$; $C(7, -4)$
7. $A(5, -1)$; $B(0, 4)$; $C(-6, -5)$; $D(0, -1)$
8. $A(7, 5)$; $B(-3, 2)$; $C(-6, -5)$; $D(4, -2)$
9. $A(8, 0)$; $B(2, 6)$; $C(-5, 0)$; $D(0, -7)$; $E(6, -5)$
10. $A(5, 6)$; $B(-7, 0)$; $C(6, -2)$; $D(-3, 7)$; $E(-1, -6)$
11. Connect the following points by a smooth curve instead of straight line segments: $A(12, 8)$; $B(0, 4)$; $C(-3, 2)$; $D(-4, 0)$; $E(-3, -2)$; $F(0, -4)$; $G(12, -8)$.
12. Connect the following points by a smooth curve: $A(5, 0)$; $B(4, 3)$; $C(3, 4)$; $D(0, 5)$; $E(-3, 4)$; $F(-4, 3)$; $G(-5, 0)$; $H(-4, -3)$; $I(-3, -4)$; $J(0, -5)$; $K(3, -4)$; $L(4, -3)$; $A(5, 0)$.

27.3. GRAPH OF AN EQUATION

We have seen that an equation containing x and y may have many solutions, or pairs of numbers that make the equation true. Consider the equation

$$x + y = 8$$

x	y
6	2
4	4
8	0
0	8
9	-1
-3	11

This equation is true for the pair of numbers, $x = 6$ and $y = 2$. However, it is true also if $x = 4$ and $y = 4$. The equation has many solution pairs, some of which are shown in the table at the left. Any one of these pairs of numbers will satisfy this equation.

GRAPHING

To find a pair of numbers that will satisfy an equation involves algebra. It has nothing to do with geometry.

Now, here is one of the most surprising stories in mathematics. For hundreds of years people had been working with algebraic equations, finding solutions to equations such as the one we have shown. Yet no one thought of connecting algebra with geometry. Up to this point there is no connection. But then, along came Rene Descartes, a Frenchman, born in 1596. While a young man in his teens, he had a brilliant idea, an idea that marks one of the major advances in mathematics.

This was Descartes's idea: if the numbers, $x = 6$ and $y = 2$, satisfy the equation, $x + y = 8$, let us call this pair of numbers the coordinates of a point on a rectangular coordinate system. That is, *we take a pair of numbers from an algebraic equation, and then use this pair of numbers to represent a point in geometry.* The result was the uniting of algebra and geometry into one of the most powerful tools in mathematics.

Let us follow through with the equation, $x + y = 8$. We have seen that several pairs of numbers will satisfy the equation. We shall call each pair of numbers the coordinates of a point. As a result, we shall have several points.

Descartes discovered an important fact, which we can also discover. If the points we get from the x and y values of an equation are connected by straight line segments, the result is one straight line. All the points will lie on the same straight line (Fig. 27.5). On the other hand, if we take any point on the line, it will have coordinates that satisfy the equation.

There are two facts we should note here.

1. *If we take any pair of numbers that satisfy the equation, this pair of numbers will be the coordinates of a point on the line.*

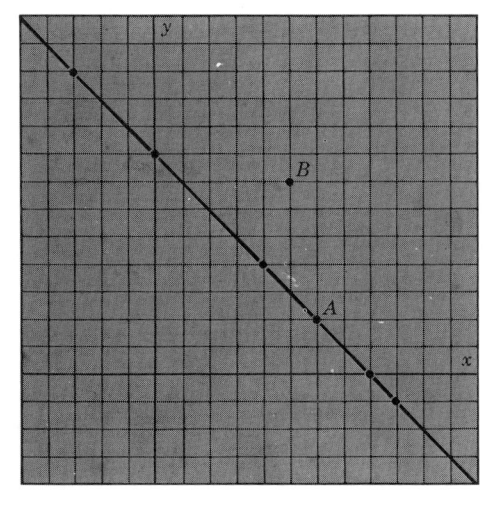

FIGURE 27.5

2. *If we take any point on the line, such as point A, and find its coordinates from the graph, this pair of numbers will satisfy the equation.*

The point B has coordinates that do not satisfy the equation because the point does not lie on the line.

Since both the above conditions are true, the straight line in Figure 27.5 is the graph or picture of the equation, $x+y=8$. An equation such as $x+y=8$, is called a *linear equation* because its graph is a straight line. It is known that *the graph of any first degree equation is a straight line*. A *first-degree equation* is an equation containing only first degree terms. It has no terms containing x^2, y^2, or xy.

In finding values of x and y and in constructing the graph of a linear equation, we should keep a few facts in mind.

1. Since we know the graph is a straight line, it would be sufficient to find only two points because two points determine a straight line. However, one or two extra points should be plotted as a check on the accuracy of the work.

2. If only two points are used, they should not be too close together. The straight line can be drawn more accurately if the two points are some distance apart.

3. The easiest points to find are usually those where the line cuts the x-axis and the y-axis. These points are easily found by setting each letter equal to zero, and then solving for the value of the other letter. First, we let $x=0$, and then find the value of y. Then we set $y=0$, and solve for the value of x. However, in some equations, such as, $2x=3y$, when $x=0$, then $y=0$. Then we need at least one other point.

4. The straight line should be drawn *through* the points and should not end at any of the points plotted. It should be understood that the line is unlimited in extent, not limited by the size of the paper.

Example 1. Graph the equation $2x-3y=18$.

Solution. To find any pair of values that will satisfy the equation we take any convenient value for one letter, and then find the corresponding value of the other letter. First we find the zero values. Several sets of values may be found, as shown here:

if $x=$	0	9	3	6	−3
$y=$	−6	0	−4	−2	−8

Now these corresponding values are plotted as points. The points are then connected by a straight line. The graph is shown in Figure 27.6.

Example 2. Graph the equation $2x=5y$.

GRAPHING 427

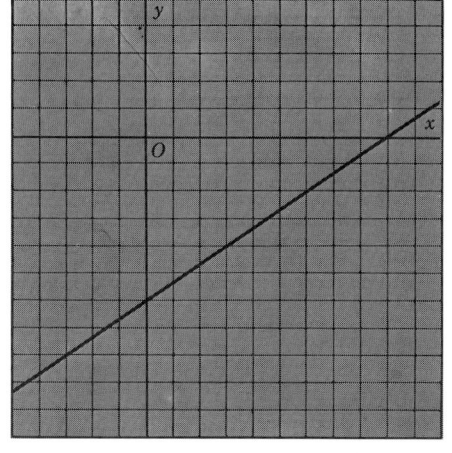

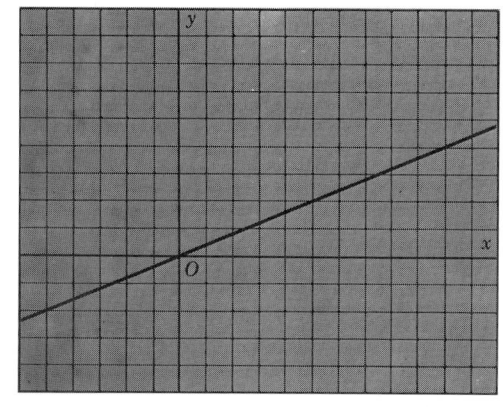

FIGURE 27.6

FIGURE 27.7

Solution. For pairs of corresponding values, we have the following:

$x=0$	5	$7\frac{1}{2}$	-5	3	10
$y=0$	2	3	-2	$1\frac{1}{5}$	4

The graph is shown in Figure 27.7.

Exercise 27.2

Graph the following linear equations:

1. $x+2y=8$
2. $2x-y=6$
3. $3x+2y=12$
4. $2x-3y=18$
5. $5x+2y=20$
6. $3x-4y=12$
7. $x-2y=9$
8. $x+3y=10$
9. $2x+y=7$
10. $3x+7y=21$
11. $5x+6y=30$
12. $7y-4x=28$
13. $4x-5y=40$
14. $3y-2x=24$
15. $3x-5y=30$
16. $2x+3y=20$
17. $5x+4y=28$
18. $5x+3y=25$
19. $2x-3y=13$
20. $3x-5y=19$
21. $3x+4y=23$
22. $3x=5y$
23. $4x-3y=0$
24. $5x+2y=0$
25. $4x-3y=2$
26. $2x+3y=1$
27. $3x-5y=4$
28. $3x+4y=5$

27.4. SOLVING SYSTEMS OF EQUATIONS BY GRAPHING

Graphing can be used to solve a system of equations. Consider the system:

$$x + y = 8$$
$$x - y = 2$$

We have seen in Figure 27.5 that the graph of the first equation is a straight line. Now let us graph the second equation, $x - y$, on the same graph paper. We find several sets of numbers that satisfy the equation. We can use the following pairs of values:

$x =$	0	8	2	10
$y =$	-2	6	0	8

Next, we plot the points corresponding to these pairs of values. Connecting these points, we get another straight line. Figure 27.8 shows the lines for both equations: $x + y = 8$ and $x - y = 2$.

Note that point A lies on the first line. Therefore, its coordinates will satisfy the first equation, $x + y = 8$. Since point B lies on the second line, its coordinates will satisfy the second equation, $x - y = 2$. The coordinates of point C will not satisfy either equation, since it does not lie on either line.

Only one point will have coordinates that satisfy both equations. That is the point where the two lines intersect. From the graph, it appears to be the

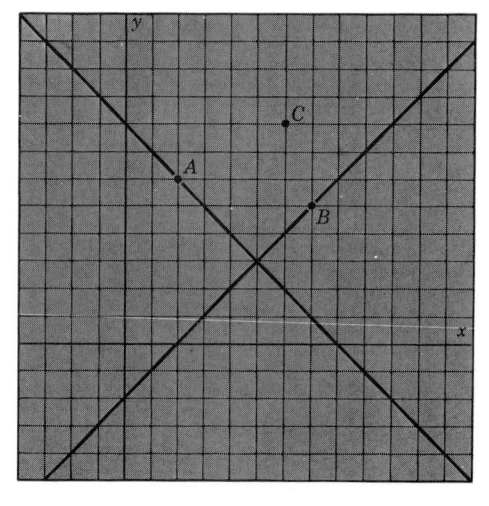

FIGURE 27.8

GRAPHING 429

point whose coordinates are approximately as follows: $x = 5$; $y = 3$. These values should satisfy both equations.

This system of two equations could be solved by some algebraic method, such as addition or subtraction, or by substitution. However, there is sometimes an advantage in the graphical method. In some more advanced types of equations, algebraic methods cannot be used. Then we must resort to the graphical method.

In some problems, the point of intersection of the two lines is rather easily determined. However, in many cases, it is not easy to read the exact coordinates from the graph. This is especially true when the values are fractions. It should always be remembered that *any answers read from a graph must be assumed to be only approximate.* If answers do not appear to be whole numbers, they can be estimated in fractions, even in tenths.

Example 1. Solve the following system by graphing:

$$x + 2y = 9$$
$$2x - y = 8$$

Solution. We first find pairs of numbers that satisfy each equation:

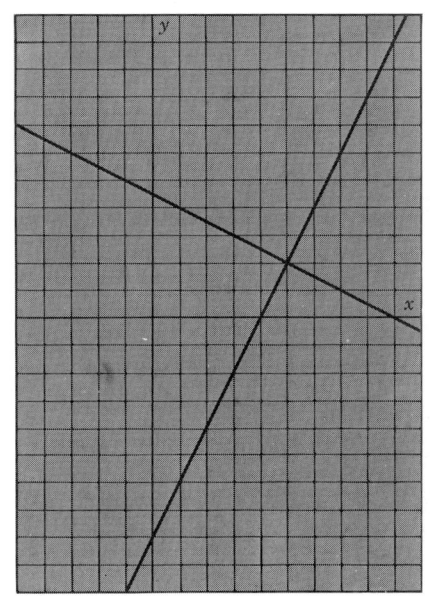

For $x + 2y = 9$				For $2x - y = 8$			
$x = 0$	9	7		$x =$ 0	4	8
$y = 4\frac{1}{2}$	0	1		$y = -8$	0	8

FIGURE 27.9

430 ALGEBRA

Plotting the points for each equation, we get the graph of each equation (Fig. 27.9). The point of intersection appears to be approximately: $x = 5$; and $y = 2$; that is (5, 2).

Example 2. Graph the system: $2x + y = 4$
$x - 3y = 9$

Solution. We take the following pairs of values:

For $2x + y = 4$

$x =$	0	2	6
$y =$	4	0	−8

For $x - 3y = 9$

$x =$	0	9	12
$y =$	−3	0	1

The graphs of the equations are shown in Figure 27.10. The point of intersection appears to be approximately (3, −2).

Example 3. Solve the system by graphing:

$$3x - 2y = 0$$
$$2x + 5y = 15$$

Solution. We might use the following values:

For $3x - 2y = 0$

$x =$	0	2	4
$y =$	0	3	6

For $2x + 5y = 15$

$x =$	0	$7\frac{1}{2}$	5
$y =$	3	0	1

Plotting the points for corresponding values, we get the two lines in Figure

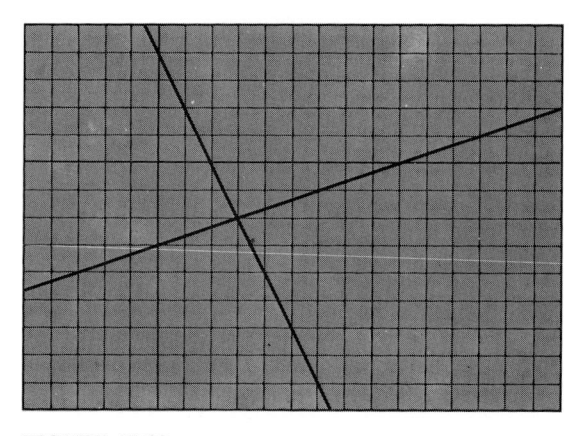

FIGURE 27.10

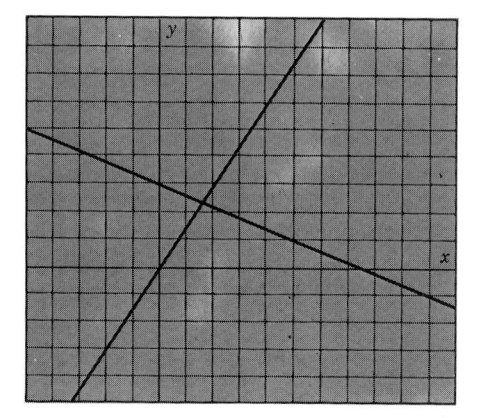

FIGURE 27.11

27.11. The point of intersection appears to have the following coordinates: $x = 1\frac{3}{5}$; $y = 2\frac{2}{5}$; or, in decimals: $x = 1.6$; $y = 2.4$.

If the equations are solved by algebraic methods, the answers are found to be exactly: $x = 1\frac{11}{19}$; $y = 2\frac{7}{19}$.

If the example calls for a graphical solution, the correct answer is the approximate solution as read from the graph, since it is impossible to read a value such as $1\frac{11}{19}$ from a graph.

Exercise 27.3

Solve the following systems graphically:

1. $x + 2y = 10$
   $3x - y = 9$

2. $2x - 3y = 12$
   $3x + y = 7$

3. $x - 2y = -8$
   $2x + y = 9$

4. $3x + 2y = -6$
   $4x - 5y = 25$

5. $5x - 2y = 12$
   $2x - 5y = -6$

6. $3x + 4y = 12$
   $5x + 3y = 15$

7. $2x - 3y = 6$
   $5x - 4y = 20$

8. $3x - 5y = 30$
   $2x + y = 10$

9. $2x + 5y = 20$
   $3x - 2y = 18$

10. $5x + 4y = 20$
    $2x - 5y = -10$

11. $5x - 6y = 30$
    $3x + y = 12$

12. $3x + 4y = 12$
    $2x - 5y = -10$

13. $3x + 4y = 8$
    $2x + 5y = 3$

14. $4x - 5y = 7$
    $3x - 4y = 6$

15. $4x + 3y = -4$
    $3x - 2y = 2$

16. $4x + 3y = 11$
    $3x + 2y = 7$

17. $5x - 7y = 8$
    $4x + 5y = 17$

18. $5x + 4y = 7$
    $3x - 5y = -19$

19. $3x + y = 9$
    $4x - 3y = 12$

20. $5x - 2y = 15$
    $2x + 3y = 6$

21. $2x - 3y = 12$
    $5x - 2y = 8$

22. $3x - 2y = 0$
    $5x + 4y = 15$

23. $2x - 5y = 0$
    $3x + 4y = 20$

24. $3x - y = 6$
    $5x - 3y = 0$

28 EXPONENTS, POWERS, AND ROOTS

28.1. MULTIPLICATION

At this time it is necessary to review carefully the rules to be observed in operations involving exponents. We have seen how to find the product of two quantities expressed with exponents. Consider the example

$$(x^2) \cdot (x^3)$$

In multiplying the two quantities, x^2 and x^3, we note that the base is the same. Then we get the product by adding the exponents:

$$(x^2) \cdot (x^3) = x^5$$

The quantity, x^2, means: $x \cdot x$; the quantity, x^3, means: $x \cdot x \cdot x$. Then the product, $(x^2)(x^3)$, means:

$$(x \cdot x) \cdot (x \cdot x \cdot x) = x^5.$$

This example is an illustration of the rule for multiplication.

RULE 1. In multiplying quantities expressed with exponents, we add the exponents of the same base.

To apply this rule, the base must be the same. We do not multiply the exponents. We add the exponents when we multiply the quantities.

To emphasize the rule, consider the problem in multiplication:

$$(5^2) \cdot (5^4) = 5^6$$

To show that the rule is true, let us consider each quantity in parentheses: $5^2 = 25$; $5^4 = 625$; then the problem means:

$$(25)(625) = 15625 = 5^6$$

433

ALGEBRA

Exponents on different bases cannot be added. For example,

$$(x^2) \cdot (y^4) = x^2 y^4$$

In a problem involving only arithmetic numbers, we can expand each factor and then find the product. For example,

$$(3^2) \cdot (5^4) = (9)(625) = 5625$$

The rule for multiplication can be stated as a formula:

$$(x^a) \cdot (x^b) = x^{a+b}$$

The rule for multiplication involving exponents holds true for all kinds of exponents. If no exponent is shown, the exponent is understood to be 1. The rule can be extended to cover several factors. The following examples show the application of the rule:

$(x^6)(x^2) = x^8;$
$(n^3)(n^0) = n^3;$
$(x^6)(x^{-2}) = x^4;$
$(x^{1/2})(x^{1/3}) = x^{5/6};$
$(10^4)(10^{-4}) = 10^0;$
$(10^{3.124})(10^{1.413}) = 10^{4.537}$
$(4^3)(5^2) = (64)(25) = 1600$

$(y^4)(y) = y^5;$
$(5^3)(5^5) = 5^8;$
$(2^4)(2^6) = 2^{10}$
$(n^{-3})(n^4) = n;$
$(10^5)(10^3)(10) = 10^9;$
$(x^n)(x^c)(x) = x^{n+c+1};$

Exercise 28.1

Find the powers:

1. $(-5)^2$
2. $(-4)^3$
3. $-(-4)^3$
4. $(-3)^4$
5. $-(8)^2$
6. $-(-2)^5$
7. $-(-7)^2$
8. $(-\frac{1}{5})^2$
9. $(-0.02)^3$
10. $-(0.03)^4$
11. $-(-0.4)^3$
12. -6^2

Multiply as indicated:

13. $(x^7)(x^3)$
14. $(n^{-5})(n^4)$
15. $(y^3)(y)$
16. $(n^{-2})(n^{-4})$
17. $(y^{-2})(y^7)$
18. $(x^6)(x^0)$
19. $(y^{1.4})(y^{2.3})$
20. $(x^{3.21})(x^{-1.12})$

EXPONENTS, POWERS, AND ROOTS 435

21. $(n^{2.1})(n^{-4.5})$
22. $(x^{2/3})(x^{3/4})$
23. $(n^{3a})(n^{2a})(n)$
24. $(x^3)(x^7)(x^{-4})$
25. $(10^3)(10^4)(10)(10^0)$
26. $(6^{-3})(6^4)(6^0)(6)(6^{-1})$
27. $(2)(2^6)(2^{-2})(2^0)(2^3)$
28. $(10^{4.31})(10^{3.14})(10^{1.68})$
29. $(10^{1.234})(10^{2.121})(10^{0.468})$
30. $(10^{2.34})(10^{1.414})(10^{-1.012})$
31. $(2^4)(3^3)(5^2)$
32. $(6^2)(5^3)(4^2)$
33. $(10^3)(5^2)(2^5)$
34. $(2^3)(4^3)(10^4)$

28.2. DIVISION

We have seen and used the following rule for the division of quantities expressed with exponents.

RULE 2. In division involving quantities expressed with exponents on the same base, we subtract the exponent in the divisor from the exponent in the dividend.

The rule may be stated as a formula

$$x^a \div x^b = x^{a-b}$$

To show the reason for the rule, consider the following problem:

$$x^6 \div x^2$$

We can write the same problem in the form of a fraction:

$$\frac{x^6}{x^2}$$

If we show the factors of the numerator and the denominator, we have

$$\frac{x^6}{x^2} = \frac{x \cdot x \cdot x \cdot x \cdot x \cdot x}{x \cdot x} = x^4$$

Now we see that numerator and denominator can be divided by $(x)(x)$, and we get the reduced fraction, x^4. Note that the answer can be found by subtracting the exponents. This example shows the reason for the rule. The answer in division can be checked by multiplication, as we usually check division. For example, $(x^2)(x^4) = x^6$.

436 ALGEBRA

The rule is true for all kinds of exponents. The following examples show the application of the rule for division involving exponents:

$$x^8 \div x^2 = x^6$$
$$n^3 \div n^0 = n^3$$
$$x^4 \div x = x^3$$
$$10^7 \div 10^2 = 10^5$$
$$x^c \div x^d = x^{c-d}$$
$$y^2 \div y^6 = y^{-4}$$

$$y^{10} \div y^2 = y^8$$
$$x^4 \div x^{-3} = x^7$$
$$n^5 \div n^4 = n$$
$$n^5 \div n^5 = n^0$$
$$n^{-3} \div n^{-5} = n^2$$
$$10^{5.2} \div 10^{3.8} = 10^{1.4}$$

Exercise 28.2

Perform the indicated divisions:

1. $x^8 \div x^2$
2. $x^{10} \div x^5$
3. $x^{12} \div x^5$
4. $n^4 \div n^{-3}$
5. $n^{-3} \div n^5$
6. $n^5 \div n$
7. $x^{-3} \div x^{-5}$
8. $y^{-5} \div y^{-6}$
9. $y^{-4} \div y^{-3}$
10. $y^5 \div y^5$
11. $x^9 \div x^3$
12. $x^4 \div x^6$
13. $n \div n^4$
14. $n^6 \div n^6$
15. $n^{-3} \div n^{-3}$
16. $x \div x$
17. $8^4 \div 8^3$
18. $2^{10} \div 2^5$
19. $x^{4.31} \div x^{2.15}$
20. $n^{5.46} \div n^{-1.24}$
21. $x^{2.13} \div x^{5.34}$
22. $10^{3.89} \div 10^{1.21}$
23. $10^{1.46} \div 10^{5.32}$
24. $10^{2.36} \div 10^{-1.54}$

28.3. POWER OF A POWER

By the expression "power of a power" we mean a problem such as the following example:

$$(x^5)^3$$

This example means the third power of the fifth power of x. The exponent 3 means that we are to use the quantity x^5 three times as a factor. That is,

$$(x^5)^3 \text{ means: } (x^5) \cdot (x^5) \cdot (x^5)$$

Then we apply the rule for multiplication and get the answer.

$$(x^5)^3 \text{ means } (x^5) \cdot (x^5) \cdot (x^5) = x^{15}$$

EXPONENTS, POWERS, AND ROOTS

The exponent on the base x in the answer could be found by multiplying the two exponents; that is, the power times the power: $(3)(5) = 15$. This example is an illustration of the rule for finding the power of a power.

RULE 3. To find a power of a power, we multiply the powers to get the new exponent on the given base. The base remains the same.

The rule may be stated as a formula:

$$(x^a)^b = x^{ab}$$

This is *not* the same as the rule for multiplication: $(x^a)(x^b) = x^{a+b}$.

The rule is true for all kinds of exponents. The following examples show the application of this rule:

$$(x^4)^5 = x^{20}; \quad (y^3)^{-2} = y^{-6}; \quad (n^4)^0 = n^0; \quad (x^6)^{2/3} = x^4$$

$$(5^6)^{1/2} = 5^3; \quad (8^3)^{1/3} = 8; \quad (10^{1.5})^4 = 10^6; \quad (10^4)^{1.2} = 10^{4.8}$$

$$(4^{-2})^{-3} = 4^6 = 4096; \quad (3^3)^2 = 3^6 = 729; \quad [(x^2)^3]^5 = x^{30}$$

Exercise 28.3

Find the indicated powers:

1. $(x^2)^6$
2. $(y^3)^{-2}$
3. $(n^{-5})^3$
4. $(10^{-3})^{-4}$
5. $(n^8)^{1/2}$
6. $(x^6)^{2/3}$
7. $(x^{-4})^{-1/2}$
8. $(n^{-8})^{-3/4}$
9. $(y^{2/3})^{3/2}$
10. $(n^{-2})^{-1/2}$
11. $(10^5)^2$
12. $(2^4)^3$
13. $(x^{1.2})^{3.5}$
14. $(x^{-2.6})^{2.1}$
15. $(y^{3.5})^{-2}$
16. $(x^{1.32})^{4.5}$
17. $(2^4)^{2.25}$
18. $(3^{2.4})^{2.5}$
19. $(x^{3.12})^0$
20. $(y^0)^{2.314}$
21. $(10^{2.143})^3$
22. $(10^{3.121})^4$
23. $(x^{4.2})^{-0.51}$
24. $(n^{-2.3})^{-1.2}$
25. $(10^{4.32})^{1/2}$
26. $(10^{5.43})^{1/3}$
27. $(10^{-3.2})^{1/4}$
28. $(10^{1.45})^{1/5}$
29. $[(x^2)^2]^3$
30. $[(n^2)^4]^{1/2}$
31. $[(x^4)^6]^{1/3}$
32. $[(x^{-2})^{-3}]^{-2}$

28.4. POWER OF A PRODUCT

A power of a product is shown in the following example:

$$(xy)^3$$

Here we have the product of two factors, x and y. In this example, the product, xy, is the be raised to the third power. For the meaning, we have

$$(xy)^3 \text{ means } (xy)(xy)(xy) = x \cdot y \cdot x \cdot y \cdot x \cdot y$$

Since multiplication can be done in any order of factors, we can write

$$x \cdot y \cdot x \cdot y \cdot x \cdot y = x \cdot x \cdot x \cdot y \cdot y \cdot y = x^3 y^3$$

Notice that the *exponent on the product becomes the exponent on each factor.*
This example is an illustration of the rule for the power of a product.

RULE 4. **The power of a product of several factors is equal to the product of the several factors raised to the same power as that of the product.**

The rule may be stated as a formula:

$$(xyz)^n = x^n y^n z^n$$

That is, the exponent on a product applies to each factor.
To emphasize the rule, consider a problem using arithmetic numbers:

$$(2 \cdot 6)^3 = (2^3) \cdot (6^3) = (8) \cdot (216) = 1728$$

If we first expand the quantity in parentheses, we get the same answer:

$$(2 \cdot 6)^3 = (12)^3 = 1728$$

If we have a product of several factors, each with separate powers, then an exponent applies only to the factor on which it is placed. For example if $x = 3$ and $y = 4$, then

$$xy^2 = (3)(4^2) = 48; \quad \text{but} \quad (xy)^2 = (12)^2 = 144$$

We must be very careful when negative signs are involved. For example, in $-x^2$, the exponent 2 does *not* apply to the minus sign. If $x = 5$, then

$$-x^2 = -5^2 = -25; \quad \text{but} \quad (-x)^2 = (-5)^2 = +25$$

EXPONENTS, POWERS, AND ROOTS

Example 1. Simplify the expression $(2x^2yz^3)^4$.

Solution. The expression is the power of a product. The exponent 4 is placed on each of the factors within the parentheses; we get

$$(2^4) \cdot (x^2)^4 \cdot (y)^4 \cdot (z^3)^4$$

Now the problem involves also the power of a power. Applying the rule, we get

$$(2^4)(x^8)(y^4)(z^{12}) = 16x^8y^4z^{12}$$

Example 2. Simplify the expression $(3x^2y)^4 \cdot (5xy^2)^3 \cdot (xy)^{-2}$.

Solution. This problem involves powers of products; powers of powers; and multiplication. Applying the first two of these, we get

$$(81x^8y^4) \cdot (125x^3y^6) \cdot (x^{-2}y^{-2})$$

Now we apply the rule for multiplication, and get

$$10125x^9y^8$$

Exercise 28.4

Simplify the following expressions:

1. $(ab)^4$
2. $(rst)^5$
3. $(xyz)^{-3}$
4. $(5x^5)^2$
5. $(-5x^4)^3$
6. $(3x^3)^4$
7. $(10x^{10})^2$
8. $(x^3y^5)^2$
9. $(a^2b^4c^3)^{-2}$
10. $(4a^2b^4c)^3$
11. $(x^2y^{-4}z^{-1})^{-2}$
12. $(5a^3bx^{-2})^4$
13. $(-4x^3y^{-2})^4$
14. $(-3x^3y^2)^3$
15. $(-2x^2yz)^5$
16. $-(-3x^2y)^4$
17. $-(-6x^{-2}y)^3$
18. $-5(-2x^3y^5)^4$
19. $(x^4y^6z^8)^{1/2}$
20. $(a^6b^9c^3)^{2/3}$
21. $(x^3y^5z^2)^{1/2}$
22. $(2^4 \cdot 3^2 \cdot 5)^2$
23. $(2^2 \cdot 3^2 \cdot 5)^3$
24. $(10^4 \cdot 8^2 \cdot 5^6)^{1/2}$
25. $(4x^3y^2)^2 \cdot (-2x^2y)^3$
26. $(5x^5y)^3 \cdot (-6xy^3)^2$
27. $(-2a^2b)^3 \cdot (-5a^3b)^2$
28. $(-2^4)(-3)^2(-3x^4y^3)^3$

28.5. POWER OF A FRACTION

An exponent on a fraction applies to the entire fraction, numerator and denominator. For example,

$$\left(\frac{4}{5}\right)^3 \text{ means } \left(\frac{4}{5}\right) \cdot \left(\frac{4}{5}\right) \cdot \left(\frac{4}{5}\right) = \frac{4 \cdot 4 \cdot 4}{5 \cdot 5 \cdot 5} = \frac{4^3}{5^3}$$

Note that the exponent 3 applies to numerator and denominator. This example is an illustration of the rule involving powers of fractions.

RULE 5. **An exponent on a fraction as a quantity indicates that both numerator and denominator are to be raised to the indicated power.**

The rule may be stated as a formula:

$$\left(\frac{x}{y}\right)^n = \frac{x^n}{y^n}$$

Example 1. Simplify

$$\left(\frac{2x}{3y}\right)^5$$

Solution. The exponent 5 applies to both factors of the numerator and both factors of the denominator, and we get

$$\left(\frac{2x}{3y}\right)^5 = \frac{32x^5}{243y^5}$$

Example 2. Simplify

$$\left(\frac{5x^2 y^3 z^0}{6ab^2 c}\right)^4.$$

Solution. The exponent 4 applies to numerator and denominator. However, both numerator and denominator consist of products and powers. Therefore, we also apply Rule 4 and Rule 3. We get

$$\frac{625 x^8 y^{12} z^0}{1296 a^4 b^8 c^4}$$

Example 3. Simplify

$$\left(\frac{3x^{-2} y^3 z^{-1}}{7a^4 b^{-1} c^{-3}}\right)^2$$

EXPONENTS, POWERS, AND ROOTS 441

Solution. The exponent 2 applies to each factor in both numerator and denominator. Applying Rule 4 and Rule 5, we get

$$\frac{9x^{-4}y^6z^{-2}}{49a^8b^{-2}c^{-6}}$$

Exercise 28.5

Simplify the following expressions:

1. $\left(\dfrac{1}{5}\right)^3$
2. $\left(\dfrac{2}{3}\right)^5$
3. $\left(\dfrac{x}{y}\right)^{-3}$
4. $\left(\dfrac{x^2}{a^3}\right)^4$
5. $\left(\dfrac{4x^3}{3y^2}\right)^3$
6. $\left(\dfrac{2a^2b^3c}{3xy^5}\right)^4$
7. $\left(\dfrac{x^6y^4}{a^8b^2}\right)^{1/2}$
8. $\left(\dfrac{a^6b^3c^9}{x^{12}}\right)^{1/3}$
9. $\left(\dfrac{m^{15}}{n^9}\right)^{2/3}$
10. $\left(\dfrac{x^4y^{-8}}{a^{12}b^{-4}}\right)^{-1/4}$
11. $\left(\dfrac{2x^2y^3z^4}{3ab^3c^{-1}}\right)^4$
12. $\left(\dfrac{3xy^{-2}z^{-1}}{5a^2bc^{-3}}\right)^3$
13. $\left(\dfrac{6x^2y^4}{4ab^3c}\right)^3$
14. $\left(\dfrac{x^0y^6z^2}{a^4b^8c^{-2}}\right)^{1/2}$
15. $\left(\dfrac{x^{-3}y^{-6}z^0}{a^0b^9c^{-12}}\right)^{1/3}$
16. $\left(\dfrac{a^4b^{-2}c^0}{x^{-6}y^8}\right)^{3/2}$

28.6. ZERO EXPONENT

In mathematics, we sometimes get a zero exponent, as in the expressions:

$$x^0; \quad 5^0; \quad y^0; \quad \text{and so on}$$

Such expressions must be given some logical meaning. We cannot explain the meaning of a zero exponent in the same way as we do for positive integral exponents. For example,

$$5^2 \text{ means } 5 \cdot 5; \quad 5^3 \text{ means } 5 \cdot 5 \cdot 5; \quad 5^0 \text{ means ?}$$

To understand the meaning of a zero exponent, let us see how they come about. Consider the following problem in division:

$$x^4 \div x^4$$

442 ALGEBRA

If we apply the rule for division involving exponents, we get

$$x^4 \div x^4 = x^{4-4} = x^0$$

Whether we like it or not, we get a zero exponent.

Now let us look at the same problem in another way. We write the problem as a fraction:

$$\frac{x^4}{x^4} = 1$$

We know that any quantity (except zero) divided by itself is equal to 1. When we use the law of exponents in division, we get

$$x^4 \div x^4 = x^0$$

Yet we know that

$$x^4 \div x^4 = 1$$

In one way, we get the answer, x^0. In the other way, we get 1. Therefore, we can say that

$$x^0 = 1$$

Whatever example we use, we get the same result:

$$n^7 \div n^7 = n^0; \quad \text{but} \quad n^7 \div n^7 = 1; \quad \text{then} \quad n^0 = 1$$

Concerning a zero exponent, we have the following rule.

RULE 6. Any quantity (except 0) expressed with a zero exponent is equal to 1.

Examples. $(3x)^0 = 1$; $(5x^3y^2)^0 = 1$; $(3x^2 + 5y^3)^0 = 1$.
In the expression, $3x^0$, the $x^0 = 1$; then $3x^0 = 3(1) = 3$.

Exercise 28.6

Simplify the following expressions:

1. 4^0
2. $(-3)^0$
3. $(2abc)^0$
4. $(6x^2y)^0$
5. $-(5x)^0$
6. $7(x^2)^0$
7. $8x^0y^0$
8. $-6a^0b^0$
9. -5^0
10. $(7x^2y^3)^0$

EXPONENTS, POWERS, AND ROOTS 443

11. $(-8n^3x^4)^0$
12. $-(-2^3ab^2)^0$
13. $4(x^2+y^2)^0$
14. $-(a^3b-c^2)^0$
15. $-3(4x^3-5)^0$
16. $5x^0y^0z^0$
17. $-5a^0b^0c^0$
18. $-6r^0s^0t^0$
19. $(5)(3^0)(x^0)$
20. $(-5)^0(-3)^0(-2)^0$
21. $(-3)(-2)^0(6x^0)$
22. $6n^0+(3n)^0$
23. $(8x^0)-8x^0$
24. $5a^0+(3a)^0+2^0$
25. $\dfrac{6x^0}{8y^0}$
26. $\dfrac{10n^0}{-2n^0}$
27. $\dfrac{-4x^0}{(-4x)^0}$
28. $\dfrac{(8x^0)(3y^0)}{6(2xy)^0}$
29. $\dfrac{5x^0+(2x)^0}{(3x)^0-4x^0}$
30. $\dfrac{(3x)^0+7x^0}{(4x)^0-3x^0}$

28.7. NEGATIVE EXPONENT

In mathematics we sometimes get a negative exponent, as in the expression, 5^{-2}. A negative exponent cannot be explained as we explain the meaning of a positive integral exponent. For example, in an expression such as 5^3, we have the meaning: $5 \cdot 5 \cdot 5 = 125$. A negative exponent cannot be explained in this way. Yet it must be given some logical meaning.

To understand the meaning of a negative exponent, let us see how it comes about. Consider the following example in division:

$$x^3 \div x^5$$

Now we recall the rule for division involving exponents. In division, we subtract the exponents. Then, in this example, we get

$$x^3 \div x^5 = x^{-2}$$

Whether we like it or not, we get a negative exponent.

Now, if we write the division as a fraction, and reduce the fraction, we get

$$\frac{x^3}{x^5} = \frac{x \cdot x \cdot x}{x \cdot x \cdot x \cdot x \cdot x} = \frac{1}{x^2}$$

Therefore, we conclude that the answers mean the same, and we have

$$x^{-2} = \frac{1}{x^2}$$

The foregoing example is an illustration of the rule.

RULE 7. Any quantity with a negative exponent can be written as the reciprocal of the same quantity with a positive exponent.

The rule may be stated in another way:

Any factor in the numerator of a fraction may be transferred to the denominator provided the sign of its exponent is changed; any factor in the denominator may be transferred to the numerator provided the sign of its exponent is changed.

Warning. Separate terms of polynomial numerators or denominators may not be shifted in this way. The rule says factors, not terms.

Example 1. Change this expression so that all exponents are positive:

$$\frac{x^2 y^{-3}}{a^{-1} b^4}$$

Solution. We transfer the factor, a^{-1}, to the numerator, and the factor y^{-3} to the denominator, and then change the signs of their exponents. We get

$$\frac{x^2 y^{-3}}{a^{-1} b^4} = \frac{ax^2}{b^4 y^3}$$

The expression can be written without showing a denominator by transferring all factors to the numerator:

$$\frac{x^2 y^{-3}}{a^{-1} b^4} = \frac{ab^{-4} x^2 y^{-3}}{1} = ab^{-4} x^2 y^{-3}$$

Example 2. Simplify each of the following expressions, writing each without negative exponents:

(A) $\dfrac{a}{b^{-2} c^{-2}}$ (B) $\dfrac{a}{b^{-2} + c^{-2}}$

Solution. In example (A), the two quantities, b^{-2} and c^{-2}, are *factors*. Therefore they can be transferred to the numerator if the signs of the exponents are changed:

(A) $\dfrac{a}{b^{-2} c^{-2}} = ab^2 c^2$ (denominator of 1)

However, in example (B), the denominator consists of two *terms*. The separate terms cannot be moved to the numerator. Example (B) can be simplified as follows:

$$(B) \quad \frac{a}{b^{-2}+c^{-2}} = \frac{a}{\frac{1}{b^2}+\frac{1}{c^2}} = \frac{a}{\frac{c^2+b^2}{b^2c^2}} = \frac{ab^2c^2}{c^2+b^2}$$

Another method of simplifying example (b) is to multiply numerator and denominator of the original fraction by the quantity, b^2c^2. Then we get

$$\frac{(b^2c^2)}{(b^2c^2)} \cdot \frac{a}{b^{-2}+c^{-2}} = \frac{ab^2c^2}{c^2+b^2}$$

Exercise 28.7

Express each of the following without negative exponents:

1. x^{-3}
2. $a^{-2}b^{-4}$
3. $a^2x^{-2}y^3z^{-1}$
4. $5x^2y^{-3}z^{-2}$
5. $-3a^3b^{-3}c^{-1}$
6. $4^{-2}x^{-5}y^{-3}$
7. $-3^{-1}a^{-2}b^2c$
8. $(-2^{-3})(5^2)a^{-1}b^3$
9. $(6^{-2})(3^2)x^{-2}y^{-1}$
10. $\dfrac{a^2b^{-1}}{x^{-1}y^3z}$
11. $\dfrac{1}{6x^2y^{-3}z^{-1}}$
12. $\dfrac{4^{-1}a^2b^{-2}}{6^{-1}a^{-1}b^2}$
13. $\dfrac{1}{-3ab^3c^{-4}}$
14. $\dfrac{8^{-1}a^{-1}b^3c}{2^{-3}x^{-1}y^2z^0}$
15. $\dfrac{9^{-1}x^{-3}y^2z^{-2}}{3^{-2}x^{-1}y^{-2}z}$
16. $\dfrac{1}{x^{-1}+y^{-1}}$
17. $\dfrac{x+y}{x^{-1}-y^{-1}}$
18. $\dfrac{a^{-2}-b^{-2}}{a^{-1}b^{-1}}$

19–21. Write each of the expressions in No. 10–12 with no expressed denominator; that is, with a denominator of +1, understood.

Evaluate each of the following expressions:

22. $(-4)^{-3}$
23. $(6^2)(2^{-5})$
24. $4^{-2}+3^{-2}$
25. $5^{-2}-2^{-3}$
26. $(6^{-2})(2^{+6})$
27. $50^{-1}+40^{-1}$
28. $(3^{-3})(9)(6^2)$
29. $2^{-3}+3^{-2}-6^{-1}$
30. $4^{-1}+6^{-1}-3^{-1}$
31. $\dfrac{1}{4^{-2}+2^{-3}}$
32. $\dfrac{3^{-2}+2^0}{2^{-3}+3^0}$
33. $\dfrac{5^{-2}-3^{-2}}{3^{-1}+5^{-1}}$

28.8. ROOTS OF NUMBERS

Before we explain the meaning of a frictional exponent, we must review briefly the meaning of a *root* of a number. A root of a number is one of the equal factors of the number. The *square root* of a number is one of the *two equal factors* of the number. For example the square root of 169 is 13 because 169 is the product of the two equal factors: 13 and 13.

The *cube root* of a number is one of the three equal factors of the number. The cube root of 125 is 5 because 125 is equal to (5)(5)(5). The *fourth root* of a number is one of the four equal factors. The fourth root of 81 is 3 because $(3)(3)(3)(3) = 81$.

The symbol for a root is the *radical sign* ($\sqrt{}$). The number under the radical sign is called the *radicand*. The particular root to be found is indicated by a small number placed in the notch of the radical sign. This number is called the *index* of the root. The expression

$$\sqrt[3]{125} = 5$$

means that the cube root of 125 is to be found. The cube root of 125 is 5. The entire expression is called a *radical*.

The *square root* is so often used that the index 2 is usually omitted. For example, $\sqrt{16} = 4$. The following examples show the roots of some numbers:

$$\sqrt{361} = 19 \qquad \sqrt[3]{27} = 3 \qquad \sqrt[4]{16} = 2 \qquad \sqrt[5]{32} = 2$$

$$\sqrt[4]{10{,}000} = 10 \qquad \sqrt{9x^2} = 3x \qquad \sqrt[3]{64x^6} = 4x^2$$

28.9. FRACTIONAL EXPONENTS

In mathematics we sometimes get a fractional exponent such as $5^{1/2}$. When this happens, the expression must be given some logical meaning. The expression does not mean one-half of 5. If we take $(\tfrac{1}{2})(5)$, then the $(\tfrac{1}{2})$ is a factor, not an exponent. Our problem now is to determine the meaning when an exponent is a fraction.

We have seen that one of the two equal factors of a number is called the square root of a number. For example,

$$\text{since} \quad (13)(13) = 169, \quad \text{then} \quad 13 = \sqrt{169}$$

That is, if two equal factors are multiplied together, one of these two equal factors is the square root of the product. Now consider the following example:

$$(x^{1/2})(x^{1/2}) = x^{(1/2 + 1/2)} = x$$

EXPONENTS, POWERS, AND ROOTS 447

Here we have two equal factors whose product is x. Then one of the factors must be the square root of x; that is,

$$x^{1/2} = \sqrt{x}$$

Therefore, the fractional exponent, $\frac{1}{2}$, means the same as the square root of the base. Other examples:

$$25^{1/2} = \sqrt{25} = 5; \quad 36^{1/2} = 6; \quad 5^{1/2} = \sqrt{5} = 2.236$$

For the exponent $\frac{1}{3}$, we can get the meaning in the same way. We have seen that the cube root of a number is one of the three equal factors of the number. For example,

$$\text{since} \quad (6)(6)(6) = 216; \quad \text{then} \quad 6 = \sqrt[3]{216}$$

Now consider the following example:

$$(x^{1/3})(x^{1/3})(x^{1/3}) = x^{(1/3+1/3+1/3)} = x$$

Therefore, one of these three equal factors is the cube root of x: that is,

$$x^{1/3} = \sqrt[3]{x}$$

Other fractional exponents have similar meanings:

$$8^{1/3} = \sqrt[3]{8} = 2 \quad 64^{1/3} = 4 \quad 16^{1/4} = 2$$

$$64^{1/6} = 2 \quad 243^{1/5} = 3 \quad (10{,}000{,}000)^{1/7} = 10$$

In the foregoing example, we had exponents with a numerator of 1, such as $\frac{1}{2}, \frac{1}{3}, \frac{1}{4}, \frac{1}{5}$, and so on. Now let us consider a fractional exponent in which the numerator is something other than 1, as in $x^{2/3}$. The meaning can be shown in the same way as in the previous examples. Consider the multiplication:

$$(x^{2/3}) \cdot (x^{2/3}) \cdot (x^{2/3}) = x^{6/3} = x^2$$

Here, again, we have three equal factors whose product is x^2. Therefore, one of the factors must be the cube root of x^2; that is,

$$x^{2/3} = \sqrt[3]{x^2}$$

Notice the following facts.

1. *In the fractional exponent, $\frac{2}{3}$, on the base x, the numerator 2 is the power on the base x in the radicand.*
2. *The denominator of the fractional exponent is the index of the root.*

This example is an illustration of the general rule for the meaning of fractional exponents.

RULE 8. In a fractional exponent, such as p/r, the numerator p of the exponent indicates a power, and the denominator r indicates a root.

The rule may be stated as a formula:

$$N^{p/r} = \sqrt[r]{N^p}$$

That is, a number N raised to the p/r power is equal to the rth root of the pth power of N.

Example 1. Find the value of $8^{2/3}$.

Solution. The expression means the cube root of 8^2, which is equal to 4. However, we can first find the cube root of 8, which is 2. Then this value can be squared. We get the same answer either way.

Example 2. Find the value of $25^{3/2}$.

Solution. The expression means the square root of 25^3, or $\sqrt{15{,}625}$. However, it is simpler to take the square root of 25 first, which is 5. Then we find 5^3, which is equal to 125.

Example 3. Find the value of $16^{-3/4}$.

Solution. First we apply Rule 7 concerning negative exponents. Then we apply the rule for fractional exponents, and get

$$16^{-3/4} = \frac{1}{16^{3/4}} = \tfrac{1}{8}$$

Example 4. Find the value of $7^{3/2}$.

Solution. In this example, if we first find the square root of 7, we get the approximate value, 2.646. Then this number must be raised to the third power. In this instance, it is better to find the third power of 7, and then find the square root of the result.

$$7^3 = 343; \quad \text{then} \quad \sqrt{343} = 18.52 \text{ (approx.)}$$

Exercise 28.8

Express the following with fractional exponents:

1. $\sqrt{n}$
2. $\sqrt{xy}$
3. $\sqrt[3]{35}$
4. $\sqrt[3]{x^2}$
5. $\sqrt{a^2-b^2}$
6. $\sqrt[4]{x^3+y^3}$
7. $\sqrt[5]{n^3+4}$
8. $\sqrt[3]{x^2y^2}$
9. $\sqrt{(x-2)^3}$
10. $\sqrt{a^2-b^2}$
11. $\sqrt[3]{x^2y^2z^4}$
12. $\sqrt{x^5yz^3}$

Evaluate each of the following expressions:

13. $16^{1/2}$
14. $81^{1/4}$
15. $27^{2/3}$
16. $16^{5/4}$
17. $32^{3/5}$
18. $9^{5/2}$
19. $4^{7/2}$
20. $8^{5/3}$
21. $\left(\dfrac{4}{9}\right)^{1/2}$
22. $\left(\dfrac{25}{49}\right)^{1/2}$
23. $\left(\dfrac{9}{16}\right)^{3/2}$
24. $\left(\dfrac{36}{49}\right)^{1/2}$
25. $16^{-1/4}$
26. $9^{-1/2}$
27. $125^{-2/3}$
28. $64^{-1/6}$
29. $(-8)^{1/3}$
30. $(-32)^{4/5}$
31. $(-27)^{2/3}$
32. $(-64)^{4/3}$
33. $5^{3/2}$
34. $10^{2/3}$
35. $2^{5/2}$
36. $3^{4/3}$
37. $(4^2+3^2)^{-1/2}$
38. $(5^2-3^2)^{3/2}$
39. $(12^2+5^2)^{1/2}$
40. $(6^2+8^2)^{3/2}$
41. $(15^2+8^2)^{1/2}$
42. $(3^2+7)^{3/2}$
43. $(11^2+4)^{2/3}$
44. $(4)(25)^{3/2}(4^{-1/2})$
45. $(20^2-16^2)^{1/2}$
46. $(13^2-12^2)^{1/2}$

28.10. SCIENTIFIC NOTATION

In scientific work we often use numbers that are very large or very small. For example, the number of electrons in one coulomb is the large number approximately equal to 6,280,000,000,000,000,000. The wave length of yellow light is approximately equal to 0.000023 of an inch. When such numbers are written in the usual way, they are sometimes awkward to use in computation. Such numbers are often written in a form called *scientific notation*.

450 ALGEBRA

Numbers are expressed in scientific notation by using powers of 10. For example, consider the number 93,200,000. To express this number in scientific notation, we first place the decimal point just to the right of the first significant digit. *The first significant digit of a number is the first digit that is not zero, starting at the left.* In the number 93,200,000, the first significant digit is 9. Then we place the decimal point between the 9 and the 3:

$$9.32$$

This position is called the *standard position of the decimal point.*

Now we multiply the number, 9.32, by the proper power of 10 that will make the value equal to the original value. To determine the proper power of 10, we count the number of places from standard position to the decimal point of the original number. In the number 93,200,000, the decimal point is understood to be at the right of the number, since it is a whole number. From the standard position to the decimal point of the original number, we count 7 places to the right. The power of 10 to be used in this example is, therefore, 7. The number, written in scientific notation, becomes

$$(9.32)(10^7)$$

The result is equal in value to the original number.

In the case of a decimal fraction, the number can be written in a similar manner. However, the power of 10 will be *negative*. For example, take the number

$$0.00000245$$

The first significant digit in this number is 2. Therefore, the standard position for the decimal point is between the 2 and the 4. Then we write

$$2.45$$

Now we count the number of places from standard position to the decimal point. In doing so, we move *from right to left.* The decimal point is six places to the *left* of standard position.

$$0.0\,0\,0\,0\,0\,2\,4\,5$$
$$\uparrow$$
standard position

Therefore we multiply the number 2.45 by 10^{-6}. In scientific notation, we have

$$0.00000245 = (2.45)(10^{-6})$$

EXPONENTS, POWERS, AND ROOTS 451

Scientific notation is especially convenient in connection with very large or very small numbers. However, any number can be written in scientific notation.

The number 36 can be written as $(3.6)(10^1)$, or simply $(3.6)(10)$.
The number 8.2 can be written as $(8.2)(10^0)$, which is $(8.2)(1)$.
The number 0.047 can be written as $(4.7)(10^{-2})$.

Scientific notation is convenient for expressing the relations between metric measurements:

1 meter = 10 decimeters	then	1 decimeter = 10^{-1} meter
1 meter = 10^2 centimeters		1 centimeter = 10^{-2} meter
1 meter = 10^3 millimeters		1 millimeter = 10^{-3} meter
1 meter = 10^6 microns		1 micron = 10^{-6} meter
1 kilometer = 10^3 meters		1 meter = 10^{-3} kilometer

In multiplication and division, it is often convenient to express the numbers in scientific notation.

Example 1. Multiply the following: $(460,000)(0.035)$.

Solution. In scientific notation, we have $(4.6)(10^5)(3.5)(10^{-2})$
Rearranging factors we have $(4.6)(3.5)(10^5)(10^{-2})$
Multiplying the first two factors, $(4.6)(3.5) = 16.1$
Multiplying the powers of 10, $(10^5)(10^{-2}) = 10^3$
The power of 10 in the product determines the position of the decimal point. Then
$$(460,000)(0.035) = (16.1)(10^3) = 16,100$$

Example 2. Multiply $(0.000000245)(32900)$.

Solution. In scientific notation: $(2.45)(10^{-7})(3.29)(10^4)$
Rearranging factors, $(2.45)(3.29)(10^{-7})(10^4)$
Multiplying the first two factors, $(2.45)(3.29) = 8.06$ (rounded off)
Combining the powers of 10, $(10^{-7})(10^4) = 10^{-3}$
For the product, we get

$$(8.06)(10^{-3}) = 0.00806$$

Example 3. Divide as indicated: $(45,000) \div (0.0018)$.

Solution. In scientific notation, we have $(4.5)(10^4) \div (1.8)(10^{-3})$
First we divide the following: $(4.5) \div (1.8) = 2.5$
Dividing the powers of 10, $(10^4) \div (10^{-3}) = 10^7$
For the quotient, we have

$$(2.5)(10^7) = 25,000,000$$

452 ALGEBRA

Example 4. Divide as indicated: $(0.0078) \div (25000)$.

Solution. In scientific notation we have $(7.8)(10^{-3}) \div (2.5)(10^4)$
First we divide the following: $(7.8) \div (2.5) = 3.12$
Dividing the powers of 10, $(10^{-3}) \div (10^4) = 10^{-7}$
For the quotient, we have

$$(3.12)(10^{-7}) = 0.000000312$$

Exercise 28.9

Write the number in each of the following examples in scientific notation:

1. The sun is about 92,800,000 miles from the earth.
2. Light travels about 186,200 miles per second.
3. The wavelength of red light is approximately 0.00063 millimeter.
4. The wavelength of yellow light is about 0.0000228 inch.
5. One angstrom unit is equal to one ten-millionth of a millimeter.
6. An atom weighs about 0.000000000000000000000166 gram.
7. An atom is approximately 0.000000005 inch in diameter.
8. One radio station broadcasts on a frequency of 1260000 cycles per second.
9. Light travels about 5,872,000,000,000 miles in one year.
10. One coulomb is equal to about 6,280,000,000,000,000,000 electrons.

Evaluate the following, using scientific notation. Express the answer in scientific notation and then in expanded form. Round answers to three significant digits. If an answer begins with the digit 1, round the answer off to four significant digits.

11. $(125000)(260)$
12. $(0.000142)(3400)$
13. $(0.000234)(0.00412)$
14. $(36000000)(42.3)$
15. $(63000)(0.0000214)$
16. $(0.00000072)(9300)$
17. $(7200000) \div (450)$
18. $(480000) \div (0.0015)$
19. $(0.00084) \div (2400)$
20. $(0.00096) \div (750)$
21. $(0.0036) \div (16000)$
22. $(15600) \div (65000)$
23. $(280000) \div (610000)$
24. $(320000) \div (0.0071)$
25. $(29) \div (640000)$
26. $(4.7) \div (6300)$

29 RADICALS

29.1. SQUARE ROOTS

In arithmetic we have already had the problem of finding the square roots of numbers. For example, if a square floor has an area of 169 square feet, then the length of one side is found by taking the square root of 169, which is 13. We usually write the problem:

$$\sqrt{169} = 13$$

In some actual problems, as the one mentioned here, we do not consider negative values. It would be meaningless to say that the side of a square is -13 feet. We cannot measure length by negative numbers.

Yet in mathematics, we do find that negative numbers force themselves into the solution of problems. For example, we get negative numbers as the roots of equations:

$$\text{if} \quad 5 - 3x = 17; \quad \text{then} \quad x = -4$$

Negative numbers also appear as the square roots of numbers. To be able to work with square roots correctly, it is necessary to understand exactly what is meant by the square root of any number. We have already mentioned the meaning in Section 28.8, but it is well to emphasize the meaning here.

The square root of any number is one of the two equal factors of the number. For example, the square root of 1225 is 35, because 35 can be multiplied by itself to make 1225; that is, $(35)(35) = 1225$.

Now, if we multiply the number, -35, by itself, we get $+1225$. Therefore, we must say that the square root of 1225 can also be -35. Consider another example. The square root of 169 is $+13$ because $(+13)(+13) = +169$. However, the square root of $+169$ is also -13, because $(-13)(-13) = +169$.

Every number has two square roots, one positive, the other negative. The

453

root indicated by the positive radical sign is called the *principal square root*. For example, +3 is the principal square root of 9.

When we use the symbol, $\sqrt{}$, to indicate a root, we must be sure to use it correctly. The symbol means the *principal root*. If we wish to indicate the negative square root, we use the symbol, $-\sqrt{}$, with a minus sign before the radical. To indicate both square roots by use of the symbol, we must write $\pm\sqrt{}$. For example,

$$\sqrt{9} = +3; \qquad -\sqrt{9} = -3; \qquad \pm\sqrt{9} = \pm 3$$

29.2. ANY ROOT OF A NUMBER

As we have previously mentioned (Sec. 28.8), *the cube root of a number is one of the three equal factors of a number.* For example, the cube root of 216 is 6 because $6^3 = 216$. The cube root is indicated by the expression

$$\sqrt[3]{216} = 6$$

The *index* 3 indicates that the cube root is to be found. In this problem the number 216 is called the *radicand*. The radicand is the number of which a root is to be found.

In the same way, we indicate the fourth root of 16 as $\sqrt[4]{16} = 2$. The fourth root of 16 is 2 because $2^4 = 16$. In general, any root r is indicated by the expression:

$$\sqrt[r]{N}$$

The expression means that we are to find the rth root of N.

We have said that the principal root is the positive value of the radical. However, in some cases, *the principal root may actually be negative.* For example,

$$\sqrt[3]{-216} = -6, \qquad \text{because} \qquad (-6)^3 = -216$$

In fact, *the principal odd root of a negative number is negative.*

The following additional examples show principal roots:

$$\sqrt[4]{81} = 3; \quad \sqrt[6]{64} = 2; \quad \sqrt[5]{-32} = -2; \quad \sqrt[3]{-125} = -5; \quad \sqrt[7]{-1} = -1;$$

In algebra, the radical sign has the same meaning as in arithmetic.

$$\sqrt{x^6} = x^3, \qquad \text{because} \qquad (x^3)(x^3) = x^6$$

$$\sqrt[3]{x^{12}} = x^4, \qquad \text{because} \qquad (x^4)(x^4)(x^4) = x^{12}$$

$$\sqrt{x^2 - 6x + 9} = x - 3, \qquad \text{because} \qquad (x-3)^2 = x^2 - 6x + 9$$

RADICALS 455

The foregoing examples are illustrations of the following rule.

RULE. To find the indicated root of a given power expressed with exponents, divide the exponents by the index of the root.

The rule may be stated as a formula:

$$\sqrt[r]{x^n} = x^{n/r}$$

The following examples show the application of the rule:

$$\sqrt[3]{10^6} = 10^2; \qquad \sqrt[3]{x^5} = x^{5/3}; \qquad \sqrt[4]{n^3} = n^{3/4};$$

$$\sqrt[3]{-64x^6} = -4x^2; \qquad \sqrt{25x^8y^4} = 5x^4y^2; \qquad \sqrt[4]{625x^{12}} = 5x^3$$

In algebra, as in arithmetic, *any expression has two square roots, one positive, the other negative.* If no sign (+ or −) is shown before the radical sign, then the principal root is indicated. For example, the square roots of x^2 are $+x$ and $-x$; but

$$\sqrt{x^2} = +x$$

Exercise 29.1

Find the indicated root in each radical:

1. $\sqrt{x^8}$
2. $\sqrt{y^4}$
3. $\sqrt{a^{10}}$
4. $\sqrt{5^{1.6}}$
5. $\sqrt[3]{7^{2.4}}$
6. $\sqrt[4]{10^{3.6}}$
7. $\sqrt[3]{64x^9}$
8. $\sqrt[3]{27x^{15}}$
9. $\sqrt[4]{16x^{16}}$
10. $\sqrt[3]{-8n^6}$
11. $\sqrt[5]{-32n^5}$
12. $\sqrt[3]{-343n^{18}}$
13. $\sqrt{36x^5}$
14. $\sqrt[3]{512x^7}$
15. $\sqrt[5]{243x^4}$
16. $\sqrt{(x-2)^2}$
17. $\sqrt[3]{(a+b)^3}$
18. $\sqrt{(x+y)^4}$

19. $\sqrt{n^2+8n+16}$
20. $\sqrt{a^2-2a+1}$
21. $\sqrt{n^2-24n+144}$
22. $-\sqrt{16x^8}$
23. $-\sqrt[3]{-27x^3}$
24. $-\sqrt[4]{256x^7}$

29.3. RATIONAL AND IRRATIONAL NUMBERS

A rational number is defined as a number that can be expressed as the quotient of two integers. That is, *a rational number can be written as a common fraction with numerator and denominator as integers.*

1. All common fractions are rational:

    2/3 is the quotient: $2 \div 3$;   13/23 means $13 \div 23$

2. All integers are rational:

    5 can be written 5/1; -3 can be written $-3/1$

3. All decimal fractions are rational:

    0.7 can be written 7/10;   3.47 can be written 347/100

Such numbers as $\sqrt{2}$, $\sqrt{3}$, $\sqrt{5}$, and π, are called *irrational numbers* to distinguish them from rational numbers. The number $\sqrt{2}$ is irrational because the exact square root of 2 cannot be written as a common fraction or as an exact decimal fraction. The same is true regarding such numbers as $\sqrt{3}$, $\sqrt{5}$, $\sqrt{6}$, and many other radicals, such as $\sqrt[3]{4}$. The number π is also irrational because it cannot be expressed as a common fraction or as an exact decimal.

We often use approximations for irrational numbers, such as the following:

$\sqrt{2} = 1.4142$ (approx.); $\sqrt{3} = 1.732$ (approx.)

$\sqrt{5} = 2.236$ (approx.); $\pi = 3.14159$ (approx.)

Let us consider a little more carefully the irrational number, $\sqrt{2}$. The number is *irrational*, and is only approximately equal to 1.414. However, if we multiply this number, 1.414, by itself, we get 1.999396, which is less than 2. The square root of 2 is more nearly equal to 1.414213. Yet if we multiply this number by itself, we do not get exactly 2. The square root of 2 will never come out even, as a common fraction or as a decimal, no matter how far we carry out the decimal part.

However, we can express the *exact square root of 2 by a symbol*:

$$\sqrt{2}$$

Now, if we say that this symbol is to represent the *exact square root of 2*, then this symbol *multiplied by itself* must be equal to 2. That is,

$$(\sqrt{2})(\sqrt{2}) = 2$$

Remember. The square root of any given number means some other number that can be multiplied by itself to produce the given number.

The following examples show the meaning of square root:

$$\sqrt{3} \cdot \sqrt{3} = 3; \quad (\sqrt{41})(\sqrt{41}) = 41; \quad (\sqrt{7})(\sqrt{7}) = 7$$

$$(\sqrt{10})(\sqrt{10}) = 10; \quad (\sqrt{x})(\sqrt{x}) = x; \quad (\sqrt{-5})(\sqrt{-5}) = -5$$

29.4. SIMPLIFYING RADICALS

In any radical, the root should be stated as a rational number if possible. For example, the expression, $\sqrt{49}$, should be stated as 7. The following expression should be simplified as shown:

$$\sqrt[3]{125x^6} = 5x^2$$

If a radical represents an irrational number, it can often be simplified to a form that is easier to use in a problem. In simplifying radicals, we have two objectives:

1. *Make the radicand as small as possible.*
2. *Eliminate fractions in the radicand.*

To see how the radicand is made as small as possible, consider the following example:

Example 1. Simplify the radical $\sqrt{32}$.

Solution. First we separate the radicand 32 into two *factors* such that one of the factors is the largest possible perfect square:

$$\sqrt{32} = \sqrt{(16)(2)}$$

We still have only one radical. But now we write this as the product of two radicals:

$$(\sqrt{16})(\sqrt{2})$$

Now we simplify the first radical, a factor, and get $\sqrt{16}=4$.
Then the 4 becomes the coefficient of the other radical, $\sqrt{2}$. The entire process is shown by the following steps:

$$\sqrt{32} = \sqrt{(16)(2)} = (\sqrt{16})(\sqrt{2}) = 4\sqrt{2}$$

Now, if we wish, we can easily find the approximate decimal value:

$$4\sqrt{2} = 4(1.4142) = 5.6568 \text{ (approx.)}$$

Let us summarize the steps in making the radicand as small as possible.

1. *First, we separate the radicand, if possible, into two factors so that one factor is a perfect power (square or other power) corresponding to the index of the root. If a square root is indicated, we find a perfect square. If a cube root is indicated, we find a perfect cube.*
2. *Second, we show the one radical as the product of two radicals.*
3. *Finally, we take the root of the perfect power and place this root as the coefficient of the other radical, which is irrational.*

The procedure depends on the following principle:

PRINCIPLE. A radicand may be factored and the radical may be written as the product of two radicals.

The following examples show how radicals are simplified.

Example 2. $3\sqrt{98} = 3\sqrt{(49)(2)} = 3(\sqrt{49})(\sqrt{2}) = 3(7)\sqrt{2} = 21\sqrt{2}.$

Example 3. $4\sqrt{\dfrac{5}{9}} = 4\sqrt{\dfrac{1}{9} \cdot 5} = 4\left(\dfrac{1}{3}\right)(\sqrt{5}) = \dfrac{4}{3}\sqrt{5}.$

Example 4. Simplify $\sqrt{\dfrac{2}{3}}.$

Solution. To eliminate the fraction in the radicand, we multiply the numerator and denominator by some number that will *make the denominator a perfect square*. Here we multiply numerator and denominator by 3:

$$\sqrt{\dfrac{2}{3}} = \sqrt{\dfrac{6}{9}} = \sqrt{\dfrac{1}{9} \cdot 6} = \dfrac{1}{3}\sqrt{6} \quad \text{(taking out the square root of 1/9)}$$

Example 5. Simplify $\sqrt{\dfrac{5}{32}}.$

RADICALS 459

Solution. To eliminate the fraction in the radicand, we multiply numerator and denominator by 2 to make the denominator a perfect square. We get

$$\sqrt{\frac{5}{32}} = \sqrt{\frac{10}{64}} = \sqrt{\frac{1}{64} \cdot 10} = \frac{1}{8}\sqrt{10}$$

Example 6. $\sqrt{12x^3y^6} = \sqrt{4x^2y^6 \cdot 3x} = 2xy^3\sqrt{3x}.$

In this example, we take out the square root of the perfect square, $4x^2y^6$.

Example 7. Simplify $\sqrt[3]{40}$.

Solution. Since a cube root is indicated, we look for a factor of 40 that is a perfect cube. It is the number 8. Then we have

$$\sqrt[3]{40} = \sqrt[3]{(8)(5)} = \sqrt[3]{8} \cdot \sqrt[3]{5} = 2\sqrt[3]{5}$$

Warning. You cannot remove roots of separate terms of a polynomial radicand.

For example, suppose we have the problem: $\sqrt{13}$. Suppose we write

$$\sqrt{13} = \sqrt{9+4}$$

So far, the statement is correct. However, now we *cannot* take the square root of separate terms. If we did so, we would get: $3+2=5$, as the $\sqrt{13}$. Roots may be taken only of *factors* of the radicand, not of separate terms.

Examples. Is this correct: $\sqrt{49-25} = 7-5$? No.

Does $\sqrt{x^2+y^2} = x+y$? No. Does $\sqrt{x^2-y^2} = x-y$? No.

Exercise 29.2

Simplify each of the following radicals as much as possible. Find the approximate numerical value (decimal value) of the first 8.

1. $\sqrt{48}$
2. $2\sqrt{75}$
3. $4\sqrt{18}$
4. $3\sqrt{50}$
5. $2\sqrt{45}$
6. $3\sqrt{20}$
7. $5\sqrt{24}$
8. $4\sqrt{63}$
9. $\frac{1}{2}\sqrt{72}$
10. $\frac{1}{5}\sqrt{300}$

11. $\dfrac{1}{4}\sqrt{128}$

12. $\dfrac{3}{5}\sqrt{75}$

13. $\dfrac{1}{3}\sqrt{90}$

14. $\dfrac{3}{4}\sqrt{80}$

15. $\dfrac{2}{3}\sqrt{54}$

16. $\dfrac{5}{3}\sqrt{99}$

17. $\sqrt{\dfrac{3}{4}}$

18. $\sqrt{\dfrac{7}{9}}$

19. $\sqrt{\dfrac{25}{32}}$

20. $\sqrt{\dfrac{49}{18}}$

21. $2\sqrt{\dfrac{3}{8}}$

22. $6\sqrt{\dfrac{16}{27}}$

23. $5\sqrt{\dfrac{4}{5}}$

24. $8\sqrt{\dfrac{2}{3}}$

25. $\sqrt[3]{16}$

26. $\sqrt[3]{54}$

27. $\sqrt[4]{48}$

28. $\sqrt[5]{64}$

29. $\sqrt{18x^3}$

30. $\sqrt{75n^7}$

31. $\sqrt{45x^9}$

32. $\sqrt{147x^3y^3}$

33. $\sqrt[3]{24x^7}$

34. $\sqrt[3]{32n^5}$

35. $\sqrt[4]{32n^9}$

36. $\sqrt[5]{96x^7}$

37. $\sqrt{16 \cdot 9}$

38. $\sqrt{16+9}$

39. $\sqrt{64+36}$

40. $\sqrt{64 \cdot 36}$

41. $\sqrt{x^6+x^4}$

42. $\sqrt{x^4-x^2}$

43. $\sqrt{36+9+4}$

44. $\sqrt{81+36+4}$

29.5. ADDITION AND SUBTRACTION OF RADICALS

If radicals are to be combined by addition or subtraction, they must be *like* or *similar radicals*. Similar radicals are those that have the *same radicand* and the same indicated root; that is, the *same index*, such as

$$\sqrt{5}; \quad 3\sqrt{5}; \quad -2\sqrt{5}; \quad 7\sqrt{5}$$

or
$$6\sqrt[3]{5xy}; \quad -4\sqrt[3]{5xy}; \quad 8\sqrt[3]{5xy}; \quad -\sqrt[3]{5xy}$$

Like or similar radicals are added or subtracted by adding or subtracting their coefficients, just as in combining any algebraic terms for example,

$$6\sqrt{2}+3\sqrt{2}-5\sqrt{2}=4\sqrt{2}$$

Sometimes unlike radicals can be transformed into like radicals and the results combined.

Example 1. Combine the terms $\sqrt{3}+7\sqrt{2}-5\sqrt{3}-\sqrt{2}+6\sqrt{3}$. Combining like radicals, we get

$$6\sqrt{2}+2\sqrt{3}$$

For the approximate decimal value, we have $6(1.414)+2(1.732)=11.948$.

Example 2. Combine $6\sqrt{2}+\sqrt{12}+5\sqrt{32}-6\sqrt{18}-7\sqrt{3}$.

Solution. When the radicals are simplified, the expression becomes

$$6\sqrt{2}+2\sqrt{3}+20\sqrt{2}-18\sqrt{2}-7\sqrt{3}=8\sqrt{2}-5\sqrt{3}$$

If required, the decimal value is $8(1.414)-5(1.732)=2.652$ (approx.).

Exercise 29.3

Simplify the radicals in each of the following examples, and combine like radicals. Finally, find the approximate decimal value of each example.

1. $\sqrt{2}+5\sqrt{72}+4\sqrt{50}-\sqrt{128}$
2. $\sqrt{5}+\sqrt{45}-\sqrt{80}+\sqrt{180}$
3. $\sqrt{3}-\sqrt{12}+3\sqrt{8}+\sqrt{200}$
4. $\sqrt{75}-\sqrt{125}-2\sqrt{108}+\sqrt{320}$
5. $\sqrt{27}+\sqrt{288}-\sqrt{48}-\sqrt{162}$
6. $6\sqrt{32}-\sqrt{98}+3\sqrt{96}-2\sqrt{54}$
7. $3\sqrt{40}+5\sqrt{490}-\sqrt{250}+\sqrt{60}$
8. $\sqrt{300}+\sqrt{800}-\sqrt{45}+\sqrt{450}$
9. $\sqrt{18}+\sqrt{12}+\sqrt{20}+\sqrt{24}+\sqrt{28}$
10. $\sqrt{10}+\sqrt{100}+\sqrt{1000}+\sqrt{10000}$

29.6. MULTIPLICATION OF RADICALS

We have seen that one radical may be written as the product of two radicals. For example,

$$\sqrt{18} \text{ can be written } \sqrt{9} \cdot \sqrt{2}$$

462 ALGEBRA

In the same way,

$$\sqrt{15} \text{ can be written } \sqrt{5} \cdot \sqrt{3}$$

Then the reverse procedure, multiplication, is possible. That is,

$$\sqrt{5} \cdot \sqrt{3} = \sqrt{15}$$

In general, if two radicals have the same index, then the radicands can be multiplied together. That is,

$$\sqrt{x} \cdot \sqrt{y} = \sqrt{xy}$$

This operation is always permitted, *with one exception*. If the two radicands are *both negative*, the product of the two radicands is *not positive*. This special case involves imaginary numbers, which require special treatment. However, at this time we do not include examples of this kind.

If two or more radicals have the same index, we follow these steps in multiplication.

1. *Multiply the coefficients for the coefficient of the product, observing the rule for multiplication of signed numbers.*
2. *Multiply the radicands for the radicand of the product.*
3. *Simplify the resulting radical if possible.*

Example 1. Multiply $(-4\sqrt{2})(-5\sqrt{3})$.

Solution. Multiplying coefficients, $\qquad (-4)(-5) = +20$
Multiplying radicands, $\qquad\qquad\qquad\qquad (2)(3) = 6$
For the complete answer, we have: $\quad (-4\sqrt{2})(-5\sqrt{3}) = +20\sqrt{6}$.

Example 2. Multiply $(3\sqrt{2})(4\sqrt{2})$.

Solution. Multiplying coefficients: $(3)(4) = +12$
For the radicands, $\sqrt{2} \cdot \sqrt{2} = 2$
For the product, we have $(3\sqrt{2})(4\sqrt{2}) = (12)(2) = 24$.

Note. In Example 2, notice that the radicands are equal. Whenever two radicands are equal, they should not be multiplied together.

In Example 2, we do not multiply the radicands and get $\sqrt{4}$. This is not only unnecessary, but it could actually lead to error in some cases. Instead, whenever we have a product such as $(\sqrt{2})(\sqrt{2})$, the product of the radicals should be immediately stated as

$$(\sqrt{2})(\sqrt{2}) = 2$$

RADICALS 463

In the same way,
$$(\sqrt{3})(\sqrt{3}) = 3$$

Example 3. Multiply $(-5\sqrt{3})(4\sqrt{3})$.

Solution. Multiplying coefficients:
$$(-5)(4) = -20$$

For the product of the radicals, we have
$$(\sqrt{3})(\sqrt{3}) = 3$$

For the entire product, we have
$$(-5\sqrt{3})(4\sqrt{3}) = (-20)(3) = -60$$

Example 4. Multiply $(2\sqrt{3})(-\sqrt{7})(-4\sqrt{2})$.

Solution. Multiplying coefficients:
$$(2)(-1)(-4) = +8$$

For the product of the radicands, we get
$$(3)(7)(2) = 42$$

Then the entire product becomes
$$(2\sqrt{3})(-\sqrt{7})(-4\sqrt{2}) = +8\sqrt{42}$$

Example 5. Multiply $(7\sqrt{3})(-2\sqrt{6})$.

Solution. Multiplying the coefficients, $(7)(-2) = -14$
For the product of the radicands, we get
$$(3)(6) = 18$$

Then the product becomes
$$(7\sqrt{3})(-2\sqrt{6}) = -14\sqrt{18}$$

Now we can simplify the radical:
$$-14\sqrt{18} = -14\sqrt{9 \cdot 2} = (-14)(3)\sqrt{2} = -42\sqrt{2}$$

Example 6. Multiply $(\sqrt{2})(\sqrt{8})$.

Solution. In most cases, the product of two irrational numbers is another irrational number. However, in this example, it happens that the product is rational:

$$(\sqrt{2})(\sqrt{8}) = \sqrt{16} = 4$$

Example 7. Multiply $(-5)(-2\sqrt{7})$

Solution. In this example, we multiply the coefficients, and get

$$(-5)(-2\sqrt{7}) = 10\sqrt{7}$$

Example 8. Multiply $2\sqrt{3}(5\sqrt{3} + 4\sqrt{5} - 7\sqrt{2})$.

Solution. This example involves the *multiplication of a monomial times a polynomial*. It is similar to the problem:

$$2x(5x + 4y - 7z)$$

We multiply each term of the polynomial by the monomial. For the product, we have

first term: $(2\sqrt{3})(5\sqrt{3}) = (10)(3) = 30$
second term: $(2\sqrt{3})(4\sqrt{5}) = 8\sqrt{15}$
third term: $(2\sqrt{3})(-7\sqrt{2}) = -14\sqrt{6}$

The complete product is

$$30 + 8\sqrt{15} - 14\sqrt{6}$$

Example 9. Multiply $(4\sqrt{6} + 3\sqrt{2})(3\sqrt{6} - 5\sqrt{2})$.

Solution. This example involves the *multiplication of two binomials*. It is similar in form to the problem: $(4x + 3y)(3x - 5y)$. We show the steps in the multiplication by writing one binomial below the other as shown at the left. Note especially the multiplications:

$$4\sqrt{6} + 3\sqrt{2}$$
$$3\sqrt{6} - 5\sqrt{2}$$
$$\overline{12 \cdot 6 + 9\sqrt{12}}$$
$$ -20\sqrt{12} - 15 \cdot 2$$
$$\overline{72 - 11\sqrt{12} - 30}$$

$$(\sqrt{6})(\sqrt{6}) = 6 \quad \text{and} \quad (\sqrt{2})(\sqrt{2}) = 2$$

Note that $\sqrt{12} = 2\sqrt{3}$. Then we get the simple answer: $42 - 22\sqrt{3}$. The answer can now be expressed as an approximate decimal: 3.896. If we were to express the original binomials in decimal form and then multiply, the work would involve much computation with decimals.

Example 10. Multiply $(2\sqrt{3}+3\sqrt{5})(2\sqrt{3}-3\sqrt{5})$.

Solution. Note that the *two binomials are exactly the same except for the middle sign*. Then one binomial is called the *conjugate form* of the other. They differ only in the middle sign. The problem is similar in form to

$$(a+b)(a-b) = a^2 - b^2$$

Then all we need to do is to take the square of the first term minus the square of the second term.

$$(2\sqrt{3})^2 = 12; \quad (3\sqrt{5})^2 = 45$$

For the product, we have

$$(2\sqrt{3}+3\sqrt{5})(2\sqrt{3}-3\sqrt{5}) = 12 - 45 = -33$$

Note that the product is a rational number. This is always true when a binomial involving radicals is multiplied by its conjugate form.

If two radicals do not have the same index, their product can only be indicated. However, the product may be simplified by reducing the radicals to the same order; that is, with the same index.

Example 11. Multiply $(\sqrt{3})(\sqrt[3]{5})$.

Solution. Here we change each radical so that they have the *same index*. First, *we write each radical in exponential form*. Then we change the fractional exponents to new fractions with the same denominator:

$$\sqrt{3} = 3^{1/2} = 3^{3/6} = \sqrt[6]{3^3} = \sqrt[6]{27}; \quad \sqrt[3]{5} = 5^{1/3} = 5^{2/6} = \sqrt[6]{5^2} = \sqrt[6]{25}$$

Now the two radicals have the same index; then we multiply the radicands:

$$(\sqrt[6]{27})(\sqrt[6]{25}) = \sqrt[6]{27 \cdot 25} = \sqrt[6]{675}$$

Exercise 29.4

Multiply; find approximate values for No. 3, 8, 13, 18, 23:

1. $(5\sqrt{2})(4\sqrt{3})$
2. $(-2\sqrt{5})(3\sqrt{2})$
3. $(-4\sqrt{2})(-3\sqrt{7})$
4. $(-3\sqrt{2})(5\sqrt{2})$
5. $(4\sqrt{5})(-2\sqrt{5})$
6. $(7\sqrt{3})(-2\sqrt{3})$

7. $(5\sqrt{18})(3\sqrt{2})$
8. $(-3\sqrt{8})(-4\sqrt{2})$
9. $(2\sqrt{32})(5\sqrt{2})$
10. $(3\sqrt{7})(\sqrt{7})(4\sqrt{3})$
11. $(4\sqrt{5})(2\sqrt{5})(\sqrt{5})$
12. $(-2\sqrt{3})(-\sqrt{3})(4\sqrt{3})$
13. $(2\sqrt{15})(3\sqrt{15})(\sqrt{15})$
14. $(4\sqrt{5})(-3\sqrt{5})(-5\sqrt{5})$
15. $3\sqrt{2}(5\sqrt{3}-2\sqrt{5}+4\sqrt{2})$
16. $-2\sqrt{5}(3\sqrt{2}-4\sqrt{5}+2\sqrt{10})$
17. $2\sqrt{3}(7\sqrt{3}+\sqrt{6}-\sqrt{12})$
18. $4\sqrt{2}(3\sqrt{5}+2\sqrt{3}-5\sqrt{8})$
19. $(3\sqrt{2}-2\sqrt{7})(4\sqrt{2}+5\sqrt{7})$
20. $(3\sqrt{5}+4\sqrt{2})(\sqrt{5}-3\sqrt{2})$
21. $(3\sqrt{3}+5\sqrt{2})(3\sqrt{3}-5\sqrt{2})$
22. $(3\sqrt{5}-4\sqrt{2})(3\sqrt{5}+4\sqrt{2})$
23. $(3\sqrt{2}-2\sqrt{6})^2$
24. $(4\sqrt{3}+5\sqrt{6})^2$
25. $(\sqrt{6}-5)^2$
26. $(8-3\sqrt{2})^2$
27. $(\sqrt[3]{2})(\sqrt[4]{3})$
28. $(\sqrt{5})(\sqrt[3]{6})$

29.7. DIVISION INVOLVING RADICALS

Division is the inverse of multiplication. Recall that *in multiplication, if the index of the radicals is the same*, then we follow the steps:

1. *Multiply the coefficients together.*
2. *Multiply the radicands together.*

Then, *in division, we follow the reverse process:* To divide one radical by another, we have the following steps: (provided the radicals have the same index.)

1. *Divide the coefficients.*
2. *Divide the radicands.*

However, in division we often run into some difficulties, as we shall see in some of the following examples.

Example 1. Divide $15\sqrt{14} \div 3\sqrt{2}$.

Solution. Dividing the coefficients, $15 \div 3 = 5$.
Dividing the radicands, $14 \div 2 = 7$
For the quotient, we have $15\sqrt{14} \div 3\sqrt{2} = 5\sqrt{7}$.

Example 2. $12\sqrt{15} \div 8\sqrt{3} = \dfrac{3}{2}\sqrt{5}$, or $\dfrac{3\sqrt{5}}{2}$

Example 3. $4 \div \sqrt{3}$

Solution. First, we write the division as a fraction:

$$\frac{4}{\sqrt{3}}$$

When division is indicated as a fraction, our objective is *to eliminate the radical in the denominator*. To do so, *we multiply numerator and denominator by a quantity that will make the denominator rational*. Such a quantity is called a *rationalizing factor*. In this example, we multiply numerator and denominator by $\sqrt{3}$. We get

$$\frac{(4)(\sqrt{3})}{(\sqrt{3})(\sqrt{3})} = \frac{4\sqrt{3}}{3}$$

Notice that *the denominator has been changed to a rational number*. The answer can be left as shown. However, if we wish, we can find the approximate decimal value:

$$\frac{4\sqrt{3}}{3} = \frac{4(1.732)}{3} = \frac{6.928}{3} = 2.309 \text{ (approx.)}$$

In a problem such as Example 3, note that when the division is written in the form of a fraction, we *eliminate the radical in the denominator*. That is, we make the denominator a rational number. There is a good reason for doing so. If we were to find the approximate value of the expression in the first form, we should have

$$\frac{4}{\sqrt{3}} = \frac{4}{1.732}$$

Notice that the division would then involve long division by a decimal fraction. On the other hand, when the denominator is changed to a rational number, the division is usually much easier. That is why we rationalize the denominator.

Example 4. Rationalize the denominator in the fraction, $\dfrac{5}{3\sqrt{2}}$

Solution. We multiply the numerator and denominator by $\sqrt{2}$, and get

$$\frac{5}{3\sqrt{2}} = \frac{(5)(\sqrt{2})}{(3\sqrt{2})(\sqrt{2})} = \frac{5\sqrt{2}}{6}$$

Now the expression can easily be reduced to an approximate decimal:

$$\frac{5\sqrt{2}}{6} = \frac{5(1.4142)}{6} = 1.1785 \text{ (approx.)}$$

Example 5. Rationalize the denominator of the fraction: $\dfrac{8}{4+\sqrt{3}}$.

Solution. When the denominator is a binomial containing a radical, then *the rationalizing factor must be the conjugate form of the denominator.* Since the denominator is the binomial, $4+\sqrt{3}$, the conjugate form is $4-\sqrt{3}$. The conjugate form is the same binomial with the middle sign changed. We get

$$\frac{8}{4+\sqrt{3}} = \frac{(8)(4-\sqrt{3})}{(4+\sqrt{3})(4-\sqrt{3})} = \frac{8(4-\sqrt{3})}{16-3} = \frac{8(4-\sqrt{3})}{13}$$

Example 6. Rationalize the denominator in the fraction:

$$\frac{\sqrt{5}+\sqrt{10}}{3\sqrt{5}+\sqrt{10}}$$

Solution. We multiply numerator and denominator by the conjugate form of the denominator:

$$\frac{(\sqrt{5}+\sqrt{10})(3\sqrt{5}-\sqrt{10})}{(3\sqrt{5}+\sqrt{10})(3\sqrt{5}-\sqrt{10})} = \frac{3\cdot 5 + 2\sqrt{50} - 10}{45 - 10} = \frac{15 - 10 + 2\sqrt{50}}{35}$$

The radical, $2\sqrt{50}$, can be simplified to $10\sqrt{2}$. The answer reduces to

$$\frac{5+10\sqrt{2}}{35} = \frac{1+2\sqrt{2}}{7}$$

Exercise 29.5

Rationalize the denominator in each example, and simplify if possible:

1. $\dfrac{1}{\sqrt{3}}$
2. $\dfrac{2}{\sqrt{5}}$
3. $\dfrac{5}{\sqrt{2}}$
4. $\dfrac{12}{\sqrt{3}}$
5. $\dfrac{20}{\sqrt{5}}$
6. $\dfrac{14}{\sqrt{2}}$
7. $\dfrac{8}{5\sqrt{2}}$
8. $\dfrac{15}{4\sqrt{3}}$

9. $\dfrac{35}{3\sqrt{5}}$

10. $\dfrac{3}{2\sqrt{15}}$

11. $\dfrac{5}{3\sqrt{10}}$

12. $\dfrac{12}{5\sqrt{6}}$

13. $\dfrac{2\sqrt{5}}{5\sqrt{3}}$

14. $\dfrac{3\sqrt{2}}{4\sqrt{5}}$

15. $\dfrac{5\sqrt{3}}{3\sqrt{2}}$

16. $\dfrac{3}{3-\sqrt{5}}$

17. $\dfrac{\sqrt{3}}{4-\sqrt{7}}$

18. $\dfrac{\sqrt{2}}{4+\sqrt{6}}$

19. $\dfrac{1}{5\sqrt{2}-2\sqrt{3}}$

20. $\dfrac{1}{4\sqrt{3}-\sqrt{5}}$

21. $\dfrac{1}{2\sqrt{5}+3\sqrt{2}}$

22. $\dfrac{2\sqrt{3}}{7-4\sqrt{3}}$

23. $\dfrac{5\sqrt{2}}{8-3\sqrt{7}}$

24. $\dfrac{6\sqrt{5}}{5-2\sqrt{6}}$

25. $\dfrac{3-\sqrt{5}}{3+\sqrt{5}}$

26. $\dfrac{4+\sqrt{7}}{4-\sqrt{7}}$

27. $\dfrac{3+\sqrt{2}}{8-5\sqrt{2}}$

28. $\dfrac{\sqrt{3}+\sqrt{5}}{4\sqrt{3}-2\sqrt{5}}$

29. $\dfrac{\sqrt{3}-\sqrt{6}}{2\sqrt{3}+\sqrt{6}}$

30. $\dfrac{\sqrt{6}-\sqrt{2}}{\sqrt{6}+3\sqrt{2}}$

30 QUADRATIC EQUATIONS

30.1. DEFINITIONS

A *quadratic equation* is an equation containing the second power, but no higher power, of the letter, such as x^2. The following is a quadratic equation: $x^2 - 5x + 6 = 0$.

An equation is often named according to the highest power of the letter it contains. Let us first consider the various kinds of equations.

An equation containing only the *first power* of x or other letter is called a *linear equation*. We have already solved equations of this kind.

Linear equations: $\quad 3x = 15; \quad 5x + 2 = 19$.

An equation containing a term in the *second power* but no higher power is called a *quadratic equation*.

Quadratic equations: $\quad 2x^2 - 7x + 3 = 0; \quad 4x^2 = 3x - 2$

An equation containing the *third power* of a letter but no higher power is called a *cubic equation*.

Cubic equations: $\quad x^3 - 5x^2 + 2x = 4; \quad 3y^3 - 5y^2 + 6 = 0$

An equation containing the *fourth power* but no higher power of a letter is called a *quartic equation*.

Quartic equations: $\quad 2x^4 - 5x^2 - 3 = 0; \quad 3x^4 - 5x^3 + 2x = 7$

Equations of a degree higher than the second are often named simply after the term of highest degree.

Third degree equation: $\quad x^3 - 4x^2 + 5x + 8 = 0$.
Fourth degree equation: $\quad 2x^4 + 5x^3 + 6x = 9$.
Fifth degree equation: $\quad n^5 - 2n^3 + 4n^2 + 3 = 0$.

A *root* or *solution* of any equation is any number that satisfies the equation; that is, any number that makes the equation true. We shall find that a *quadratic equation has two roots, a cubic equation has three roots*, and *a quartic equation has four roots*. In general, the number of roots of any equation is equal to the degree of the equation.

In this chapter, our concern is with quadratic equations. The general form of a quadratic equation is

$$ax^2 + bx + c = 0$$

This means:

1. The x^2 term has some coefficient we call a.
2. The x term has some coefficient we call b.
3. The constant term, which we call c, is the term or part that does not contain x in any form. (Of course, this term could be called cx^0.)

One or both of the coefficients, b and c, may be zero. However, the coefficient a cannot be zero for then we should not have a quadratic equation.

Here are some examples of quadratic equations. Notice that each must have a term containing a letter squared.

$$3x^2 + 5x + 2 = 0; \quad 4x^2 + 7x = 0; \quad 6x^2 + 8 = 0; \quad 5y^2 = 8$$

In the first equation, $a = 3$; $b = 5$; $c = 2$.
In the second equation, $a = 4$; $b = 7$; $c = 0$.
In the third equation, $a = 6$; $b = 0$; $c = 8$.

In a quadratic equation, if b is equal to zero, then the equation contains no first degree term. The equation is then called a *pure quadratic*, such as the following:

$$x^2 - 9 = 0; \quad 3x^2 + 7 = 0; \quad 4x^2 - 25 = 0$$

If a quadratic equation contains the first degree term, x, in addition to the x^2 term, it is called a *complete quadratic*, such as the following:

$$x^2 + x - 6 = 0; \quad 4x^2 + 7x = 0$$

We have said that a root of an equation is any number that satisfies the equation. Consider the following quadratic equation.

$$x^2 - 5x + 6 = 0$$

Let us try to find the roots by trial and error:

We try $x = 1$; does $1^2 - 5(1) + 6 = 0$? No. Then 1 is not a root.
We try $x = 2$; does $2^2 - 5(2) + 6 = 0$? Yes. Then 2 is a root.
We try $x = 3$; does $3^2 - 5(3) + 6 = 0$? Yes. Then 3 is a root.
We try $x = 4$; does $4^2 - 5(4) + 6 = 0$? No. Then 4 is not a root.

To *solve an equation* means to find the roots. In the foregoing equation, $x^2 - 5x + 6 = 0$, we found the roots by guessing. However, we need a more systematic method for solving quadratic equations. There are several methods, which will now be explained.

30.2. SOLVING A PURE QUADRATIC EQUATION

We have said that *a pure quadratic contains no first degree term*. To solve such an equation, we proceed as follows:

(1) *Isolate the term containing x^2, transposing other terms to the other side of the equation.*
(2) *If the x^2 has some coefficient other than 1, divide both sides of the equation by this coefficient, so that the coefficient of x^2 becomes 1.*
(3) *Take the square roots of both sides of the equation.*

Example 1. Solve the pure quadratic: $5x^2 - 45 = 0$.

Solution. Transposing the 45, $\quad 5x^2 = 45$
Dividing both sides of the equation by 5, $\quad x^2 = 9$
Taking the square roots of both sides, $\quad x = \begin{cases} +3 \\ -3 \end{cases}$

The two answers are often written: $x = \pm 3$; which means: $x = +3$; and $x = -3$.

Note. Actually it should be understood that the square roots of the left side of the equation are $+x$ and $-x$, just as the square roots of the right side are $+3$ and -3. This would lead to the following condition:

$$\begin{rcases} +x \\ -x \end{rcases} = \begin{cases} +3 \\ -3 \end{cases}$$

This statement implies four equations:

$$+x = +3; \quad +x = -3; \quad -x = +3; \quad -x = -3$$

However, the first and the fourth of these equations have the same meaning. This is also true for the second and the third. For this reason, the "+" and "−" are omitted before the x on the left side of the equation.

Example 2. Solve the equation $\qquad x^2 - 36 = 0$.

Solution. Transposing, $\qquad x^2 = 36$

Taking the square roots of both sides, $\qquad x = \begin{cases} +6 \\ -6 \end{cases}$, or $\quad x = \pm 6$

Example 4. Solve the equation $\qquad 4x^2 - 25 = 0$.

Solution. Transposing the 25, $\qquad 4x^2 = 25$

Dividing both sides by 4, $\qquad x^2 = \dfrac{25}{4}$

Taking the square roots of both sides, $\qquad x = \pm \dfrac{5}{2}$, or $\quad \pm 2.5$

Example 5. Solve the equation $\qquad 9x^2 - 20 = 0$.

Solution. Transposing the 20, $\qquad 9x^2 = 20$

Dividing both sides by 9, $\qquad x^2 = \dfrac{20}{9}$

Taking the square root of both sides, $\qquad x = \pm\sqrt{\dfrac{20}{9}} = \pm\dfrac{2}{3}\sqrt{5}$

Example 6. Solve the equation $\qquad 18x^2 - 5 = 0$.

Solution. Transposing the 5, $\qquad 18x^2 = 5$

Dividing both sides by 18, $\qquad x^2 = \dfrac{5}{18}$

Taking square root of both sides, $\qquad x = \pm\sqrt{\dfrac{5}{18}} = \pm\dfrac{\sqrt{10}}{6}$

Example 7. Solve the equation $\qquad 3x^2 - 16 = 0$.

Solution. Transposing the 16, $\qquad 3x^2 = 16$

Dividing both sides by 3, $\qquad x^2 = \dfrac{16}{3}$

Taking square root of both sides, $\qquad x = \pm\sqrt{\dfrac{16}{3}} = \pm\dfrac{4\sqrt{3}}{3}$

QUADRATIC EQUATIONS

Example 8. Solve the equation $\qquad x^2+9=0.$

Solution. Transposing the 9, $\qquad x^2=-9$
Taking square root of both sides, $\qquad x=\pm\sqrt{-9}$; the roots are imaginary.

Note. The square roots of negative numbers are called *imaginary numbers.*

Exercise 30.1

Solve the following pure quadratic equations by the foregoing method.

1. $x^2-16=0$
2. $x^2-25=0$
3. $x^2-81=0$
4. $4y^2-49=0$
5. $9y^2-100=0$
6. $16t^2-121=0$
7. $4n^2-1=0$
8. $16t^2-9=0$
9. $36y^2-25=0$
10. $9x^2-32=0$
11. $4x^2-45=0$
12. $25x^2-12=0$
13. $I^2-10^6=0$
14. $9E^2-10^8=0$
15. $V^2-10^{-6}=0$
16. $8x^2-3=0$
17. $32x^2-5=0$
18. $50x^2-7=0$
19. $3x^2-4=0$
20. $2x^2-9=0$
21. $5x^2-16=0$
22. $2n^2-5=0$
23. $3n^2-7=0$
24. $5x^2-3=0$
25. $x^2+16=0$
26. $x^2+25=0$
27. $n^2+144=0$
28. $9n^2-20=0$
29. $6n^2-169=0$
30. $16t^2-225=0$
31. $s-16t^2=0$ (solve for t)
32. $h^2-k^2=0$ (solve for h)
33. $A-\pi r^2=0$ (solve for r)
34. $V-\pi r^2 h=0$ (solve for r)

30.3. SOLVING QUADRATIC EQUATIONS BY FACTORING

Some quadratic equations, such as the following, can be solved by factoring:

$$x^2-5x+6=0$$

Factoring the left side, we have

$$(x-2)(x-3)=0$$

Now we have two factors whose product is zero. Let us see what this implies.

476 ALGEBRA

If we multiply two or more factors together and one of them is zero, then the product must be zero. For example,

$$(4) \cdot (0) = 0; \quad 0 \cdot 5 = 0; \quad 16 \cdot 35 \cdot 0 \cdot 48 = 0$$

Moreover, we know that *if the product of two factors is zero, then one of the factors must be zero.* We cannot get a product of zero unless one factor is zero. For example,

if $a \cdot b = 0$, then either a or b is zero.

Now, let us consider the example,

$$x^2 - 5x + 6 = 0$$

Factoring the left side, we have

$$(x-2)(x-3) = 0$$

Here the product of two factors is zero. Therefore, one of the factors must be zero. Either factor may be equal to zero. Then we take each factor and set it equal to zero:

if $x - 2 = 0$, $x = 2$; then one root is 2, which we can call r_1.
if $x - 3 = 0$, $x = 3$; then one root is 3, which we can call r_2.

Each of these roots will make the equation true:

if $x = 2$, we have $2^2 - 5(2) + 6 = 0$, which is true.
if $x = 3$, we have $3^2 - 5(3) + 6 = 0$, which is also true.

To solve a quadratic equation by factoring *depends on having zero as the product of two or more factors;* that is, one side of the equation must be zero. However, factoring cannot be used if the expression cannot be factored.

The following examples show how some quadratic equations can be solved by factoring. Observe the following steps:

1. *See that one side of the equation is zero, transposing if necessary.*
2. *Factor the other side of the equation.*
3. *Set each factor equal to zero and solve for x.*

Example 1. Solve by factoring: $\quad x^2 + 12 = 7x$

Solution. Transposing, $\quad x^2 - 7x + 12 = 0$
Factoring the left side, $\quad (x-3)(x-4) = 0$

Now we set each factor equal to zero and solve each for x:

if $x - 3 = 0$, $x = 3$, which we call r_1.
if $x - 4 = 0$, $x = 4$, which we call r_2.

Either root may be called r_1; then the other root is called r_2.

Example 2. Solve by factoring: $x^2 = x + 20$.

Solution. Transposing, $x^2 - x - 20 = 0$
Factoring the left side, $(x + 4)(x - 5) = 0$

We set each factor equal to zero and solve for x:

if $x + 4 = 0$, $x = -4$, which we call root r_1.
if $x - 5 = 0$, $x = 5$, which we call r_2.

Example 3. Solve by factoring: $y^2 - 5y + 6 = 12$

Solution. If we factor, we get $(y - 2)(y - 3) = 12$

However, factoring at this point does not help. If the product is not zero (0), then we know nothing about either factor. We must have zero on one side.

Transposing, $y^2 - 5y + 6 - 12 = 0$
Combining like terms, $y^2 - 5y - 6 = 0$
Factoring, $(y + 1)(y - 6) = 0$
if $y + 1 = 0$, $y = -1$; if $y - 6 = 0$, $y = 6$.

Then the two roots are -1 and $+6$.

Example 4. Solve by factoring: $3x^2 + 7x + 2 = 0$

Solution. Factoring the left side: $(x + 2)(3x + 1) = 0$
If $x + 2 = 0$, $x = -2$; if $3x + 1 = 0$, $x = -\frac{1}{3}$.

Example 5. Solve by factoring: $4x^2 = 5x$.

Solution. Transposing, $4x^2 - 5x = 0$
Factoring the left side, $x(4x - 5) = 0$
If $x = 0$, then one root is 0; if $4x - 5 = 0$, $x = \frac{5}{4}$.

Note. In solving equations like Example 5, we must be careful to solve for both roots. A common error in this type of equation is to divide both sides of the equation by x. If this is done, we get $4x - 5 = 0$, from which we get only one root, $x = \frac{5}{4}$. By dividing both sides of the equation by x, we lose the one root, $x = 0$.

Example 6. Solve by factoring: $\qquad x^2 + 9 = 6x$

Solution. Transposing, $\qquad x^2 - 6x + 9 = 0$
Factoring the left side, $\qquad (x-3)(x-3) = 0$
If the first factor is equal to zero, we have $x - 3 = 0$; $x = 3$.
If the second factor is equal to zero, we have $x - 3 = 0$; $x = 3$.

Note that in this example, *the roots are equal.* We might be inclined to say that the equation has only one root. However, remember a quadratic equation always has *two* roots. Here, the root 3, is often called a *double root*.

Example 7. Solve by factoring: $\qquad x^2 = 16$.

Solution. This pure quadratic equation can be solved by factoring.

Transposing, $\qquad x^2 - 16 = 0$
Factoring, $\qquad (x+4)(x-4) = 0$
If $x + 4 = 0$, $x = -4$; $\quad$ if $x - 4 = 0$, $x = 4$.
The roots are numerically equal but opposite in sign.

Example 8. Given the equation: $\qquad x^2 + 6x + 2 = 0$.

Note. This equation cannot be solved by factoring.

Exercise 30.2

Solve the following quadratic equations by factoring:

1. $x^2 + 7x + 10 = 0$
2. $x^2 + 15 = 8x$
3. $x^2 + 8 = 6x$
4. $y^2 = y + 12$
5. $y^2 = 6 + y$
6. $y^2 = 20 - y$
7. $n^2 = 9n - 18$
8. $n^2 + 9n + 20 = 0$
9. $n^2 + 16 = 10n$
10. $x^2 + 16 = 8x$
11. $x^2 = 12x - 36$
12. $x^2 + 10x + 25 = 0$
13. $2x^2 - 7x + 3 = 0$
14. $3x^2 - 5x = 2$
15. $4x^2 + 2 = 9x$
16. $4n^2 - 20n + 25 = 0$
17. $9n^2 + 4 = 12n$
18. $9 = 24n - 16n^2$
19. $3y^2 = 5y$
20. $4y^2 + 7y = 0$
21. $5t^2 = 8t$
22. $9x^2 = 5x$
23. $25n^2 = 36$
24. $25 = 4x^2$
25. $4x^2 - 9 = 5x$
26. $12x^2 - 12 = 7x$
27. $6x^2 + 5x = 6$
28. $6x^2 - 5x - 4 = 0$
29. $2x^2 - 5x - 6 = 0$
30. $4x^2 - 15 - 4x = 0$

30.4. SOLVING QUADRATIC EQUATIONS BY COMPLETING A SQUARE

The method of solving a quadratic equation by *completing a square* can be used to solve all types of quadratic equations. However, the method is a rather long, complicated process. For this reason, it is not much used for this purpose. However, the technique of completing a square is a useful device in much work in mathematics. Our chief purpose here is to derive a general formula for solving all types of quadratic equations. We illustrate the method by two examples.

Example 1. Solve by completing a square: $\quad x^2 - 8x - 9 = 0$

Solution. Transpose the 9 to the right side: $\quad x^2 - 8x = 9$

Now we complete the square on the left side by adding 16. Then we must add 16 also to the right side:

$$x^2 - 8x + 16 = 9 + 16$$

We write the left side as a square and at the same time combine the terms on the right:

$$(x-4)^2 = 25$$

Taking the square roots of both sides, $\quad x - 4 = \pm 5$

The result represents two equations; we solve each for x:

If $x - 4 = +5$, $x = 9$; if $x - 4 = -5$, $x = -1$.

The two roots are 9, and -1. This equation could have been solved by factoring.

Summary of steps in solving by completing a square.

1. See that the coefficient of x^2 is 1. If it is some number other than 1, divide both sides of the equation by the coefficient of x^2.
2. Transpose the constant term to the right side of the equation.
3. Add some quantity to both sides that will make the left side a perfect square. That is, complete the square on the left side. This quantity will be the square of one-half the coefficient of x.
4. Write the left side as a square of a binomial, and at the same time combine the terms on the right side.
5. Take the square roots of both sides, using "+" and "−" on the right side.
6. Solve the resulting equations for x.

Note. *Steps 1 and 2 may be reversed.*

Example 2. Solve by completing the square: $3x^2 + 7x + 2 = 0$.

Solution. Dividing both sides by 3, $\quad x^2 + \dfrac{7}{3}x + \dfrac{2}{3} = 0$

Transposing the $\frac{2}{3}$ to the right side, $\quad x^2 + \dfrac{7}{3}x = -\dfrac{2}{3}$

To find the quantity to be added to the left side to make a perfect square, we take $\frac{1}{2}$ of $\frac{7}{3}$, which is $\frac{7}{6}$. We square the $\frac{7}{6}$ and get $\frac{49}{36}$. This is the quantity to be added. It must be added to both sides of the equation. Then we get

$$x^2 + \frac{7}{3}x + \frac{49}{36} = -\frac{2}{3} + \frac{49}{36}$$

Now we write the left side as a square, combining the terms on the right:
$$\left(x + \frac{7}{6}\right)^2 = \frac{25}{36}$$

We take the square roots of both sides: $\quad x + \dfrac{7}{6} = \pm\dfrac{5}{6}$

Solving the resulting equations:

$$\text{if } x + \frac{7}{6} = +\frac{5}{6};\ x = -\frac{1}{3}; \quad \text{if } x + \frac{7}{6} = -\frac{5}{6},\ x = -2.$$

Note. *This equation can also be solved by factoring.*

Exercise 30.3

Solve the following equations by completing the square:

1. $x^2 - 4x - 5 = 0$
2. $x^2 + 6x - 7 = 0$
3. $x^2 - 6x + 8 = 0$
4. $x^2 + 7x + 10 = 0$
5. $x^2 - 2x - 8 = 0$
6. $x^2 + 3x - 10 = 0$
7. $x^2 - 5x - 14 = 0$
8. $x^2 + x - 6 = 0$
9. $x^2 - x - 2 = 0$
10. $2x^2 + 3x - 2 = 0$
11. $4x^2 - 5x - 6 = 0$
12. $3x^2 - 2x - 5 = 0$
13. $x^2 - 4x + 1 = 0$
14. $x^2 - 6x + 2 = 0$
15. $x^2 + 2x - 4 = 0$
16. $x^2 + 4x - 6 = 0$

30.5. SOLVING QUADRATIC EQUATIONS BY FORMULA

We have solved equations such as the following by factoring:

$$3x^2 + 7x + 2 = 0$$

In this equation, the roots are determined by the numbers: 3, 7, and 2. If these numbers are placed in the proper formula, the roots will be found.

We have said that the general quadratic equation has the form

$$ax^2 + bx + c = 0$$

In this equation: a represents the coefficient of x^2, the square term,
b represents the coefficient of x, the first-degree term,
c represents the constant term.

In every quadratic equation, the roots are determined by the numbers, a, b and c. If these numbers are placed in the proper formula, the roots will be obtained. Here is the formula, sometimes called the *"root machine."*

$$x = \frac{-b \pm \sqrt{b^2 - 4ac}}{2a}$$

This is the famous quadratic formula. If the numbers represented by a, b, and c are placed properly in the formula, the roots can be computed. We shall solve the following equation by use of the formula.

Example 1. Solve by formula: $3x^2 + 7x + 2 = 0$.

Solution. First, we write the formula:

$$x = \frac{-b \pm \sqrt{b^2 - 4ac}}{2a}$$

In this example, $a = 3$; $b = 7$; $c = 2$.
Inserting the constants in the formula,

$$x = \frac{-7 \pm \sqrt{49 - (4)(3)(2)}}{(2)(3)}$$

Then

$$x = \frac{-7 \pm \sqrt{49 - 24}}{6} = \frac{-7 \pm \sqrt{25}}{6} = \frac{-7 \pm 5}{6}$$

482 ALGEBRA

Then x has the two values:

$$x = \frac{-2}{6} = -\frac{1}{3}; \quad \text{and} \quad x = \frac{-12}{6} = -2$$

At this point, we should see how the formula is derived. To get the formula, we solve the general equation, $ax^2 + bx + c = 0$, by completing the square. We begin with the equation

$$ax^2 + bx + c = 0$$

Dividing both sides of the equation by a,

$$x^2 + \frac{b}{a}x + \frac{c}{a} = 0$$

Transposing the constant term,

$$x^2 + \frac{b}{a}x = -\frac{c}{a}$$

Add $\left(\frac{b}{2a}\right)^2$ to both sides,

$$x^2 + \frac{b}{a}x + \frac{b^2}{4a^2} = -\frac{c}{a} + \frac{b^2}{4a^2}$$

We write the left side as a square and combine the terms on the right:

$$\left(x + \frac{b}{2a}\right)^2 = \frac{b^2 - 4ac}{4a^2}$$

Taking the square roots of both sides,

$$x + \frac{b}{2a} = \frac{\pm\sqrt{b^2 - 4ac}}{2a}$$

Solving for x,

$$x = -\frac{b}{2a} \pm \frac{\sqrt{b^2 - 4ac}}{2a} = \frac{-b \pm \sqrt{b^2 - 4ac}}{2a}$$

QUADRATIC EQUATIONS 483

In using the formula, it is important that we identify correctly the numbers, a, b, and c, especially with regard to sign.

The equation should always be written so that a is positive. That is, *the x^2 term should always be made positive. One side must be zero.*

Example 2. Solve by formula $\quad 5x+6=4x^2$.

Solution. Transposing,
$$-4x^2+5x+6=0$$

Multiplying both sides by -1,
$$4x^2-5x-6=0$$

In this equation, $a=4$, $b=-5$, $c=-6$.
We shall rewrite the formula:
$$x=\frac{-b\pm\sqrt{b^2-4ac}}{2a}$$

Inserting the values of a, b, and c,
$$x=\frac{+5\pm\sqrt{25-(4)(4)(-6)}}{8}$$
$$=\frac{+5\pm\sqrt{25+96}}{8}$$
$$x=\frac{+5\pm\sqrt{121}}{8}=\frac{+5\pm 11}{8}$$

Then for one root, $\quad x=\dfrac{5+11}{8}=2$

for the other root, $\quad x=\dfrac{5-11}{8}=-\dfrac{3}{4}$

Example 3. Solve by formula $\quad x^2-6x+9=0$.

Solution. In this equation, $a=1$, $b=-6$, $c=9$.
Inserting the constants in the formula,
$$x=\frac{6\pm\sqrt{36-(4)(1)(9)}}{2}$$

Then

$$x = \frac{6 \pm \sqrt{36-36}}{2} = \frac{6 \pm \sqrt{0}}{2} = \frac{6 \pm 0}{2} = \begin{cases} +3 \\ +3 \end{cases}$$

In this equation, the two roots are equal. This will always be the result when the quantity under the radical sign is equal to zero.

Example 4. Solve by formula $x^2 = 4x+1$.

Solution. Transposing to make one side zero,

$$x^2 - 4x - 1 = 0$$

In this equation, $a = 1$, $b = -4$, $c = -1$.
Inserting these numbers in the formula,

$$x = \frac{+4 \pm \sqrt{16 - (4)(1)(-1)}}{2}$$

Then
$$x = \frac{+4 \pm \sqrt{16+4}}{2} = \frac{+4 \pm \sqrt{20}}{2}$$

At this point, we see that the roots are irrational since we have $\sqrt{20}$. However, the radical can be simplified and the roots reduced to lower terms:

$$x = \frac{+4 \pm 2\sqrt{5}}{2} = +2 \pm \sqrt{5}; \quad \text{or} \quad x = \begin{cases} 2+2.236 \\ 2-2.236 \end{cases} = \begin{cases} 4.236 \\ -0.236 \end{cases}$$

Example 5. Solve by formula: $3x^2 - 4x + 2 = 0$.

Solution. In this equation, $a = 3$, $b = -4$, $c = 2$.
Inserting these values in the formula,

$$x = \frac{+4 \pm \sqrt{16 - (4)(3)(2)}}{6}$$

Then
$$x = \frac{+4 \pm \sqrt{16-24}}{6} = \frac{+4 \pm \sqrt{-8}}{6}$$

Since we get a negative number under the radical sign, we simply state at this time that the roots of the equation are *imaginary*.

Example 6. Solve by formula: $4x^2 - 3x = 0$.

Solution. In this equation, $a = 4$, $b = -3$, $c = 0$. Inserting these values in the formula,

$$x = \frac{+3 \pm \sqrt{9 - (4)(4)(0)}}{8}$$

Then

$$x = \frac{+3 \pm \sqrt{9}}{8} = \frac{+3 \pm 3}{8}$$

The roots are: 0, and $\frac{3}{4}$.

Exercise 30.4

Solve the following quadratic equations by formula:

1. $3x^2 - 7x + 2 = 0$
2. $4x^2 + 9x + 2 = 0$
3. $2x^2 + 5x + 3 = 0$
4. $2x^2 + 3 = 5x$
5. $5x + 2 = 3x^2$
6. $4x + 3 = 4x^2$
7. $5x^2 = x + 4$
8. $2x^2 = x + 3$
9. $x^2 + 6 = 5x$
10. $x^2 = 10 - 3x$
11. $x^2 - 20 = x$
12. $x^2 + 3x = 18$
13. $4y^2 - 4y - 3 = 0$
14. $5y^2 - 3y - 2 = 0$
15. $6x^2 + 5x - 4 = 0$
16. $9x^2 + 12x + 4 = 0$
17. $4x^2 + 20x + 25 = 0$
18. $9x^2 + 16 = 24x$
19. $x^2 - 4x + 1 = 0$
20. $x^2 + 6x + 4 = 0$
21. $n^2 - 2n - 7 = 0$
22. $n^2 + 4n + 3 = 0$
23. $4n^2 - 2 = 7n$
24. $3t^2 + 5 = 8t$
25. $x^2 - 2x + 4 = 0$
26. $x^2 - 4x + 8 = 0$
27. $x^2 + 6x + 40 = 0$
28. $5x^2 = 6x - 4$
29. $3x^2 + 5x = 0$
30. $4x^2 - 9x = 0$
31. $3x^2 - 16 = 0$
32. $2x^2 - 9 = 0$
33. $5x^2 - 7 = 0$
34. $3x^2 = 5$

30.6. WORD PROBLEMS INVOLVING QUADRATIC EQUATIONS

In solving word problems involving quadratic equations, we begin in the same way as with any word problem. We let some letter represent one of the unknowns. Then we express the other unknown in terms of the same letter.

The equation we set up may turn out to be a quadratic. Then we solve the quadratic equation by any convenient method.

To solve a quadratic equation that appears in a word problem, we try factoring as a first choice. However, if the factors are not easily seen, we use the quadratic formula. In some problems we may get a pure quadratic equation, which is then solved by the pure quadratic method.

In most practical problems, any negative solution can usually be discarded because it often has no practical meaning. For example, the side of a rectangle can not be a negative number. However, if a problem involves only numbers, then all values must be considered, negative as well as positive.

Example 1. Find two consecutive odd numbers whose product is 63.

Solution. Let $x =$ the smaller odd number.
Then $x+2 =$ the next odd number.
For the product, we multiply one number by the other. For the equation, we have
$$x(x+2) = 63$$

Removing parentheses, $\quad x^2 + 2x = 63$
Transposing, $\quad x^2 + 2x - 63 = 0$
Factoring, $\quad (x-7)(x+9) = 0$
If $x-7 = 0$, $x = 7$; $\quad$ if $x+9 = 0$; $x = -9$.

Taking 7 as the smaller of the consecutive odd numbers, the next odd number is 9. Then one set of answers is the set of odd numbers: 7 and 9.

However, if we take the second answer, $x = -9$, for the smaller number, then the larger odd number is $-9+2 = -7$. Then we have the second set of consecutive odd numbers: -9 and -7. The product is still $+63$.

Example 2. The length of a rectangle is 6.5 inches greater than the width. The area of the rectangle is 360 square inches. Find the width and length.

Solution. Let $x =$ the width (in inches).
Then $x + 6.5 =$ the length (in inches).
For the area, we take the width times the length. Then we have the equation
$$x(x+6.5) = 360$$

Removing parentheses, $\quad x^2 + 6.5x = 360$
Transposing, $\quad x^2 + 6.5x - 360 = 0$
Multiplying both sides by 2, $\quad 2x^2 + 13x - 720 = 0$
Factoring, $\quad (2x+45)(x-16) = 0$
If $2x+45 = 0$, $x = -22.5$; $\quad$ if $x - 16 = 0$, $x = 16$.

In this example, we discard the negative answer as having no meaning, since the width can not be negative. Then, for the answer, we have

$$\text{width} = 16 \text{ inches}; \quad \text{length} = 22.5 \text{ inches, (that is, } 16+6.5)$$

Note. *In an example of this kind, it may be a little difficult to find the factors. If so, the quadratic formula can be used.*

Example 3. One square field has a side that is 12 rods longer than the side of a smaller square field. The total area of the two fields is 1224 square rods. Find the size of each field.

Solution. Let $x =$ the side of the smaller field (in rods).
Then $x + 12 =$ the side of the larger field (in rods).
For the areas of the fields, we have (in square rods):
$\quad x^2 =$ area of smaller field; $\quad (x+12)^2 =$ area of larger field.
For the equation, we have
$$x^2 + (x+12)^2 = 1224$$

Removing parentheses, $\quad x^2 + x^2 + 24x + 144 - 1224 = 0$
Combining like terms, $\quad 2x^2 + 24x - 1080 = 0$
Dividing both sides by 2, $\quad x^2 + 12x - 540 = 0$
Factoring, $\quad (x-18)(x+30) = 0$

Then, $x = 18$; and $x = -30$. We discard the negative value. Then the side of the smaller field is 18 rods; the larger field, $18 + 12 = 30$ rods.

Example 4. The hypotenuse of a right triangle is 45 inches long. One leg of the right triangle is 9 inches longer than the other leg. Find the perimeter and the area of the triangle.

Solution. Let $x =$ the length of the shorter leg (in inches).
Then $x + 9 =$ the length of the longer leg (in inches).
Now we use the Pythagorean rule: the square of the hypotenuse of a right triangle is equal to the sum of the squares of the legs. Then we have the equation:

$$x^2 + (x+9)^2 = 45^2$$

Removing parentheses, $\quad x^2 + x^2 + 18x + 81 = 2025$
Transposing and combining, $\quad 2x^2 + 18x - 1944 = 0$
Dividing both sides by 2, $\quad x^2 + 9x - 972 = 0$
Factoring, $\quad (x-27)(x+36) = 0$
Solving, $\quad x = +27$; and $x = -36$.

We discard the negative as having no meaning in this problem. Then, taking 27 as the length of the short leg, the long leg is $27+9=36$.

For the perimeter, we have $\quad 27+36+45=108$ inches.
For the area we have $\quad\quad\quad \frac{1}{2}(36)(27)=486$ (square inches).

Example 5. The sum of a number and its square is 90. Find the number.

Solution. Let $x =$ the number.
Then $x^2 =$ the square of the number.
For the equation, we have,
$$x^2 + x = 90$$

Transposing, $\quad\quad\quad x^2 + x - 90 = 0$
Factoring, $\quad\quad\quad (x-9)(x+10) = 0$
Solving for x, $\quad\quad x = +9$; and $x = -10$.

If we take the first answer, $x = +9$, we have the square: $9^2 = 81$. Now, if we add the number to its square, we have: $81 + 9 = 90$. Therefore, one answer is $x = 9$, as the number. Now, let us take the second answer: $x = -10$. The square of -10 is $+100$. If we add the square of -10 to -10 itself, we have: $(+100) + (-10) = 90$. Therefore, the second answer is also correct.

Example 6. A bus driver has a regular run of 384 miles at a regular rate of travel. One day, he had to reduce his speed by 16 miles per hour, and as a result, he was 2 hours late. Find his regular rate of travel and the usual time of the trip.

Solution. We make use of the distance formula: $d = rt$. Then $t = d/r$. That is, *the time is equal to the distance divided by the rate.*
  Let $x =$ his regular rate (in miles per hour)
Then $x - 16 =$ the slower rate (in miles per hour) for the particular day.
For the time required in each case, we divide the distance by the rate:

Then $\quad\quad\quad \dfrac{384}{x} =$ the usual time (in hours)

and

$\quad\quad\quad \dfrac{384}{x-16} =$ the time required at the slower rate (in hours)

Since the time was 2 hours greater at the slower rate, we have the equation
$$\frac{384}{x-16} = \frac{384}{x} + 2$$

We multiply both sides of the equation by the quantity: $x(x-16)$, and get

$$384x = 384(x-16) + 2x(x-16)$$

Removing parentheses, $\qquad 384x = 384x - 6144 + 2x^2 - 32x$
Transposing and combining, $\qquad 2x^2 - 32x - 6144 = 0$
Dividing both sides by 2, $\qquad x^2 - 16x - 3072 = 0$
Factoring, $\qquad (x-64)(x+48) = 0$

Then $x = 64$, and $x = -48$; representing the rates of speed. In this problem, we discard the negative value for the rate. The regular rate is therefore 64 miles per hour. The regular time of the trip was 6 hours.

Exercise 30.5

1. Find two consecutive even numbers whose product is 168.
2. The sum of a number and its square is 42. Find the number.
3. The sum of the squares of two consecutive odd numbers is 290. What are the numbers?
4. The length of a rectangle is 4 inches greater than the width. The area of the rectangle is 60 square inches. Find the width, length, and perimeter.
5. The length of a rectangle is 3.5 inches greater than the width. The area of the rectangle is 186 square inches. Find the width and the length.
6. The length of a rectangle is 3 inches less than twice the width. The area of the rectangle is 104 square inches. Find the length and width.
7. The length of a rectangle is 3.25 inches greater than the width. The area of the rectangle is 308 square inches. Find the width and length.
8. A farmer has two square fields. The larger field has a side that is 8 rods greater than the side of the smaller field. The total area of the two fields is 424 square rods. Find the size of each field.
9. One square has a side that is 10 inches greater than the side of another square. The total area of the two squares is 1850 square inches. Find the size and the area of each square.
10. The length of a rectangle is twice the width and the area is 450 sq. in. Find the width and the length.
11. The length of a rectangle is 8 inches greater than the width. If the width is increased by 2 inches and the length is increased by 7 inches, the area is doubled. Find the width and length.
12. The hypotenuse of a right triangle is 17 inches long. One leg is 7 inches longer than the other leg. Find the perimeter and area of the triangle.
13. The hypotenuse of a right triangle is 22.5 inches long. One leg is 4.5 inches longer than the other leg. Find the perimeter of the triangle.
14. A lawn has a length that is 1.5 times the width. If the width and the length are each increased by 8 yards, the area is doubled. Find the size of the lawn.
15. A bus driver has a regular run of 360 miles at a regular rate of travel. One day

490 ALGEBRA

due to bad weather, his rate was reduced by 15 miles per hour, and as a result he was 4 hours late. Find his regular rate and time.

16. Two people made a trip of 54 miles by bicycle. One traveled at a rate of 1.5 miles per hour faster than the other, and as a result he made the trip in 3 hours less time. How fast did each travel?

17. Two men start on a hike of 18 miles. One walks 0.75 miles per hour faster than the other and as a result, he travels the distance in 2 hours less time. Find the rate of travel for each.

18. A bus driver on a regular run of 312 miles found that he had to reduce his regular speed by 9 miles per hour. Then the trip took 1.5 hours longer than the usual time. What was his usual speed?

19. Two boys agree to mow a lawn that is 120 feet long and 90 feet wide. Each boy is to mow half the lawn. How wide a strip of uniform width should the first boy mow around the two sides and two ends until half the lawn is mowed?

20. An orchard has 525 fruit trees planted in rows of uniform length. The number of trees in each row is 4 more than the number of rows. Find the number of rows and the number of trees in each row.

21. A total of 192 tomato plants are planted in rows of uniform length. The number of plants in each row is 3 times the number of rows. Find the number of rows and the number of plants in each row.

22. Some plants are planted in rows of uniform length. The number of plants in each row is 18 more than the number of rows. If there are a total of 175 plants, find the number of rows and the number of plants in each.

23. A variable electric current, i, is given by the formula: $i = t^2 - 7t + 12$. If t represents the number of seconds, at what two times is the current equal to 2 amperes?

30.7. IMAGINARY NUMBERS*

In mathematics it often happens that we have a problem involving the square root of a negative number, as in the equation

$$x^2 + 9 = 0$$

Transposing,

$$x^2 = -9$$

To solve the equation, we must find a number whose square is -9. Now, we know that the square root of -9 is neither $+3$ nor -3. If we multiply $+3$ by itself, we get $+9$. If we multiply -3 by itself, we get $+9$. That is,

$$(+3)^2 = +9 \quad \text{and} \quad (-3)^2 = +9$$

Then the square root of -9 must be a new kind of number.

* Optional.

Although we cannot express the square root of -9 as an ordinary number, we can represent it by the symbol:

$$\sqrt{-9}$$

If we use this symbol, $\sqrt{-9}$, to represent the square root of -9, then this symbol multiplied by itself must equal -9. This is the meaning of square root. That is,

$$(\sqrt{-9})(\sqrt{-9}) = -9$$

The symbol, $\sqrt{-9}$, is called an *imaginary number*. *The square root of any negative number is called an imaginary number*. All other numbers, such as those we have used up to this time, are called *real numbers*.

Imaginary numbers: $\sqrt{-4}$; $\sqrt{-7}$; $\sqrt{-25}$; $\sqrt{-\frac{1}{2}}$; $\sqrt{-\frac{3}{4}}$; $\sqrt{-.25}$; $\sqrt{-\pi}$.

Real numbers: 3; -5; -9; $\frac{2}{3}$; π; $\sqrt{2}$; $\sqrt{5}$; 2.13; $\sqrt{\pi}$; $-\sqrt{3}$.

Imaginary numbers can be simplified in a way that is similar to simplifying some radicals. For example, we have seen that a radicand can be separated into factors:

$$\sqrt{32} = \sqrt{(16)(2)} = \sqrt{16} \cdot \sqrt{2} = 4\sqrt{2}$$

In a similar way, we can separate the radicand of an imaginary number so that (-1) appears by itself as one of the factors:

$$\sqrt{-4} = \sqrt{4(-1)} = \sqrt{4} \cdot \sqrt{-1} = 2\sqrt{-1}, \quad \text{or 2 of these: } \sqrt{-1}$$

$$\sqrt{-9} = \sqrt{9} \cdot \sqrt{-1} = 3\sqrt{-1}; \quad \text{or 3 of these: } \sqrt{-1}$$

$$\sqrt{-32} = \sqrt{32} \cdot \sqrt{-1} = \sqrt{16} \cdot \sqrt{2} \cdot \sqrt{-1} = 4\sqrt{2} \cdot \sqrt{-1}$$

Note that each imaginary number can be written with $\sqrt{-1}$ as one factor, having a real number as its coefficient. An imaginary number can always be simplified so that $\sqrt{-1}$ appears as a factor. This symbol is called the imaginary unit. That is

$$\sqrt{-1} = i \text{ (imaginary) unit}$$

This definition is often shortened to read: $\sqrt{-1} = i$. The i is used because it is the first letter of the word *imaginary*. In electrical engineering, the unit, $\sqrt{-1}$, is usually represented by j because the letter i is often used to represent electric current.

Remember, we do not get rid of the radical, $\sqrt{-1}$, simply by calling it i or j. The i or j is used only for convenience to represent the imaginary unit, $\sqrt{-1}$.

Imaginary numbers are apt to be confusing to a student when he is faced with them for the first time. You may wonder, "What good are such numbers?" It is true we cannot count with them. But neither can we count with negative numbers. Yet we have seen that negative numbers have important uses. Imaginary numbers have important uses in mathematics, especially in electrical engineering. The first thing to do is to learn how to work with imaginary numbers. Then you will find that they are not so confusing as you might think.

To work with imaginary numbers, we first simplify them so each one shows the imaginary unit, $\sqrt{-1}$, or i or j. Then we can add and subtract such numbers.

Examples. (Use either i or j for the imaginary unit, $\sqrt{-1}$.)
1. Add: $\sqrt{-9} + \sqrt{-25} + \sqrt{-49} = 3i + 5i + 7i = 15i$.
2. Subtract: $\sqrt{-81} - \sqrt{-16} = 9j - 4j = 5j$.

In multiplying imaginary numbers, we must remember that $\sqrt{-1}$ means the square root of -1. Therefore, this number, $\sqrt{-1}$, multiplied by itself is equal to -1. Then

$$\sqrt{-1} \cdot \sqrt{-1} = -1; \text{ that is, } (i^2) = (i)(i) = -1, \text{ a \textit{real} number}$$

Example 3. Multiply $(5i)(7i) = (5)(7)(i)(i) = 35i^2 = 35(-1) = -35$.

Example 4. Multiply $(-3)(6j)(2j)(4j)$.

Solution. The products of the coefficients is -144. Then we also have

$$(j)(j)(j) = (j^2)(j) = (-1)(j) = -j$$

The complete product becomes: $(-144)(-j) = +144j$.

Example 5. Multiply $(\sqrt{-9})(\sqrt{-16})$.

Solution. First we simplify each imaginary number and get: $(3i)(4i)$. Then the product becomes: $(3)(4)(i^2) = 12i^2 = 12(-1) = -12$. In this example, if we first multiply the radicands, we get the wrong answer.

Exercise 30.6

Simplify the following:

1. $\sqrt{-81}$
2. $\sqrt{-100}$
3. $3\sqrt{-144}$
4. $5\sqrt{-36}$
5. $\sqrt{-169}$
6. $\sqrt{-40}$
7. $\sqrt{-75}$
8. $\sqrt{-0.16}$
9. $\sqrt{-\dfrac{4}{9}}$
10. $\sqrt{-25x^2}$

Add or subtract as indicated:

11. $\sqrt{-49}+\sqrt{-64}+\sqrt{-1}+5\sqrt{-1}$
12. $3\sqrt{-196}-\sqrt{-9}-2\sqrt{-25}$

Multiply:

13. $3\sqrt{-1}\cdot 5\sqrt{-1}$
14. $(4i)(7i)$
15. $(6j)(8j)$
16. $(-6j)(5j)$
17. $(\sqrt{-25})(\sqrt{-9})$
18. $(\sqrt{-4})(\sqrt{-100})$
19. $(\sqrt{-32})(\sqrt{-18})$

Solve these quadratic equations:

20. $x^2+16=0$
21. $x^2+4=0$
22. $x^2+400=0$
23. $4x^2+25=0$

30.8. COMPLEX NUMBERS*

Sometimes the root of a quadratic equation consists of two parts. As an example, let us solve the following equation by formula:

$$x^2-4x+13=0$$

By the formula, we get

$$x = \frac{4\pm\sqrt{16-52}}{2} = \frac{4\pm\sqrt{-36}}{2} = \frac{4\pm 6i}{2} = 2\pm 3i$$

One root is the expression, $2+3i$; the other root is $2-3i$.

An entire expression, such as $2+3i$, is to be considered as one number. The number, $2+3i$, is partly real and partly imaginary. The real part is 2; the imaginary part is $3i$. The entire number, $2+3i$, is called a *complex number*.

* Optional.

494 ALGEBRA

A complex number is a number that is partly real and partly imaginary. It has the general form: $a + bi$. That is, a represents the real part, and b represents the coefficient of i in the imaginary part. The two parts of a complex number are separated by a plus (+) or a minus (−). The two parts *cannot* be combined into a single term.

Examples. $3 + 5i$; $-7 + 2i$; $8 - 3i$; $-4 - i$; $-\frac{3}{4} - \frac{5}{4}i$.

If we use j to represent the unit, $\sqrt{-1}$, then a complex number has the form: $a + bj$. In engineering, the j factor is often written before its coefficient. For example, the number, $3 + 7j$, is often written: $3 + j7$. The number, $-5.32 + 2.45j$, is often written: $-5.32 + j2.45$.

If two complex numbers are identically the same except for opposite signs of the imaginary parts, then the two complex numbers are called *conjugates*. Then *one of the numbers is called the conjugate of the other*. For example, the number, $3 + 5i$, is the conjugate of the number, $3 - 5i$. The number, $-4 - j7$, is the conjugate of the number, $-4 + j7$.

In the general form, $a + bi$, of a complex number, if the real part a is zero, then the number is called a pure imaginary, such as the number: $0 + 5i$. If the coefficient b is zero, the number is entirely real, such as in the number, $3 + 0i$. Any real number can be written as a complex number. For example, the number 6 can be written: $6 + 0i$. This is often done in electrical engineering.

To add or subtract complex numbers, we must add or subtract separately the real parts and the imaginary parts, since the two parts are different kinds of numbers.

Examples. (a) Add: $\begin{array}{r} 1 - 5i \\ -6 + 3i \\ \hline -5 - 2i \end{array}$ (b) Subtract: $\begin{array}{r} 1 - 5i \\ -6 + 3i \\ \hline +7 - 8i \end{array}$

(c) Add: $\begin{array}{r} -2.71 - j3.25 \\ 5.12 - j1.41 \\ \hline 2.41 - j4.66 \end{array}$ (d) Subtract: $\begin{array}{r} -2.71 - j3.25 \\ 5.12 - j1.41 \\ \hline -7.83 - j1.84 \end{array}$

In the multiplication of complex numbers, we treat the i as we would any other letter in algebra. However, whenever i^2 appears, it is immediately changed to -1. In the final result, the real part of a complex number is always written first.

Example 1. Multiply $-3(4 - 5i) = -12 + 15i$.

Example 2. Multiply $2i(3 + 7i) = 6i + 14i^2 = 6i - 14 = -14 + 6i$.

Example 3. Multiply $(-5 - 2i)(4 - 3i)$.

QUADRATIC EQUATIONS

Solution. $-20-8i+15i+6i^2=-20+7i-6=-26+7i.$

Example 4. Using any example, show that the product of two conjugate complex numbers is a real number.

Solution. Let us use the conjugates: $-3+5i$ and $-3-5i$. Multiplying, we get

$$+9-25i^2=+9+25=34$$

In division involving complex numbers, if the divisor is a real number, we divide as with any algebraic expression involving letters.

Example 5. Divide $(12-20i) \div 4$.

Solution. As with any polynomial, such as $(12-20x) \div 4$, we divide the 4 into both terms of the polynomial. Then we get $3-5i$.

Example 6. Divide $(-5+j3) \div 7 = -\dfrac{5}{7}+j\dfrac{3}{7}.$

If the divisor is an imaginary number or a complex number, we make the divisor a real number as shown in the following examples.

Example 7. Divide $(8-3j) \div (5j)$.

Solution. First we indicate the division as a fraction:

$$\frac{8-3j}{5j}$$

Now, if we multiply both numerator and denominator by j, the denominator becomes a real number.

$$\frac{(8-3j)(j)}{(5j)(j)} = \frac{8j-3j^2}{5j^2} = \frac{8j+3}{-5} = -\frac{8j}{5}-\frac{3}{5} = -\frac{3}{5}-\frac{8}{5}j$$

Example 8. Divide $(2+5i) \div (4-3i)$.

Solution. First we indicate the division as a fraction:

$$\frac{2+5i}{4-3i}$$

To make the denominator a real number, we multiply it by the conjugate of the denominator; that is, we multiply by the complex number, $4+3i$. Then we must *also multiply the numerator by the same quantity*. We get

$$\frac{(2+5i)(4+3i)}{(4-3i)(4+3i)} = \frac{8+26i+15i^2}{16-9i^2} = \frac{8+26i-15}{16+9} = \frac{-7+26i}{25} = -\frac{7}{25}+\frac{26}{25}i$$

Exercise 30.7

Solve each of the following equations by formula. Notice that the two roots of each equation are conjugate complex numbers.

1. $x^2 - 6x + 13 = 0$
2. $x^2 + 10x + 34 = 0$
3. $y^2 - 2y + 2 = 0$
4. $n^2 + 6n + 10 = 0$
5. $2t^2 - 2t + 5 = 0$
6. $5x^2 - 2x + 1 = 0$
7. $n^2 - 6n + 45 = 0$
8. $x^2 + 2x + 9 - 0$
9. $x^2 + 4x + 16 = 0$
10. $3x^2 + 4x + 3 = 0$
11. $5x^2 - 10x + 9 = 0$
12. $7x^2 - 10x + 5 = 0$

Add the following complex numbers.

13. $4 + 3i$
    $5 + 2i$
14. $-5 - i$
    $-2 - 3i$
15. $-5 + 4i$
    $4 - 3i$
16. $5 + 4i$
    $5 - 4i$
17. $-7 + 3i$
    $7 + 3i$
18. $4 + 5i$
    $-6 + 0i$
19. $-3 - 2j$
    $-3 - 2j$
20. $-5 + j4$
    $-3 - j$
21. $2.1 + j1.4$
    $5.8 + j3.2$
22. $-3.3 - j4.5$
    $3.1 + j2.1$

23–32. In Nos. 13–22, subtract the bottom numbers from the top in each.

To each of the following, add its conjugate:

33. $2 + 5i$
34. $-3 + 4i$
35. $5 - 3i$
36. $-2 - j5$
37. $3 - i$

38–42. From each number in Nos. 33–37, subtract its conjugate.

Multiply as indicated:

43. $-4(3 - 2i)$
44. $5(-4 - 3i)$
45. $-3i(5 - 2i)$
46. $-4j(-3 + 5j)$

47–50. Multiply the two numbers given in Nos. 13–16 above.

51. Using an example, show that the product of two conjugate complex numbers is a real number.

Divide as indicated:

52. $(20+10i) \div 5$

53. $(-24+18i) \div 6$

54. $(-15-9i) \div 12$

55. $(8-7i) \div 3$

56. $(10-7i) \div 3i$

57. $(7-5i) \div 4i$

58. $(-9-5i) \div (-2i)$

59. $(6) \div (5i)$

60. $(4) \div (-3j)$

61. $\dfrac{5+4i}{2+3i}$

62. $\dfrac{-6+3i}{1-2i}$

63. $\dfrac{3-i}{-4+3i}$

64. $\dfrac{-4+5i}{-3-5i}$

65. $\dfrac{2+5i}{2-5i}$

66. $\dfrac{-3-2i}{-3+2i}$

67. $\dfrac{1}{3+4i}$

68. $\dfrac{1}{-5-2i}$

69. $\dfrac{5-3i}{5+3i}$

31 RATIO AND PROPORTION

31.1. RATIO

We often wish to compare quantities in everyday life. For example, we say that one object is heavier than another or that one board is longer than another. If we wish to compare 8 feet with 6 feet, we can use subtraction:

$$8 \text{ feet} - 6 \text{ feet} = 2 \text{ feet}$$

That is, 8 feet is 2 feet greater than 6 feet.

Whenever we say one quantity is some amount *greater than* or *less than* another, we are comparing the two by *subtraction*. When we say, "John is 2 inches taller than James," we are comparing by *subtraction*.

We can also compare two quantities by *division*. A comparison by division is called the *ratio* of the two quantities. For example, we can compare 12 pounds with 4 pounds by dividing:

$$12 \text{ pounds} \div 4 \text{ pounds} = 3$$

Then we say the ratio of 12 pounds to 4 pounds is 3. That is, 12 pounds is 3 times as much as 4 pounds. The ratio "3" is an abstract number and has no denomination.

When we compare numbers by subtraction, the *difference has the same name or denomination* as the quantities compared. For example,

$$12 \text{ feet} - 9 \text{ feet} = 3 \text{ feet}$$

$$12 \text{ hours} - 9 \text{ hours} = 3 \text{ hours}$$

When we find the *ratio* of two quantities (that is, comparing by division), *the answer has no denomination.* For example,

$$12 \text{ ft.} \div 9 \text{ ft.} = \frac{4}{3}; \quad 12 \text{ hours} \div 9 \text{ hours} = \frac{4}{3}$$

500 ALGEBRA

A ratio is often indicated by a colon (:). For example, the ratio of 6 feet to 9 feet is often written

$$6 \text{ ft.} : 9 \text{ ft.}$$

This is read, "6 feet is to 9 feet." The colon represents the words "*is to.*"

A ratio can be written as a fraction. For example, the ratio of 6 feet to 9 feet can be written

$$\frac{6 \text{ ft.}}{9 \text{ ft.}}, \text{ which reduces to } \frac{2}{3}$$

The ratio, $\frac{2}{3}$, has no name or denomination. It is an abstract number.

When we wish to compare two quantities in any way, the quantities must be of the same kind. We cannot compare quantities that are of different kinds, such as 12 feet and 4 hours. However, quantities of the same kind can be compared even though they may be measured in different units. For example, we can compare 4 feet with 2 yards by first expressing the two measurements in the same units. The ratio of 4 feet to 2 yards can be stated as the ratio of 4 feet to 6 feet (that is, 2 yards). Then we have

$$4 \text{ ft.} : 6 \text{ ft.} = \frac{4 \text{ ft.}}{6 \text{ ft.}} = \frac{2}{3}$$

That is, 4 feet is $\frac{2}{3}$ as much as 2 yards.

Examples. State the ratio indicated in each of the following, and reduce the ratio to its simplest form: (a) 10 feet to 15 feet; (b) 1 yard to 2 feet; (c) 20 pounds to 5 pounds; (d) 2640 feet to 1 mile.

Solutions.

(a) 10 feet : 15 feet = 10 feet/15 feet = $\frac{2}{3}$;
(b) 1 yard : 2 feet = 3 feet : 2 feet = $\frac{3}{2}$;
(c) 20 pounds : 5 pounds = 20 pounds/5 pounds = 4;
(d) 2640 feet : 1 mile = 2640 feet : 5280 feet = $\frac{1}{2}$

Exercise 31.1

Express the ratio indicated in each of the following examples, and then reduce the ratio to its simplest form:

1. The ratio of 12 in. to 16 in.
2. The ratio of $15 to $25
3. The ratio of 20 lb to 30 lb
4. The ratio of 8 ft. to 4 yd.

5. The ratio of 8 miles to 6 miles.
6. The ratio of 5 yd. to 6 ft.
7. The ratio of 6 hours to 1 day
8. The ratio of 1 mile to 1320 ft.
9. The ratio of 12 in. to 1 ft.
10. The ratio of 5 minutes to $\frac{1}{2}$ hr.
11. The ratio of 32 lb. to 20 lb.
12. The ratio of 1 sq. in. to 1 sq. ft.
13. The ratio of 14 oz. to 1 lb.
14. The ratio of 80¢ to $1.
15. The ratio of 18 qts. to 1 gal.
16. The ratio of 15 sq. ft. to 1 sq. yd.
17. One circle has a radius of 10 inches, and another circle has a radius of 5 inches. What is the ratio of the diameter of the first circle to the diameter of the second? What is the ratio of their circumferences? What is the ratio of their areas?
18. One square has a side 6 inches, and a second square has a side 18 inches. What is the ratio of the sides of the two squares? What is the ratio of their perimeters? Of their diagonals? Of their areas?
19. On a certain plan, $\frac{1}{2}$ inch represents an actual length of 1 foot. What is the ratio of the distance on the plan to the actual distance on the ground?
20. On a certain map, 1 inch represents 5 miles. What is the ratio of the distance on the map to the actual distance on the ground?

31.2. PROPORTION

Let us consider the following two ratios:

(a) The ratio of 4 feet to 6 feet is $\frac{2}{3}$.
(b) The ratio of $8 to $12 is $\frac{2}{3}$.

Here we have two ratios that have the same value. The ratio in each case is $\frac{2}{3}$. Since the ratios are equal, we can state this fact with an equation:

$$4 \text{ ft.} : 6 \text{ ft.} = \$8 : \$12$$

That is, the ratio of 4 feet to 6 feet is the same as the ratio of $8 to $12.

Definition. *A statement of equality between two equal ratios is called a proportion.*

Both ratios in a proportion may refer to the same kind of quantity, as in the proportion:

$$4 \text{ ft.} : 6 \text{ ft.} = 10 \text{ ft.} : 15 \text{ ft.}$$

However, the ratios may refer to different kinds of quantities:

$$4 \text{ ft.} : 6 \text{ ft.} = \$8 : \$12$$

Even though the quantities compared are of different kinds, the ratios are the same, and the two ratios can be stated as a proportion.

A proportion has four terms, called the *first*, *second*, *third*, and *fourth*. The first and the fourth terms are called the *extremes* of the proportion. The second and the third terms are called the *means*. The fourth term is often called the *fourth proportional* to the other three terms.

We have seen that a ratio can be written as a fraction. For example, in the following proportion, the ratios can be written as fractions:

$$6:8 = 9:12, \quad \text{can be written} \quad \frac{6}{8} = \frac{9}{12}$$

In general terms, the proportion: $a:b = c:d$ can be written

$$\frac{a}{b} = \frac{c}{d}$$

If we multiply both sides of this equation by bd, we get

$$ad = bc$$

Note that the term ad is the *product of the extremes* of the proportion. The term bc is the *product of the means*. From this example, we have the following principle of a proportion:

PRINCIPLE. In any proportion, the product of the means is equal to the product of the extremes.

This principle enables us to find any missing term of a proportion.

Example 1. Find the value of the term x of the proportion:

$$5:x = 4:7$$

Solution. The product of the means is $4x$; the product of the extremes is 35. Then we can write the equation

$$4x = 35$$

Solving the equation, we get

$$x = 8\tfrac{3}{4}, \quad \text{or} \quad 8.75$$

Example 2. Find the value of the term x in the following proportion:

$$x:4 = (x-5):6$$

RATIO AND PROPORTION

Solution. We first state that the product of the means is equal to the product of the extremes. Then

$$4(x-5) = 6x$$

Removing parentheses, $\quad 4x - 20 = 6x$

Solving the equation, $\quad -10 = x$

To check the answer, we put -10 for x in the proportion:

$$-10:4 = (-10-5):6$$

or $-10:4 = -15:6$; that is,

$$\frac{-10}{4} = \frac{-15}{6}$$

Each ratio is equal to $-5/2$, so the proportion is true.

Example 3. Find the value of k in the proportion: $3:(k-2) = 5:(k+4)$.

Solution. For the equation, we have

$$5(k-2) = 3(k+4)$$

Removing parentheses, we get $\quad 5k - 10 = 3k + 12$

Solving for k, $\quad k = 11$

As a check, we find that if 11 is put in place of k in the proportion, each ratio is equal to $\frac{1}{3}$.

31.3. MEAN PROPORTIONAL

It is possible that the second term of a proportion is the same as the third term, as in the proportion

$$2:6 = 6:18$$

If the second and the third terms of a proportion are the same, that quantity is called the *mean proportional* between the first and fourth terms. In such a proportion, we have only three different quantities. The fourth term is then called the *third proportional* to the first and second terms.

504 ALGEBRA

In the proportion, $2:6=6:18$, the number 6 is the *mean proportional* between 2 and 18; the number 18 is the *third proportional* to the 2 and 6.
Consider the proportion:

$$a:b=b:c$$

Here the mean proportional is b. For the equation, we can write

$$b^2 = ac$$

If we solve this equation for b, we get $\quad b = \pm\sqrt{ac}.$

Note that the mean proportional, b, may be positive or negative.

Example 4. Find the mean proportional between 4 and 25.

Solution. Let $x =$ the mean proportional. We set up the proportion:

$$4:x = x:25$$

We get the equation, $\quad x^2 = 100$

Solving for x, we get $\quad x = \pm 10$

The mean proportional can be either $+10$ or -10.
Check. Using $+10$, we have the proportion:

$$\frac{4}{+10} = \frac{+10}{25}; \text{ which is } \frac{2}{5} = \frac{2}{5}$$

Using the -10, we have the proportion:

$$\frac{4}{-10} = \frac{-10}{25}; \text{ which is } -\frac{2}{5} = -\frac{2}{5}$$

Exercise 31.2

Find the value of the unknown in each of these proportions.
1. $4:x = 3:12$
2. $2:7 = x:21$
3. $x:9 = 6:2$
4. $3:2.4 = 8:x$

5. $x:10 = 21:4$
6. $5:x = 3.2:7.2$
7. $9.2:x = 15:3.5$
8. $2.4:12 = 10.8:x$
9. $4:5 = x:8.3$
10. $6:5 = x:11$
11. $3:x = 4:13$
12. $2:7 = 9:x$
13. $36:x = x:4$
14. $4:x = x:8$
15. $3:x = x:25$
16. $x:9 = 34:7$
17. $x:5 = 10:3$
18. $2x:5 = 3:4$
19. $4:(x-2) = 13:(2x+1)$
20. $(2x+3):3 = (x-2):5$
21. $(3n-1):7 = (2n+1):3$
22. $(4n-3):6 = (3n+1):5$
23. $k:(k-4) = (k+4):(k-6)$
24. $(k-3):k = (k+6):(k+3)$

25. Find the fourth proportional to 2, 3, and 7.
26. Find the fourth proportional to 5, 2, and 9.
27. Find the mean proportional between 2 and 128.
28. Find the mean proportional between 4 and 18.
29. Find the mean proportional between $\frac{1}{16}$ and 9.
30. Find the mean proportional between -3 and -20.
31. In an electric circuit, the current I is proportional to the voltage E. If $I = 1.5$ amperes when $E = 25$ volts, what voltage is needed for a current of 4.2 amperes?
32. A certain antifreeze mixture calls for 4 quarts of antifreeze in 3.5 gallons of water. How much antifreeze is needed for 5 gallons of water?
33. According to Hooke's law, in any elastic body, the distortion is proportional to the distorting force, if kept within the "elastic limit." If a force of 6 pounds will stretch a spring $2\frac{1}{2}$ inches, what force will be required to stretch the spring 1 inch?
34. A certain recipe calls for $2\frac{1}{4}$ cups of flour for 6 servings. How many cups of flour are needed for 15 servings?

31.4. PROPORTIONAL DIVISION

Sometimes we wish to divide a given number into two or more parts such that the parts have certain ratios to each other.

Example 5. If a board 20 feet long is to be cut into two pieces so that the two pieces have the ratio of 2 to 3, what should be the length of each piece?

Solution. The problem may be solved in two ways. As a *first method*, we
Let $x =$ the length (in feet) of the smaller piece.
Then $20 - x =$ length (in feet) of the larger piece.
Since the ratio of the smaller to the larger is to be $\frac{2}{3}$, we write the

equation:

$$\frac{x}{20-x} = \frac{2}{3}$$

Multiply both sides by $3(20-x)$: $3x = 2(20-x)$
Removing parentheses, $3x = 40 - 2x$
Transposing and combining, $5x = 40$
Dividing both sides by 5, $x = 8$,   length (in feet) of smaller piece

For the length of the larger piece, we have $20 - 8 = 12$.

The two pieces have lengths of 8 feet and 12 feet, respectively, which have the desired ratio: 8 ft. : 12 ft. $= \frac{2}{3}$.

The problem can also be solved by the following method.

We let $2x =$ the length (in feet) of the smaller piece
Then $3x =$ the length (in feet) of the larger piece.

For the equation, we have

$$2x + 3x = 20 \text{ (total length, in feet)}$$

Solving this equation, we get

$$x = 4$$

For the length of the smaller piece, we have

$$2x = 8 \text{ (feet)}$$

For the length of the larger piece, we have

$$3x = 12 \text{ (feet)}$$

Example 6. A board 12 feet long is to be cut into three pieces so that the pieces have the ratio of 2 to 3 to 4; that is, the ratio of the three pieces is to be the same as the numbers, 2, 3, and 4. Find the length of each piece.

Solution. If we let x represent some number, we can say the lengths can be represented as follows (changing 12 feet to 144 inches):

$2x =$ the length (in inches) of the smallest piece,
$3x =$ the length (in inches) of the next piece,
$4x =$ the length (in inches) of the longest piece.

Since the total length of the board is 144 inches, we have the equation

$$2x + 3x + 4x = 144$$

Combining like terms, $9x = 144$
Dividing both sides by 9, $x = 16$

For the lengths of the pieces, we have

$$2x = 32; \quad 3x = 48; \quad 4x = 64 \text{ (inches)}$$

Exercise 31.3

1. A board 14 feet long is to be cut into two pieces whose lengths have the ratio: 3 to 5. Find the length of each piece. (14 feet = 168 inches.)
2. A wire 30 inches long is to be cut into two pieces whose lengths have the ratio of 5 to 7. Find the length of each piece.
3. Brass is an alloy of zinc and copper, containing 2 parts of zinc to 3 parts of copper. Find the amount of each in 8 pounds of brass.
4. Pewter metal is an alloy of lead and tin, containing 1 part of lead for every 4 parts of tin. Find the amount of each in 12 pounds of pewter.
5. Eighteen-carat gold contains 1 part of copper to 3 parts of gold. Find the amount of each in 30 ounces of 18-carat gold.
6. A picture is 32 inches long and 24 inches wide. In a reproduction of the picture, the length is 12 inches. What is the width?
7. In a mixture of antifreeze and water, the mixture contains 2 parts of antifreeze to 7 parts of water. Find the amount of each in 30 gallons of the mixture.
8. A feed mixture consisting of wheat and oats contains 4 parts of wheat and 5 parts of oats. Find the amount of each grain in 240 pounds of the mixture.
9. The number of degrees in the angles of a triangle have the ratio, 4 to 5 to 6. Find the size of each angle. (Total number of degrees is 180.)
10. A farmer has 320 acres of land. Part of the land is used for pasture, part is planted in corn, and the rest in oats. The amounts in pasture, corn, and oats, have the ratio of 4 to 5 to 7. Find the number of acres in each.
11. A concrete mixture consists of cement, sand, and gravel, in the ratio of 1, 2, and 6. Find the amount of each in 1 ton of concrete (2000 lb.).
12. A concrete mixture consists of cement, sand, and gravel in the ratio of 3, 4, and 8. Find the amount of each in 1 cubic yard of concrete (27 cu. ft.).
13. A man has $12,000 invested in stocks, bonds, and savings, in the ratio of 2, 4, and 9. Find the amount of each investment.
14. A store has 300 cans of canned vegetables, consisting of beans, corn, and peas. The number of cans of the three kinds are in the ratio of 3, 5, and 7. Find the number of cans of each kind.
15. An automobile dealer has a stock of 108 cars, some small cars, some medium size, and some large expensive cars, the number of each kind being in the ratio of 9 to 7 to 2. Find the number of cars of each kind.

16. A contractor agrees to build 60 houses, some low-priced, some medium-priced, and some high-priced, the numbers of the three kinds being in the ratio of 7, 4, and 1. How many of each kind does he build?

17. Chrome steel contains chromium, nickel, and steel, in the proportion of 1 part chromium, 1 part nickel, and 23 parts steel. Find the amount of each kind in 1 ton of chromium steel.

18. A university library contains 15,120 books, consisting of books on mathematics, science, history, and literature. In the order given here, the numbers have the ratio of 2, 3, 5, and 8. Find the number of books of each kind in the library.

IV TRIGONOMETRY

32 THE TRIGONOMETRIC RATIOS

32.1. IMPORTANCE OF TRIGONOMETRY

Trigonometry is one of the most useful forms of mathematics. It has many applications in engineering and other sciences. One important use is in finding distances that are difficult or impossible to measure directly, such as the height of a flagpole or the distance across a lake. Moreover, it is not a difficult subject when the basic ideas are once understood.

32.2. DIRECT AND INDIRECT MEASUREMENTS

By *direct measurement*, we mean applying a measuring instrument directly to the thing measured. When you measure the length of a room with a yardstick or a tape measure, you are using *direct measurement*. That is, you are measuring the length directly. However, it would be very difficult to measure the height of a flagpole by direct measurement. To measure the distance across a lake, it would be practically impossible to use direct measurement.

Instead of trying to measure some distances directly, we measure them *indirectly* by means of trigonometry. For example, to measure the height of a flagpole, we make a couple of measurements on the ground, and then compute the height by using trigonometry. To measure the distance across a lake, we look at a point on the opposite shore from two different positions on our side and then use trigonometry to compute the distance across the lake.

32.3. ANGLES

Trigonometry involves the study of angles and triangles. To see how to use trigonometry to solve problems, it is first necessary to understand the

512 TRIGONOMETRY

meaning of an angle. Most people would take an angle to mean the figure formed by two straight lines drawn from the same point. This is true. However, let us carefully define a few terms in connection with an angle.

In geometry, we learn that a straight line is understood to have no end but instead unlimited extent in both directions. However, a straight line drawn in one direction from a point is understood to have one end point and unlimited extent in one direction. Such a line with one end point is called a *half-line* or a *ray*.

Now, if two rays, *AB* and *AC*, are drawn from the same point, they form an angle *BAC* (Fig. 32.1). The two rays, *AB* and *AC*, are called the *sides* of the angle. The point *A* from which they are drawn is called the *vertex* of the angle. The angle *BAC* is the figure formed. The *size* of the angle is determined by the amount of *opening* between the rays.

FIGURE 32.1

An angle can also be defined as the result of motion. Much engineering involves motion like the circular motion of a dynamo or generator. Suppose we begin with the ray *OA* (Fig. 32.2) with the end point at *O*. Now we turn or *rotate* the ray about its end point *O* to the new position *OB*, as shown by the curved arrow. Then the angle is formed as a result of motion. The curved arrow shows the direction and the amount of rotation of the ray. Instead of thinking of two different rays, we can think of only *one ray* but in *two different positions*. Then the angle is formed by the two *different positions of the same ray*. This is like the angle formed by two different positions of the long hand of a clock as it moves from one position to another.

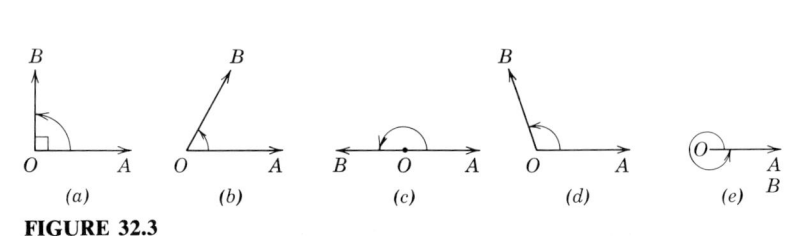

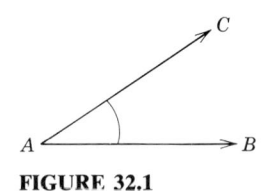

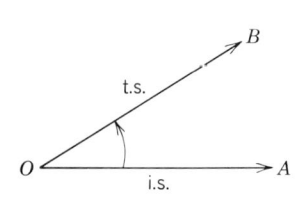

FIGURE 32.2

If we think of an angle as the amount of rotation of a ray, then there is no limit to the size of an angle. The first position of the ray is called the *initial side* of the angle. The final position of the ray is called the *terminal side* of the angle. The point *O* is the *vertex* of the angle.

i.s. = initial side;   t.s. = terminal side

32.4. KINDS OF ANGLES

In geometry, we learn the different kinds of angles, as shown in Figure 32.3. In each case, the curved arrow shows the amount of rotation of the ray. If the sides of the angle form a square corner, the angle is called a *right angle*

FIGURE 32.3

THE TRIGONOMETRIC RATIOS

(Fig. 32.3a). An angle smaller than a right angle is called an *acute angle* (b). If the two sides of the angle point in opposite directions, the angle is called a *straight angle* (c) because the two sides form a straight line. An angle greater than a right angle but less than a straight angle is called an *obtuse angle* (d). If the ray rotates until it reaches its original position, the angle formed is called one *revolution* (e).

32.5. MEASUREMENT OF ANGLES

An angle is often measured in degrees. One revolution is called 360 degrees (360°). A straight angle contains 180°. A right angle contains 90°. An acute angle is less than 90°. An obtuse angle contains more than 90° but less than 180°.

Angles can be measured by means of an instrument called a *protractor*. A protractor consists essentially of a semicircle showing numbers of degrees. To measure an angle, the protractor is placed on the angle with the center O of the semicircle at the vertex of the angle and with one side of the angle along the diameter of the circle. Then the other side of the angle lies along a point on the semicircle that shows the number of degrees in the angle. By using the same idea, we can draw an angle of any desired size. (For more complete explanation of the use of protractor, see Section 6.6, of geometry.)

32.6. TRIANGLES

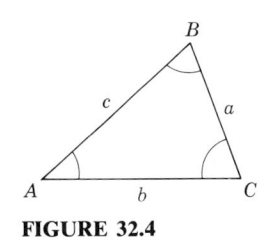

FIGURE 32.4

In trigonometry we deal with triangles as well as with angles, especially with the *right triangle*. In geometry we learn certain facts about triangles. A triangle is a figure bounded by three straight line segments (Fig. 32.4). The triangle has three sides and three angles. The three sides and three angles are called the six *elements* of the triangle. A point where two sides meet is called a *vertex*. Then the triangle has *three vertices*.

A triangle is usually named by capital letters at the vertices, as the triangle ABC. Each side is often named with a small letter that corresponds to the opposite angle. In Figure 32.4, the angles are A, B, and C. The side opposite the angle A is called side a, the side b is opposite angle B, and the side c is opposite angle C.

From geometry we learn the following facts about triangles.

1. *In any triangle the sum of the three angles is* 180°.
2. *If the three sides of a triangle are equal, then the three angles are also equal. The converse is also true.*
3. *If two sides of a triangle are equal, then the angles opposite these sides are also equal. The converse is also true.*

514 TRIGONOMETRY

4. *If one side of a triangle is greater than another side, then the angle opposite the greater side is greater than the angle opposite the smaller side. The converse is also true.*

32.7. THE RIGHT TRIANGLE

A triangle that has a right angle is called a *right triangle*. (It was once called a *rectangled triangle*.) Since the right angle contains 90°, the sum of the other two angles is 90°, because the sum of the three angles is 180°. Therefore, the other two angles are acute angles. They are also complementary. (*Complementary angles* are two angles whose sum is 90°.)

In a right triangle, the side opposite the right angle is called the *hypotenuse*. The other two sides are called the *legs* of the triangle.

Let us construct a particular right triangle ABC (Fig. 32.5). We make AC 8 inches long, and we make CB 6 inches long, with a right angle at C. Now we draw the line segment AB and we have a right triangle. Angle C is the right angle; angle A and angle B are the acute angles. The hypotenuse of the right triangle is the line segment AB. The legs of the triangle are AC and CB.

Now we can find the length of the hypotenuse AB without measuring. We find the square of each leg:

$$8^2 = 64; \qquad 6^2 = 36$$

Now we add the squares of the legs: $8^2 + 6^2 = 64 + 36 = 100$. Then the square of the hypotenuse is 100. For the length of the hypotenuse, we take the square root of 100: $\sqrt{100} = 10$. Then the hypotenuse is 10 inches long.

In any right triangle, if we *square the two legs* and *add these squares*, we get the *square of the hypotenuse*. This is one of the most useful rules in mathematics. It is called the *Pythagorean* rule.

PYTHAGOREAN RULE. The square of the hypotenuse of a right triangle is equal to the sum of the squares of the legs.

The pythagorean rule is true for all right triangles whatever the size or shape. The rule is named after Pythagoras, a Greek mathematician and philosopher who proved the truth of the statement over 2000 years ago.

Example 1. Find the length of the hypotenuse of a right triangle DEF, with legs equal to 5 inches and 12 inches, respectively, and the right angle at F.

Solution. The triangle is shown in Figure 32.6. Squaring the legs, we have

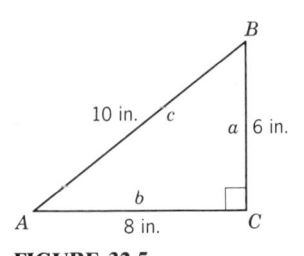

$$5^2 = 25; \qquad 12^2 = 144$$

THE TRIGONOMETRIC RATIOS

Adding the squares, we get $25+144=169$. Then the length of the hypotenuse is found by taking the square root of 169: $\sqrt{169}=13=$ length of hypotenuse DE.

Example 2. Find the length of the hypotenuse of a right triangle RST, with legs equal to 4 inches and 8 inches, respectively, and T the right angle.

Solution. The triangle is shown in Figure 32.7. Squaring the legs, we have

$$4^2 = 16; \qquad 8^2 = 64$$

FIGURE 32.7

Adding the squares, we get $16+64=80$. For the length of the hypotenuse, we have

$$\sqrt{80} = 4\sqrt{5} = 4(2.236) = 8.944 \text{ (approx.)}$$

Example 3. The length of the hypotenuse of a right triangle is 17 inches, and the length of one leg is 8 inches. Find the length of the other leg.

Solution. The triangle is shown in Figure 32.8, right angle at C. If we know the length of the hypotenuse and one leg, we use the Pythagorean rule in reverse. In this case we *subtract* the squares.

FIGURE 32.8

$$\begin{aligned}
\text{square of the hypotenuse} &= 17^2 = 289 \\
\text{square of one leg} &= 8^2 = \underline{64} \\
\text{subtracting the squares} & 225
\end{aligned}$$

Then the length of the other leg is $\sqrt{225}=15$.

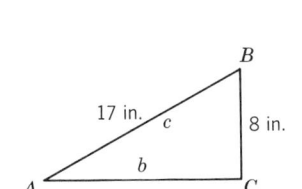

Exercise 32.1

Each of the following examples represents a right triangle ABC with c as the hypotenuse, and a and b as the legs. Two of these lengths are given. Find the unknown side or hypotenuse. Leave in radical form.

1. $a = 24$ cm; $b = 32$ cm.
2. $a = 21$ inches; $b = 72$ inches.
3. $a = 35$ in; $c = 65$ in.
4. $b = 16$ cm; $c = 34$ cm.
5. $a = 8$ in; $b = 16$ in.
6. $a = 4$ in; $b = 4$ in.
7. $b = 8$ cm; $c = 12$ cm.
8. $b = 6$ ft; $c = 8$ ft.
9. $a = 4$ ft; $c = 10$ ft.
10. $a = 3$ in; $b = 5$ ft.
11. $b = 15$ in; $c = 21$ in.
12. $a = 8$ in; $b = 1$ ft.
13. $a = 33$ in; $b = 56$ in.
14. $a = 4$ ft; $c = 73$ in.
15. $a = 39$ cm; $c = 89$ cm.
16. $b = 11$ ft; $c = 13$ ft 1 in.

32.8. THE TRIGONOMETRIC RATIOS

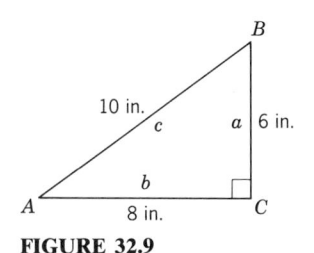

Trigonometry is concerned chiefly with certain ratios between the sides of a right triangle. In triangle *ABC* (Fig. 32.9) suppose $AC = 8$ inches, $BC = 6$ inches, and $AB = 10$ inches. Now suppose we want to know the ratio of side *BC* to side *AC*. The ratio between the two sides is written as a fraction:

$$\frac{a}{b} = \frac{BC}{AC} = \frac{6 \text{ inches}}{8 \text{ inches}} = \frac{3}{4}$$

FIGURE 32.9

The ratio has the value $\frac{3}{4}$. This number, $\frac{3}{4}$, is not inches or degrees. It is simply a relation between the two sides. It is a comparison between the two. A *ratio is a comparison by division.* That is, *BC* is $\frac{3}{4}$ of *AC*.

In this triangle, the angle *A* is approximately 37°, or more nearly 36.9°. In any right triangle, if the ratio of the two legs is $\frac{3}{4}$, the smaller acute angle is approximately 36.9°, whatever the lengths of the sides. In this triangle, the ratio, $\frac{3}{4}$, is called the *tangent of angle A*.

In any right triangle, we can write six ratios between the sides and the hypotenuse, taking two lengths at a time. For example, in the triangle in Figure 32.9, we can write the following ratios: (omitting the word *inches*):

$$\frac{6}{10}; \quad \frac{8}{10}; \quad \frac{6}{8}; \quad \frac{8}{6}; \quad \frac{10}{8}; \quad \frac{10}{6}$$

These ratios can be reduced to lower terms or to decimals if we wish. The ratios do not mean inches, feet, or degrees. They are abstract numbers. For example, the ratio

$$\frac{6 \text{ inches}}{10 \text{ inches}} = \frac{3}{5}$$

These six ratios are given names. The names of the ratios refer to a particular one of the acute angles. If we refer to angle *A*, then we have

The ratio 6/10 is called the *sine of angle A*
The ratio 8/10 is called the *cosine of angle A*
The ratio 6/8 is called the *tangent of angle A*
The ratio 8/6 is called the *cotangent of angle A*.
The ratio 10/8 is called the *secant of angle A*.
The ratio 10/6 is called the *cosecant of angle A*.

These ratios are called the six *trigonometric ratios of an angle.*

32.9. DEFINITIONS OF THE TRIGONOMETRIC RATIOS

Now we need to define exactly what is meant by each of the trigonometric ratios. To do so, we must name the three sides of the right triangle. Whatever acute angle we are considering, we define the sides as follows:

1. The *hypotenuse* is the side opposite the right angle.
2. One leg of the right triangle is the side *opposite* the angle.
3. The other leg is the side *adjacent* to the angle; that is, the side that helps form the angle with the hypotenuse.

For any one of the acute angles, we have these definitions; we state the ratios as fractions. Then, with reference to a particular angle,

		Abbreviations
$\dfrac{\text{side opposite}}{\text{hypotenuse}}$	is called the *sine* of the angle	(sin)
$\dfrac{\text{side adjacent}}{\text{hypotenuse}}$	is called the *cosine* of the angle	(cos)
$\dfrac{\text{side opposite}}{\text{side adjacent}}$	is called the *tangent* of the angle	(tan)
$\dfrac{\text{side adjacent}}{\text{side opposite}}$	is called the *cotangent* of the angle	(cot)
$\dfrac{\text{hypotenuse}}{\text{side adjacent}}$	is called the *secant* of the angle	(sec)
$\dfrac{\text{hypotenuse}}{\text{side opposite}}$	is called the *cosecant* of the angle	(csc)

To shorten the definitions, we abbreviate as follows:

For "side opposite" we use "opp"
For "side adjacent" we use "adj"
For "hypotenuse", we use "hyp"

We also abbreviate the names of the ratios and omit the word "of." The names of the ratios should always be pronounced as though written out in full.

To illustrate these ratios, suppose we have the triangles *DEF* in Figure 32.10, with sides lettered *d*, *e*, and *f* as the hypotenuse and θ (*theta*) as the

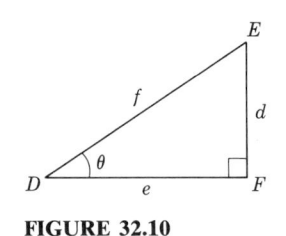

FIGURE 32.10

angle opposite the side d. Then, from the foregoing definitions, we have

$$\sin \theta = \frac{\text{opp}}{\text{hyp}} = \frac{d}{f} \qquad \cos \theta = \frac{\text{adj}}{\text{hyp}} = \frac{e}{f}$$

$$\tan \theta = \frac{\text{opp}}{\text{adj}} = \frac{d}{e} \qquad \cot \theta = \frac{\text{adj}}{\text{opp}} = \frac{e}{d}$$

$$\sec \theta = \frac{\text{hyp}}{\text{adj}} = \frac{f}{e} \qquad \csc \theta = \frac{\text{hyp}}{\text{opp}} = \frac{f}{d}$$

Example 1. In a certain right triangle, ABC, side $a = 5$ inches, side $b = 12$ inches. Find the length of the hypotenuse c, and then state the numerical values of the sine, cosine, tangent, and cotangent of angle A.

Solution. First we find the length of the hypotenuse c. Squaring the sides, we have

$$5^2 = 25; \qquad 12^2 = 144$$

Adding the squares, we get 169. Then the hypotenuse is: $\sqrt{169} = 13$. Now we sketch the triangle (Fig. 32.11) showing the lengths of the sides and the hypotenuse. For angle A, we have the following ratios:

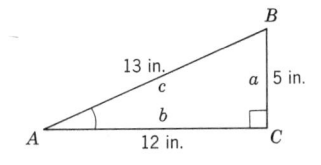

FIGURE 32.11

$$\sin A = \frac{\text{opp}}{\text{hyp}} = \frac{5}{13}; \qquad \cos A = \frac{\text{adj}}{\text{hyp}} = \frac{12}{13}$$

$$\tan A = \frac{\text{opp}}{\text{adj}} = \frac{5}{12}; \qquad \cot A = \frac{\text{adj}}{\text{opp}} = \frac{12}{5}$$

As decimals, we have:
$\sin A = 0.385$; $\cos A = 0.923$; $\tan A = 0.417$; $\cot A = 2.4$.
In this triangle, it happens that angle $A = 22.6°$ (approximately).

Example 2. In a certain right triangle, DEF, side $d = 12$ inches, side $e = 6$ inches. Find the length of the hypotenuse and then find the sine, cosine, and the tangent of angle D.

Solution. First we find the length of the hypotenuse f. Squaring the sides, we have $12^2 = 144$; $6^2 = 36$; adding the squares, $144 + 36 = 180$. Then the hypotenuse is: $\sqrt{180} = 6\sqrt{5}$. Now we sketch the triangle (Fig. 32.12) showing the lengths of the sides and the hypotenuse f.

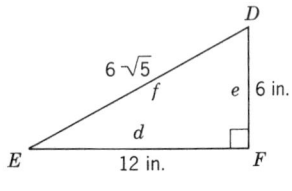

FIGURE 32.12

THE TRIGONOMETRIC RATIOS **519**

For angle D, we have the following ratios: (stated as decimals)

$$\sin D = \frac{\text{opp}}{\text{hyp}} = \frac{12}{6\sqrt{5}} = \frac{2}{\sqrt{5}} = 0.894 \text{ (approx.)}$$

$$\cos D = \frac{\text{adj}}{\text{hyp}} = \frac{6}{6\sqrt{5}} = \frac{1}{\sqrt{5}} = 0.447 \text{ (approx.)}$$

$$\tan D = \frac{\text{opp}}{\text{hyp}} = \frac{12}{6} = 2.000$$

In this example, it happens that angle D is approximately 63.4°.

Example 3. In a certain right triangle, RST, the right angle is angle T, side $s = 4$ inches, and the hypotenuse $= 8$ inches. Find the sine, cosine, and tangent of angle R.

Solution. In this example, the hypotenuse t is known. Then we find the following squares: $t^2 = 8^2 = 64$; $s^2 = 4^2 = 16$. To find the side r we subtract the squares: $64 - 16 = 48$. Then the square of side r is 48; and $r = \sqrt{48} = 4\sqrt{3}$. Now we sketch the triangle (Fig. 32.13) showing the lengths of the sides and the hypotenuse, t. For angle R, we have the following ratios:

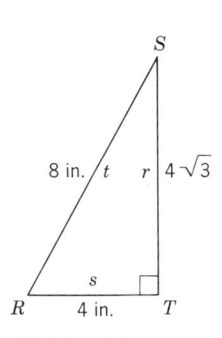

FIGURE 32.13

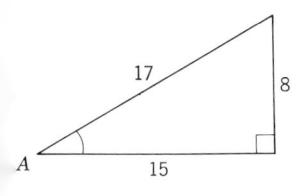

$$\sin R = \frac{\text{opp}}{\text{hyp}} = \frac{4\sqrt{3}}{8} = \frac{\sqrt{3}}{2} = 0.866 \text{ (approx.)}$$

$$\cos R = \frac{\text{adj}}{\text{hyp}} = \frac{4}{8} = 0.5$$

$$\tan R = \frac{\text{opp}}{\text{adj}} = \frac{4\sqrt{3}}{4} = \sqrt{3} = 1.732 \text{ (approx.)}$$

In this example, it happens that angle R is exactly 60°.

Note. As a student you should memorize the trigonometric ratios so well that you can identify them instantly. If you want to be able to work problems in trigonometry quickly, you should know the trigonometric ratios as well as you know the multiplication table or the names of your friends. Whenever you see the following ratios, you should be able to name them at once:

$$\frac{\text{opp}}{\text{hyp}} \qquad \frac{\text{adj}}{\text{hyp}} \qquad \frac{\text{opp}}{\text{adj}} \qquad \frac{\text{adj}}{\text{opp}} \qquad \frac{\text{hyp}}{\text{adj}} \qquad \frac{\text{hyp}}{\text{opp}}$$

For angle A in the figure, name the following ratios instantly (right angle is shown):

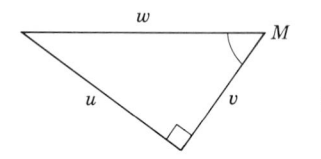

$$\frac{8}{15} \quad \frac{15}{17} \quad \frac{17}{8} \quad \frac{15}{8} \quad \frac{17}{15} \quad \frac{8}{17}$$

For angle M in the figure, name the following ratios instantly (right angle is shown):

$$\frac{u}{v} \quad \frac{u}{w} \quad \frac{w}{u} \quad \frac{v}{w} \quad \frac{v}{u} \quad \frac{w}{v}$$

Exercise 32.2

Each of the following examples represents a right triangle with the right angle at C, hypotenuse $= c$, and legs a and b. Sketch each triangle and find the unknown length. Then write the numerical value of the sine, cosine, and tangent of the angle indicated for each example. The angle A is opposite side a, and the angle B is opposite side b. Leave answers in radical form.

1. $a = 21$ cm; $b = 26$ cm (angle A).
2. $b = 36$ cm; $c = 39$ cm (angle B).
3. $a = 24$ in; $b = 32$ in (angle B).
4. $b = 24$ in; $c = 25$ in (angle A).
5. $a = 2$ cm; $c = 6$ cm (angle A).
6. $a = 6$ cm; $b = 2$ cm (angle B).
7. $a = 5$ in; $b = 10$ in (angle B).
8. $a = 6$ in; $c = 9$ in (angle A).
9. $b = 8$ cm; $c = 16$ cm (angle A).
10. $b = 6$ cm; $c = 8$ cm (angle B).
11. $b = 5$ in; $c = 7$ in (angle A).
12. $a = 5$ cm; $b = 5$ cm (angle A).
13. $a = 7$ cm; $b = 1$ cm (angle A).
14. $b = 1$ cm; $c = 4$ cm (angle B).
15. $a = 1$ in; $b = 3$ in (angle B).
16. $a = 3$ in; $b = 2$ in (angle B).

33 TABLES OF TRIGONOMETRIC RATIOS

33.1. THE TRIGONOMETRIC RATIOS OF AN ANGLE

To see the meaning of the trigonometric ratios of any angle, consider the triangle ABC, with angle C as the right angle (Fig. 33.1). Suppose $AC = 4$ inches, and $CB = 3$ inches. To find the hypotenuse, we square the legs:

$$4^2 = 16; \quad 3^2 = 9$$

Adding the squares, we get

$$4^2 + 3^2 = 16 + 9 = 25$$

Then the hypotenuse AB is equal to $\sqrt{25} = 5$. Then we have the following ratios:

$$\sin A = \frac{3}{5}; \quad \cos A = \frac{4}{5}; \quad \tan A = \frac{3}{4}$$

In this triangle, it happens that A is equal to approximately $36.9°$.

Now let us double the sides and the hypotenuse of the triangle, so that

$$AE = 8 \text{ inches}; \quad DE = 6 \text{ inches}; \quad AD = 10 \text{ inches}$$

We get the new triangle, ADE, and angle A remains unchanged at $36.9°$.

Using the triangle ADE, we have

$$\sin A = \frac{ED}{AD} = \frac{6}{10} = \frac{3}{5}; \quad \cos A = \frac{AE}{AD} = \frac{8}{10} = \frac{4}{5}; \quad \tan A = \frac{ED}{AE} = \frac{6}{8} = \frac{3}{4}$$

The sine, cosine, and the tangent of angle A have the same values as in the smaller triangle. *Increasing the lengths of the sides of the right triangle does*

522 TRIGONOMETRY

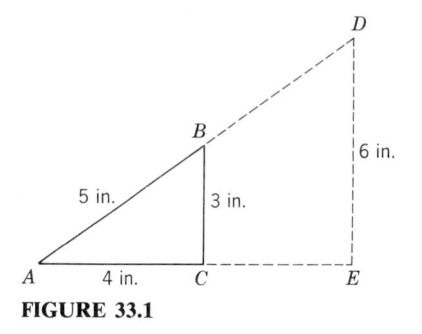

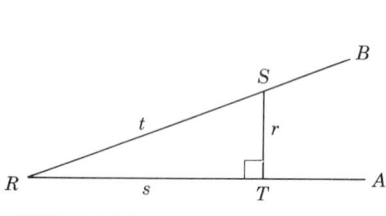

FIGURE 33.1

FIGURE 33.2

not change the value of the trigonometric ratios as long as the angle A remains unchanged.

The foregoing example shows one important fact. The trigonometric ratios of an angle do not depend on the lengths of the sides of the triangle, but only on the size of the angle. *The ratio values will change only if the angle itself changes in size.* Therefore, the ratios are called *trigonometric functions of the angle.*

We have said that angle A (Fig. 33.1) is approximately 36.9°. Since the values of the ratios will not change if angle A is 36.9°, whatever the lengths of the sides of the triangle, we can say

$$\sin 36.9° = \frac{3}{5} = 0.6; \quad \cos 36.9° = \frac{4}{5} = 0.8; \quad \tan 36.9° = \frac{3}{4} = 0.75$$

Let us consider another example. We begin with an angle of 20°, having sides RA and RB (Fig. 33.2). Now we take any point T on the side RA. We draw a line through T perpendicular to RA, meeting RB at S. Now we have a right triangle RST with the right angle at T. Let us call the sides r, s, and t, to correspond to the opposite angles.

For the sine of angle R, we have

$$\sin R = \frac{\text{opp}}{\text{hyp}} = \frac{r}{t}$$

Notice in the figure that the side r is approximately $\frac{1}{3}$ as long as the hypotenuse t. This ratio is the same no matter how long we make the sides of the right triangle. Then we can say

$$\sin 20° = \text{approximately } \frac{1}{3}$$

The actual ratio for a 20° angle has been computed to be approximately

0.3420. That is,

$$\sin 20° = 0.3420 \text{ (approx.)}$$

Also, for the 20° angle, notice that the side s is almost as long as the hypotenuse t. Then the cosine of 20° is almost 1. The cosine ratio has been computed to be approximately 0.9397 for a 20° angle.

For the tangent ratio in the figure, we have

$$\tan 20° = \frac{\text{opp}}{\text{adj}} = \frac{r}{s}$$

which appears to be a little over $\frac{1}{3}$. For a 20° angle, the tangent ratio has been computed to be approximately 0.3640. To summarize:

$$\sin 20° = 0.3420; \quad \cos 20° = 0.9397; \quad \tan 20° = 0.3640$$

33.2. TABLES OF TRIGONOMETRIC RATIOS

The numerical values of the trigonometric ratios have been computed for angles from zero to 90°. The values are usually listed in tabular form. One of the most useful tables is one showing the values of the ratios for degrees and tenths of a degree. Most tables show the values of the sine, the cosine, the tangent, and the cotangent values because these ratios are most used (see Appendix, Table 3).

For example, to find the values of the ratios for an angle of 21.4°, we follow down the first column with the heading "Degrees" to the angle 21.4°. Then, opposite this angle, we find the value of each trigonometric ratio under the proper heading. There we find

$$\sin 21.4° = 0.3649; \quad \cos 21.4° = 0.9311;$$

$$\tan 21.4° = 0.3919; \quad \cot 21.4° = 2.552$$

We must not lose sight of the meaning of these ratios. For example, when we say that the sine of 21.4° is 0.3649, we mean that if we have an angle of 21.4° in a right triangle, then the ratio of the side opposite to the hypotenuse would be 0.3649; that is,

$$\frac{\text{opp}}{\text{hyp}} = 0.3649$$

In most tables, the values of the ratios are arranged so that the angles may be read downward at the left side and upward at the right. This kind of arrangement avoids *the* unnecessary repetition of values. For angles from zero to 45° we read downward at the left. For angles from 45° to 90°, we read upward at the right and then read the *footings* instead of the headings.

For example, to find the values of the ratios for 14.5°, we look in the left-hand column until we come to 14.5°. Opposite this angle we find the following values under the proper headings:

$$\sin 14.5° = 0.2504; \qquad \cos 14.5° = 0.9681;$$

$$\tan 14.5° = 0.2586; \qquad \cot 14.5° = 3.867$$

To find the values for 64.2°, we read upwards in the right-hand column until we come to the angle. Opposite this angle, we find these values above the footings:

$$\sin 64.2° = 0.9003; \qquad \cos 64.2° = 0.4352;$$

$$\tan 64.2° = 2.069; \qquad \cot 64.2° = 0.4834$$

The tables of the trigonometric ratios can also be used in reverse. If a certain ratio value is known, the angle itself may be found in the table. For example, if we know that the sine of an angle is 0.3190, we look for this value under the heading "sine." There we find that this sine value corresponds to the angle 18.6°.

In using the table in reverse, we must sometimes use *footings* instead of headings. For example, suppose we know that the sine of an angle is 0.9041. We find this value in the sine column if we read the footings. Then we find that this value for the sine corresponds to an angle of 64.7°.

Other examples:
If the tangent of angle A is 2.513, then angle $A = 68.3°$.
If $\cot B = 0.7080$, then $B = 54.7°$.
If $\cos D = 0.3272$, then $D = 70.9°$.

If the exact value of a ratio is not found in the table, then we usually take the nearest angle. For example, if $\sin T = 0.5420$, then $T = 32.8°$ (approx.).

Exercise 33.1

Find the value of each of the following:

1. $\sin 10.5°$
2. $\cos 14.4°$
3. $\tan 18.8°$
4. $\cot 23.9°$
5. $\cos 35.6°$
6. $\sin 30.7°$
7. $\cot 44.2°$
8. $\tan 39.3°$
9. $\tan 54.5°$
10. $\cot 59.4°$

11. sin 45.6°
13. cot 76.2°
15. cos 71.8°
17. 10 sin 12°
19. 30 tan 20°
21. 3 cos 42.1°
23. 5 tan 33.9°
25. 4 tan 83.8°
27. 8 sin 79.1°

12. cos 50.1°
14. tan 81.3°
16. sin 66.9°
18. 20 cos 23°
20. 40 cot 29°
22. 4 sin 30.4°
24. 6 cot 38.2°
26. 7 cot 73.3°
28. 9 cos 82.9°

Find the angle indicated by the letter corresponding to each of the following function values:

29. sin A = 0.2840
31. tan C = 0.4706
33. tan E = 2.135
35. cos G = 0.6280
37. cos J = 0.8440
39. sin M = 0.7340

30. cos B = 0.9403
32. cot D = 1.760
34. sin F = 0.8320
36. cot H = 0.2180
38. tan K = 1.805
40. cot N = 2.870

33.3. ARC-FUNCTIONS, OR INVERSE FUNCTIONS

At this point, we introduce the terms *arcsine, arccosine, arctangent,* and so on. These terms are often convenient to use. The term *arc* can be used as a prefix to the names of all the trigonometric functions. These expressions mean *angles.* For example, an arcsine is an angle.

For these expressions, we use the following abbreviations:

    arcsin (pronounced *ark-sine*)
    arccos (pronounced *ark-cosine*)
    arctan (pronounced *ark-tangent*)
    arccot (pronounced *ark-cotangent*)
    arcsec (pronounced *ark-secant*)
    arccsc (pronounced *ark-cosecant*)

To understand the term *arcsin,* note these two statements:

If
$$\sin A = 0.5$$

then
$$A = \text{the angle whose sine is } 0.5$$

The two statements are equivalent. For the second statement, we can write

$$A = \arcsin 0.5$$

TRIGONOMETRY

Here, the word *arcsin* means "the angle whose sine is." Now, of course, if

$$\sin A = 0.5$$

then $$A = 30°$$

Then we can say $$\text{arcsin } 0.5 = 30°$$

In the same way, we can say:

$$\text{arcsin } 0.3420 = 20°$$

As another example, suppose we want to find: arctan 0.9004. The expression, "arctan 0.9004" means the angle whose tangent is 0.9004. Using the table, we find that the tangent ratio 0.9004 corresponds to an angle of 42°. Then we can say

$$\text{arctan } 0.9004 = 42°$$

To see one advantage of the use of such a term as *arcsine*, consider the following example, which requires *two* statements:

If $$\sin B = 0.6820$$

then $$B = 43°$$

Instead of using two statements, we can say the same thing in one statement:

$$\text{arcsin } 0.6820 = 43°$$

The arc-functions are often called *inverse functions*. They are sometimes indicated by a "−1" written in the position of an exponent on the word *sine*. That is,

"arcsin x" is sometimes written: "$\sin^{-1} x$"

The following examples show the meaning of these terms. The student should check each example to be sure the angle is correct and the meaning is clear.

(1) arcsin 0.4067 = 24°, because sin 24° = 0.4067
(2) arctan 0.6494 = 33°, because tan 33° = 0.6494
(3) arccos 0.7880 = 38°   (4) arccot 1.804 = 29°
(5) arcsin 0.8192 = 55°   (6) arccos 0.4540 = 63°
(7) arctan 2.097 = 64.5°   (8) arccot 0.3404 = 71.2°

Exercise 33.2

Find the following angles:

1. arcsin 0.1908
2. arccos 0.9690
3. arctan 0.4327
4. arccot 1.905
5. arcsin 0.8607
6. arccos 0.7230
7. arctan 1.196
8. arccot 0.5402
9. arccos 0.8643
10. arcsin 0.6280
11. arccot 1.865
12. arctan 2.840
13. arctan 0.6346
14. arccot 0.7590
15. arcsin 0.9355
16. 2 arcsin 0.4540
17. 3 arccos 0.9848
18. 2 arctan 1

Find the following:

19. arcsin 0.1736 + arcsin 0.2588
20. arcsin 0.9397 − arcsin 0.3420
21. arcsin 0.8660 − arccos 0.9063
22. arctan 1.1918 + arctan 0.3640
23. arccot 3.732 + arcsin 0.5000
24. arctan 4.705 − arcsin 0.3090

34 SOLVING RIGHT TRIANGLES

34.1. SOLVING STATED OR WORD PROBLEMS BY TRIGONOMETRY

All the problems we consider here involve a right triangle. In solving stated problems, the following steps will serve as a guide.

(1) Sketch a figure showing the right triangle involved.
(2) Label all parts, showing given values and the parts to be found.
(3) Set up a ratio between two sides, at least one known side.
(4) Tell what function this ratio represents with reference to an acute angle.
(5) Solve the resulting equation for the unknown value.
(6) If more than one unknown is to be found, find the other unknowns.

Example 1. Suppose we wish to measure the height of a flagpole. We assume the pole is perpendicular to the ground, forming a right angle. From a point 90 feet from the foot of the pole, it is possible by means of an instrument to measure the angle of elevation to the top of the pole. Suppose this angle is 35.3°. How high is the pole?

Solution. First we sketch a figure showing the right triangle involved. We assume the ground is level and the pole is upright (Fig. 34.1). Now we use a letter to represent the height to be found.
   Let h = the height of the pole (in feet).
   As a second step, we set up the ratio

$$\frac{h}{90}$$

(This step should always be done.) Now we notice that this ratio represents the tangent of the angle 35.3°. Then we write the equation stating

530 TRIGONOMETRY

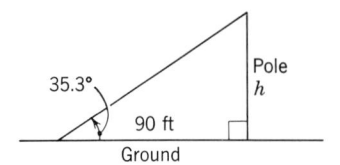

FIGURE 34.1

this fact:

$$\frac{h}{90} = \tan 35.3°$$

From the table, we find that the tangent of 35.3° is 0.708. Then we have

$$\frac{h}{90} = 0.708$$

Solving,

$$h = (90)(0.708) = 63.7$$

The height of the pole is 63.7 feet, rounded to three significant digits.

Note. In Example 1, if the measurement of angle A (angle of elevation) is made from some distance, such as 5 feet, above the ground level, this distance must be added to the value of h, the height, to get the true height of the pole. In some problems, this extra distance is negligible and can be omitted.

Example 2. An electric light pole is braced with a wire cable fastened 4 feet from the top of the pole and to a stake in the ground 18 feet from the foot of the pole. If the brace makes an angle of 73° with the ground, find the height of the pole and the length of the brace.

Solution. We sketch the figure (Fig. 34.2), showing the right triangle, with parts labeled. Now we set up the ratio, using h for the height to the top of the brace:

$$\frac{h}{18}$$

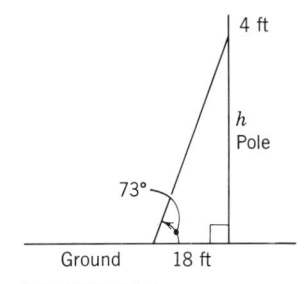

FIGURE 34.2

Some students have much trouble in working stated problems because they start to look for a trigonometric function. The ratio itself will select the correct function. Now we notice that the ratio we have set up is the tangent of the angle 73°. Then we write

$$\frac{h}{18} = \tan 73°$$

SOLVING RIGHT TRIANGLES 531

From the table we find that $\tan 73° = 3.271$. Then

$$\frac{h}{18} = 3.271$$

Solving for h, we get

$$h = (18)(3.271) = 58.9 \text{ (rounded off)}$$

We add the 4 extra feet at the top and get the height of the pole: 62.9 feet.

To find the length of the brace, c, we set up the ratio

$$\frac{18}{c}$$

Now we recognize this ratio as the cosine of the angle, 73°. Then we have

$$\frac{18}{c} = \cos 73°$$

Then

$$\frac{18}{c} = 0.2924$$

Solving this equation for c, we get $c = 61.6$ ft (rounded off).

Example 3. A 36-foot ladder leans up against a vertical wall. If the ladder makes an angle of 68.5° with the ground, how high on the wall will the ladder reach? How far is the foot of the ladder from the wall?

Solution. Figure 34.3 shows the right triangle in the problem.

Let w = the height of the wall to the top of the ladder. Then we set up the ratio

$$\frac{w}{36}$$

Notice that this fraction represents the ratio which is the sine of 68.5°. Then we write the equation:

$$\frac{w}{36} = \sin 68.5°$$

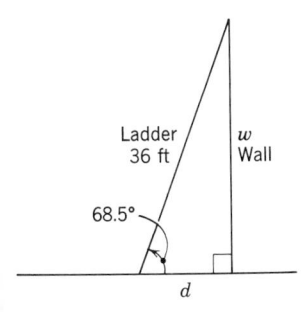

FIGURE 34.3

From the table, we get $\sin 68.5° = 0.9304$. Then we have the equation

$$\frac{w}{36} = 0.9304; \qquad w = (36)(0.9304) = 33.5 \text{ (approx.)}$$

Therefore, the ladder reaches a height of 33.5 feet on the wall.

Let d = the distance from the foot of the ladder to the wall. Then we set up the ratio

$$\frac{d}{36}$$

Now we note that this ratio represents the cosine of the angle 68.5°. Then

$$\frac{d}{36} = \cos 68.5°; \quad \text{or} \quad \frac{d}{36} = 0.3665; \quad d = (36)(0.3665) = 13.2 \text{ (approx.)}$$

Example 4. A man stands on the top of a cliff that is 300 feet above the level of the water below. He sees a small boat out on the water. The angle of depression to the boat is 14.7°. How far is the boat from the cliff?

Solution. We make a sketch of the right triangle involved in the problem and label all parts (Fig. 34.4). The man's angle of depression is the same as the angle of elevation from the boat to the man.

Let d = the distance from the boat to the cliff. Then we write the ratio

$$\frac{d}{300} = \cot 14.7° \qquad (\cot 14.7° = 3.812)$$

Then we have

$$\frac{d}{300} = 3.812$$

Solving for d, we get

$$d = (300)(3.812) = 1143.6 \text{ (feet)}$$

The answer can be reasonably rounded off to 1140 feet.

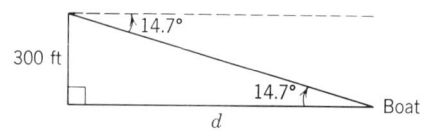

FIGURE 34.4

Example 5. Find the angle of elevation of the sun when a flagpole 48 feet high casts a shadow 56 feet long.

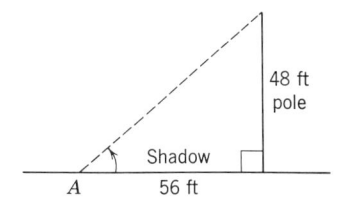

FIGURE 34.5

Solution. Figure 34.5 shows the right triangle involved in the problem. Here we set up the ratio between two sides:

$$\frac{48}{56}$$

Now we note that this ratio represents the tangent of A, the angle of elevation of the sun. Then we write

$$\frac{48}{56} = \tan A$$

Reducing the fraction to a decimal, we have $0.8571 = \tan A$. From the table, we find that the angle A is $40.6°$.

Note. In this problem it happens that the tangent value, 0.8571 appears in the table. If we do not find the exact value of a function in the table, we take the nearest angle.

Exercise 34.1

1. The shadow of a flagpole is 73 feet long when the sun is at an elevation of $39.7°$. Find the height of the flagpole.
2. A telephone pole is braced with a wire fastened 3 feet from the top of the pole. The other end of the brace is fastened to a stake in the ground some distance from the foot of the pole. If the wire brace is 50 feet long and makes an angle of $56°$ with the ground, how high is the pole?
3. A television tower is braced by a wire cable. One end of the cable is fastened to the tower 20 feet from the top. The other end of the brace is fastened to a stake in the ground 120 feet from the foot of the tower. The cable makes an angle of $74.5°$ with the ground. How high is the tower? How long is the cable?
4. The shadow of a tree is 82 feet when the sun is at an elevation of $27°$. How high is the tree?
5. A ladder 40 feet long leans up against a vertical wall. The angle between the wall and the ladder is $15.6°$. How high on the wall does the ladder reach? How far is the foot of the ladder from the wall?

6. The Washington Monument is 555 feet high. How long is its shadow when the elevation of the sun is 65°?

7. An inclined railway is built to the top of a hill that is 200 feet above the level of the bottom of the railway. If the angle of elevation of the railway is 42°, how long is the railway?

8. An escalator from the first floor to the second floor of a building is 42 feet long. If the escalator makes an angle of 28.4° with the floors, find the vertical distance between the floors.

9. To find the angle of elevation of the sun, a 6-foot pole is set up vertically. Then the shadow is found to be 11 feet 3 inches. Find the angle of elevation of the sun.

10. Find the angle of elevation of the sun when a 60-foot flagpole casts a shadow 72 feet long.

11. A road has a uniform elevation of 6°. Find the increase in elevation in driving $\frac{1}{4}$ mile along the road.

12. A railroad track 500 feet long has a rise of 40 feet. What is the angle of elevation of the track?

13. A telephone pole 40 feet high is braced by a wire fastened 4 feet from the top of the pole. The other end of the brace is fastened to a stake in the ground 15 feet from the foot of the pole. What angle does the brace make with the ground? How long is the brace?

14. A plank 18 feet long is used to roll a barrel onto a truck. If the platform of the truck is 5.2 feet above the ground, what angle does the plank make with the ground?

15. Two positions on level land, *A* and *B*, are 800 feet apart. At a certain instant a helicopter is directly above point *B*. At that instant, the angle of elevation of the helicopter from *A* is 73.2°. Find the height of the helicopter.

16. From the top of a cliff 200 feet above a lake, the angle of depression to a small boat on the lake is 18°. How far is the boat from the cliff? Some time later, the angle of depression of the boat is 32°. How far has the boat moved toward shore?

17. From the top of an 80-foot building, a man looks at the near side of a street below and finds the angle of depression is 65°. How far is the street from the building. Then the man finds that the angle of depression to the far side of the street is 41°. How wide is the street?

34.2. SOLVING A RIGHT TRIANGLE

In working a word problem, we are usually asked to find only one or two of the unknowns. However, to "solve" a triangle means to find all the unknown elements.

The three sides and three angles of a triangle are called the six *elements* of the triangle. To solve a triangle, we must know at least *three* of the elements, *one of which must be a side*. In a right triangle, we know that one angle is a right angle. Then two more elements must be known, one of which must be a side.

SOLVING RIGHT TRIANGLES

In finding the unknown elements of a triangle, we make use of the trigonometric ratios, just as we have done in stated word problems. We also recall the fact that the sum of the angles of a triangle is 180°. In a right triangle, the sum of the two acute angles is therefore 90°.

Note. *The Pythagorean rule should not be used in finding the length of a side or the hypotenuse. That rule should be reserved to be used as a check on the accuracy of the work.*

The angles of a triangle are usually named with capital letters placed at the vertices. Each side is named with a small letter that corresponds to the opposite angle. In naming the angles of a triangle, the letters A, B, and C, are often used, with the letter C as the right angle. However, any letters may be used to name the angles.

Example 1. In a certain right triangle ABC, angle C is the right angle, angle $A = 24.3°$, and side $b = 46$ inches. Find a, and c, and angle B.

Solution. We make a sketch of the triangle, labeling the parts (Fig. 34.6) and showing the right angle as the square corner. Now we set up a ratio between an unknown side and a known side:

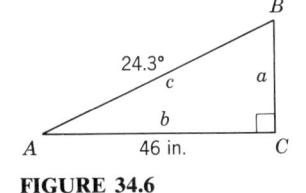

$$\frac{a}{b}$$

which is

$$\frac{a}{46}$$

FIGURE 34.6

We notice that this ratio represents the tangent of angle A. Then we write the equation

$$\frac{a}{46} = \tan 24.3° \qquad (\tan 24.3° = 0.4515, \text{ from table})$$

Solving for a,

$$a = (46)(0.4515) = 20.8 \text{ (rounded off)}$$

To find c, we set up the ratio,

$$\frac{46}{c}$$

This ratio represents the cosine of the angle 24.3°. From the table, we get

$$\cos 24.3° = 0.9114$$

536 TRIGONOMETRY

Then we have the equation:

$$\frac{46}{c} = 0.9114$$

If we multiply both sides of this equation by c, and then divide by the cosine of 24.3°, which is 0.9114, we get the equation

$$c = \frac{46}{\cos 24.3°}; \quad \text{or} \quad c = \frac{46}{0.9114} = 50.5 \text{ (approx.)}$$

For angle B, we take
$$90° - 24.3° = 65.7°$$

We now have the answers:

$$a = 20.8; \quad c = 50.5; \quad B = 65.7°$$

The answers can be checked by the Pythagorean rule; that is,

$$a^2 + b^2 \text{ should equal } c^2; \quad \text{or} \quad (20.8)^2 + (46)^2 = (50.5)^2$$

$$(20.8)^2 + (46)^2 = 2548.64; \quad (50.5)^2 = 2550.25$$

If these answers are nearly the same, the work can be considered correct.

Note. To find c in this example, if we set up the ratio, $c/46$, this ratio represents the secant of angle A. Without a table of secants, we write instead, $46/c$, which is the cosine of A. We must use this ratio, even though the solution involves division.

Example 2. In a certain right triangle ABC, angle C is the right angle; angle $A = 42.3°$, and $a = 25$ inches. Find b, c, and B.

Solution. We sketch the triangle (Fig. 34.7). To find b, we set up the ratio

$$\frac{b}{25}$$

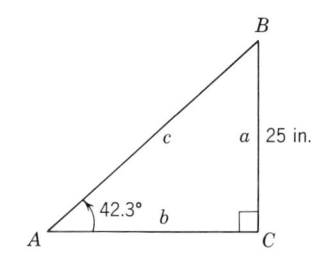

FIGURE 34.7

Now we see that this ratio represents the cotangent of the angle A, or cot 42.3°. Then we write the equation

$$\frac{b}{25} = \cot 42.3°$$

SOLVING RIGHT TRIANGLES **537**

In the table we find that cot 42.3° = 1.099. Then we have the equation,

$$\frac{b}{25} = 1.099$$

Solving for b, we get

$$b = (25)(1.099) = 27.5 \text{ (approx.)}$$

To find c, we set up the ratio

$$\frac{25}{c}$$

Now we see that this ratio represents the sine of angle A, which is 0.673. Then we have

$$\frac{25}{c} = \sin A$$

Solving for c, we get

$$c = \frac{25}{\sin 42.3°} = \frac{25}{0.673} = 37.1 \text{ (approx.)}$$

To find angle B, we have

$$B = 90° - 42.3° = 47.7°$$

As a check, we find that $(25)^2 + (27.5)^2$ is very nearly equal to $(37.1)^2$.

Example 3. In a certain right triangle DEF, angle F is the right angle, angle $D = 34.8°$, and $f = 52$ inches. Find d, e, and E.

Solution. We sketch the figure (Fig. 34.8). To find d, we have

$$\frac{d}{52} = \sin 34.8°$$

Then

$$d = (52)(\sin 34.8°)$$

$$= (52)(0.5707) = 29.7 \text{ (approx.)}$$

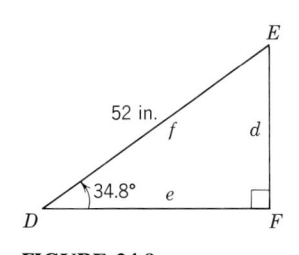

FIGURE 34.8

538 TRIGONOMETRY

To find e, we have

$$\frac{e}{52} = \cos 34.8°$$

Then

$$e = (52)(\cos 34.8°) = (52)(0.8211) = 42.7 \text{ (approx.)}$$

To find angle E, we take

$$E = 90° - 34.8° = 55.2°$$

As a check, we find that $d^2 + e^2 = 2705.38$; $f^2 = 2704.00$.

Example 4. In a certain right triangle ABC, angle C is the right angle, angle $B = 65.1°$, and $b = 53$ inches. Find a, c, and A.

Solution. We sketch the figure (Fig. 34.9). To find a, we write

$$\frac{a}{53} = \cot B$$

or

$$\frac{a}{53} = \cot 65.1° = 0.4642$$

Then

$$a = (53)(0.4642) = 24.6 \text{ (approx.)}$$

For c, we write the ratio

$$\frac{53}{c} = \sin 65.1°$$

Multiplying both sides by c and then dividing both sides by $\sin 65.1°$, we get

$$c = \frac{53}{\sin 65.1°} = \frac{53}{0.907} = 58.4 \text{ (approx.)}$$

For angle A, we have

$$A = 90° - 65.1° = 24.9°$$

Checking by the Pythagorean rule, we find that

$$(24.6)^2 + (53)^2 \text{ is very nearly equal to } (58.4)^2$$

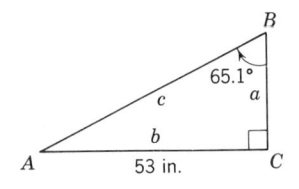

FIGURE 34.9

SOLVING RIGHT TRIANGLES 539

Example 5. In a certain right triangle ABC, angle C is the right angle, the side $a = 18$ inches, and $b = 25$ inches. Find c, A, and B.

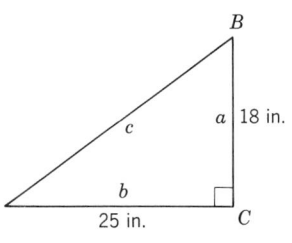

Solution. We sketch the triangle (Fig. 34.10). We could find the hypotenuse c immediately by the Pythagorean rule, but we save that rule for the check. Instead, we first find angle A:

We set up the ratio
$$\frac{18}{25}$$

FIGURE 34.10

Now we see this ratio represents the tangent of A. Then we write
$$\frac{18}{25} = \tan A$$

As a decimal, we get
$$\tan A = 0.7200; \quad \text{then} \quad A = 35.8° \text{ (approx.)}$$

Now we use angle A to find c:
$$\frac{18}{c} = \sin 35.8°$$

Solving for c,
$$c = \frac{18}{\sin 35.8°}; \quad \text{or} \quad c = \frac{18}{0.585} = 30.8 \text{ (approx.)}$$

For angle B, we have
$$B = 90° - 35.8° = 54.2°$$

Example 6. In a certain triangle RST, angle T is a right angle, $s = 32$ inches, and $t = 39$ inches. Find R, r, and S.

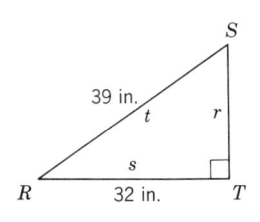

Solution. We sketch the triangle (Fig. 34.11). To find R, we set up the ratio
$$\frac{32}{39}$$

FIGURE 34.11

Now we see this ratio represents the $\cos R$. Reducing the fraction to a decimal, we have
$$0.8205 = \cos R$$

540 TRIGONOMETRY

Then
$$R = 34.9°$$

From the table we find this cosine value is opposite the angle 34.9° (approx.). Now we use angle R to find r. We set up the ratio

$$\frac{r}{39} = \sin 34.9°$$

Since $\sin 34.9° = 0.5721$, we have

$$\frac{r}{39} = 0.5721$$

Solving for r, we get

$$r = (39)(0.5721) = 22.3 \text{ (approx.)}$$

For angle S, we have
$$S = 90° - 34.9° = 55.1°$$

Exercise 34.2

Solve the following triangles. Each triangle is understood to be lettered ABC, with angle C as the right angle and the sides named with small letters that correspond to the opposite angles. Round off answers to *three significant digits* (or to *four* if the first digit is 1).

1. $A = 19.4°$; $c = 42$ in.
2. $A = 61.9°$; $c = 33$ cm
3. $B = 24.2°$; $c = 60$ cm
4. $B = 68.7°$; $c = 510$ in.
5. $A = 33.8°$; $b = 36$ ft
6. $A = 59.2°$; $b = 18$ ft
7. $B = 41.5°$; $a = 74$ cm
8. $B = 53.7°$; $a = 16$ cm
9. $A = 18.3°$; $a = 24$ in.
10. $A = 63.9°$; $a = 52$ in.
11. $B = 16.5°$; $b = 25$ cm
12. $B = 76.1°$; $b = 90$ cm
13. $a = 11$ cm; $b = 20$ cm
14. $a = 13.2$ cm; $c = 25$ cm
15. $a = 42$ in.; $b = 16$ in.
16. $b = 24$ in.; $c = 52$ in.
17. $a = 23$ cm; $c = 26$ cm
18. $b = 33$ cm; $c = 38$ cm
19. $A = 11.2°$; $c = 240$ cm
20. $A = 60.6°$; $c = 760$ cm
21. $A = 13.1°$; $b = 68$ cm
22. $B = 69.8°$; $c = 86$ cm
23. $B = 14.3°$; $b = 25.4$ in.
24. $A = 12.3°$; $a = 30.8$ in.
25. $B = 68.1°$; $a = 28.5$ cm
26. $A = 72.9°$; $b = 95$ cm
27. $A = 79.1°$; $a = 256$ cm
28. $B = 73.8°$; $h = 78.5$ cm
29. $B = 22.9°$; $c = 72.5$ cm
30. $A = 39.4°$; $c = 1200$ cm
31. $A = 16.3°$; $b = 24.8$ in.
32. $A = 56.2°$; $a = 32.2$ cm
33. $B = 38.8°$; $a = 126$ in.
34. $B = 68.3°$; $b = 52.4$ in.

35 INTRODUCTION TO SET THEORY*

35.1. INTRODUCTION

At some point in your study of mathematics, you will be faced with problems involving *sets*. Then it will be well to have some idea of the meaning of a set, as well as a knowledge of some of the elementary operations with sets. The purpose of this chapter is to present an introduction to set theory as a preparation for future study involving sets.

35.2. MEANING OF A SET

One of the basic concepts in mathematics is the idea of a set. Although the term *set*, strictly speaking, is undefined, it can be considered as a collection of objects or things. The term *set* is common in everyday use, as when we speak of a set of books, a set of dishes, a set of furniture, and so on. We can also speak of a set of ideas, a set of qualities, or a set of conditions.

The individual objects or ideas in a set are called the *elements* of the set. For example, consider a class of forty students in a mathematics class. The set is the entire class or collection of students. The set may have the name "Mathematics I, Section A." Each student in the class is an element of the set. There are then forty elements in this set.

The elements of a set are usually of a certain kind having something in common with one another, such as in the following sets:

(a) The set of students in a class,
(b) The set of letters in the English alphabet,
(c) The set of the three students: John, Mary, and Robert,
(d) The set of vowels in the English alphabet,
(e) The set of people in the United States,
(f) The set of counting numbers: 1, 2, 3, 4, and so on.

* Optional

In most cases, the word *set* refers to elements that do have some common characteristic, even though the elements may be only concepts or ideas, such as the set of qualities: kindness, friendship, and joy. However, a set may consist of elements that have no common characteristic. For example, a set may consist of the following elements: Hoover Dam, Charles Dickens, Mount Hood, and Yale University. As another example, a speaker may announce that he will speak on the set of four topics: the Constitution, modern music, Mexico, and the Twenty-third Psalm.

35.3. NOTATION FOR SETS AND ELEMENTS

A set is often named with a capital letter of the alphabet. The elements of the set are usually enclosed in a pair of braces: { }. If there are only a few elements in the set, the elements may be listed. Listing of the elements of a set is called the *roster* or *tabulation* method for indicating the elements. For example, we may have the following set:

$$N = \{1, 2, 3, 4, 5\}$$

The expression is read, "N is the set of numbers, 1, 2, 3, 4, and 5." This set contains only five elements. It is called a *finite set*. A *finite set* is a set having a limited number of elements.

The elements of a finite set may be indicated in other ways. In some cases, they are shown by what is called *partial listing*. If the number of elements is so great that it would be inconvenient to list all, then they may be indicated by listing the first few, followed by a series of dots, and then perhaps the final element in the set. For example, the set of twenty-six letters of the alphabet may be indicated as follows:

$$A = \{a, b, c, d, \ldots, z\}$$

The capital letter A names the set. The dots should be read "and so on." This set is finite and contains twenty-six elements.

If the elements of a set have some recognizable common characteristic, they may be indicated in another way. Then the elements may be represented by a letter followed by a descriptive of determining phrase. For example, the set of letters of the alphabet may be indicated in the following manner:

$$A = \{x \mid x \text{ is a letter of the English alphabet}\}$$

The vertical line "|" is read, "such that." Then this set is read: "A is the set of elements x such that x is a letter of the English alphabet." Note that x

represents any letter of the alphabet, and A is the name of the set. This is a *finite* set.

As another example of this kind of notation, we may have the set

$$P = \{x \mid x \text{ has been a president of the United States}\}$$

The set is read: "P is the set of people x such that x has been a president of the United States."

A set may contain a great number of elements, yet be finite. For example, the following is a finite set:

$$H = \{x \mid x \text{ is a living human being}\}$$

This set is read: "H is the set of elements such that x is a living human being." The set contains many elements, but it is *finite*.

A set may contain an infinite number of elements. Then it is called an *infinite set*. An example of an infinite set is the set of all the natural numbers; that is, all the positive integers, beginning with 1. Such a set may be indicated in two ways:

$$N = \{x \mid x \text{ is a positive integer}\}$$

$$N = \{1, 2, 3, 4, 5, \ldots\}$$

A set may have only a single element, such as the following sets:

$$C = \{5\}$$

This set contains the single element 5.

$$D = \{x \mid x + 3 = 7\}$$

This set contains the single element 4.

It is important to distinguish between a set containing only one element and the element itself. For example, we may have a class containing only one student. A student, Tom, may be the only student taking a certain course in Math-12. The class consisting of the student Tom is not the same as Tom himself. The class may be cancelled, but Tom still exists.

The following set contains two elements:

$$R = \{x \mid x^2 = 9\}$$

This set has two elements: $+3$ and -3.

A set may have *no* element. Such a set is called a *null* set or the *empty set*. The empty set is often denoted by the Greek letter *phi* (ϕ). The following is the empty set; it contains no element.

$$W = \{x \mid x \text{ is a past woman president of the United States}\}$$

For this set we can write:

$$W = \phi$$

Warning. Do not enclose ϕ in braces. Also, do not put zero (0) in the empty set. To do so would make zero (0) an element of the set. Instead the empty set may be indicated by a pair of braces alone.

In the foregoing example, we can write

$$w = \{\ \};\quad \text{or}\quad \phi = \{\ \}$$

Note. A set must have well-defined elements. There must be no question about whether or not a particular element belongs to the set. It must not be a matter of opinion whether an element is to be included. The set of the ten greatest living people in the world is not a true set because there might well be disagreement whether a particular person belongs in the set.

To indicate that a particular element belongs to a set, we use the symbol $\in$, which resembles a capital E with a rounded back. For example, suppose we have the set of four elements:

$$S = \{a, b, c, d\}$$

We indicate that a particular element, say, c, is an element of the set S in the following manner:

$$c \in S$$

The statement is read: "c is an element of S", or "c is in S." For the set S as shown, we also have

$$a \in S;\quad b \in S;\quad d \in S$$

However, $g \notin S$. This is read: "g is not a member of S," or "g is not in S". Note that the symbol, $\notin$, recalls the symbol $\neq$ for "does not equal."

35.4. EQUAL SETS

Two sets, A and B, are said to be equal if every element in A is also in B; and every element in B is also in A. The order of listing of the elements is immaterial. For example, consider the two sets:

$$A = \{c, d, e\} \quad \text{and} \quad B = \{e, c, d\}$$

Note that the elements in set A are exactly the same as the elements in B. Then we can say

$$A = B$$

The order in which the elements are listed in a set does not affect the meaning of a set. If a committee consists of a set of three people, they may be seated in any order, and the committee will still consist of the same people; that is, the elements of the set will be the same.

35.5. SUBSETS

A subset can be thought of as a set that is part of another set. For example, consider the two sets:

$$A = \{1, 2, 3, 4, 5, 6\} \quad B = \{2, 4, 5\}$$

Note that set B contains only elements that are also in set A. Then we say that set B is a subset of set A.

To indicate that one set is a subset of another set, we use the symbol "$\subset$" Then we can write:

$$B \subset A$$

The expression is read: "B is a subset of A." Note that all the elements of B are also in A.

If the symbol is reversed, it is read "contains." For example,

$$A \supset B \text{ is read "} A \text{ contains } B\text{."}$$

As another example of a subset, let us denote the set of the letters of the alphabet by L, the set of vowels by V, and the set of consonants by C. Then we can say:

$$V \subset L, \quad \text{which is read "} V \text{ is a subset of } L\text{."}$$

Also: $\quad L \supset V, \quad$ read "L contains V."

Also: $\quad C \subset L, \quad$ and $\quad L \supset C.$

In this example, the sets V and C are called *disjoint* sets. Two disjoint sets are sets such that neither set has any element that is contained in the other. That is, they have no element in common.

Definition of Subset. One set S is a subset of another set T if and only if every element in S is also in T.

Note that, according to this definition, the two sets S and T may actually be equal. For example, consider the two sets:

$$S = \{c, d, e, f\} \quad \text{and} \quad T = \{d, f, e, c\}$$

Note that every element in S is also in T. Therefore, by the definition, S is a subset of T; that is $S \subset T$. We can also say $S = T$.

To indicate that S is a subsct of T and is also equal to T, we use the symbol: $\subseteq$. That is, $S \subseteq T$; also $T \subseteq S$.

In most instances, we think of a subset as being only a part or portion of a larger set. Then the subset is called a *proper* subset. For example, if

$$A = \{a, b, c, d, e\} \quad \text{and} \quad B = \{d, a, c\}$$

then we say that B is a proper subset of A. That is, $B \subset A$.

35.6. ALL POSSIBLE SUBSETS OF A GIVEN SET

A subset of a given set may contain any number of the elements of the given set. For example, if a set contains, say, three elements, then the subsets of this set may contain *none*, *one*, *two*, or *three* of the three elements of the given set. The empty set is also considered a subset of every set. Moreover, it is a *proper subset* of every set except of the empty set itself.

Let us consider all the possible subsets of the set: $S = \{x, y, z\}$. They are:

$$\{x, y\}; \{x, z\}; \{y, z\}; \{x\}; \{y\}; \{z\}; \{x, y, z\}; \{\ \}.$$

Note that there are eight subsets of the set: $S = \{x, y, z\}$. All of these are proper subsets except the subset $\{x, y, z\}$.

It can be shown that the total number of subsets in any given set can be determined by the following rule. The number of elements in the given set can be taken as a power of 2. For example, the foregoing set S has 3 elements. Then we take 3 as the exponent on 2. That is, for the total number of subsets, we take 2^3, which is equal to 8.

As another example, if a given set has 5 elements, the total number of subsets of this set is equal to 2^5, which is equal to 32. This number will include the empty set as well as the subset containing all of the elements of the given set. All of these are *proper subsets* except the subset containing all the elements of the given set.

A question may arise here. Why do we call the empty set a subset of a given set? Or why do we say one possible subset of a given set contains all the elements of the given set?

To answer these questions, consider the following problem. Suppose we let P represent the set of all people who have been president of the United States. This set contains 37 elements. We can write

$$P = \{x \mid x \text{ is a person who has been president of the U.S.}\}$$

Now suppose we wish to list the elements of this set in classes or subsets as follows according to age at the time of inauguration. Each class is a subset of P.

Subset	Age at Inauguration	Number of Elements in Subset
A	20–34	0
B	35–70	37

The subset A contains no elements. It is the empty set. The subset B contains 37 elements. Then we can write

$$A \subset P \quad \text{and} \quad B \subseteq P$$

In this case, the set A, the empty set, is a *proper subset of P*.

35.7. UNION AND INTERSECTION OF SETS; OVERLAPPING OF SETS

It sometimes happens that only some of the elements of one set are contained in another set. For example, let us consider a mathematics class of 40 students.

Now we ask this question of the class: How many of you are studying physics? The answer, let us say, is 28. Call this set P. Then we ask this question of the class: How many of you are studying biology? The answer, let us say, is 13. Call this set B.

One conclusion is certain. With only 40 people in the class, some of them must be studying both physics and biology. The two sets, P and B, are not mutually exclusive. There must be some overlapping of the two sets. Some of the elements in P must also be in B.

A second point must be considered. There are 40 students in all in the class. There are 28 in set P and 13 in set B. However, there may be some students among the 40 who are not taking either physics or biology. The students who are not taking physics or biology make up what is called the *complementary* set of this class.

At this point we are interested in discovering three facts:

(1) How many students are taking *both physics and biology*? That is, how many elements are in both sets, P and B. The answer is called the *intersection* of the two sets, indicated by the symbol: $P \cap B$, read "P intersection B."

(2) How many of the 40 students are taking *either physics or biology or both*? That is, we wish to know the total number of individual students in both sets, P and B, counting each student just once. The answer is called the *union* of the sets, shown by the symbol: $P \cup B$, read "P union B."

(3) How many students are not taking either physics or biology? The answer to this question is called the *complementary set*, denoted by T.

Definitions.

1. The *intersection* of two or more sets includes only those elements that are found in all of the two or more sets.

2. The *union* of two or more sets contains all the separate individual elements contained in all the sets, counting each separate element just once.

3. The *universal set*, denoted by U, contains all the elements under consideration in a problem. In the foregoing example, the universal set consists of the 40 students in the class.

4. The *complementary set*, denoted by T, includes the elements in the universal set but not contained in the other given sets.

5. *Disjoint sets* are sets that have no element in common.

Let us suppose that the complementary set T contains 7 elements. Then it is possible to find the number of elements in the union of the two sets, P and B, and also in their intersection. Since the total number of elements in the universal set U is 40, and T contains 7 elements, then the total number of elements in the union of the two sets, P and B, must be 33. Then we can write

$$P \cup B \text{ contains 33 elements}$$

To find the number of elements in the intersection, $P \cap B$, we proceed as follows. The number of elements in P alone is 28, and the number in B alone is 13. If we add the number of elements in P and B, we get 41. But the union of the two sets contains only 33 elements. Therefore, there must be an overlapping equal to the difference, which is 8. This is the number of elements in the intersection. That is

$$P \cap B \text{ contains 8 elements}$$

35.8. THE VENN DIAGRAM

The situation in the foregoing example can be represented graphically by what is called a *Venn diagram.* Let us represent set *P* by one circle and set *B* by another circle (Fig. 35.1).

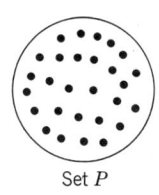

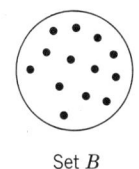

Set *P*   Set *B*

FIGURE 35.1

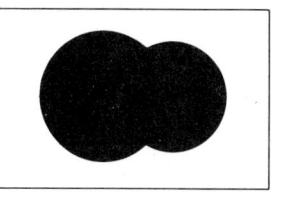

FIGURE 35.2

In the diagram (Fig. 35.2), the rectangle represents the entire class, which we call the universal set *U*. The sets, *P* and *B*, are subsets of the universal set *U*. The overlapping of sets *P* and *B* is shown by the overlapping of the circles.

In Figure 35.2, set *P* (28 elements) is shown by one circle, shaded in one direction. Set *B* (13 elements) is shown by the other circle, shaded in another direction.

The heavily shaded portion, shaded in both directions, represents the overlapping of the two sets. This portion represents the number of students taking both physics and biology; that is, the *intersection* of the two sets. Note that the intersection of the two sets contains elements belonging to both sets, or, in symbols, $P \cap B$. Any element in the intersection must be in both sets.

The entire shaded portion represents the *union* of the two sets; that is, the number of students taking either physics or biology or both. The union includes all students taking one or both of these courses, counting each student just once. For the union, we write: $P \cup B$.

The unshaded portion of the rectangle represents the complementary set *T*, which contains the students taking neither physics nor biology. Note that the union of the three sets, *P*, *B*, and *T*, contains the entire class of 40 students. That is, $P \cup B \cup T$ contains 40 elements.

Exercise 35.1

Given the three sets:

$$X = \{a, b, c, d, e, f\}; \quad Y = \{d, e, f\}; \quad Z = \{b, c\}$$

Taking the universal set U as the entire alphabet, find the following:

1. $X \cup Y$
2. $Y \cup Z$
3. $Y \cap Z$
4. $Z \cup X$
5. $X \cap Z$
6. $Y \cap U$
7. $Y \cap X$
8. $Y \cup U$
9. $X \cup U$
10. $T \cap U$

Tell whether or not each of the following is true:

11. $b \in X$
12. $d \in X$
13. $c \in Y$
14. $k \in U$
15. $a \in U$
16. $i \in X$
17. $\phi \in X$
18. $d \in Y$
19. $\phi \subset X$
20. $h \in T$
21. $T \subset U$
22. $c \subset X$
23. $c \in Y$
24. $\phi \subset T$
25. $X \supset U$
26. $Y \subset U$
27. $U \supset T$
28. $Y \supset Z$
29. $Z \subset U$
30. $T \subset U$

APPENDIX

TABLE 1. AMOUNT AT COMPOUND INTEREST $(1+i)^n$

Rate i

Periods	$\frac{5}{8}$%	$\frac{3}{4}$%	$\frac{7}{8}$%	1%	$1\frac{1}{8}$%	Periods
4	1.02524	1.03034	1.03546	1.04060	1.04577	4
8	1.05111	1.06160	1.07218	1.08286	1.09362	8
12	1.07763	1.09381	1.11020	1.12683	1.14367	12
16	1.10483	1.12699	1.14957	1.17258	1.19601	16
20	1.13271	1.16118	1.19034	1.22019	1.25075	20
24	1.16129	1.19641	1.23255	1.26973	1.30799	24
28	1.19060	1.23271	1.27626	1.32129	1.36785	28
32	1.22064	1.27011	1.32152	1.37494	1.43045	32
36	1.25145	1.30865	1.36838	1.43077	1.49592	36
40	1.23803	1.34835	1.41691	1.48886	1.56438	40
44	1.31540	1.38926	1.46716	1.54932	1.63597	44
48	1.34860	1.43141	1.51918	1.61223	1.71084	48
52	1.38263	1.47483	1.57306	1.67769	1.78914	52
56	1.41752	1.51958	1.62884	1.74581	1.87102	56
60	1.45329	1.56568	1.68660	1.81670	1.95665	60
64	1.48997	1.61318	1.74641	1.89046	2.04619	64
68	1.52757	1.66213	1.80834	1.96722	2.13984	68
72	1.56612	1.71255	1.87247	2.04710	2.23777	72
76	1.60564	1.76451	1.93887	2.13022	2.34018	76
80	1.64616	1.81804	2.00763	2.21672	2.44727	80
84	1.68770	1.87320	2.07883	2.30672	2.55927	84
88	1.73029	1.93003	2.15254	2.40038	2.67640	88
92	1.77395	1.98859	2.22888	2.49785	2.79889	92
96	1.81872	2.04892	2.30792	2.59927	2.92698	96
100	1.86462	2.11108	2.38976	2.70481	3.06093	100
104	1.91167	2.17513	2.47451	2.81464	3.20101	104
108	1.95991	2.24112	2.56226	2.92893	3.34751	108
112	2.00937	2.30912	2.65312	3.04785	3.50071	112
116	2.06008	2.37917	2.74721	3.17161	3.66092	116
120	2.11206	2.45136	2.84463	3.30039	3.82846	120
124	2.16536	2.52573	2.94551	3.43440	4.00367	124
128	2.22001	2.60236	3.04996	3.57385	4.18690	128
132	2.27603	2.68131	3.15812	3.71896	4.37851	132
136	2.33347	2.76266	3.27011	3.86996	4.57890	136
140	2.39235	2.84648	3.38608	4.02710	4.78845	140
144	2.45272	2.93284	3.50615	4.19062	5.00759	144
148	2.51462	3.02182	3.63049	4.36077	5.23677	148
152	2.57808	3.11350	3.75923	4.53784	5.47643	152
156	2.64314	3.20796	3.89254	4.72209	5.72706	156
160	2.70984	3.30528	4.03058	4.91383	5.98915	160

TABLE 1. (*Continued*)

			Rate *i*			
Periods	**1¼%**	**1⅜%**	**1½%**	**1⅝%**	**1¾%**	**Periods**
2	1.02516	1.02769	1.03022	1.03276	1.03531	2
4	1.05095	1.05614	1.06136	1.06660	1.07186	4
6	1.07738	1.08539	1.09344	1.10155	1.10970	6
8	1.10449	1.11544	1.12649	1.13764	1.14888	8
10	1.13227	1.14633	1.16054	1.17491	1.18944	10
12	1.16075	1.17807	1.19562	1.21341	1.23144	12
14	1.18995	1.21069	1.23176	1.25316	1.27492	14
16	1.21989	1.24421	1.26899	1.29422	1.31993	16
18	1.25058	1.27866	1.30734	1.33663	1.36653	18
20	1.28204	1.31407	1.34686	1.38042	1.41478	20
22	1.31429	1.35045	1.38756	1.42565	1.46473	22
24	1.34735	1.38784	1.42950	1.47236	1.51644	24
26	1.38125	1.42627	1.47271	1.52060	1.56998	26
28	1.41599	1.46576	1.51722	1.57042	1.62541	28
30	1.45161	1.50635	1.56308	1.62187	1.68280	30
32	1.48813	1.54806	1.61032	1.67501	1.74221	32
34	1.52557	1.59092	1.65900	1.72989	1.80372	34
36	1.56394	1.63498	1.70914	1.78657	1.86741	36
38	1.60329	1.68025	1.76080	1.84511	1.93334	38
40	1.64362	1.72677	1.81402	1.90556	2.00160	40
42	1.68497	1.77458	1.86885	1.96799	2.07227	42
44	1.72735	1.82372	1.92533	2.03247	2.14543	44
46	1.77081	1.87422	1.98353	2.09906	2.22118	46
48	1.81535	1.92611	2.04348	2.16784	2.29960	48
50	1.86102	1.97944	2.10524	2.23887	2.38079	50
52	1.90784	2.03425	2.16887	2.31222	2.46485	52
54	1.95583	2.09058	2.23443	2.38798	2.55187	54
56	2.00503	2.14847	2.30196	2.46622	2.64197	56
58	2.05547	2.20796	2.37154	2.54702	2.73525	58
60	2.10718	2.26909	2.44322	2.63047	2.83182	60
64	2.21453	2.39649	2.59314	2.80566	3.03531	64
68	2.32735	2.53104	2.75227	2.99253	3.25342	68
72	2.44592	2.67314	2.92116	3.19183	3.48721	72
76	2.57053	2.82323	3.10041	3.40441	3.73780	76
80	2.70148	2.98174	3.29066	3.63115	4.00639	80
84	2.83911	3.14915	3.49259	3.87299	4.29429	84
88	2.98375	3.32595	3.70691	4.13094	4.60287	88
92	3.13576	3.51269	3.93438	4.40607	4.93363	92
96	3.29551	3.70991	4.17580	4.69952	5.28815	96
100	3.46340	3.91820	4.43205	5.01252	5.66816	100

TABLE 1. (*Continued*)

Periods	2%	2¼%	2½%	2¾%	3%	Periods
1	1.02000	1.02250	1.02500	1.02750	1.03000	1
2	1.04040	1.04551	1.05062	1.05576	1.06090	2
3	1.06121	1.06903	1.07689	1.08479	1.09273	3
4	1.08243	1.09308	1.10381	1.11462	1.12551	4
5	1.10408	1.11768	1.13141	1.14527	1.15927	5
6	1.12616	1.14283	1.15969	1.17677	1.19405	6
7	1.14869	1.16854	1.18869	1.20913	1.22987	7
8	1.17166	1.19483	1.21840	1.24238	1.26677	8
9	1.19509	1.22171	1.24886	1.27655	1.30477	9
10	1.21899	1.24920	1.28008	1.31165	1.34392	10
11	1.24337	1.27731	1.31209	1.34772	1.38423	11
12	1.26824	1.30605	1.34489	1.38478	1.42576	12
13	1.29361	1.33544	1.37851	1.42287	1.46853	13
14	1.31948	1.36548	1.41297	1.46199	1.51259	14
15	1.34587	1.39621	1.44830	1.50220	1.55797	15
16	1.37279	1.42762	1.48451	1.54351	1.60471	16
17	1.40024	1.45974	1.52162	1.58596	1.65285	17
18	1.42825	1.49259	1.55966	1.62957	1.70243	18
19	1.45681	1.52617	1.59865	1.67438	1.75351	19
20	1.48595	1.56051	1.63862	1.72043	1.80611	20
22	1.54598	1.63152	1.72157	1.81635	1.91610	22
24	1.60844	1.70577	1.80873	1.91763	2.03279	24
26	1.67342	1.78339	1.90029	2.02455	2.15659	26
28	1.74102	1.86454	1.99650	2.13743	2.28793	28
30	1.81136	1.94939	2.09757	2.25660	2.42726	30
32	1.88454	2.03810	2.20376	2.38242	2.57508	32
34	1.96068	2.13085	2.31532	2.51526	2.73191	34
36	2.03989	2.22782	2.43254	2.65550	2.89828	36
38	2.12230	2.32920	2.55568	2.80356	3.07478	38
40	2.20804	2.43519	2.68506	2.95987	3.26204	40
42	2.29724	2.54601	2.82100	3.12491	3.46070	42
44	2.39005	2.66186	2.96381	3.29914	3.67145	44
46	2.48661	2.78300	3.11385	3.48309	3.89504	46
48	2.58707	2.90964	3.27149	3.67729	4.13225	48
50	2.69159	3.04205	3.43711	3.88232	4.38391	50
52	2.80033	3.18048	3.61111	4.09879	4.65089	52
54	2.91346	3.32521	3.79392	4.32732	4.93412	54
56	3.03117	3.47653	3.98599	4.56859	5.23461	56
58	3.15362	3.63473	4.18778	4.82332	5.55340	58
60	3.28103	3.80013	4.39979	5.09225	5.89160	60

TABLE 1. (*Continued*)

			Rate *i*			
Periods	**3½%**	**4%**	**4½%**	**5%**	**5½%**	**Periods**
1	1.03500	1.04000	1.04500	1.05000	1.05500	1
2	1.07122	1.08160	1.09202	1.10250	1.11302	2
3	1.10872	1.12486	1.14117	1.15762	1.17424	3
4	1.14752	1.16986	1.19252	1.21551	1.23882	4
5	1.18769	1.21665	1.24618	1.27628	1.30696	5
6	1.22926	1.26532	1.30226	1.34010	1.37884	6
7	1.27228	1.31593	1.36086	1.40710	1.45468	7
8	1.31681	1.36857	1.42210	1.47746	1.53469	8
9	1.36290	1.42331	1.48610	1.55133	1.61909	9
10	1.41060	1.48024	1.55297	1.62889	1.70814	10
11	1.45997	1.53945	1.62285	1.71034	1.80209	11
12	1.51107	1.60103	1.69588	1.79586	1.90121	12
13	1.56396	1.66507	1.77220	1.88565	2.00577	13
14	1.61869	1.73168	1.85194	1.97993	2.11609	14
15	1.67535	1.80094	1.93528	2.07893	2.23248	15
16	1.73399	1.87298	2.02237	2.18287	2.35526	16
17	1.79468	1.94790	2.11338	2.29202	2.48480	17
18	1.85749	2.02582	2.20848	2.40662	2.62147	18
19	1.92250	2.10685	2.30786	2.52695	2.76565	19
20	1.98979	2.19112	2.41171	2.65330	2.91776	20
21	2.05943	2.27877	2.52024	2.78596	3.07823	21
22	2.13151	2.36992	2.63365	2.92526	3.24754	22
23	2.20611	2.46472	2.75217	3.07152	3.42615	23
24	2.28333	2.56330	2.87601	3.22510	3.61459	24
25	2.36324	2.66584	3.00543	3.38635	3.81339	25
26	2.44596	2.77247	3.14068	3.55567	4.02313	26
27	2.53157	2.88337	3.28201	3.73346	4.24440	27
28	2.62017	2.99870	3.42970	3.92013	4.47784	28
29	2.71188	3.11865	3.58404	4.11614	4.72412	29
30	2.80679	3.24340	3.74532	4.32194	4.98395	30
32	3.00671	3.50806	4.08998	4.76494	5.54726	32
34	3.22086	3.79432	4.46636	5.25335	6.17424	34
36	3.45027	4.10393	4.87738	5.79182	6.87209	36
38	3.69601	4.43881	5.32622	6.38548	7.64880	38
40	3.95926	4.80102	5.81636	7.03999	8.51331	40
42	4.24126	5.19278	6.35162	7.76159	9.47553	42
44	4.54334	5.61652	6.93612	8.55715	10.54650	44
46	4.86694	6.07482	7.57442	9.43426	11.73852	46
48	5.21359	6.57053	8.27146	10.40127	13.06526	48
50	5.58493	7.10668	9.03264	11.46740	14.54196	50

TABLE 1. (*Continued*)

Rate i

Periods	6%	6½%	7%	7½%	8%	Periods
1	1.06000	1.06500	1.07000	1.07500	1.08000	1
2	1.12360	1.13422	1.14490	1.15562	1.16640	2
3	1.19102	1.20795	1.22504	1.24230	1.25971	3
4	1.26248	1.28647	1.31080	1.33547	1.36049	4
5	1.33823	1.37009	1.40255	1.43563	1.46933	5
6	1.41852	1.45914	1.50073	1.54330	1.58687	6
7	1.50363	1.55399	1.60578	1.65905	1.71382	7
8	1.59385	1.65500	1.71819	1.78348	1.85093	8
9	1.68948	1.76257	1.83846	1.91724	1.99900	9
10	1.79085	1.87714	1.96715	2.06103	2.15892	10
11	1.89830	1.99915	2.10485	2.21561	2.33164	11
12	2.01220	2.12910	2.25219	2.38178	2.51817	12
13	2.13293	2.26749	2.40984	2.56041	2.71962	13
14	2.26090	2.41487	2.57853	2.75244	2.93719	14
15	2.39656	2.57184	2.75903	2.95888	3.17217	15
16	2.54035	2.73901	2.95216	3.18079	3.42594	16
17	2.69277	2.91705	3.15882	3.41935	3.70002	17
18	2.85434	3.10665	3.37993	3.67580	3.99602	18
19	3.02560	3.30859	3.61653	3.95149	4.31570	19
20	3.20714	3.52365	3.86968	4.24785	4.66096	20
21	3.39956	3.75268	4.14056	4.56644	5.03383	21
22	3.60354	3.99661	4.43040	4.90892	5.43654	22
23	3.81975	4.25639	4.74053	5.27709	5.87146	23
24	4.04893	4.53305	5.07237	5.67287	6.34118	24
25	4.29187	4.82770	5.42743	6.09834	6.84848	25
26	4.54938	5.14150	5.80735	6.55572	7.39635	26
27	4.82235	5.47570	6.21387	7.04739	7.98806	27
28	5.11169	5.83162	6.64884	7.57595	8.62711	28
29	5.41839	6.21067	7.11426	8.14414	9.31727	29
30	5.74349	6.61437	7.61225	8.75496	10.06266	30
32	6.45339	7.50218	8.71527	10.11745	11.73708	32
34	7.25103	8.50916	9.97811	11.69197	13.69013	34
36	8.14725	9.65130	11.42394	13.51154	15.96817	36
38	9.15425	10.94675	13.07927	15.61427	18.62528	38
40	10.28572	12.41607	14.97446	18.04424	21.72452	40
42	11.55703	14.08262	17.14426	20.85237	25.33948	42
44	12.98548	15.97286	19.62846	24.09752	29.55597	44
46	14.59049	18.11682	22.47262	27.84770	34.47409	46
48	16.39387	20.54855	25.72891	32.18150	40.21057	48
50	18.42015	23.30668	29.45703	37.18975	46.90161	50

TABLE 2. SQUARES, SQUARE ROOTS, CUBES, CUBE ROOTS

No.	Square	Square Root	Cube	Cube Root	No.	Square	Square Root	Cube	Cube Root
0	0	0.000	0	0.000	50	2500	7.071	125000	3.684
1	1	1.000	1	1.000	51	2601	7.141	132651	3.708
2	4	1.414	8	1.260	52	2704	7.211	140608	3.732
3	9	1.732	27	1.442	53	2809	7.280	148877	3.756
4	16	2.000	64	1.587	54	2916	7.348	157464	3.780
5	25	2.236	125	1.710	55	3025	7.416	166375	3.803
6	36	2.449	216	1.817	56	3136	7.483	175616	3.826
7	49	2.646	343	1.913	57	3249	7.550	185193	3.849
8	64	2.828	512	2.000	58	3364	7.616	195112	3.871
9	81	3.000	729	2.080	59	3481	7.681	205379	3.893
10	100	3.162	1000	2.154	60	3600	7.746	216000	3.915
11	121	3.317	1331	2.224	61	3721	7.810	226981	3.936
12	144	3.464	1728	2.289	62	3844	7.874	238328	3.958
13	169	3.606	2197	2.351	63	3969	7.937	250047	3.979
14	196	3.742	2744	2.410	64	4096	8.000	262144	4.000
15	225	3.873	3375	2.466	65	4225	8.062	274625	4.021
16	256	4.000	4096	2.520	66	4356	8.124	287496	4.041
17	289	4.123	4913	2.571	67	4489	8.185	300763	4.062
18	324	4.243	5832	2.621	68	4624	8.246	314432	4.082
19	361	4.359	6859	2.668	69	4761	8.307	328509	4.102
20	400	4.472	8000	2.714	70	4900	8.367	343000	4.121
21	441	4.583	9261	2.759	71	5041	8.426	357911	4.141
22	484	4.690	10648	2.802	72	5184	8.485	373248	4.160
23	529	4.796	12167	2.844	73	5329	8.544	389017	4.179
24	576	4.899	13824	2.884	74	5476	8.602	405224	4.198
25	625	5.000	15625	2.924	75	5625	8.660	421875	4.217
26	676	5.099	17576	2.962	76	5776	8.718	438976	4.236
27	729	5.196	19683	3.000	77	5929	8.775	456533	4.254
28	784	5.292	21952	3.037	78	6084	8.832	474552	4.273
29	841	5.385	24389	3.072	79	6241	8.888	493039	4.291
30	900	5.477	27000	3.107	80	6400	8.944	512000	4.309
31	961	5.568	29791	3.141	81	6561	9.000	531441	4.327
32	1024	5.657	32768	3.175	82	6724	9.055	551368	4.344
33	1089	5.745	35937	3.208	83	6889	9.110	571787	4.362
34	1156	5.831	39304	3.240	84	7056	9.165	592704	4.380
35	1225	5.916	42875	3.271	85	7225	9.220	614125	4.397
36	1296	6.000	46656	3.302	86	7396	9.274	636056	4.414
37	1369	6.083	50653	3.332	87	7569	9.327	658503	4.431
38	1444	6.164	54872	3.362	88	7744	9.381	681472	4.448
39	1521	6.245	59319	3.391	89	7921	9.434	704969	4.465
40	1600	6.325	64000	3.420	90	8100	9.487	729000	4.481
41	1681	6.403	68921	3.448	91	8281	9.539	753571	4.498
42	1764	6.481	74088	3.476	92	8464	9.592	778688	4.514
43	1849	6.557	79507	3.503	93	8649	9.644	804357	4.531
44	1936	6.633	85184	3.530	94	8836	9.695	830584	4.547
45	2025	6.708	91125	3.557	95	9025	9.747	857375	4.563
46	2116	6.782	97336	3.583	96	9216	9.798	884736	4.579
47	2209	6.856	103823	3.609	97	9409	9.849	912673	4.595
48	2304	6.928	110592	3.634	98	9604	9.899	941192	4.610
49	2401	7.000	117649	3.659	99	9801	9.950	970299	4.626
50	2500	7.071	125000	3.684	100	10000	10.000	1000000	4.642

TABLE 3. NATURAL TRIGONOMETRIC FUNCTIONS FOR DECIMAL FRACTIONS OF A DEGREE

Deg.	Sin	Tan	Cot	Cos	Deg.	Deg.	Sin	Tan	Cot	Cos	Deg.
0.0	0.00000	0.00000	∞	1.0000	90.0	4.5	0.07846	0.07870	12.706	0.9969	85.5
0.1	0.00175	0.00175	573.0	1.0000	89.9	4.6	0.08020	0.08046	12.429	0.9968	85.4
0.2	0.00349	0.00349	286.5	1.0000	89.8	4.7	0.08194	0.08221	12.163	0.9966	85.3
0.3	0.00524	0.00524	191.0	1.0000	89.7	4.8	0.08368	0.08397	11.909	0.9965	85.2
0.4	0.00698	0.00698	143.24	1.0000	89.6	4.9	0.08542	0.08573	11.664	0.9963	85.1
0.5	0.00873	0.00873	114.59	1.0000	89.5	5.0	0.08716	0.08749	11.430	0.9962	85.0
0.6	0.01047	0.01047	95.49	0.9999	89.4	5.1	0.08889	0.08925	11.205	0.9960	84.9
0.7	0.01222	0.01222	81.85	0.9999	89.3	5.2	0.09063	0.09101	10.988	0.9959	84.8
0.8	0.01396	0.01396	71.62	0.9999	89.2	5.3	0.09237	0.09277	10.780	0.9957	84.7
0.9	0.01571	0.01571	63.66	0.9999	89.1	5.4	0.09411	0.09453	10.579	0.9956	84.6
1.0	0.01745	0.01746	57.29	0.9998	89.0	5.5	0.09585	0.09629	10.385	0.9954	84.5
1.1	0.01920	0.01920	52.08	0.9998	88.9	5.6	0.09758	0.09805	10.199	0.9952	84.4
1.2	0.02094	0.02095	47.74	0.9998	88.8	5.7	0.09932	0.09981	10.019	0.9951	84.3
1.3	0.02269	0.02269	44.07	0.9997	88.7	5.8	0.10106	0.10158	9.845	0.9949	84.2
1.4	0.02443	0.02444	40.92	0.9997	88.6	5.9	0.10279	0.10334	9.677	0.9947	84.1
1.5	0.02618	0.02619	38.19	0.9997	88.5	6.0	0.10453	0.10510	9.514	0.9945	84.0
1.6	0.02792	0.02793	35.80	0.9996	88.4	6.1	0.10626	0.10687	9.357	0.9943	83.9
1.7	0.02967	0.02968	33.69	0.9996	88.3	6.2	0.10800	0.10863	9.205	0.9942	83.8
1.8	0.03141	0.03143	31.82	0.9995	88.2	6.3	0.10973	0.11040	9.058	0.9940	83.7
1.9	0.03316	0.03317	30.14	0.9995	88.1	6.4	0.11147	0.11217	8.915	0.9938	83.6
2.0	0.03490	0.03492	28.64	0.9994	88.0	6.5	0.11320	0.11394	8.777	0.9936	83.5
2.1	0.03664	0.03667	27.27	0.9993	87.9	6.6	0.11494	0.11570	8.643	0.9934	83.4
2.2	0.03839	0.03842	26.03	0.9993	87.8	6.7	0.11667	0.11747	8.513	0.9932	83.3
2.3	0.04013	0.04016	24.90	0.9992	87.7	6.8	0.11840	0.11924	8.386	0.9930	83.2
2.4	0.04188	0.04191	23.86	0.9991	87.6	6.9	0.12014	0.12101	8.264	0.9928	83.1
2.5	0.04362	0.04366	22.90	0.9990	87.5	7.0	0.12187	0.12278	8.144	0.9925	83.0
2.6	0.04536	0.04541	22.02	0.9990	87.4	7.1	0.12360	0.12456	8.028	0.9923	82.9
2.7	0.04711	0.04716	21.20	0.9989	87.3	7.2	0.12533	0.12633	7.916	0.9921	82.8
2.8	0.04885	0.04891	20.45	0.9988	87.2	7.3	0.12706	0.12810	7.806	0.9919	82.7
2.9	0.05059	0.05066	19.74	0.9987	87.1	7.4	0.12880	0.12988	7.700	0.9917	82.6
3.0	0.05234	0.05241	19.081	0.9986	87.0	7.5	0.13053	0.13165	7.596	0.9914	82.5
3.1	0.05408	0.05416	18.464	0.9985	86.9	7.6	0.13226	0.13343	7.495	0.9912	82.4
3.2	0.05582	0.05591	17.886	0.9984	86.8	7.7	0.13399	0.13521	7.396	0.9910	82.3
3.3	0.05756	0.05766	17.343	0.9983	86.7	7.8	0.13572	0.13698	7.300	0.9907	82.2
3.4	0.05931	0.05941	16.832	0.9982	86.6	7.9	0.13744	0.13876	7.207	0.9905	82.1
3.5	0.06105	0.06116	16.350	0.9981	86.5	8.0	0.13917	0.14054	7.115	0.9903	82.0
3.6	0.06279	0.06291	15.895	0.9980	86.4	8.1	0.14090	0.14232	7.026	0.9900	81.9
3.7	0.06453	0.06467	15.464	0.9979	86.3	8.2	0.14263	0.14410	6.940	0.9898	81.8
3.8	0.06627	0.06642	15.056	0.9978	86.2	8.3	0.14436	0.14588	6.855	0.9895	81.7
3.9	0.06802	0.06817	14.669	0.9977	86.1	8.4	0.14608	0.14767	6.772	0.9893	81.6
4.0	0.06976	0.06993	14.301	0.9976	86.0	8.5	0.14781	0.14945	6.691	0.9890	81.5
4.1	0.07150	0.07168	13.951	0.9974	85.9	8.6	0.14954	0.15124	6.612	0.9888	81.4
4.2	0.07324	0.07344	13.617	0.9973	85.8	8.7	0.15126	0.15302	6.535	0.9885	81.3
4.3	0.07498	0.07519	13.300	0.9972	85.7	8.8	0.15299	0.15481	6.460	0.9882	81.2
4.4	0.07672	0.07695	12.996	0.9971	85.6	8.9	0.15471	0.15660	6.386	0.9880	81.1
Deg.	Cos	Cot	Tan	Sin	Deg.	Deg.	Cos	Cot	Tan	Sin	Deg.

TABLE 3. (*Continued*)

Deg.	Sin	Tan	Cot	Cos	Deg.	Deg.	Sin	Tan	Cot	Cos	Deg.
9.0	0.15643	0.15838	6.314	0.9877	81.0	13.5	0.2334	0.2401	4.165	0.9724	76.5
9.1	0.15816	0.16017	6.243	0.9874	80.9	13.6	0.2351	0.2419	4.134	0.9720	76.4
9.2	0.15988	0.16196	6.174	0.9871	80.8	13.7	0.2368	0.2438	4.102	0.9715	76.3
9.3	0.16160	0.16376	6.107	0.9869	80.7	13.8	0.2385	0.2456	4.071	0.9711	76.2
9.4	0.16333	0.16555	6.041	0.9866	80.6	13.9	0.2402	0.2475	4.041	0.9707	76.1
9.5	0.16505	0.16734	5.976	0.9863	80.5	14.0	0.2419	0.2493	4.011	0.9703	76.0
9.6	0.16677	0.16914	5.912	0.9860	80.4	14.1	0.2436	0.2512	3.981	0.9699	75.9
9.7	0.16849	0.17093	5.850	0.9857	80.3	14.2	0.2453	0.2530	3.952	0.9694	75.8
9.8	0.17021	0.17273	5.789	0.9854	80.2	14.3	0.2470	0.2549	3.923	0.9690	75.7
9.9	0.17193	0.17453	5.730	0.9851	80.1	14.4	0.2487	0.2568	3.895	0.9686	75.6
10.0	0.1736	0.1763	5.671	0.9848	80.0	14.5	0.2504	0.2586	3.867	0.9681	75.5
10.1	0.1754	0.1781	5.614	0.9845	79.9	14.6	0.2521	0.2605	3.839	0.9677	75.4
10.2	0.1771	0.1799	5.558	0.9842	79.8	14.7	0.2538	0.2623	3.812	0.9673	75.3
10.3	0.1788	0.1817	5.503	0.9839	79.7	14.8	0.2554	0.2642	3.785	0.9668	75.2
10.4	0.1805	0.1835	5.449	0.9836	79.6	14.9	0.2571	0.2661	3.758	0.9664	75.1
10.5	0.1822	0.1853	5.396	0.9833	79.5	15.0	0.2588	0.2679	3.732	0.9659	75.0
10.6	0.1840	0.1871	5.343	0.9829	79.4	15.1	0.2605	0.2698	3.706	0.9655	74.9
10.7	0.1857	0.1890	5.292	0.9826	79.3	15.2	0.2622	0.2717	3.681	0.9650	74.8
10.8	0.1874	0.1908	5.242	0.9823	79.2	15.3	0.2639	0.2736	3.655	0.9646	74.7
10.9	0.1891	0.1926	5.193	0.9820	79.1	15.4	0.2656	0.2754	3.630	0.9641	74.6
11.0	0.1908	0.1944	5.145	0.9816	79.0	15.5	0.2672	0.2773	3.606	0.9636	74.5
11.1	0.1925	0.1962	5.097	0.9813	78.9	15.6	0.2689	0.2792	3.582	0.9632	74.4
11.2	0.1942	0.1980	5.050	0.9810	78.8	15.7	0.2706	0.2811	3.558	0.9627	74.3
11.3	0.1959	0.1998	5.005	0.9806	78.7	15.8	0.2723	0.2830	3.534	0.9622	74.2
11.4	0.1977	0.2016	4.959	0.9803	78.6	15.9	0.2740	0.2849	3.511	0.9617	74.1
11.5	0.1994	0.2035	4.915	0.9799	78.5	16.0	0.2756	0.2867	3.487	0.9613	74.0
11.6	0.2011	0.2053	4.872	0.9796	78.4	16.1	0.2773	0.2886	3.465	0.9608	73.9
11.7	0.2028	0.2071	4.829	0.9792	78.3	16.2	0.2790	0.2905	3.442	0.9603	73.8
11.8	0.2045	0.2089	4.787	0.9789	78.2	16.3	0.2807	0.2924	3.420	0.9598	73.7
11.9	0.2062	0.2107	4.745	0.9785	78.1	16.4	0.2823	0.2943	3.398	0.9593	73.6
12.0	0.2079	0.2126	4.705	0.9781	78.0	16.5	0.2840	0.2962	3.376	0.9588	73.5
12.1	0.2096	0.2144	4.665	0.9778	77.9	16.6	0.2857	0.2981	3.354	0.9583	73.4
12.2	0.2113	0.2162	4.625	0.9774	77.8	16.7	0.2874	0.3000	3.333	0.9578	73.3
12.3	0.2130	0.2180	4.586	0.9770	77.7	16.8	0.2890	0.3019	3.312	0.9573	73.2
12.4	0.2147	0.2199	4.548	0.9767	77.6	16.9	0.2907	0.3038	3.291	0.9568	73.1
12.5	0.2164	0.2217	4.511	0.9763	77.5	17.0	0.2924	0.3057	3.271	0.9563	73.0
12.6	0.2181	0.2235	4.474	0.9759	77.4	17.1	0.2940	0.3076	3.251	0.9558	72.9
12.7	0.2198	0.2254	4.437	0.9755	77.3	17.2	0.2957	0.3096	3.230	0.9553	72.8
12.8	0.2215	0.2272	4.402	0.9751	77.2	17.3	0.2974	0.3115	3.211	0.9548	72.7
12.9	0.2233	0.2290	4.366	0.9748	77.1	17.4	0.2990	0.3134	3.191	0.9542	72.6
13.0	0.2250	0.2309	4.331	0.9744	77.0	17.5	0.3007	0.3153	3.172	0.9537	72.5
13.1	0.2267	0.2327	4.297	0.9740	76.9	17.6	0.3024	0.3172	3.152	0.9532	72.4
13.2	0.2284	0.2345	4.264	0.9736	76.8	17.7	0.3040	0.3191	3.133	0.9527	72.3
13.3	0.2300	0.2364	4.230	0.9732	76.7	17.8	0.3057	0.3211	3.115	0.9521	72.2
13.4	0.2317	0.2382	4.198	0.9728	76.6	17.9	0.3074	0.3230	3.096	0.9516	72.1
Deg.	Cos	Cot	Tan	Sin	Deg.	Deg.	Cos	Cot	Tan	Sin	Deg.

TABLE 3. (*Continued*).

Deg.	Sin	Tan	Cot	Cos	Deg.	Deg.	Sin	Tan	Cot	Cos	Deg.
18.0	0.3090	0.3249	3.078	0.9511	72.0	22.5	0.3827	0.4142	2.414	0.9239	67.5
18.1	0.3107	0.3269	3.060	0.9505	71.9	22.6	0.3843	0.4163	2.402	0.9232	67.4
18.2	0.3123	0.3288	3.042	0.9500	71.8	22.7	0.3859	0.4183	2.391	0.9225	67.3
18.3	0.3140	0.3307	3.024	0.9494	71.7	22.8	0.3875	0.4204	2.379	0.9219	67.2
18.4	0.3156	0.3327	3.006	0.9489	71.6	22.9	0.3891	0.4224	2.367	0.9212	67.1
18.5	0.3173	0.3346	2.989	0.9483	71.5	23.0	0.3907	0.4245	2.356	0.9205	67.0
18.6	0.3190	0.3365	2.971	0.9478	71.4	23.1	0.3923	0.4265	2.344	0.9198	66.9
18.7	0.3206	0.3385	2.954	0.9472	71.3	23.2	0.3939	0.4286	2.333	0.9191	66.8
18.8	0.3223	0.3404	2.937	0.9466	71.2	23.3	0.3955	0.4307	2.322	0.9184	66.7
18.9	0.3239	0.3424	2.921	0.9461	71.1	23.4	0.3971	0.4327	2.311	0.9178	66.6
19.0	0.3256	0.3443	2.904	0.9455	71.0	23.5	0.3987	0.4348	2.300	0.9171	66.5
19.1	0.3272	0.3463	2.888	0.9449	70.9	23.6	0.4003	0.4369	2.289	0.9164	66.4
19.2	0.3289	0.3482	2.872	0.9444	70.8	23.7	0.4019	0.4390	2.278	0.9157	66.3
19.3	0.3305	0.3502	2.856	0.9438	70.7	23.8	0.4035	0.4411	2.267	0.9150	66.2
19.4	0.3322	0.3522	2.840	0.9432	70.6	23.9	0.4051	0.4431	2.257	0.9143	66.1
19.5	0.3338	0.3541	2.824	0.9426	70.5	24.0	0.4067	0.4452	2.246	0.9135	66.0
19.6	0.3335	0.3561	2.808	0.9421	70.4	24.1	0.4083	0.4473	2.236	0.9128	65.9
19.7	0.3371	0.3581	2.793	0.9415	70.3	24.2	0.4099	0.4494	2.225	0.9121	65.8
19.8	0.3387	0.3600	2.778	0.9409	70.2	24.3	0.4115	0.4515	2.215	0.9114	65.7
19.9	0.3404	0.3620	2.762	0.9403	70.1	24.4	0.4131	0.4536	2.204	0.9107	65.6
20.0	0.3420	0.3640	2.747	0.9397	70.0	24.5	0.4147	0.4557	2.194	0.9100	65.5
20.1	0.3437	0.3659	2.733	0.9391	69.9	24.6	0.4163	0.4578	2.184	0.9092	65.4
20.2	0.3453	0.3679	2.718	0.9385	69.8	24.7	0.4179	0.4599	2.174	0.9085	65.3
20.3	0.3469	0.3699	2.703	0.9379	69.7	24.8	0.4195	0.4621	2.164	0.9078	65.2
20.4	0.3486	0.3719	2.689	0.9373	69.6	24.9	0.4210	0.4642	2.154	0.9070	65.1
20.5	0.3502	0.3739	2.675	0.9367	69.5	25.0	0.4226	0.4663	2.145	0.9063	65.0
20.6	0.3518	0.3759	2.660	0.9361	69.4	25.1	0.4242	0.4684	2.135	0.9056	64.9
20.7	0.3535	0.3779	2.646	0.9354	69.3	25.2	0.4258	0.4706	2.125	0.9048	64.8
20.8	0.3551	0.3799	2.633	0.9348	69.2	25.3	0.4274	0.4727	2.116	0.9041	64.7
20.9	0.3567	0.3819	2.619	0.9342	69.1	25.4	0.4289	0.4748	2.106	0.9033	64.6
21.0	0.3584	0.3839	2.605	0.9336	69.0	25.5	0.4305	0.4770	2.097	0.9026	64.5
21.1	0.3600	0.3859	2.592	0.9330	68.9	25.6	0.4321	0.4791	2.087	0.9018	64.4
21.2	0.3616	0.3879	2.578	0.9323	68.8	25.7	0.4337	0.4813	2.078	0.9011	64.3
21.3	0.3633	0.3899	2.565	0.9317	68.7	25.8	0.4352	0.4834	2.069	0.9003	64.2
21.4	0.3649	0.3919	2.552	0.9311	68.6	25.9	0.4368	0.4856	2.059	0.8996	64.1
21.5	0.3665	0.3939	2.539	0.9304	68.5	26.0	0.4384	0.4877	2.050	0.8988	64.0
21.6	0.3681	0.3959	2.526	0.9298	68.4	26.1	0.4399	0.4899	2.041	0.8980	63.9
21.7	0.3697	0.3979	2.513	0.9291	68.3	26.2	0.4415	0.4921	2.032	0.8973	63.8
21.8	0.3714	0.4000	2.500	0.9285	68.2	26.3	0.4431	0.4942	2.023	0.8965	63.7
21.9	0.3730	0.4020	2.488	0.9278	68.1	26.4	0.4446	0.4964	2.014	0.8957	63.6
22.0	0.3746	0.4040	2.475	0.9272	68.0	26.5	0.4462	0.4986	2.006	0.8949	63.5
22.1	0.3762	0.4061	2.463	0.9265	67.9	26.6	0.4478	0.5008	1.997	0.8942	63.4
22.2	0.3778	0.4081	2.450	0.9259	67.8	26.7	0.4493	0.5029	1.988	0.8934	63.3
22.3	0.3795	0.4101	2.438	0.9252	67.7	26.8	0.4509	0.5051	1.980	0.8926	63.2
22.4	0.3811	0.4122	2.426	0.9245	67.6	26.9	0.4524	0.5073	1.971	0.8918	63.1
Deg.	Cos	Cot	Tan	Sin	Deg.	Deg.	Cos	Cot	Tan	Sin	Deg.

TABLE 3. (*Continued*)

Deg.	Sin	Tan	Cot	Cos	Deg.	Deg.	Sin	Tan	Cot	Cos	Deg.
27.0	0.4540	0.5095	1.963	0.8910	63.0	31.5	0.5225	0.6128	1.6319	0.8526	58.5
27.1	0.4555	0.5117	1.954	0.8902	62.9	31.6	0.5240	0.6152	1.6255	0.8517	58.4
27.2	0.4571	0.5139	1.946	0.8894	62.8	31.7	0.5255	0.6176	1.6191	0.8508	58.3
27.3	0.4586	0.5161	1.937	0.8886	62.7	31.8	0.5270	0.6200	1.6128	0.8499	58.2
27.4	0.4602	0.5184	1.929	0.8878	62.6	31.9	0.5284	0.6224	1.6066	0.8490	58.1
27.5	0.4617	0.5206	1.921	0.8870	62.5	32.0	0.5299	0.6249	1.6003	0.8480	58.0
27.6	0.4633	0.5228	1.913	0.8862	62.4	32.1	0.5314	0.6273	1.5941	0.8471	57.9
27.7	0.4648	0.5250	1.905	0.8854	62.3	32.2	0.5329	0.6297	1.5880	0.8462	57.8
27.8	0.4664	0.5272	1.897	0.8846	62.2	32.3	0.5344	0.6322	1.5818	0.8453	57.7
27.9	0.4679	0.5295	1.889	0.8838	62.1	32.4	0.5358	0.6346	1.5757	0.8443	57.6
28.0	0.4695	0.5317	1.881	0.8829	62.0	32.5	0.5373	0.6371	1.5697	0.8434	57.5
28.1	0.4710	0.5340	1.873	0.8821	61.9	32.6	0.5388	0.6395	1.5637	0.8425	57.4
28.2	0.4726	0.5362	1.865	0.8813	61.8	32.7	0.5402	0.6420	1.5577	0.8415	57.3
28.3	0.4741	0.5384	1.857	0.8805	61.7	32.8	0.5417	0.6445	1.5517	0.8406	57.2
28.4	0.4756	0.5407	1.849	0.8796	61.6	32.9	0.5432	0.6469	1.5458	0.8396	57.1
28.5	0.4772	0.5430	1.842	0.8788	61.5	33.0	0.5446	0.6494	1.5399	0.8387	57.0
28.6	0.4787	0.5452	1.834	0.8780	61.4	33.1	0.5461	0.6519	1.5340	0.8377	56.9
28.7	0.4802	0.5475	1.827	0.8771	61.3	33.2	0.5476	0.6544	1.5282	0.8368	56.8
28.8	0.4818	0.5498	1.819	0.8763	61.2	33.3	0.5490	0.6569	1.5224	0.8358	56.7
28.9	0.4833	0.5520	1.811	0.8755	61.1	33.4	0.5505	0.6594	1.5166	0.8348	56.6
29.0	0.4848	0.5543	1.804	0.8746	61.0	33.5	0.5519	0.6619	1.5108	0.8339	56.5
29.1	0.4863	0.5566	1.797	0.8738	60.9	33.6	0.5534	0.6644	1.5051	0.8329	56.4
29.2	0.4879	0.5589	1.789	0.8729	60.8	33.7	0.5548	0.6669	1.4994	0.8320	56.3
29.3	0.4894	0.5612	1.782	0.8721	60.7	33.8	0.5563	0.6694	1.4938	0.8310	56.2
29.4	0.4909	0.5635	1.775	0.8712	60.6	33.9	0.5577	0.6720	1.4882	0.8300	56.1
29.5	0.4924	0.5658	1.767	0.8704	60.5	34.0	0.5592	0.6745	1.4826	0.8290	56.0
29.6	0.4939	0.5681	1.760	0.8695	60.4	34.1	0.5606	0.6771	1.4770	0.8281	55.9
29.7	0.4955	0.5704	1.753	0.8686	60.3	34.2	0.5621	0.6796	1.4715	0.8271	55.8
29.8	0.4970	0.5727	1.746	0.8678	60.2	34.3	0.5635	0.6822	1.4659	0.8261	55.7
29.9	0.4985	0.5750	1.739	0.8669	60.1	34.4	0.5650	0.6847	1.4605	0.8251	55.6
30.0	0.5000	0.5774	1.7321	0.8660	60.0	34.5	0.5664	0.6873	1.4550	0.8241	55.5
30.1	0.5015	0.5797	1.7251	0.8652	59.9	34.6	0.5678	0.6899	1.4496	0.8231	55.4
30.2	0.5030	0.5820	1.7182	0.8643	59.8	34.7	0.5693	0.6924	1.4442	0.8221	55.3
30.3	0.5045	0.5844	1.7113	0.8634	59.7	34.8	0.5707	0.6950	1.4388	0.8211	55.2
30.4	0.5060	0.5867	1.7045	0.8625	59.6	34.9	0.5721	0.6976	1.4335	0.8202	55.1
30.5	0.5075	0.5890	1.6977	0.8616	59.5	35.0	0.5736	0.7002	1.4281	0.8192	55.0
30.6	0.5090	0.5914	1.6909	0.8607	59.4	35.1	0.5750	0.7028	1.4229	0.8181	54.9
30.7	0.5105	0.5938	1.6842	0.8599	59.3	35.2	0.5764	0.7054	1.4176	0.8171	54.8
30.8	0.5120	0.5961	1.6775	0.8590	59.2	35.3	0.5779	0.7080	1.4124	0.8161	54.7
30.9	0.5135	0.5985	1.6709	0.8581	59.1	35.4	0.5793	0.7107	1.4071	0.8151	54.6
31.0	0.5150	0.6009	1.6643	0.8572	59.0	35.5	0.5807	0.7133	1.4019	0.8141	54.5
31.1	0.5165	0.6032	1.6577	0.8563	58.9	35.6	0.5821	0.7159	1.3968	0.8131	54.4
31.2	0.5180	0.6056	1.6512	0.8554	58.8	35.7	0.5835	0.7186	1.3916	0.8121	54.3
31.3	0.5195	0.6080	1.6447	0.8545	58.7	35.8	0.5850	0.7212	1.3865	0.8111	54.2
31.4	0.5210	0.6104	1.6383	0.8536	58.6	35.9	0.5864	0.7239	1.3814	0.8100	54.1
Deg.	Cos	Cot	Tan	Sin	Deg.	Deg.	Cos	Cot	Tan	Sin	Deg.

TABLE 3. (*Continued*)

Deg.	Sin	Tan	Cot	Cos	Deg.	Deg.	Sin	Tan	Cot	Cos	Deg.
36.0	0.5878	0.7265	1.3764	0.8090	54.0	40.5	0.6494	0.8541	1.1708	0.7604	49.5
36.1	0.5892	0.7292	1.3713	0.8080	53.9	40.6	0.6508	0.8571	1.1667	0.7593	49.4
36.2	0.5906	0.7319	1.3663	0.8070	53.8	40.7	0.6521	0.8601	1.1626	0.7581	49.3
36.3	0.5920	0.7346	1.3613	0.8059	53.7	40.8	0.6534	0.8632	1.1585	0.7570	49.2
36.4	0.5934	0.7373	1.3564	0.8049	53.6	40.9	0.6547	0.8662	1.1544	0.7559	49.1
36.5	0.5948	0.7400	1.3514	0.8039	53.5	41.0	0.6561	0.8693	1.1504	0.7547	49.0
36.6	0.5962	0.7427	1.3465	0.8028	53.4	41.1	0.6574	0.8724	1.1463	0.7536	48.9
36.7	0.5976	0.7454	1.3416	0.8018	53.3	41.2	0.6587	0.8754	1.1423	0.7524	48.8
36.8	0.5990	0.7481	1.3367	0.8007	53.2	41.3	0.6600	0.8785	1.1383	0.7513	48.7
36.9	0.6004	0.7508	1.3319	0.7997	53.1	41.4	0.6613	0.8816	1.1343	0.7501	48.6
37.0	0.6018	0.7536	1.3270	0.7986	53.0	41.5	0.6626	0.8847	1.1303	0.7490	48.5
37.1	0.6032	0.7563	1.3222	0.7976	52.9	41.6	0.6639	0.8878	1.1263	0.7478	48.4
37.2	0.6046	0.7590	1.3175	0.7965	52.8	41.7	0.6652	0.8910	1.1224	0.7466	48.3
37.3	0.6060	0.7618	1.3127	0.7955	52.7	41.8	0.6665	0.8941	1.1184	0.7455	48.2
37.4	0.6074	0.7646	1.3079	0.7944	52.6	41.9	0.6678	0.8972	1.1145	0.7443	48.1
37.5	0.6088	0.7673	1.3032	0.7934	52.5	42.0	0.6691	0.9004	1.1106	0.7431	48.0
37.6	0.6101	0.7701	1.2985	0.7923	52.4	42.1	0.6704	0.9036	1.1067	0.7420	47.9
37.7	0.6115	0.7729	1.2938	0.7912	52.3	42.2	0.6717	0.9067	1.1028	0.7408	47.8
37.8	0.6129	0.7757	1.2892	0.7902	52.2	42.3	0.6730	0.9099	1.0990	0.7396	47.7
37.9	0.6143	0.7785	1.2846	0.7891	52.1	42.4	0.6743	0.9131	1.0951	0.7385	47.6
38.0	0.6157	0.7813	1.2799	0.7880	52.0	42.5	0.6756	0.9163	1.0913	0.7373	47.5
38.1	0.6170	0.7841	1.2753	0.7869	51.9	42.6	0.6769	0.9195	1.0875	0.7361	47.4
38.2	0.6184	0.7869	1.2708	0.7859	51.8	42.7	0.6782	0.9228	1.0837	0.7349	47.3
38.3	0.6198	0.7898	1.2662	0.7848	51.7	42.8	0.6794	0.9260	1.0799	0.7337	47.2
38.4	0.6211	0.7926	1.2617	0.7837	51.6	42.9	0.6807	0.9293	1.0761	0.7325	47.1
38.5	0.6225	0.7954	1.2572	0.7826	51.5	43.0	0.6820	0.9325	1.0724	0.7314	47.0
38.6	0.6239	0.7983	1.2527	0.7815	51.4	43.1	0.6833	0.9358	1.0686	0.7302	46.9
38.7	0.6252	0.8012	1.2482	0.7804	51.3	43.2	0.6845	0.9391	1.0649	0.7290	46.8
38.8	0.6266	0.8040	1.2437	0.7793	51.2	43.3	0.6858	0.9424	1.0612	0.7278	46.7
38.9	0.6280	0.8069	1.2393	0.7782	51.1	43.4	0.6871	0.9457	1.0575	0.7266	46.6
39.0	0.6293	0.8098	1.2349	0.7771	51.0	43.5	0.6884	0.9490	1.0538	0.7254	46.5
39.1	0.6307	0.8127	1.2305	0.7760	50.9	43.6	0.6896	0.9523	1.0501	0.7242	46.4
39.2	0.6320	0.8156	1.2261	0.7749	50.8	43.7	0.6909	0.9556	1.0464	0.7230	46.3
39.3	0.6334	0.8185	1.2218	0.7738	50.7	43.8	0.6921	0.9590	1.0428	0.7218	46.2
39.4	0.6347	0.8214	1.2174	0.7727	50.6	43.9	0.6934	0.9623	1.0392	0.7206	46.1
39.5	0.6361	0.8243	1.2131	0.7716	50.5	44.0	0.6947	0.9657	1.0355	0.7193	46.0
39.6	0.6374	0.8273	1.2088	0.7705	50.4	44.1	0.6959	0.9691	1.0319	0.7181	45.9
39.7	0.6388	0.8302	1.2045	0.7694	50.3	44.2	0.6972	0.9725	1.0283	0.7169	45.8
39.8	0.6401	0.8332	1.2002	0.7683	50.2	44.3	0.6984	0.9759	1.0247	0.7157	45.7
39.9	0.6414	0.8361	1.1960	0.7672	50.1	44.4	0.6997	0.9793	1.0212	0.7145	45.6
40.0	0.6428	0.8391	1.1918	0.7660	50.0	44.5	0.7009	0.9827	1.0176	0.7133	45.5
40.1	0.6441	0.8421	1.1875	0.7649	49.9	44.6	0.7022	0.9861	1.0141	0.7120	45.4
40.2	0.6455	0.8451	1.1833	0.7638	49.8	44.7	0.7034	0.9896	1.0105	0.7108	45.3
40.3	0.6468	0.8481	1.1792	0.7627	49.7	44.8	0.7046	0.9930	1.0070	0.7096	45.2
40.4	0.6481	0.8511	1.1750	0.7615	49.6	44.9	0.7059	0.9965	1.0035	0.7083	45.1
						45.0	0.7071	1.0000	1.0000	0.7071	45.0
Deg.	Cos	Cot	Tan	Sin	Deg.	Deg.	Cos	Cot	Tan	Sin	Deg.

ANSWERS TO MOST ODD-NUMBERED EXERCISE PROBLEMS

Exercise 1.1, page 5.
1. 105; 76; 124; 117; 116; 129; 159; 168; 159; 168 **3.** 61; 102; 71; 80; 103; 70; 140; 84; 85; 93 **5.** 1162; 1190; 1172; 1470; 1295; 1486; 1372; 1754; 1793 **7.** 982; 1232; 1455; 1333; 1310; 1504; 1660; 1344; 1276 **9.** 2854; 2715; 2759; 2743 **11.** 45631 **13.** 77020 **15.** 69681

Exercise 1.2, page 8
1. 33; 43; 71; 10; 64; 45; 11; 23 **3.** 10; 51; 17; 32; 27; 38; 40; 56 **5.** 38; 25; 35; 38; 66; 12; 61; 59 **7.** 39; 28; 55; 23; 77; 59; 39; 64 **9.** 334; 645; 421; 202; 426; 420; 400; 210 **11.** 384; 464; 337; 359; 394; 278; 495; 356 **13.** 234; 117; 545; 474; 308; 326; 369; 206 **15.** 1287; 3970; 1475; 2393; 5990; 3883; 2780; 985 **17.** 6831; 231; 1285; 3164; 987; 1389; 2434; 3918 **19.** 16 **21.** 11 **23.** 284 **25.** 4190 **27.** 503

Exercise 1.3, page 14
1. 288; 282; 344; 316; 504; 760; 581; 553 **3.** 5502; 8406; 6904; 4606; 6102; 6902; 8064; 7808 **5.** 6278; 3298; 3120; 4484; 5546; 2755; 1541; 8256 **7.** 3484; 3744; 8232; 3404; 4465; 8277; 7056; 7912 **9.** 41031; 36582; 55845; 73108; 48662; 55774; 40082; 52896 **11.** 22920; 74080; 220806; 380608; 394842; 256360; 3383798; 2810880

Exercise 1.4, page 19 (Remainder, if any, is shown.)
1. 8079, R:1 **3.** 9708, R:5 **5.** 4806, R:2 **7.** 8064, R:7 **9.** 6012, R:3 **11.** 10377, R:7 **13.** 12029, R:3 **15.** 10500, R:4 **17.** 246, R:12 **19.** 536, R:4 **21.** 636, R:27 **23.** 266, R:30 **25.** 627, R:24 **27.** 786, R:28 **29.** 407, R:49 **31.** 809, R:12 **33.** 705 **35.** 412, R:24 **37.** 745, R:20 **39.** 435, R:20 **41.** 583 **43.** 505, R:36 **45.** 378, R:2 **47.** 619 **49.** 689 **51.** 12345679 **53.** 12345679 **55.** 18 **57.** 53

Exercise 1.5, page 19
1. 2368 **3.** 349 **5.** 4840 **7.** 102400 **9.** 12960 ft **11.** 5,865,696,000,000 **13.** 7,500,000 **15.** 2100

Exercise 2.1, page 21
1. Counting **3.** Counting **5.** Counting **7.** Counting **9.** Measurement

565

ANSWERS TO MOST ODD-NUMBERED EXERCISE PROBLEMS

Exercise 2.2, page 25

1. $\frac{1}{2}; \frac{1}{3}; \frac{1}{5}; \frac{1}{6}; \frac{2}{3}; \frac{7}{10}; \frac{2}{3}$ 3. $\frac{1}{4}; \frac{1}{3}; \frac{1}{8}; \frac{1}{2}; \frac{5}{8}; \frac{4}{5}; \frac{2}{3}$ 5. $\frac{3}{4}; \frac{2}{5}; \frac{4}{5}; \frac{2}{7}; \frac{5}{6}; \frac{3}{5}; \frac{4}{7}$ 7. $\frac{3}{4}; \frac{3}{8}; \frac{5}{7}; \frac{5}{11}; \frac{5}{7}; \frac{3}{10}; \frac{7}{12}$ 9. $\frac{5}{12}; \frac{7}{9}; \frac{5}{11}; \frac{5}{7}; \frac{5}{9}; \frac{9}{10}; \frac{5}{8}$ 11. $\frac{3}{8}; \frac{5}{8}; \frac{3}{4}; \frac{7}{16}; \frac{3}{5}; \frac{4}{9}; \frac{3}{5}$ 13. $\frac{24}{55}; \frac{2}{3}; \frac{3}{8}; \frac{445}{2268}; \frac{21}{22}; \frac{5}{6}; \frac{11}{13}$ 15. $\frac{36}{48}; \frac{32}{48}; \frac{18}{48}; \frac{28}{48}$ 17. $\frac{70}{105}; \frac{90}{105}; \frac{84}{105}; \frac{56}{105}$

Exercise 2.3, page 27

1. $3\frac{3}{4}$ 3. 8   5. $4\frac{3}{7}$ 7. $10\frac{5}{6}$ 9. $5\frac{5}{8}$ 11. $8\frac{3}{10}$ 13. $8\frac{1}{12}$ 15. $10\frac{3}{8}$ 17. 6   19. $12\frac{5}{12}$ 21. $12\frac{13}{16}$ 23. 15   25. $9\frac{1}{3}$ 27. $6\frac{13}{20}$ 29. $188\frac{1}{8}$ 31. $\frac{15}{3}; \frac{25}{5}; \frac{60}{12}$ 33. $\frac{27}{3}; \frac{45}{5}; \frac{108}{12}$ 35. $\frac{45}{3}; \frac{75}{5}; \frac{180}{12}$ 37. $\frac{13}{3}$ 39. $\frac{37}{5}$ 41. $\frac{60}{7}$ 43. $\frac{110}{9}$ 45. $\frac{75}{8}$ 47. $\frac{135}{8}$ 49. $\frac{150}{11}$ 51. $\frac{205}{16}$ 53. $\frac{371}{12}$ 55. $\frac{100}{3}$ 57. $\frac{870}{7}$ 59. $\frac{7405}{18}$ 61. $\frac{830}{7}$ 63. $\frac{1090}{11}$ 65. $\frac{1000}{7}$

Exercise 2.4, page 34

1. 12   3. 30   5. 30   7. 60   9. 42   11. 60   13. 72   15. 60   17. 1080   19. 2100   21. 10800   23. $1\frac{7}{20}; \frac{43}{48}; 1\frac{37}{54}; 1\frac{29}{72}$ 25. $\frac{31}{60}; \frac{35}{48}; \frac{11}{54}; \frac{11}{72}$ 27. $32\frac{1}{3}; 113\frac{3}{4}; 65\frac{5}{8}; 156; 9\frac{3}{8}; 16\frac{5}{12}; 27\frac{13}{40}; 76\frac{13}{36}$ 29. 3; $43\frac{3}{4}; 32\frac{3}{8}; \frac{1}{2}; 4\frac{1}{8}; 2\frac{11}{12}; 7\frac{37}{40}; 20\frac{19}{36}$ 31. $12\frac{11}{40}$ 33. $12\frac{11}{36}$ 35. $14\frac{19}{24}$ 37. $\frac{1}{24}$ 39. $\frac{1}{30}$ 41. $3\frac{5}{8}$ 43. $12\frac{13}{30}$ 45. $42\frac{31}{48}$ 47. 1   49. $7\frac{29}{72}$ 51. $5\frac{31}{120}$ 53. $1\frac{29}{120}$ 55. $31\frac{25}{36}$ 57. $21\frac{3}{8}$ 59. $13\frac{1}{8}$

Exercise 2.5, page 40

1. $\frac{8}{35}$ 3. $\frac{14}{27}$ 5. 8   7. $59\frac{1}{2}$ 9. $9\frac{4}{5}$ 11. $166\frac{1}{2}$ 13. $46\frac{1}{5}$ 15. $46\frac{1}{2}$ 17. $86\frac{1}{3}$ 19. $62\frac{3}{14}$ 21. $1\frac{1}{5}$ 23. $13\frac{1}{3}$ 25. $\frac{2}{9}$ 27. $\frac{1}{6}$ 29. $8\frac{3}{4}$ 31. $\frac{34}{69}$ 33. $1727\frac{509}{720}$ 35. $81\frac{459}{700}$ 37. $50\frac{10}{27}$ 39. $\frac{1}{2}$ 41. $\frac{5}{8}$ 43. $1\frac{1}{4}$ 45. $\frac{14}{15}$ 47. 1   49. $\frac{812}{1355}$

ANSWERS TO MOST ODD-NUMBERED EXERCISE PROBLEMS

Exercise 2.6, page 41
1. $12337\frac{1}{2}$ **3.** $200; 27\frac{7}{9}; 6\frac{17}{18}$ **5.** $10\frac{11}{16}$ in. **7.** $4\frac{7}{8}$ in. **9.** 7 pieces; $26\frac{1}{8}$ in. left **11.** 42 mph; $61\frac{3}{5}$ ft/sec

Exercise 3.1, page 45
1. $\frac{3}{5}$ **3.** $\frac{6}{25}$ **5.** $3\frac{1}{5}$ **7.** $\frac{3}{8}$ **9.** $9\frac{7}{8}$ **11.** $6\frac{1}{20}$ **13.** $4\frac{3}{40}$
15. $3\frac{1}{25}$ **17.** $\frac{1}{3}$ **19.** $4\frac{3}{80}$ **21.** $\frac{1}{7}$ **23.** $3\frac{1}{3}$ **25.** $\frac{3}{7}$ **27.** $9\frac{5}{8}$
29. $6\frac{19}{40}$ **31.** $8\frac{1}{8}$ **33.** $\frac{4}{125}$ **35.** $\frac{4}{9}$ **37.** $1\frac{41}{120}$ **39.** $4\frac{7}{160}$
41. $3\frac{9}{1400}$

Exercise 3.2, page 47
1. 1516.97 **3.** 295.535 **5.** 9797.365 **7.** 1022.908 **9.** 876.4649
11. 41.06 **13.** 3561.64 **15.** 578.347 **17.** 288.976 **19.** 1.11074
21. 1.643 **23.** 1.51302 **25.** 0.94131 **27.** 0.51294 **29.** 1.38175
31. 247.7

Exercise 3.3, page 50
1. 471530; 471500; 472000; 470000 **3.** 4071400; 4071000; 4070000; 4100000 **5.** 852.71; 852.7; 853; 850 **7.** 39.250; 39.25; 39.2; 39
9. 0.18297; 0.1830; 0.183; 0.18 **11.** 688540; 688500; 689000; 690000
13. 7290300; 7290000; 7290000 7300000 **15.** 3.2592; 3.259; 3.26; 3.3
17. 0.0057405; 0.005740; 0.00574; 0.0057 **19.** 87.005; 87.00; 87.0; 87

Exercise 3.4, page 55
1. 440.16 **3.** 8.1528 **5.** 5.8875 **7.** 0.65504 **9.** 2.9172
11. 0.0015822 **13.** 0.85695 **15.** 28.77123 **17.** 284.71056
19. 0.05836572 **21.** 0.04364091 **23.** 1012.9008 **25.** 0.08094
27. 0.002579 **29.** 3623 **31.** 0.008464 **33.** 747900 **35.** 1.371
37. 29.63 **39.** 680.8 **41.** 0.00006919

Exercise 3.5, page 56
1. 0.625 **3.** 0.3636 **5.** 3.1875 **7.** 6.875 **9.** 8.4 **11.** 3.4375
13. 4.35 **15.** 2.0667 **17.** 9.1562 **19.** 1.4453 **21.** 4.2778
23. 3.2812

Exercise 3.6, page 59 (Answers are omitted for short method.)
39. 42.9545 **41.** 0.036641

Exercise 3.7, page 59
1. 0.54 in. **3.** Total: 29.3; average: 5.87 pounds **5.** 27.9 in. **7.** 7 pieces
9. 1.375 in. **11.** 16.67 **13.** 18.175 **15.** 53 mph **17.** 454.9 miles
19. 4.8 **21.** 83.3 mph; 122.2 ft/sec **23.** 101.1 mph **25.** 733 mph
27. $1.21 **29.** Brand A: 3.833 cents per oz.

Exercise 4.1, page 65
1. $0.45 = \frac{9}{20}$ **3.** $0.06 = \frac{3}{50}$ **5.** $0.75 = \frac{3}{4}$ **7.** $0.1 = \frac{1}{10}$ **9.** $1.25 = 1\frac{1}{4}$
11. $0.375 = \frac{3}{4}$ **13.** $0.28\frac{4}{7} = \frac{2}{7}$ **15.** $5.2 = 5\frac{1}{5}$ **17.** $0.008 = \frac{1}{125}$
19. $0.225 = \frac{9}{40}$ **21.** $0.042 = \frac{21}{500}$ **23.** $0.01875 = \frac{3}{160}$ **25.** $0.005 = \frac{1}{200}$
27. $0.01\frac{1}{3} = \frac{1}{75}$ **29.** 40% **31.** $14\frac{1}{16}$% **33.** 462.5% **35.** 425%
37. $66\frac{2}{3}$% **39.** $444\frac{4}{9}$% **41.** 7.5% **43.** 336% **45.** 108%
47. 8% **49.** 0.5% **51.** 200%

Exercise 4.2, page 70
1. $184 **3.** $700 **5.** $33.60 **7.** 59.4 ft. **9.** 4.8 pounds **11.** $156
13. 1.5 **15.** $3120 **17.** $320 **19.** $9 **21.** 70 miles **23.** $6750;
$38,250 **25.** 325; 350; 300; 275 miles **27.** 3920 **29.** $12,880; $19,600;
$23,520 **31.** 15% **33.** 112.5% **35.** 2.5% **37.** 5.25%
39. 32.1% **41.** 64 **43.** 95.6 **45.** $8.50 **47.** 315 miles **49.** 15%
51. $620 **53.** 137.5%

Exercise 4.3, page 73
1. 25,155 **3.** $10,800 **5.** 2.5% **7.** 6.4% **9.** $57; 12.5%

Exercise 4.4, page 75
1. $9600 **3.** 165 lb **5.** $176; $13.20 **7.** $16,800; $3780

Exercise 4.5, page 77
1. $52.50; $402.50 **3.** $450; $2950 **5.** $400; $2800 **7.** $3.50;
$52.50 **9.** 25% gain **11.** 15% gain **13.** 20% loss **15.** 50% gain

Exercise 4.6, page 81 (Some answers are approximate.)
1. $980 **3.** $39,150 **5.** $224 **7.** $27.62 **9.** $319.20 **11.** $63.18
13. $345.80 **15.** $79.62 **17.** $4.14 **19.** 17.5% **21.** 22%
23. 58.6% **25.** 17.4% **27.** 50.3% **29.** 29.7% **31.** 36.1%
33. 35.7% **35.** $72.90, gain: 21.5% **37.** $55.25; gain: 10.5%
39. $31.32; loss: 13% **41.** 44% **43.** 37% **45.** 36%
47. 46.8%

Exercise 4.7, page 85
1. $48; $248 **3.** $2; $82 **5.** $138.67; $3338.67 **7.** $1.20; $61.20
9. $21.75; $381.75 **11.** $9; $2409 **13.** $56.25; $4556.25
15. $8.40; $308.40 **17.** $7.39; $567.39 **19.** $43.35; $723.35

Exercise 4.8, page 87
1. $6036.60; $3036.60 **3.** $6130.44; $3130.44 **5.** $3842.47; $1442.47
7. $977.33; $377.33 **9.** $986.17; $386.17 **11.** $159.18; $79.18
13. $4.04893; $8.1473; $32.9880 **15.** $33.822558 (approx.)

Exercise 49, page 90
1. $12; $288 **3.** $5.10; $234.90 **5.** $15; $785 **7.** $300; $2200
9. $32; $368; 8.7% **11.** $84; $716; 7.8% **13.** $1.36; $80.24
15. $31.65; $2078.35 **17.** $1.44; $60.06 **19.** $612.24 **21.** $1290.32
23. $1914.89 **25.** $50.35

ANSWERS TO MOST ODD-NUMBERED EXERCISE PROBLEMS

Exercise 4.10, page 92
1. $315   **3.** $1352   **5.** $31.25   **7.** 22%   **9.** 28%   **11.** $237.50
13. $150

Exercise 4.11, page 94
1. $673.20   **3.** $1116.50   **5.** $378   **7.** 4.3%   **9.** 3.5%; $16,200,00; $1015.88

Exercise 5.1, page 103 (See table for No. 13–23.)
1. 6084   **3.** 0.005329   **5.** 232,324   **7.** 0.690561   **9.** 5806.44
11. 0.00426409   **25.** $2\sqrt{6}$ **27.** $4\sqrt{2}$ **29.** $3\sqrt{6}$ **31.** $5\sqrt{3}$ **33.** $3\sqrt{10}$
35. $7\sqrt{2}$ **37.** $7\sqrt{3}$ **39.** $9\sqrt{2}$ **41.** 14   **43.** $11\sqrt{2}$ **45.** $5\sqrt{10}$
47. $30\sqrt{2}$ **49.** $60\sqrt{5}$

Exercise 6.1, page 114 (No answers are shown.)

Exercise 6.2, page 119 (some answers are omitted.)
1. 122°; 58°; 122°   **3.** 106°   **5.** 57°   **7.** 71°   **9.** 56°   **15.** 2; 5; 9; none   **17.** Yes   **23.** Yes   **25.** No   **27.** Yes

Exercise 7.1, page 126
1. 165 in.^{2}; 53 in.   **3.** 540 in.^{2}; 102 in.   **5.** 59 ft; 216 ft^2 **7.** 84,480 ft^2
9. 171 ft^2 **11.** 342.25 ft^2; 74 ft   **13.** 21 in.^{2}; 22 in.   **15.** 12 in.; 39 in.
17. 14 ft;   67 ft   **19.** 24 ft;   96 ft   **21.** 13 in.;   169 in.^{2} **23.** 8100 ft^2; 0.186 acre   **25.** 180 ft^2 **27.** $84,480

Exercise 7.2, page 132
1. 135 in.^{2} **3.** 477 in.^{2} **5.** 76 in.^{2} **7.** 84 in.^{2} **9.** 8 in.   **11.** 11.25 in.
13. 6 in.   **15.** 700 ft^2 **17.** 148.5 ft^2 **19.** 897 rd^2 **21.** 35 in.^{2}
23. 17.32 in.^{2} **25.** 30 in.^{2} **27.** 96 in.; 104 in.; 146 in.; 204 in.; 390 in.; 1154 in.

Exercise 8.1, page 145
1. 40 in.; 96 in.; 384 in.^{2} **3.** 32.5 cm; 75 cm; 187.5 cm^2 **5.** 28.6 yd; 68.6 yd; 195.5 yd^2 **7.** 41.9 in.;   100.3 in.;   414.3 in.^{2} **9.** 71.4 in.;   203.2 in.^{2}
11. 100 cm; 375 cm^2 **13.** 60.6 cm; 157.4 cm^2 **15.** 18.7 in.   **17.** 4.98 ft
19. 6462 yd   **21.** 509 rd   **23.** 7.78 yd   **25.** 13.4 in.   **27.** 47.4 in.^{2}; 35 in.   **29.** 49,884 ft^2; 1135.7 ft   **31.** 9.81 in.^{2}; 15.9 in.   **33.** 140.3 in.^{2}
35. 84.9 in.^{2} **37.** 1188.9 ft^2 **39.** 41.9 ft^2 **41.** 16.8 ft   **43.** 31.2 ft
45. 28.2 miles   **47.** 1379 cm^2; 149.8 cm

Exercise 9.1, page 155
1. 157.08 in.; 1963.5 in.^{2} **3.** 81.68 in.; 530.9 in.^{2} **5.** 87.96 in.; 615.75 in.^{2}
7. 36.13 in.; 103.87 in.^{2} **9.** 39.58 in.   **11.** 775.7 turns   **13.** 1885 ft
15. 420.2 ft; 138655 ft^2 **17.** 3204 ft^2 **19.** 17.4 in.   **21.** 41.055 mph
23. 9.03 in.   **25.** 12 in.   **27.** 9 in.; 96.75 in.^{2} **29.** 31.4 min

Exercise 10.1, page 164
1. 2900 in.^{3}; 980 in.^{2} **3.** 80 in.^{3}; 124 in.^{2} **5.** 304.9 yd^3; 2156 ft^2 **7.** 90 yd^3; $48   **9.** 465.ft^3; 17.22 yd^3 **11.** 277.5 in.^{3} **13.** 352.35 ft^3 **15.** 3456 in.^{3}; 2 ft^3; 848.64 lb.   **17.** 360 in.^{3} **19.** 11.88 gal; 1176 in.^{2} **21.** 400 yd^3
23. 5.5 in.   **25.** 13.5 in.   **27.** 7 in.; 294 in.^{2} **29.** 125 in.^{3} **31.** 3 in.; 72 in.^{3} **33.** 512 in.^{3}; 432 in.^{3} **35.** 252 in.^{2}; 300 in.^{2}

Exercise 11.1, page 170
1. 40   **3.** 34,404   **5.** 5.1   **7.** 16 in.   **9.** 52.35   **11.** 212.1 in.^{3}
13. 8.51

ANSWERS TO MOST ODD-NUMBERED EXERCISE PROBLEMS

Exercise 11.2, page 174
1. 351.9 in.² **3.** 1260.5 turns **5.** 339.3 ft³; 3 ft; 282.7 ft² **7.** 706.9 in.³; 0.409 ft³ **9.** 11.69 ft; 165.9 ft² **11.** 569.5 ft² more for 3-inch diameter **13.** 1 ft³; 8.5 ft²; 6 ft²; 192 ft²

Exercise 12.1, page 181
1. Prism: 2880 in.³; pyramid: 960 in.³ **3.** 1280 lb **5.** 498.8 in.³ **7.** 112 yd² **9.** 10.67 in.; 11.08 in.; 169 in.² **11.** 6048 in.³; 2086 in². **13.** 528 in.² **15.** 11955 lb.

Exercise 12.2, page 185
1. 3.154 yd³ **3.** 32.17 in.³ **5.** 13.3 bu **7.** 297 ft² **9.** 2078.2 in.³; 751.6 in.² **11.** 6.708 in.; 344.2 in.³; 204.2 in.² **13.** $1.78

Exercise 13.1, page 192
1. 9.42 in.; 28.27 in.²; 14.14 in.³ **3.** 57.8 in.; 1063.6 in.²; 3261.8 in.³ **5.** 2.626 in.; 1.313 in.; 21.66 in.²; 9.48 in.³ **7.** 2.825 in.; 1.4125 in.; 25.07 in.²; 11.80 in.³ **9.** 13.22 **11.** 6 in.; 904.8 in.³ **13.** 14 in.; 11494 in.³ **15.** 5 in.; 314.16 in.² **17.** 6 in.; 452.4 in.² **19.** 397 lb. **21.** Shell, 5.236 in.³ more **23.** $(2.37241)(10^{10})$ square miles **25.** 512 in.³; 432 in.³; 707.7 in.³

Exercise 14.1, page 199
1. 88.9 cm **3.** 820.2 ft **5.** 951 cm **7.** 2125 m **9.** 228.6 cm **11.** 487.7 m **13.** 0.25 in. **15.** 140.1 km **17.** 66.04 cm by 121.92 cm **19.** 16.08 ft **21.** 483.1 km **23.** 128.02 m **25.** 36 miles; 0.914 mm **27.** ft/sec: 22; 44; 66; 88; 117.3; meters/sec: 6.7; 13.4; 20.1; 26.8; 35.8

Exercise 14.2, page 204
1. 657.9 cm² **3.** 28896 cm² **5.** 633.6 m²; 104.85 m, or 344 ft **7.** 20,520 sq. ft. more than 2 acres **9.** 1415.8 cm³ **11.** 75.71 liters

Exercise 14.3, page 205
1. 81.28 cm; 406.4 cm² **3.** 27.13 m; 44.69 m² **5.** 14.17 in. **7.** 24.84 miles **9.** 2.047 in. **11.** 310.5 miles **13.** 215.3 ft² **15.** 310.1 in.² **17.** 7.752 in.² **19.** 353 ft³ **21.** 38.1 cm **23.** 0.9525 cm **25.** 59.44 m **27.** 1.006 km **29.** 173.5 m **31.** 1.609 km **33.** 4047 m² **35.** 2.589 km² **37.** 0.9144 m; or 91.44 cm **39.** 0.4536 kg **41.** 37.85 liters **43.** Difference: 410 miles or 660,000 m

Exercise 15.1, page 212 (Any letters may be used.)
1. $n+2$ **3.** $x-6$ **5.** $n+8$ **7.** $18-x$ **9.** $7n$ **11.** $\frac{11}{x}$ **13.** $8n+11$ **15.** $5n-6$ **17.** $\frac{2x}{3}$ **19.** xy **21.** $\frac{m}{n}$ **23.** t^2 **25.** x^3 **27.** $\frac{x+y}{xy}$ **29.** $3(n+6)$

Exercise 15.2, page 214
1. 29 **3.** 7 **5.** 8 **7.** 24 **9.** 30 **11.** 4.5 **13.** 20 **15** 16.5 **17.** 0 **19.** 19 **21.** 43 **23.** 34 **25.** 18 **27.** 8 **29.** 30 **31.** 39 **33.** 0 **35.** 12 **37.** 11 **39.** 198

Exercise 16.1, page 220
1. 8 **3.** 8 **5.** 7 **7.** 8 **9.** 9 **11.** 1 **13.** 9 **15.** 20 **17.** 8.5 **19.** 3.6 **21.** 6 **23.** 2.5 **25.** 9 **27.** 3.5

ANSWERS TO MOST ODD-NUMBERED EXERCISE PROBLEMS

Exercise 16.2, page 221
1. 8   **3.** 21   **5.** 18   **7.** 12.5   **9.** 2.6   **11.** 7   **13.** 4
15. 3   **17.** 8   **19.** 8   **21.** 0   **23.** 7   **25.** 0   **27.** 0   **29.** 5

Exercise 17.1, page 226
1. 13; −11; −5; +6; 30; −28; +7   **3.** −1; 5; 12; −10; −13; 0; −1   **5.** +10; −10; +15; −9; −18; 0; +10   **7.** −37; −43; 0   **9.** +743; −407; −419; 37.4; 35.5; −5.38

Exercise 17.2, page 228
1. +7; +10; −6; +4   **3.** −3; −13; 0; −5   **5.** 8; 14; −7; 5   **7.** −65; 0; −60; −1   **9.** 21; 57; −8; 13   **11.** −14; 23; 334; 7.38

Exercise 17.3, page 230
1. 6   **3.** −11   **5.** −7   **7.** 18   **9.** 7   **11.** −1   **13.** 0   **15.** 0
17. −13

Exercise 17.4, page 233
1. −14   **3.** 90   **5.** 120   **7.** −42   **9.** 80   **11.** −300   **13.** −10.2
15. 10080   **17.** $\dfrac{1}{12}$ **19.** −120   **21.** 72   **23.** 800   **25.** −234

Exercise 17.5, page 234
1. 5   **3.** −5   **5.** 1   **7.** −12   **9.** 6   **11.** −10   **13.** −15   **15.** 8
17. 4   **19.** −2.5   **21.** $-\dfrac{15}{7}$ **23.** −0.8   **25.** 92   **27.** −22
29. −0.8   **31.** 230   **33.** $-\dfrac{8}{3}$ **35.** 1.6

Exercise 18.1, page 240
1. $9x$; $-9x^2$; $-5xy$; $4n$; $-4ab$; 0; $6x^2y^2+x^2y$ **3.** $4a$; $-4n$; $-6x^3$; $-5xy^2-x^2y$; $-7x^3y$, $-10xy^3$; $-8b^2c-2bc$ **5.** $7xy^2$; $-9st^2$; $-ab^3$; 0; $2rs$; $-12x^2y^2$; $8xyz$
7. x^3; ax^3; $6bx$; $16xy^2$; $2a^2b^2$; $-4yz-3x$; $-7a^2c^3d$ **9.** $-x$; $6x^2$; $-7xy$; $10n$; $4ab$; $-8x^2y$; $6x^2y^2-x^2y$ **11.** $-10a$; $-4n$; $10x^3$; $-5xy^2+x^2y$; $-11x^3y$; 0; $-8b^2c+2bc$ **13.** $-5xy^2$; $5st^2$; $-17ab^3$; $-4x^2y$; $-4rs$; 0; 0 **15.** $11x^3$; $-7ax^3$; $-8bx$; 0; $-12a^2b^2$; $-4yz+3x$; a^2c^3d **17.** $7x^2y$ **19.** $-6x^2y^3$ **21.** x^2y

Exercise 18.2, page 244
1. $6a+5b-2c+ab$ **3.** $6x-3z+2$ **5.** $xy-y^2$ **7.** $x^2-4xy+5y^2-2x$
9. $5x^2+xy+y^2+5$ **11.** $1-x^2-3x-y$ **13.** $x^2-xy+3y^2-x+3y$
15. $x^2y^2-2x^2y+xy^2-x+y$

Exercise 19.1, page 250
1. $2^2x^4y^3$ **3.** $5^4x^3y^2z$ **5.** 10^5x^2yz **7.** $6x^7$ **9.** $12n^7$ **11.** $12x^3y^5$
13. $-30ax^4y^2$ **15.** $28abcde$ **17.** $24x^4y^5$ **19.** $6x^8y^4z$
21. $-24x^6y^3z^5$ **13.** $-a^3b^4c^4d^2$ **25.** $-24x^4y^4$ **27.** 4000000   **29.** 32000

Exercise 19.2, page 252
1. $2x$ **3.** 1   **5.** $4x^2$ **7.** $3x^4$ **9.** $-10x^3y^2$ **11.** $-2a^2$
13. $-4x^3yz^3$ **15.** $3b$ **17.** 1   **19.** $-4yz$ **21.** $-4a^3b^4$ **23.** 1

ANSWERS TO MOST ODD-NUMBERED EXERCISE PROBLEMS

Exercise 20.1, page 254
1. $6x+15$ **3.** $40-15n$ **5.** $8-28a$ **7.** $35x-15y$ **9.** $12x^2+15xy$
11. $8x^2-14x+2$ **13.** $15xy-12x^2-18y^2$ **15.** $28x^3-12x^2+4x$ **17.** $2x^2+8x^3-6x^4$ **19.** $18n^5+24n^4x-30n^3x^2$ **21.** $5xy^5-10x^2y^4+20x^3y^3-15x^4y^2$
23. $3xy^3+15x^2y^4+18xy^5-12x^3y^3$ **25.** $24xy^3+12x^2y^2-18x^3y$
27. $36c^2d^3-18c^3d^2-30c^4d$

Exercise 20.2, page 256
1. $3x+4$ **3.** $3r-4s$ **5.** $2x+1.5$ **7.** $2x-0.75y$ **9.** $-5x-3$ **11.** $3n-4n^2$ **13.** $4x+1$ **15.** $-9x^2+3x-2$ **17.** $4x^2+3x-1$ **19.** $-6n^2y^3+4ny^2-3y$ **21.** $1-5a-4a^2bc$ **23.** $5x^8-4x^6+3x^4-2x^2-1$

Exercise 20.3, page 258
1. $12x^2-7x-10$ **3.** $8x^2+26x+15$ **5.** $6x^2+29x+35$ **7.** $8x^2-10xy-25y^2$ **9.** $25n^2-16$ **11.** $21x^2y^2-23xy+6$ **13.** $25a^2x^2-64$ **15.** $49n^2-14nx-15x^2$ **17.** $25x^4-36$ **19.** $144n^2x^2-25$ **21.** $63m^2n^2+mn-12$
23. $18x^3+x-6$ **25.** $15x^3-x^2-19x+12$ **27.** $18x^4-47x^2+11x+12$
29. $8x^3-27$ **31** $9x^5-25x^3-6x^2-2x+20$ **33.** $36x^3-27x^2-100x+75$
35. $96n^4-16n^3-86n^2+9n+18$ **37.** $4x^4+14x^3-62x^2+20x$

Exercise 20.4, page 262 (R = remainder.)
1. $3x-7$; R: 6   **3.** $5x+7$; R: 12   **5.** $4x^2+5x-3$; R: 4   **7.** $6x^2-4x+5$; R: -16 **9.** $3n^2+4n+5$; R: 49   **11.** $3x^3+6x^2+5x-2$; R: 9   **13.** $2x^2-4x+3$; R: 6   **15.** $2x+5$; R: 4   **17.** $5x-6$; R: 0   **19.** $3x^2+4x-2$; R: -2
21. $4h^2+h-2$; R: -7 **23.** $2v^3-4v^2+3v-5$; R: 10   **25.** $8x^3+12x^2-18x-27$; R: 0   **27.** $8x^3-12x^2+18x-27$; R: 0   **29.** $2a^3+6a^2b-4ab^2+3$; R: 0
31. $3x^2-2x+5$; R: $3-5x$ **33.** $2c^2+3c+1$; R: 0

Exercise 21.1, page 271

1. 3   **3.** 7   **5.** 3   **7.** -3 **9.** -2 **11.** -4 **13.** 4.5   **15.** $5\frac{1}{3}$
17. 3.5   **19.** 6   **21.** 4   **23.** -9 **25.** $4\frac{4}{7}$ **27.** $-4\frac{2}{3}$ **29.** 1.6
31. 0   **33.** -7.5 **35.** 3.75   **37.** $2\frac{1}{3}$ **39.** 3   **41.** -3 **43.** -1
45. -8.5 **47.** 0

Exercise 21.2, page 275
1. 2   **3.** -2 **5.** 2.5   **7.** -2 **9.** -4 **11.** 0.5   **13.** -3.5
15. -4 **17.** $-\dfrac{3}{16}$

Exercise 22.1, page 281
1. Slide rule: $17.50; drawing set; $25.50   **3.** 15.5 in. by 26 in.   **5.** $A=52.5°$; $B=22.5°$; $C=105°$ **7.** 67.5°; 37.5°; 75°   **9.** 8.5 amp; 11.5 amp

Exercise 22.2, page 283
1. 13 yr; 41 yr   **3.** 13.5 yr; 19.5 yr   **5.** 23 yr; 46 yr   **7.** 24 yr; 48 yr
9. 23 yr; 27 yr; 30 yr

Exercise 22.3, page 285
1. 38; 39; 40; 41   **3.** 29; 31; 33; 35; 37   **5.** 205; 245; 285; 325   **7.** 24; 32; 40; 48   **9.** inches: 18; 21; 24; 27; 30

ANSWERS TO MOST ODD-NUMBERED EXERCISE PROBLEMS

Exercise 22.4, page 287
1. 11 nickels; 3 dimes **3.** 31 nickels; 15 quarters **5.** 39 nickels; 17 dimes; 13 quarters **7.** 60 at 30 cents: 24 at 75 cents

Exercise 22.5, page 291
1. 4 hr; 3 P.M. **3.** 5.25 hr **5.** 54 mph **7.** 2 P.M. **9.** 2880 miles

Exercise 23.1, page 294
1. $120x^4$ **3.** $-30n^7$ **5.** $10a^5b^7$ **7.** $12x^7y^6$ **9.** $-12r^6s^3t^2$
11. $-24a^6b^9$ **13.** $24x^7y^6$ **15.** $-x^8y^6$ **17.** $-270x^{10}y^6$

Exercise 23.2, page 297 (Answers are omitted for No. 1–20)
21. $\dfrac{3x}{4y}$ **23.** $\dfrac{3a}{5b}$ **25.** $\dfrac{x}{2}$ **27.** $\dfrac{3x^2}{5y}$

Exercise 23.3, page 298
1. $6x-21$ **3.** $7n-21n^2$ **5.** $36x-60$ **7.** $48a^2x^3-18ax^2$ **9.** $80rs^2t^2-20rst$ **11.** $10x-6y+8$ **13.** $7n^3+35n^2-21n$ **15.** $20x^4-35x^3y+5x^2$ **17.** $3xy^2+15xy^3-12x^2y^2$ **19.** $12a^3b^2-20a^2b^3+28ab^4+4ab^2$ **21.** $3x^2y^3-12x^4y^4-6x^6y^5-15x^3y^2$ **23.** $24x^7-18x^8+6x^6+6x^5-12x^4$

Exercise 23.4, page 301
1. $5(x+3)$ **3.** $7(2x+5)$ **5.** $5(3t-1)$ **7.** $3x(3x+2)$ **9.** $5n(8n-5)$
11. $12n(2n^2+3)$ **13.** $9xy(2-3x)$ **15.** $11xy^3(3x-4y)$ **17.** $\pi h(R^2-r^2)$
19. $2x^3(3x^3-1)$ **21.** $6x^3(1-3x)$ **23.** $10n^3(1-3n^3)$ **25.** $2xy(4xy^2+6x^2y-3)$ **27.** $3ab(4a^2-6a+1)$ **29.** $3x(x^2+5xy-2y^2)$ **31.** $4xy(2x^2-4xy-3y^2)$ **33.** $6ab^2(2a^2+5ab-3b^2)$ **35.** $2x(5x^3+4x^2-3x+1)$ **37.** $x(x^4+x^3-x^2+x-1)$

Exercise 23.5, page 303
1. x^2-25 **3.** x^2-49 **5.** $9x^2-1$ **7.** $25n^2-4$ **9.** $16t^2-81$
11. $9y^2-64$ **13.** $4a^2-49n^2$ **15.** $400x^2-289y^2$ **17.** $1600t^2-361x^2$
19. $9n^2-\dfrac{1}{4}$ **21.** $64x^2-\dfrac{1}{16}$ **23.** n^6-49 **25.** $16r^8-81s^2$ **27.** $100a^2b^2-121c^2$ **29.** $6400-1=6399$ **31.** $90000-4=89996$ **33.** $360000-9=359991$

Exercise 23.6, page 308
1. $(2x+3)(2x-3)$ **3.** $(5y-4)(5y+4)$ **5.** $(s+t)(s-t)$ **7.** $(3x-1)(3x+1)$ **9.** $(8x+1)(8x-1)$ **11.** $(3a-b)(3a+b)$ **13.** $(6x+5)(6x-5)$
15. $(3t+8)(3t-8)$ **17.** $(4s+3t)(4s-3t)$ **19.** $(7s+12t)(7s-12t)$
21. $(14+5n)(14-5n)$ **23.** $\left(3n+\dfrac{2}{3}\right)\left(3n-\dfrac{2}{3}\right)$ **25.** $3n(n+3)(n-3)$
27. $7t(2t+x)(2t-x)$ **29.** $4(n+5t)(n-5t)$ **31.** $(5n+0.4)(5n-0.4)$
33. $9(x+0.1)(x-0.1)$ **35.** $(y^2+4x^2)(y+2x)(y-2x)$ **37.** $(x+y+n)(x+y-n)$
39. $(x+y+2)(x-y-2)$ **41.** $(x+y-5)(x-y-1)$

Exercise 23.7, page 311
1. $m^2+2mn+n^2$ **3.** $c^2-2cd+d^2$ **5.** $R^2-2Rr+r^2$ **7.** $x^2-16x+64$
9. $x^2-20x+100$ **11.** $169-26y+y^2$ **13.** $4x^2-20x+25$ **15.** $16x^2-56x+49$ **17.** $64x^2-80x+25$ **19.** $9x^2+48xy+64y^2$ **21.** $64n^2-112mn+49m^2$ **23.** $400x^2+200xy+25y^2$ **25.** $x^4+10x^2y+25y^2$ **27.** $25n^6-70n^3x+49x^2$ **29.** $49r^2s^2t^2+126rst+81$ **31.** $144x^2+264xy^2+121y^4$
33. $225m^2n^4-30mn^2t+t^2$ **35.** $25x^4y^2+110x^2y+121$ **37.** $x^2+x+\dfrac{1}{4}$

Exercise 23.7, page 311 (continued)

39. $64y^2 - 4y + \frac{1}{16}$ **41.** $y^2 - 1.2y + 0.36$ **43.** $256x^2 + 24x + 0.5625$
45. $4x^2 - 3x + \frac{9}{16}$ **47.** $3600 + 600 + 25 = 4225$ **49.** $(x-2)^2 - 2y(x-2) + y^2$
51. $x^2 - 2x(y-3) + (y-3)^2$ **53.** $25x^2 + 90x + 81$ **55.** $64y^2 - 48y + 9$
57. $16n^2 + 2n + \frac{1}{16}$ **59.** $4x^2 + x + \frac{1}{16}$ **61.** $x^2 = 25a^2 + 30a + 9$ **63.** $x - 2 = 25$; or $x = 27$

Exercise 23.8, page 313 (Only perfect squares are shown.)
1. $(y+5)^2$ **5.** $(3x-8)^2$ **7.** $(2x-7)^2$ **11.** $(7y+9)^2$ **13.** $(4t-9)^2$
15. $(6-7x)^2$ **17.** $(13y+1)^2$ **21.** $(5x+10)^2$ **23.** $\left(t+\frac{1}{2}\right)^2$

Exercise 23.9, page 314 (Middle term to be supplied is shown.)
1. $8x$; $(x+4)^2$ **3.** $-14n$; $(n-7)^2$ **5.** $-20x$; $(2x-5)^2$ **7.** $10n$; $(5n+1)^2$
9. $12y$; $(6y+1)^2$ **11.** $-66tx$; $(11t-3x)^2$ **13.** $2x$; $\left(3x+\frac{1}{3}\right)^2$ **15.** $-x$; $\left(x-\frac{1}{2}\right)^2$ **17.** $1.8x$; $(3x+0.3)^2$ **19.** $-40F$; $(F-20)^2$ **21.** $-24t$; $(12-t)^2$
23. $14n^2xy^3$; $(xy^3+7n^2)^2$

Exercise 23.10, (Number to be added is shown.)
1. 25; $(x+5)^2$ **3.** 64; $(t-8)^2$ **5.** 225; $(x+15)^2$ **7.** 121; $(y+11)^2$
9. 144; $(m-12)^2$ **11.** $\frac{1}{4}$; $\left(y+\frac{1}{2}\right)^2$ **13.** $\frac{25}{4}$; $\left(n+\frac{5}{2}\right)^2$ **15.** $\frac{81}{4}$; $\left(n-\frac{9}{2}\right)^2$
17. 25; $(n+5)^2$ **19.** 0.16; $(x+0.4)^2$ **21.** 0.01; $(t+0.1)^2$ **23.** 1; $(7x+1)^2$
25. $9y^2$; $(7x+3y)^2$ **27.** $25d^2$; $(6c-5d)^2$ **29.** $\frac{25}{36}$; $\left(3n+\frac{5}{6}\right)^2$
31. $\frac{9}{16}n^2$; $\left(4m+\frac{3}{4}n\right)^2$ **33.** $\frac{81}{16}y^2$; $\left(2x-\frac{9}{4}y\right)^2$ **35.** $\frac{1}{36}$; $\left(3x+\frac{1}{6}\right)^2$

Exercise 23.11, page 318
1. $x^2+11x+28$ **3.** $x^2+14x+48$ **5.** $n^2-8n+12$ **7.** $y^2-12y+32$
9. $n^2+17n+60$ **11.** $x^2-5x-50$ **13.** x^2-x-56 **15.** t^2-t-72
17. $x^2+15x-16$ **19.** $y^2-7y-30$ **21.** $y^2-11y-60$ **23.** t^2-36
25. $n^2+26n-120$ **27.** $n^2+45n-144$ **29.** $63+2y-y^2$ **31.** x^4+3x^2-40 **33.** x^8-10x^4+16 **35.** $50+5x-x^2$

Exercise 23.12, page 321
1. $(x+2)(x+3)$ **3.** $(x-3)(x-5)$ **5.** $(n-4)(n-5)$ **7.** $(y-4)(y-7)$
9. $(y+4)(y+9)$ **11.** $(x-2)(x-20)$ **13.** $(t+2)(t-10)$ **15.** $(t+2)(t-20)$
17. $2x^2(x+2)(x-7)$ **19.** $(n-5)(n+6)$ **21.** $(n+8)(n-9)$ **23.** $2x(y-1)(y+12)$ **25.** $(x-5)^2$ **27.** $(n+7)^2$ **29.** No factors **31.** $3(x+1)(x+18)$
33. $2x^2(x+1)(x-24)$ **35.** $(n+4)(n-18)$ **37.** $(y-2)(y+42)$ **39.** $(y+4)(y-21)$ **41.** $(x+4)(x-15)$ **43.** $y^2(x+3)(x-16)$ **45.** $n^3(x+4)(x-12)$
47. $(x^2-12)^2$

Exercise 23.13, page 324
1. $6x^2+19x+15$ **3.** $12x^2-25x+12$ **5.** $28x^2+29x+6$ **7.** $20x^2+9x-20$ **9.** $20x^2-3x-35$ **11.** $20t^2-t-12$ **13.** $8n^2+25n+3$ **15.** $4n^2-25n+6$ **17.** $24c^2+7c-6$ **19.** $18-9x-14x^2$ **21.** $24-10y-25y^2$

23. $4a^2b^2+ab-14$ **25.** x^4-8x^2-35 **27.** $2y^4-y^2-10$ **29.** $28a^2+ab-45b^2$ **31.** $4n^2-101n+25$ **33.** $25n^2-1226n+49$ **35.** $72a^2-ab-b^2$

Exercise 23.14, page 326
1. $(3x-2)(x-5)$ **3.** $(5x-2)(x-3)$ **5.** $(3n+5)(n+4)$ **7.** $(4x-3)(x-2)$
9. $(4x+3)(2x+5)$ **11.** $(4n+3)(n+2)$ **13.** $(4x-1)(2x+3)$ **15.** $(5x-6)(x-1)$ **17.** $(3y-8)^2$ **19.** $(2n-3)(5n+7)$ **21.** $(5n-6)(6n+7)$
23. $(x+2)(9x+8)$ **25.** $(3n-5)(12n-5)$ **27.** $(n+7)(4n+7)$ **29.** $(x+6)(8x-1)$

Exercise 23.15, page 327
1. $(R-r)(R^2+Rr+r^2)$ **3.** $(x+3)(x^2-3x+9)$ **5.** $(4x+1)(16x^2-4x+1)$
7. $(T-10)(T^2+10T+100)$ **9.** $(t-7)(t^2+7t+49)$ **11.** $(x^2+y^4)(x^4-x^2y^4+y^8)$ **13.** $\left(4x+\dfrac{2}{3}\right)\left(16x^2-\dfrac{8x}{3}+\dfrac{4}{9}\right)$ **15.** $\left(4x-\dfrac{1}{3}\right)\left(16x^2+\dfrac{4x}{3}+\dfrac{1}{9}\right)$
17. $(2x-y)(2x+y)(16x^4+4x^2y^2+y^4)$

Exercise 23.16, page 329
1. $4(2x-3)(2x+3)$ **3.** $5(a-6b)(a+6b)$ **5.** $9x^2(x-3y)(x+3y)$
7. $(n^2+9)(n-3)(n+3)$ **9.** $n^6(n^2+1)(n+1)(n-1)$ **11.** $x^4(2+x)(4-2x+x^2)$ **13.** $n^3(n+6)(n-7)$ **15.** $6n^2(n+3)(n-4)$ **17.** $6x^2(x+2)(x-6)$ **19.** $3x^2(3x+2)(2x-5)$ **21.** $(a-4-x)(a-4+x)$ **23.** $(y-3)(y+3)^2$ **25.** $(2y-x+3)(2y+x-3)$ **27.** $(x+2)(x-3)(x+3)$ **29.** $(x+y-2)(x-y+4)$

Exercise 24.1, page 336
1. $\dfrac{2y}{3x}$ **3.** $\dfrac{3x^2}{7y}$ **5.** $\dfrac{5s^3}{9r}$ **7.** $\dfrac{5x^2}{4z}$ **9.** $\dfrac{5y^2}{6x}$ **11.** $\dfrac{4n}{7m}$ **13.** 2 **15.** $\dfrac{3}{x}$
17. 1 **19.** $\dfrac{x-5}{x-2}$ **21.** $\dfrac{x-5}{x+3}$ **23.** $\dfrac{n+5}{n+10}$ **25.** $\dfrac{x+2}{x+6}$ **27.** $\dfrac{x-8}{x-4}$
29. $\dfrac{x(x+2)}{3(x-1)}$ **31.** $\dfrac{x^2(x+6)}{3(x-2)}$ **33.** $\dfrac{x^2(x+5)}{3(x-2)}$

Exercise 24.2, page 341
1. $\dfrac{12}{35}$ **3.** $\dfrac{ax}{by}$ **5.** $\dfrac{35mrs}{48txy}$ **7.** $\dfrac{3}{10}$ **9.** $\dfrac{3}{7}$ **11.** $\dfrac{2ac}{5b}$ **13.** $\dfrac{4a}{9c}$
15. $\dfrac{8ax^2}{15by^2}$ **17.** $\dfrac{9}{5abx^2y}$ **19.** $\dfrac{2}{5}$ **21.** $\dfrac{9}{5}$ **23.** $\dfrac{2b^2}{3r}$ **25.** $\dfrac{9a}{8b^2x}$
27. $\dfrac{14cd}{3}$ **29.** $\dfrac{24xr^2}{7z}$ **31.** $\dfrac{2y^4}{3a}$ **33.** $\dfrac{7a}{9by}$ **35.** 1 **37.** $\dfrac{16b^2x^3}{27a^5c}$
39. $\dfrac{13axy^2}{3c^3}$ **41.** $\dfrac{x+2}{x+6}$ **43.** $\dfrac{(x-2)(x+4)}{x(x+1)(x-4)}$ **45.** $\dfrac{5(x+3)}{6x^2(x+5)}$ **47.** $\dfrac{(x-1)^2}{2x(x^2-9)}$

Exercise 24.3, page 344
1. $\dfrac{15ay}{8x}$ **3.** $\dfrac{3a^2m}{16n^2}$ **5.** $\dfrac{10x}{3m}$ **7.** $\dfrac{a}{2cx^2}$ **9.** $\dfrac{x}{5}$ **11.** $\dfrac{4x(x+y)}{3(2x+3y)}$
13. $\dfrac{2(x-8)}{3(x-4)}$ **15.** $\dfrac{3(x+4)}{x+8}$ **17.** $\dfrac{3x-1}{12x^2(3x+1)}$ **19.** $\dfrac{x(2x+5)}{x+2}$
21. $\dfrac{(x+4)(x-3)^2}{2(x-1)(x+2)(x-4)}$ **23.** $\dfrac{3x+1}{4(x+1)}$

ANSWERS TO MOST ODD-NUMBERED EXERCISE PROBLEMS

Exercise 24.4, page 351

1. $\dfrac{5x}{4}$ 3. $\dfrac{t}{4}$ 5. $\dfrac{3n}{4}$ 7. $\dfrac{1}{x}$ 9. $\dfrac{5}{3x}$ 11. $\dfrac{5x}{st}$ 13. $\dfrac{43n}{36}$ 15. $\dfrac{7n}{6}$
17. $\dfrac{7x-11}{18}$ 19. $\dfrac{7x+9}{36}$ 21. $\dfrac{1}{6}$ 23. $\dfrac{19}{36}$ 25. $\dfrac{5n}{3}$ 27. $\dfrac{n}{2}$ 29. $\dfrac{13r}{30}$
31. $\dfrac{18n-7nx}{6x^2}$ 33. $\dfrac{15tx-44t}{12x^2}$ 35. $\dfrac{2n-26}{15}$ 37. $\dfrac{11xy+6x-6y}{6xy}$
39. $\dfrac{-10m-3n}{6mn}$

Exercise 24.5, page 358

1. $\dfrac{x+23}{(x-2)(x+3)}$ 3. $\dfrac{8x+25}{(x-3)(x+4)}$ 5. $\dfrac{4x^2+10x-4}{(x-2)(x+2)}$ 7. $\dfrac{7}{x-3}$ 9. $\dfrac{14}{x-4}$
11. $\dfrac{2y-11}{(y-3)(y-4)}$ 13. $\dfrac{1-n-n^2}{n^2-9}$ 15. $\dfrac{n^2-3n-5}{n^2-4}$ 17. $\dfrac{9x+12y}{(3x-y)(3x+2y)}$
19. $\dfrac{3n^2-10n+1}{n^2+n-6}$ 21. $\dfrac{4+2x-x^2}{x^2-5x+6}$ 23. $\dfrac{-n-5}{n^2+3n}$ 25. $\dfrac{x^2-8x+6}{x^2-3x+2}$
27. $\dfrac{-4x^3+20x^2+15}{x(x-5)}$ 29. $\dfrac{4x^2-3x^3+x+8}{x(x-2)}$

Exercise 25.1, page 363

1. $5x$ 3. $4x$ 5. $6n$ 7. $35t$ 9. $40t$ 11. $3t$ 13. $2x-5$ 15. $4-3x$ 17. $3(5x-3)$ 19. $4(3x+2)$ 21. $15(2x-7)$ 23. $6(2a-1)$

Exercise 25.2, page 367

1. 21 3. 42 5. 2.4 7. 10 9. 8 11. -8 13. 8.75
15. $10\dfrac{2}{3}$ 17. $6\dfrac{2}{3}$ 19. 9 21. 3 23. -23 25. 7.5 27. 2.4
29. 160 31. 31.25 33. 7.5 and 17.5

Exercise 25.3, page 371

1. 2 3. -10.5 5. $-19/14$ 7. 7 9. 3 11. 9.5 23. $\dfrac{57}{16}$
15. -8 17. -6 19. 3 21. 1.3 23. -2.2 25. 80 27. 60

Exercise 25.4, page 374

1. 5 3. -2 5. -0.5 7. 3.5 9. 8 11. -3 13. -8 15. -7
17. -6

Exercise 25.5, page 377

1. 5 3. 6 5. 2.5 7. -1 9. -2 11. 12/7 13. 5 15. $10\dfrac{2}{3}$
17. $\dfrac{28}{3}$ 19. 1.6 21. -15 23. $\dfrac{5}{8}$ 25. -5 27. 1.5

Exercise 25.6, page 380

1. $W=\dfrac{V}{LH}$ 3. $r=\dfrac{A}{2\pi h}$ 5. $b=\dfrac{2A}{a}$ 7. $a=\dfrac{bf}{b-f}$ 9. $C=\dfrac{5}{9}(F-32)$

ANSWERS TO MOST ODD-NUMBERED EXERCISE PROBLEMS 577

11. $r=\dfrac{S-a}{S-l}$ **13.** $B=\dfrac{2A}{h}-b$ **15.** $F=\dfrac{9}{5}C+32$ **17.** $r=\dfrac{eR}{E-e}$ **19.** $n=\dfrac{l-a+d}{d}$ **21.** $h=\dfrac{V}{\pi(R^2-r^2)}$ **23.** 8.4   **25.** 6   **27.** 40.86   **29.** 12   **31.** 20°   **33.** 8.5 cm   **35.** 25   **37.** 17.5 in.

Exercise 25.7, page 386
1. $120   **3.** Length: 22.5 in.; width: 15 in.; area: 337.5 in.^{2} **5.** 18 in. by 30 in.; perimeters: 96 in.; 94 in.   **7.** $\dfrac{13}{19}$ **9.** 3.5   **11.** 27   **13.** $4800 at 6%; $3200 at 8%   **15.** $3\dfrac{3}{7}$ hr   **17.** 3.83   **19.** Total: 120 A; corn: 32 A; wheat: 20 A; oats: 40 A   **21.** 20 qt   **23.** 37.5 qt of 12%; 7.5 qt of 60%

Exercise 26.1, page 395 (Answers are in alphabetical order of letters.)
1. 12; 8   **3.** 8; 7   **5.** 7; −2   **7.** 5; 2   **9.** 5; 1   **11.** 4; −1   **13.** $\dfrac{42}{11}$; $-\dfrac{4}{11}$ **15.** 3; −1   **17.** 3; 2   **19.** 7; −1   **21.** $\dfrac{3}{7}$; $-\dfrac{23}{7}$ **23.** −2; −6   **25.** $\dfrac{3}{2}$; $\dfrac{1}{2}$ **27.** −1.5; −4.5   **29.** 2; 3   **31.** $\dfrac{14}{13}$; $\dfrac{20}{13}$ **33.** $\dfrac{48}{17}$; $-\dfrac{29}{17}$ **35.** −15; −10.5

Exercise 26.2, page 399 (Answers are in alphabetical order of letters.)
1. 2; 3   **3.** $\dfrac{18}{7}$; $-\dfrac{26}{7}$ **5.** 4; −2   **7.** $\dfrac{5}{3}$; $\dfrac{1}{3}$ **9.** 1.5; 0.5   **11.** $\dfrac{1}{3}$; $\dfrac{4}{3}$ **13.** 2; −2   **15.** $-\dfrac{44}{7}$; $-\dfrac{40}{7}$ **17.** $\dfrac{1}{25}$; $\dfrac{7}{25}$

Exercise 26.3, page 401 (Answers are in alphabetical order of letters.)
1. $\dfrac{34}{19}$; $-\dfrac{6}{19}$ **3.** $\dfrac{13}{7}$; $\dfrac{4}{7}$ **5.** $\dfrac{26}{37}$; $-\dfrac{10}{37}$ **7.** 15; 5   **9.** $-\dfrac{24}{13}$; $\dfrac{53}{26}$ **11.** $\dfrac{13}{18}$; $\dfrac{7}{3}$

Exercise 26.4, page 405 (Answers are in alphabetical order of letters.)
1. 3; 2; 1   **3.** 1; 3; 2   **5.** 1; 2; 3   **7.** 3; 2; 1   **9.** 3; −2; 1   **11.** −1; 4; 2   **13.** 1; 2; 3   **15.** 3; −4; −2   **17.** 2.5; −0.5; 2   **19.** 2.75; −3.5; −1.75   **21.** 1; −4.2; −4.4   **23.** 4; 1; −1; −2

Exercise 26.5, page 407 (Answers are given for independent equations.)
1. $\dfrac{4}{3}$; −1   **3.** Dependent   **5.** −20.5; 17   **7.** Inconsistent   **9.** $\dfrac{17}{12}$; $\dfrac{1}{4}$ **11.** Inconsistent   **13.** Inconsistent   **15.** Inconsistent   **17.** Inconsistent

Exercise 26.6, page 417
1. 33.6°; 56.4°   **3.** 6.5 amps; 11.5 amps   **5.** 57 nickels; 26 dimes   **7.** 38 dimes; 15 quarters   **9.** 315 at 20¢; 110 at 50¢   **11.** General: 1280; grandstand: 540   **23.** 3:15 P.M.; 288.75 miles   **15.** Car: 110 mi; train: 450 mi   **17.** $7350 at 5%; $5250 at 7%   **19.** $15\dfrac{5}{8}$ qt of 8%; $4\dfrac{3}{8}$ qt of 40%   **21.** 35 lb at 60¢; 15 lb at $1.10   **23.** 12 hr of B; 12 hr of C

ANSWERS TO MOST ODD-NUMBERED EXERCISE PROBLEMS

Exercise 27.1 and Exercise 27.2: graphs not shown for these.

Exercise 27.3, page 431 (Answers are approximate, usually to nearest tenth.)
1. 4; 3   **3.** 2; 5   **5.** 3.4; 2.6   **7.** 5.1; 1.4   **9.** 6.8; 1.3   **11.** 4.4; −1.3   **13.** 4; −1   **15.** −0.1; −1.2   **17.** 3; 1   **19.** 3; 0   **21.** 0; −4   **23.** 4.3; 1.7

Exercise 28.1, page 434
1. 25   **3.** 64   **5.** −64   **7.** −49   **9.** −0.000008   **11.** 0.064   **13.** x^{10} **15.** y^4 **17.** y^5 **19.** $y^{3.7}$ **21.** $n^{-2.4}$ **23.** n^{5a+1} **25.** 100,000,000   **27.** 256   **29.** $10^{3.823}$ **31.** 10,800   **33.** 800,000

Exercise 28.2, page 436
1. x^6 **3.** x^9 **5.** n^{-8} **7.** x^2 **9.** y^{-1} **11.** x^6 **13.** n^{-3} **15.** n^0 **17.** 8   **19.** $x^{2.16}$ **21.** $x^{-3.21}$ **23.** $10^{-3.86}$

Exercise 28.3, page 437
1. x^{12} **3.** n^{-15} **5.** n^4 **7.** x^2 **9.** y **11.** 10^{10} **13.** $x^{4.2}$ **15.** y^{-7} **17.** 512   **19.** x^0 **21.** $10^{6.429}$ **23.** $x^{-2.142}$ **25.** $10^{2.16}$ **27.** $10^{-0.8}$ **29.** x^{12} **31.** x^8

Exercise 28.4, page 439
1. a^4b^4 **3.** $x^{-3}y^{-3}z^{-3}$ **5.** $-125x^{12}$ **7.** $100x^{20}$ **9.** $a^{-4}b^{-8}c^{-6}$ **11.** $x^{-4}y^8z^2$ **13.** $256x^{12}y^{-8}$ **15.** $-32x^{10}y^5z^5$ **17.** $216x^{-6}y^3$ **19.** $x^2y^3z^4$ **21.** $x^{3/2}y^{5/2}z$ **23.** 5,832,000   **25.** $-128x^{12}y^7$ **27.** $-200a^{12}b^5$

Exercise 28.5, page 441
1. $\dfrac{1}{125}$ **3.** $\dfrac{x^{-3}}{y^{-3}}$ **5.** $\dfrac{64x^9}{27y^6}$ **7.** $\dfrac{x^3y^2}{a^4b}$ **9.** $\dfrac{m^{10}}{n^6}$ **11.** $\dfrac{16x^8y^{12}z^{16}}{81a^4b^{12}c^{-4}}$ **13.** $\dfrac{27x^6y^{12}}{8a^3b^9c^3}$ **15.** $\dfrac{x^{-1}y^{-2}z^0}{a^0b^3c^{-4}}$

Exercise 28.6, page 442
1. 1   **3.** 1   **5.** −1   **7.** 8   **9.** −1   **11.** 1   **13.** 4   **15.** −3   **17.** −5   **19.** 5   **21.** −3   **23.** −7   **25.** $\dfrac{3}{4}$ **27.** −4   **29.** −2

Exercise 28.7, page 445
1. $\dfrac{1}{x^2}$ **3.** $\dfrac{a^2y^3}{x^2z}$ **5.** $\dfrac{-3a^3}{b^3c}$ **7.** $\dfrac{-b^2c}{3a^2}$ **9.** $\dfrac{1}{4x^2y^{-1}}$ **11.** $\dfrac{y^3z}{6x^2}$ **13.** $\dfrac{-c^4}{3ab^3}$ **15.** $\dfrac{y^4}{x^2z^3}$ **17.** $\dfrac{xy(x+y)}{y-x}$ **19.** $a^2b^{-1}xy^{-3}z^{-1}$ **21.** $3(2^{-1})a^3b^{-4}$ **23.** $\dfrac{9}{8}$ **25.** $\dfrac{-17}{200}$ **27.** $\dfrac{9}{200}$ **29.** $\dfrac{5}{72}$ **31.** $\dfrac{16}{3}$ **33.** $\dfrac{-2}{15}$

Exercise 28.8, page 449
1. $n^{1/2}$ **3.** $35^{1/3}$ **5.** $(a^2-b^2)^{1/2}$ **7.** $(n^3+4)^{1/5}$ **9.** $(x-2)^{3/2}$ **11.** $(x^2y^2z^4)^{1/3}$ **13.** 4   **15.** 9   **17.** 8   **19.** 128   **21.** $\dfrac{2}{3}$ **23.** $\dfrac{27}{64}$ **25.** $\dfrac{1}{2}$ **27.** $\dfrac{1}{25}$ **29.** $-\dfrac{1}{2}$ **31.** 9   **33.** 11.18   **35.** 3.87   **37.** 0.2   **39.** 13   **41.** 17   **43.** 25   **45.** 12

ANSWERS TO MOST ODD-NUMBERED EXERCISE PROBLEMS

Exercise 28.9, page 452
1. $(9.28)(10^7)$ **3.** $(6.3)(10^{-4})$ **5.** 10^{-7} **7.** $5(10^{-9})$ **9.** $(5.872)(10^{12})$
11. $(3.25)(10^7) = 32{,}500{,}000$ **13.** $(9.64)(10^{-7}) = 0.000000964$ **15.** 1.348
17. $(1.6)(10^4) = 16{,}000$ **19.** $(3.5)(10^{-7}) = 0.00000035$ **21.** $(2.25)(10^{-7}) = 0.000000225$ **23.** $(4.59)(10^{-1}) = 0.459$ **25.** $(4.53)(10^{-5}) = 0.0000453$

Exercise 29.1, page 455
1. x^4 **3.** a^5 **5.** $7^{0.8}$ **7.** $4x^3$ **9.** $2x^4$ **11.** $-2n$ **13.** $6x^{2.5}$
15. $3x^{4/5}$ **17.** $a+b$ **19.** $n+4$ **21.** $n-12$ **23.** $3x$

Exercise 29.2, page 459
1. $4\sqrt{3}$ **3.** $12\sqrt{2}$ **5.** $6\sqrt{5}$ **7.** $10\sqrt{6}$ **9.** $3\sqrt{2}$ **11.** $2\sqrt{2}$ **13.** $\sqrt{10}$
15. $2\sqrt{6}$ **17.** $\tfrac{1}{2}\sqrt{3}$ **19.** $\tfrac{5}{8}\sqrt{2}$ **21.** $\tfrac{1}{2}\sqrt{6}$ **23.** $2\sqrt{5}$ **25.** $2\sqrt[3]{2}$
27. $2\sqrt[4]{3}$ **29.** $3x\sqrt{2x}$ **31.** $3x^4\sqrt{5x}$ **33.** $2x^2\sqrt[3]{3x}$ **35.** $2n^2\sqrt[4]{2n}$
37. 12 **39.** 10 **41.** $x^2\sqrt{x^2+1}$ **43.** 7

Exercise 29.3, page 461
1. $43\sqrt{2} = 60.8$ **3.** $16\sqrt{2}-\sqrt{3} = 20.9$ **5.** $3\sqrt{3}-\sqrt{2} = 3.78$
7. $36\sqrt{10}+2\sqrt{15} = 121.6$ **9.** 22.4

Exercise 29.4, page 465
1. $20\sqrt{6}$ **3.** $12\sqrt{14} = 44.9$ **5.** -40 **7.** 90 **9.** 80 **11.** $40\sqrt{5}$
13. $90\sqrt{15} = 348.6$ **15.** $15\sqrt{6}-6\sqrt{10}+24$ **17.** $30+6\sqrt{2}$ **19.** $7\sqrt{14}-46$
21. -23 **23.** $42-24\sqrt{3} = 0.432$ **25.** $31-10\sqrt{6}$ **27.** $(432)^{1/12}$

Exercise 29.5, page 468
1. $\dfrac{\sqrt{3}}{3}$ **3.** $\dfrac{5\sqrt{2}}{2}$ **5.** $4\sqrt{5}$ **7.** $\dfrac{4\sqrt{2}}{5}$ **9.** $\dfrac{7\sqrt{5}}{3}$ **11.** $\dfrac{\sqrt{10}}{6}$ **13.** $\dfrac{2\sqrt{15}}{15}$
15. $\dfrac{5\sqrt{6}}{6}$ **17.** $\dfrac{4\sqrt{3}+\sqrt{21}}{9}$ **19.** $\dfrac{5\sqrt{2}+2\sqrt{3}}{38}$ **21.** $\dfrac{2\sqrt{5}-3\sqrt{2}}{2}$ **23.** $40\sqrt{2}+15\sqrt{14}$ **25.** $\dfrac{7-3\sqrt{5}}{2}$ **27.** $\dfrac{34+23\sqrt{2}}{14}$ **24.** $\dfrac{4-3\sqrt{2}}{2}$

Exercise 30.1, page 475
1. ± 4 **3.** ± 9 **5.** $\pm\dfrac{10}{3}$ **7.** $\pm\dfrac{1}{2}$ **9.** $\pm\dfrac{5}{6}$ **11.** $\pm\dfrac{3\sqrt{5}}{2}$ **13.** $\pm 10^3$
15. ± 0.001 **17.** $\pm\dfrac{\sqrt{10}}{8}$ **19.** $\pm\dfrac{2\sqrt{3}}{3}$ **21.** $\pm\dfrac{4\sqrt{5}}{5}$ **23.** $\pm\dfrac{\sqrt{21}}{3}$
25. $\pm\sqrt{-16}$ (imag.) **27.** $\pm\sqrt{-144}$ (imag.) **29.** $\pm\dfrac{13\sqrt{6}}{6}$ **31.** $\pm\dfrac{\sqrt{s}}{4}$
33. $\dfrac{A}{\pi}$

Exercise 30.2, page 478
1. $-2; -5$ **3.** $2; 4$ **5.** $3; -2$ **7.** $3; 6$ **9.** $2; 8$ **11.** $6; 6$ **13.** $3; \dfrac{1}{2}$
15. $2; \dfrac{1}{4}$ **17.** $\dfrac{2}{3}; \dfrac{2}{3}$ **19.** $0; \dfrac{5}{3}$ **21.** $0; \dfrac{8}{5}$ **23.** $+\dfrac{6}{5}; -\dfrac{6}{5}$
25. $-1; \dfrac{9}{4}$ **27.** $\dfrac{2}{3}; -\dfrac{3}{2}$ **29.** Not factorable

ANSWERS TO MOST ODD-NUMBERED EXERCISE PROBLEMS

Exercise 30.3, page 480
1. 5; −1   **3.** 2; 4   **5.** 4; −2   **7.** 7; −2   **9.** 2; −1   **11.** 2; $-\dfrac{3}{4}$
13. $2+\sqrt{3}$; $2-\sqrt{3}$ **15.** $-1+\sqrt{5}$; $-1-\sqrt{5}$

Exercise 30.4, page 485
1. 2; $\dfrac{1}{3}$ **3.** −1; $-\dfrac{3}{2}$ **5.** 2; $-\dfrac{1}{3}$ **7.** 1; $-\dfrac{4}{5}$ **9.** 2; 3   **11.** 5; −4
13. $\dfrac{3}{2}$; $-\dfrac{1}{2}$ **15.** $\dfrac{1}{2}$; $-\dfrac{4}{3}$ **17.** $-\dfrac{5}{2}$; $-\dfrac{5}{2}$ **19.** $2\pm\sqrt{3}$ **21.** $1\pm 2\sqrt{2}$
23. 2; $-\dfrac{1}{4}$ **25.** $\dfrac{2\pm\sqrt{-12}}{2}$ (imag.)   **27.** $\dfrac{-6\pm\sqrt{-124}}{2}$ (imag.)   **29.** 0; $-\dfrac{5}{3}$
31. $\dfrac{\pm 4\sqrt{3}}{3}$ **33.** $\pm\dfrac{\sqrt{35}}{5}$

Exercise 30.5, page 489
1. 12; 14; also −14; −12   **3.** 11; 13; also −13; −11   **5.** 12 in. by 15.5 in.
7. 16 in. by 19.25 in.   **9.** 25 in.; 35 in.   **11.** 6 in. by 14 in.   **13.** 54 in.
15. 45 mph; 8 hr.   **17.** 2.25 mph; 3 mph   **19.** 15 ft
21. 8 rows; 24 plants in each   **23.** 2 sec; 5 sec

Exercise 30.6, page 493
1. $9i$ **3.** $36i$ **5.** $13i$ **7.** $3i\sqrt{5}$ **9.** $\dfrac{2}{3}i$ **11.** $21i$ **13.** −15
15. −48   **17.** −15   **19.** −24   **21.** $\pm 2i$ **23.** $\pm\dfrac{5i}{2}$

Exercise 30.7, page 496 (No answers are given for No. 1–12.)
13. $9+5i$ **15.** $-1+i$ **17.** $0+6i$ **19.** $-6-4j$ **21.** $7.9+j4.6$
23. $-1+i$ **25.** $-9+7i$ **27.** $-14+0i$ **29.** 0   **31.** $-3.7-j1.8$
33. 4   **35.** 10   **37.** 6   **39.** $8i$ **41.** $-j10$ **43.** $-12+8i$
45. $-6-15i$ **47.** $14+23i$ **49.** $-8+31i$ **53.** $-4+3i$ **55.** $\dfrac{8}{3}-\dfrac{7}{3}i$
57. $-\dfrac{5}{4}-\dfrac{7}{4}i$ **59.** $-\dfrac{6}{5}i$ **61.** $\dfrac{22-7i}{13}$ **63.** $\dfrac{-15-13i}{25}$ **65.** $\dfrac{-21+20i}{29}$
67. $\dfrac{3-4i}{25}$ **69.** $\dfrac{8-15i}{17}$

Exercise 31.1, page 500
1. $\dfrac{3}{4}$ **3.** $\dfrac{2}{3}$ **5.** $\dfrac{4}{3}$ **7.** $\dfrac{1}{4}$ **9.** 1   **11.** $\dfrac{8}{5}$ **13.** $\dfrac{7}{8}$ **15.** $\dfrac{9}{2}$ **17.** 2; 2; 4   **19.** $\dfrac{1}{24}$

Exercise 31.2, page 504
1. 16   **3.** 27   **5.** 52.5   **7.** 2.146   **9.** 6.64   **11.** 9.75   **13.** +12; −12   **15.** $+5\sqrt{3}$; $-5\sqrt{3}$ **17.** $\dfrac{50}{3}$ **19.** 6   **21.** −2   **23.** $\dfrac{8}{3}$ **25.** 10.5
27. +16; −16   **29.** +0.75; −0.75   **31.** 70   **33.** 2.4 lb

ANSWERS TO MOST ODD-NUMBERED EXERCISE PROBLEMS

Exercise 31.3, page 507
1. 63 in.; 105 in. **3.** 3.2 lb; 4.8 lb **5.** copper: 7.5; gold: 22.5 parts
7. antifreeze: $6\frac{2}{3}$ gal; water $23\frac{1}{3}$ gal **9.** 48°; 60°; 72° **11.** cement: $222\frac{2}{9}$ lb; sand: $444\frac{4}{9}$ lb; gravel: $1333\frac{1}{3}$ lb **13.** stocks: $1600; bonds: $3200; savings: $7200; **15.** 54; 42; 12 **17.** chrom: 80 lb; nickel: 80 lb; steel: 1840 lb

Exercise 32.1, page 515
1. 40 cm **3.** 54.8 in. **5.** 17.89 in. **7.** 8.94 cm **9.** 9.17 ft
11. 14.7 in. **13.** 65 in. **15.** 80 cm

Exercise 32.2, page 520 (Order of values is sine, cosine, tangent.)
1. $\frac{21}{33.4}$; $\frac{26}{33.4}$; $\frac{21}{26}$ **3.** $\frac{4}{5}$; $\frac{3}{5}$; $\frac{4}{3}$ **5.** $\frac{1}{3}$; $\frac{2\sqrt{2}}{3}$; $\frac{\sqrt{2}}{4}$ **7.** $\frac{2\sqrt{5}}{5}$; $\frac{\sqrt{5}}{5}$; 2
9. $\frac{\sqrt{3}}{2}$; $\frac{1}{2}$; $\sqrt{3}$ **11.** $\frac{2\sqrt{6}}{7}$; $\frac{5}{7}$; $\frac{2\sqrt{6}}{5}$ **13.** $\frac{7\sqrt{2}}{10}$; $\frac{\sqrt{2}}{10}$; 7 **15.** $\frac{3\sqrt{10}}{10}$; $\frac{\sqrt{10}}{10}$; 3

Exercise 33.1, page 524 (See Table No. 3 for No. 1–16.)
17. 2.079 **19.** 10.92 **21.** 2.226 **23.** 3.360 **25.** 36.82 **27.** 7.856
29. 16.5° **31.** 25.2° **33.** 64.9° **35.** 51.1° **37.** 32.4° **39.** 47.2°

Exercise 33.2, page 527
1. 11° **3.** 23.4° **5.** 60.5° **7.** 50.1° **9.** 30.2° **11.** 28.2°
13. 32.4° **15.** 69.3° **17.** 30° **19.** 25° **21.** 35° **23.** 45°

Exercise 34.1, page 533
1. 60.1 ft **3.** Tower: 452.7 ft; cable: 449.1 ft **5.** 38.5 ft; 10.76 ft
7. 298.9 ft **9.** 28.1° **11.** 138 ft **13.** 67.4°; 39 ft **15.** 2650 ft
17. 54.7 ft

Exercise 34.2, page 540
1. $B = 70.6°$; $a = 13.95$ in.; $b = 39.6$ in. **3.** $A = 65.8°$; $a = 54.7$ cm; $b = 24.6$ cm **5.** B: 56.2°; a: 24.1 ft; c: 43.3 ft **7.** A: 48.5°; b: 65.5 cm; c: 98.8 cm **9.** B: 71.7°; b: 72.6 in.; c: 76.4 in. **11.** A: 73.5°; a: 84.4 cm; c: 88.0 cm **13.** A: 28.8°; B: 61.2°; c: 22.8 cm **15.** A: 69.1°; B: 20.9°; c: 44.9 in. **17.** A: 62.2°; B: 27.8°; b: 12.13 cm **19.** B: 78.8°; a: 46.6 cm; b: 235.4 cm **21.** B: 76.9°; a: 15.82 cm; c: 69.8 cm **23.** A: 64.6°; a: 99.6 in.; c: 102.8 in. **25.** A: 21.9°; b: 70.9 cm; c: 76.4 cm **27.** B: 10.9°; b: 49.3 cm; c: 260.7 cm **29.** A: 67.1°; a: 66.8 cm; b: 28.2 cm **31.** B: 73.7°; a: 7.25 in.; c: 25.8 in. **33.** A: 51.2°; b: 101.3 in.; c: 161.7 in.

Exercise 35.1, page 549
1. {a, b, c, d, e, f} **3.** ∅ or { } **5.** {b, c} **7.** {d, e, f}
9. {a, b, c, ... x, y, z}; entire alphabet **11.** True **13.** False **15.** True
17. False **19.** True **21.** True **23.** False **25.** False **27.** False
29. True

INDEX

Abscissa, 422
Absolute value, 224
Acre, 20
Acute, angle, 112
  triangle, 116
Addend, 3
Addition, 3
  of algebraic fractions, 345
  of arithmetic fractions, 29
  associative law, 3
  axiom of, 267
  combinations, 6
  commutative law, 3
  horizontal, 229, 241
  of monomials, 239
  of polynomials, 243
  of signed numbers, 225
  of whole numbers, 3
Adjacent angles, 112
Age problems, 281
Algebra, 209
Algebraic expression, 237
  value of, 213
Ampere, 122
Angle, 110, 512
  acute, 112
  central, 148
  of depression, 532
  of elevation, 530
  inscribed, 149
  measurement of, 113, 513
  minute of, 114
  obtuse, 112
  one degree, 113
  opening, amount of, 110
  protractor use, 113
  reflex, 113
  right, 112, 512
  sides of, 110
  straight, 111
  vertex of, 110, 512
Angle degree, 148
Angles, adjacent, 112
  complementary, 114, 116
  supplementary, 114
  vertical, 112
Apex, of cone, 182
  of pyramid, 177
Apollo moon flights, 20
Arc degree, 148
Arc of circle, 148
  center of, 148
  intercepted arc, 149
  major arc, 148
  midpoint of, 148
  minor arc, 148
Arc functions, 523
  arccos, 523
  arccot, 523
  arccsc, 523
  arcsec, 523
  arcsin, 523
  arctan, 523
Arithmetic numbers, 209
Assessed valuation, 93
Associative law, addition, 3
  for multiplication, 11
Axioms, 266

Bank discount, 8
Base, in exponents, 210
  in percentage, 64
Binomial, 238
Borrowing, in subtraction, 7
Braces, 211
Brackets, 211
Buying and selling, 75

Calculator, 100
Carrying, in addition, 4
Center of arc, 148
  of circle, 147

583

of gravity, 117
of sphere, 187
Centimeter, 196
    relation to inch, 198
Central angle, 148
Centroid, 117
Chord, 148
Circle, 147
    arc of, 148
    area of, 150
    center of, 147
    central angle, 148
    chord of, 148
    circumference of, 147
    diameter of, 147
    radius of, 147
    secant line, 147
    sector of, 154
    segment of, 154
    semi-circle, 148
    tangent to, 147
Circles, concentric, 148
Coefficient, 237
    numerical, 237
Coin and money problems, 286
Commission, 91
    rate of, 91
Commutative law, addition, 3
    for multiplication, 11
Comparison, solving systems, 399
Complementary angles, 114
Complementary set, 548
Completing a square, 315
    solving quadratics, 479
Complex numbers, 493
    addition and subtraction of, 494
    conjugates, 494
    division of, 495
    multiplication of, 494

Composite number, 10
Compound interest, 85
Concentric circles, 148
Cone, 182
    apex of, 182
    base of, 182
    circular, 182
    frustum of, 184
    volume of, 183
Congruent figures, 130, 140
Conjugate complex numbers, 494
Consecutive integers, 283
Coordinate system, 421
Cosecant of angle, 516
Cosine of angle, 516
Cotangent of angle, 516
Counting number, 21
Cube, 157
    edges of, 157
    faces of, 157
    lateral area of, 161
    vertices of, 153
    volume of, 161
Cube of number, 97
Cube root, 98
Cubes, sum and difference, in factoring, 327
Cubic equation, 471
Cubic inch, 159
Cylinder, 167
    altitude of, 167
    bases of, 167
    circular, 167
    hollow, 172
    lateral area of, 171
    oblique, 167
    radius of, 167
    right cylinder, 167
    volume of, 168
Cylindrical surface, 167

Decagon, 115

Decameter, 197
Decimeter, 196
Decimal, repeating, 55
    terminating, 55
Decimal fractions, 43
    addition of, 46
    changing to common, 44
    division of, 53
    multiplication of, 50
    subtraction of, 46
Decimal point, 43
Degree, 113
    angle, 148
    arc, 148
Degree of equation, 471
Denominator, 22
Dependent equations, 406
Derived equation, 406
Descartes, Rene, 425
Diagonal, 119
Diameter, 149
    of sphere, 187
Difference, between squares, 304
    in subtraction, 6
Digits, 1
    significant, 47
Direct measurement, 511
Discount, bank, 88
    successive, 79
    trade, 78
Disjoint sets, 545, 548
Dissimilar terms, 238
Dividend, whole numbers, 15
Divisibility of numbers, 25
Division, by zero, 268
    algebraic, 259
    arithmetic, 15
    of fractions, 37, 342
    of monomials, 252
    of polynomials, 255, 259
    of signed numbers, 234

INDEX

Elements of triangle, 513
Empty set, 544
Equation, 217, 265
  cubic, 471
  degree of, 471
  dependent, 406
  derived, 406
  first degree, 426
  fractional, 361
  graph of, 424
  inconsistent, 406
  independent, 406
  linear, 426
  literal, 378
  parentheses in, 272
  quadratic, 471
  quartic, 471
  root of, 221
  as scale, 265
  solution of, 220
  transposing in, 268
  variable denominator, 374
Equations, systems of, 389
Equiangular, 116
Equilateral, 116
  triangle, 143
Equilateral triangle, 143
  altitude of, 144
  each angle of, 143
  area of, 144
Equivalent fractions, 23
Even number, 2
Extremes of proportion, 502
Exponent, 97, 239, 247
  fractional, 446
  negative, 443
  zero, 441
Exponents, in division, 251, 435
  in multiplication, 247, 443
  in power of fraction, 440
  in power of a power, 436
  in power of a product, 438
  in scientific notation, 450

Faces, of polyhedron, 157
Factor, 10, 237, 294
  prime, 296
Factoring, 294
  common factor, 298
  cubes, sum and difference, 326
  grouping for, 327
  perfect squares, 312
  to reduce fractions, 295
  to solve quadratics, 475
  squares, difference between, 303
  trinomials, 319, 322
First degree equations, 426
"Five Golden Rules for Solving Problems," 278
Formula, definition, 123
  quadratic, 481
Fourth proportional, 502
Fractional equation, 361
Fractional exponent, 446
Fractions, 22
  addition and subtraction of, 29, 345
  algebraic, 331
  decimal, 43
  division of, 37, 342
  equivalent, 23
  fundamental principle of, 24, 332
  improper, 23
  multiplication of, 35, 337
  proper, 23
  reducing, 24, 295
  signs of, 355
  terms of, 22
  variable denominators, 352
Frustum, of cone, 184
  of pyramid, 179
Fundamental principle of fractions, 24, 332

Gain and loss, in percentage, 75
Gallon, 162, 169
Geometric line, 107
  point, 107
  solid, 157
Geometry, 107
  plane and solid, 109
Gram, 205
Graphing, solving equations, 428
Graph of equation, 424
Great circle of sphere, 187

Half-line, 108, 512
Hectare, 204
Hectometer, 197
Heptagon, 115
Hero's formula, 129
Hexagon, 115
Hollow cylinder, 172
Horizontal addition, 8, 229, 241, 243
Hypotenuse, 135, 136, 142, 514

Imaginary numbers, 490
  addition and subtraction, 492
  division, 495
  multiplication, 492, 494
  simplifying, 491
Imaginary unit, 491
Improper fraction, 23
Inconsistent equations, 406

Independent equation, 406
Indeterminate equation, 390
Index of root or radical, 98, 454
Initial side of angle, 512
Inscribed angle, 149
Inspection in solving equations, 218
Integer, 2, 283
Interest, 82
   compound, 85
   formula for, 83
   maturity date, 84
   principal, 83
   rate of, 83
   simple, 85
Irrational number, 99, 456
Irregular solids, volume of, 164
Isoceles right triangle, 139
Isoceles trapezoid, 117
Isoceles triangle, 116

Kilogram, 205
Kilometer, 197
Kilowatt, 20
Kilowatt hour, 20

Leg of right triangle, 135
Light and radio waves, speed, 20, 206
Light year, 20
Line, geometric, 107
   broken, 108
   curved, 108
   segment of, 108
   straight, 108
Lines, intersecting, 110
   parallel, 110
Linear equation, 426
Linear unit, 121
Liter, 171
Literal equation, 378

Literal number, 211
Long division, arithmetic, 16
   algebra, 259
Lowest common denominator, 29
Lowest common multiple, 32

Major arc of circle, 187
Marked price, 77
Market value, 94
Maturity date, 84
   value, 89
Mean proportional, 503
Means of a proportion, 502
Measurement, 121
   unit of, 121
Median of triangle, 117
Megameter, 197
Meter, 196
   inches in, 198
Metric system, 195
   conversion to English system, 198
   cubic measure, 203
   origin, 196
   prefixes, 198
   square measure, 200
   units in, 196
   weight, 205
Micron, 197
Mil, 195
Mill, rate of tax, 93
Millimeter, 196
Minor arc of circle, 187
Minuend, 6, 227
Minute of rotation, 114
Mixed number, 23
Monomial, 238
   addition and subtraction of, 239
   division of, 252
   multiplication of, 247

Motion problems, 288
Multiple of a number, 32
Multiplicand, 10
Multiplication, 10
   associative law, 11
   axiom of, 268
   commutative law, 11
   of fractions, 35, 50, 337
   of signed numbers, 231
   of whole numbers, 10, 32

Negative exponent, 443
Negative number, 223
Null set, 544
Number, arithmetic, 209
   complex, 493
   composite, 10
   counting, 2
   divisibility of, 25
   even, 2
   general, 211
   imaginary, 490
   irrational, 100, 456
   literal, 211
   mixed, 23
   negative, 223
   odd, 2
   positive, 223
   prime, 11
   rational, 100, 456
   reciprocal of, 37
   rounding off, 47
   signed, 223
Number line, 224
Numerator, 22
Numerical coefficient, 237
Numerical value of signed number, 224

Oblique cylinder, 167
Oblique prism, 158
Oblique triangle, 116

Obtuse angle, 112
Obtuse triangle, 116
Octagon, 115
Odd number, 2
Ohm, 122
Order of operations, 39
Ordinate, 422
Origin, 422

Parallel lines, 110
Parallelogram, 117
   altitude of, 125
   base of, 125
   opposite angles of, 119
   opposite sides of, 118
Parentheses, 211, 242
   in equations, 272
Pentagon, 114
Per cent, 63
Percentage, 64
   base in, 64
   formula for, 66
   rate in, 64
   relation to fractions, 63
Perfect power of number, 99
π, (pi), 150
Place value of digits, 1
Plane figures, 109
Plane geometry, 109
Plato, 142
Point, geometric, 107
Polygon, 115
   regular, 115
   sides of, 115
   vertices of, 115
Polyhedron, 157
Polynomial, 238
   addition and subtraction of, 243
   division, 255
   horizontal addition and subtraction, 243
   multiplication, 253

Positive number, 223
Power of a number, 99
   of fraction, 440
   of power, 436
   of product, 438
Precision, 47
Prime factor, 296
Prime number, 11, 296
Principal, at interest, 83
Principal root, 454
Prism, 157
   altitude, 160
   bases, 158
   faces, 158
   lateral area, 160
   oblique, 158
   right, 158
   volume, 158
Problems, stated, 277
   age, 281
   coin and money, 286
   consecutive numbers, 283
   general, 279
   grade points, 416
   investment, 382
   involving fractions, 381
   involving quadratics, 485
   motion, 288
   rules for solving, 278
   using two or more unknowns, 408
   work problems, 383
Proceeds, of note, 88
Product, 10, 237
Proper fraction, 23
Proportion, 501
   extremes of, 502
   means of, 502
   principle of, 502
Proportional division, 505
Protractor, 113, 513
Pure quadratic, 472
Pythagorean rule, 135, 514

Pyramid, 177
   altitude, 177
   apex, 177
   base edge, 177
   frustum of, 179
   lateral area, 178
   lateral edge, 178
   lateral face, 177
   regular, 177
   slant height, 177
   volume, 177

Quadrant, 422
Quadratic equation, 274, 471
   complete quadratic, 472
   double root of, 478
   pure quadratic, 472
   solving, by completing square, 479
   by factoring, 475
   by formula, 481
Quadratic formula, 481
Quadrilateral, 115
Quotient, 15

Radian, 121
Radical, 98, 102, 454
   addition and subtraction of, 460
   division of, 466
   index of, 98, 454
   multiplication of, 461
   radicand of, 98
   rationalizing denominator, 467
   simplifying, 102, 457
Radicand, 98, 454
   separating into factors, 101, 458
Radius of circle, 147
   of sphere, 187
Rate, of commission, 91
   of discount, 88

of interest, 85
of percentage, 64
Rate in motion problems, 288
Ratio, 499
Rationalizing denominator, 467
Rationalizing factor, 467
Rational number, 100, 456
Ratios, trigonometric, 516
Ray, 107, 512
Real number, 491
Reciprocal of a number, 37
Rectangle, 118
  area of, 122
  perimeter of, 124
Rectangular coordinate system, 421
Rectangular solid, 157
Reflex angle, 113
Regular polygon, 115
Remainder, in division, 15
  in subtraction, 6
Repeating decimal, 55
Retailer, 75
Rhombus, 118
Right, angle, 112
  cylinder, 167
  prism, 158
Right triangle, 116, 135, 514
  altitude of, 117
  area of, 128
  base of, 117
  hypotenuse of, 135
  isosceles, 139
    solving, 529, 534
    30°-60° right triangle, 142
  legs of, 135
  Pythagorean rule for, 135
Ring, area of, 153

Root of equation, 221
  double root, 478
  equal roots, 478
  imaginary roots, 475, 486
Roots of numbers, 98, 454
  index of root, 98
  principal root, 450
Rounding off numbers, 47

Scale, as equation, 265
Scalene triangle, 116
Scientific notation, 449
Secant line, 147
Secant of angle, 516
Second, angle of rotation, 114
Section of land, 20
Sector of circle, 148, 154
  area of, 154
Segment of circle, 148
  area of, 155
  height of, 155
  width of, 155
Segment of sphere, 187
Semicircle, 148
Set, 541
  complementary set, 548
  elements of, 541
  empty set, 544
  notation for, 542
  null set, 544
  universal set, 548
Sets, disjoint, 546
  equal sets, 544
  intersection of, 547, 548
  union of, 547, 548
Short-cuts in arithmetic, 57
Short division, 15
Signed numbers, 223
  absolute value of, 224
  addition of, 225
  division of, 234

  multiplication of, 231
  numerical value of, 224
  subtraction of, 227
Significant digits, 47
Signs in a fraction, 355
Similar figures, 120
Similar radicals, 460
Similar or like terms, 238
Sine, 516
Simultaneous equations, 390
Skew lines, 110
Slant height of pyramid, 177
Slide rule, 100
Solid, geometric, 157
  edge of, 157
  faces of, 157
  rectangular, 157
Solid geometry, 109
Solution of equation, 220
  definition, 221
  of systems, 391
Solution problems, 384
Sound, speed of, 20
Space geometry, 109
Special products, 293
Specific gravity, 165, 169
Sphere, 187
  area of surface, 188
  center, 187
  circumference, 187
  diameter, 187
  great circle of, 187
  radius, 187
  segment of, 187
  unit sphere, 189
  volume, 189
Square, 118, 125
  area, 125
  diagonal of, 140
  perimeter, 125
Square of a binomial, 309

INDEX

Square measure, metric, 200
Square roots, table, 453
Stere, 204
Subject of a formula, 380
Subset, 545
Substitution in solving, 396
Subtraction, algebraic, 228
 of fractions, 29
 of monomials, 239
 of polynomials, 243
 of signed numbers, 227
 of whole numbers, 6
Subtrahend, 6, 227
Successive discounts, 79
Sum of numbers, 3
Supplementary angles, 114
Surface, 108
 curved, 109
 plane (flat), 109
Systems of equations, 389, 390
 solving, by addition or subtraction, 391
 by comparison, 399
 by graphing, 428
 by substitution, 396
 three or more unknowns, 401

Table of squares and cubes, 558
Tangent of angle, 516
Tangent line, 147
Taxation, 92
Terminal side of angle, 512
Terminating decimal, 55
Terms, algebraic, 238
 dissimilar, 238
 like, 238
 similar, 238
 unlike, 238
Terms of a fraction, 22
Terms of a proportion, 502
Third proportional, 503
30°-60° right triangle, 142
Transposing, 268
Transversal, 114
Trapezium, 117
Trapezoid, 117
 altitude of, 117, 130
 area of, 130
 bases of, 117, 130
 isosceles, 117
Triangle, 116
 altitude of, 128
 area of, 128
 centroid of, 117
 elements of, 513
 equiangular, 116
 equilateral, 116, 143
 isosceles, 116
 median of, 117
 obtuse, 116,
 perimeter of, 129
 right triangle, 116, 513
 scalene, 116
 sum of angles of, 116, 513
Triangular bracing, 129
Trigonometric ratios, 516, 523
 tables of, 559
Trinomial, 238

Uniform motion, 288
Unit of measure, 121
Unit sphere, 189
Universal set, 548

Variable, 390
Venn diagram, 549
Vertex, of angle, 110, 512
 of polygon, 115
Vertex angle of triangle, 117
Vertical angles, 112
Volt, 122
Volume, 158
 of cone, 183
 of cube, 161
 of cylinder, 168
 of frustum, 179, 184
 of prism, 158
 of pyramid, 177
 of sphere, 189

Water, weight of, 165
Weight, metric system, 205
Whole numbers, 1

X and Y axes, 422

Zero, division by, 268
Zero exponent, 441

SHENANDOAH COLLEGE
LIBRARY
WINCHESTER, VA.

DATE DUE			
NOV 8 1978			
APR 2 3 1979			
GAYLORD			PRINTED IN U.S.A.